HIGH FREQUENCY PROCESSES IN MAGNETIC MATERIALS

HIGH FREQUENCY PROCESSES IN MAGNETIC MATERIALS

Editors

G. Srinivasan

A. N. Slavin

Oakland University,
USA

World Scientific
Singapore • New Jersey • London • Hong Kong

Published by

World Scientific Publishing Co. Pte. Ltd.

P O Box 128, Farrer Road, Singapore 9128

USA office: Suite 1B, 1060 Main Street, River Edge, NJ 07661

UK office: 57 Shelton Street, Covent Garden, London WC2H 9HE

HIGH FREQUENCY PROCESSES IN MAGNETIC MATERIALS

ISBN 981-02-1816-8

Printed in Singapore.

PREFACE

Almost half a century has passed since the discovery of ferromagnetic resonance (FMR) by J.H.E. Griffiths (Nature **158**, 670 (1946)), but high frequency magnetic excitations still create growing interest among scientists and engineers. There are plenty of reasons for that. From the practical point of view FMR (and other magnetic resonances) provide excellent methods of non-destructive testing of materials and sophisticated imaging techniques. Propagating magnetic excitations like magnetostatic waves or spin waves can also be used for linear and nonlinear analog signal processing at microwave frequencies. At the same time, magnetic excitations at microwave frequencies provide an excellent opportunity to study both theoretically and experimentally such fundamental phenomena as linear and nonlinear diffraction of waves in anisotropic media, chaotic dynamics of waves and weak wave turbulence, formation and interaction of envelope solitons and different kinds of wave instabilities.

Recent developments in the theoretical and experimental investigations of high frequency magnetic excitations followed three major directions:

(i) study of magnetic excitations in non-traditional and man-made magnetic materials such as amorphous magnetic oxides, ortho- and hexaferrites, and magnetic multilayers and superlattices;

(ii) study of magnetic excitations in traditional materials such as yttrium-iron garnet (YIG) in the nonlinear regime or/and under non-stationary conditions;

(iii) study of magneto-optic processes involving interaction between light and microwave magnetic excitations in both linear and nonlinear regimes.

The objective of this volume is to give an account of these important new tendencies in the high frequency magnetics research.

We classified the subjects discussed in the volume into three broad categories mentioned above. Part I deals with linear microwave processes in new magnetic materials (chapters 1-5). Recent advances in traditional characterization techniques such as ferromagnetic and antiferromagnetic resonance and studies on magnetic order in noncrystalline oxides are also presented in this section. Nonstationary and nonlinear high frequency processes are discussed in Part II (chapters 6-8). Finally, the important topic of interaction of magnetic excitations with light (especially optical waveguide modes in magnetic films) is discussed in Part III (chapters 9-12).

The first chapter by Kabos deals with basic aspects of FMR, magnetostatic and spin waves in thin magnetic films and multilayers. In particular, readers would benefit from information on new FMR experimental techniques like the detection of FMR using Brillouin light scattering .

Chapter 2 provides an excellent review on the spectroscopy of rare-earth orthoferrites - magnetic materials that allow to shift FMR techniques into the submillimeter wave range. The authors of Chapter 2 also included a section on possible use of orthoferrites in signal processing devices.

In chapter 3, recent advances in studies on the magnetic structure of noncrystalline transition metal oxides are provided. In particular, studies performed within the past few years report evidence for a long-range order in oxides containing Fe_2O_3-Bi_2O_3. Some of these oxides show important properties such as high magnetization, low microwave losses, coexisting ferromagnetism and ferroelectricity, and strong magneto-optic effects.

Chapter 4 by Kalinikos and Kolodin deals with the analytic theory of dipole-exchange spin wave spectrum in magnetic multilayers coupled by a long-range dipole-dipole interaction.

Details on advances in FMR in exchange-coupled magnetic multilayers are presented in chapter 5. Zhang and Wigen describe recent studies that focus on the characterization of magnetic properties of multilayer materials consisting of alternative magnetic/nonmagnetic layers from the spectrum of collective FMR modes.

The first chapter in the second part of the book (chapter 6) describes effects associated with the propagation of magnetostatic waves (MSW) in YIG films under nonstationary conditions such as a time-dependent static magnetic field. A rich variety of new phenomena that occur under nonstationary conditions such as transformations of MSW frequency spectrum, space-time focusing of MSW pulses , spectral separation of MSW modes are studied both theoretically and experimentally.

Chapter 7 by Cherepanov and Slavin is devoted to the theoretical investigation of sample-size effects on the process of parametric excitation of short spin waves in ferrites. A comprehensive theory explaining experimentally observed sample-size effects on the threshold of parametric excitation of spin waves and on the parameters of spin wave auto-oscillations above the threshold is presented in this chapter.

Wiese and Benner in chapter 8 present a model of nonlinear interaction of spin wave modes above the threshold of their parametric excitation where the size of the sample is taken into account through the boundary conditions on the sample surfaces. This model is used to explain complicated multistability of parametric spin waves observed in YIG spheres and films.

Part III on interaction between spin waves and light contains four chapters dealing with the following topics.

In chapter 9, Stashkevich presents a new theory of multiple scattering of optical guided modes in YIG film by microwave spin waves (or magnetostatic waves) propagating in the same film, which is the case in the experiments where high efficiency of interaction is obtained. The theory is developed for the case when spin wave contains more than one harmonic component.

The importance of this theory is supported by the data presented in chapter 10 where Stancil and Bilaniuk provide an excellent review on collinear interactions between optical guided modes and magnetostatic waves in planar heterostructures based on YIG films. Practical aspects and possible applications of these interactions are covered in chapter 10. Effects of microwave and optical losses, nonlinear effects, and material requirements for devices based on this technology are also discussed.

Chapter 11 deals with Brillouin light scattering (BLS) from dipole-exchange spin waves in magnetic films and double layers. The analytical theory presented in this chapter provides explicit expressions for BLS cross-sections convenient to use for the interpretation of experiments.

Finally, a new method of space-resolved photothermal (PT) detection of FMR in magnetics is reviewed in chapter 12. The theory of PT detection, experimental scheme, results of studies on YIG and other ferrite films and spheres are discussed.

The contributors to this volume are leading authorities in their respective field, and we believe the readers would benefit immensely from the material covered in this volume.

Andrei Slavin March 1995
Gopalan Srinivasan

CONTENTS

PART III — Interaction of Microwave Magnetic Excitations with Light

PART I

LINEAR MICROWAVE PROCESSES IN NEW MAGNETIC MATERIALS

High Frequency Excitations in Thin Magnetic Films

Pavel Kaboš
Department of Theoretical and Applied Electrical Engineering,
Faculty of Electrotechnics and Information Technology, Slovak Technical University,
Bratislava, Slovakia

1. Introduction

High frequency excitations in magnetic materials have attracted considerable attention for more than five decades, starting with the discovery of ferromagnetic resonance (FMR) by Griffiths in 1946 [1]. A summary of the key theoretical and experimental results may be found in review articles and book chapters [2-6]. Although magnetic excitations continue to be of interest in a variety of materials, many recent experimental investigations have focused on properties of ultrathin films and multilayered structures [7-10]. While many of these films are prepared by molecular beam epitaxy or standard evaporation and sputtering techniques, there have also been advances in fabrication using pulsed laser deposition (PLD) [11] and low temperature non-vacuum processes [12]. These new techniques allow preparation of quality thin ferrite films, including highly anisotropic films particularly well suited for millimeter-wave applications. The key parameters characterizing a material include the saturation magnetization, anisotropy field and high frequency losses, all of which are experimentally accessible through rudimentary FMR and effective linewidth techniques [13], [14]. Several of these topics will be addressed in more detail in chapters 4, 6 and 7 of this book.

The purpose of this chapter is to provide an outline of the basic aspects of FMR processes in thin magnetic films. The approach follows a general formalism used for magnetic systems, based on the classical torque equation and Maxwell's equations.

2. Magnetic Torque Equation

The state of a magnetic material is usually described by its magnetic moment per unit volume, or magnetization, $\mathbf{M}(\mathbf{r},t)$, which in genral varies throughout the sample and is a function of time. The following study considers the dynamical motion of the magnetization when the material is subject to an applied magnetic field. Throughout most of this chapter we will assume that the applied field is uniform throughout the volume of the sample. As a model describing the dynamic behavior of the magnetization, consider a single magnetic dipole (such as a current loop) placed in applied magnetic field. As consequence of the Lorenz force law, the magnetic moment will experience a torque which tends to align the moment into the applied field direction. Since the torque is the time derivative of the angular momentum, which is proportional to the magnetic moment, the equation of motion for magnetization is

$$\frac{d\boldsymbol{M}}{dt} = -\gamma\,(\boldsymbol{M}\times\boldsymbol{H_0}), \tag{1}$$

where the factor γ, taken as positive, is given by

$$\gamma = \frac{ge}{2m_e c}. \tag{2}$$

Here, e is the absolute value of the electron charge, m_e the rest mass, and g the so-called Lande spectroscopic splitting factor (for a free electron $g = 2.0023$). The factor γ is called the magneto-mechanical ratio and has a value of $2\pi\times2.8$ MHz/Oe or 1.76×10^7 rad/(s.Oe) in Gaussian units. As mentioned above, the state of a magnetic medium is described by its magnetization, **M**(r,t). A full description of the dynamic magnetic response requires knowledge of all the forces acting on the magnetization. These forces result from the possible interactions between the magnetization and the magnetic fields in the medium. Phenomenological representation of these interactions was first formulated by Landau and Lifshitz [15] by introducing an internal effective field H_{eff}, which replaces the applied field H_0 in the torque equation (1). The action of this field is equivalent to the applied field, but includes other interactions in the magnetic material as well. The equation of motion can then be written in the form

$$\frac{d\boldsymbol{M}}{dt} = -\gamma\,(\boldsymbol{M}\times\boldsymbol{H_{eff}}) + \boldsymbol{R}. \tag{3}$$

Here, the term **R** represents the damping in the system, the specific form of which will be discussed later.

Before proceeding with the solution of equation (3), it is necessary to specify the exact form of the effective field $\mathbf{H}_{eff}$. This will be done in the next section.

3. Effective Fields

The amplitude and the direction of the internal effective field can be found to a sufficiently good approximation from the variation principle. This leads to an expression for the effective field of the form

$$\boldsymbol{H}_{eff} = -\frac{\partial F}{\partial \boldsymbol{M}} + \sum_{p} \frac{\partial}{\partial x_p} \cdot \left[\frac{\partial F}{\partial \left(\frac{\partial \boldsymbol{M}}{\partial x_p} \right)} \right]. \qquad \textbf{(4)}$$

The term F represents, for temperature T > 0, the free energy of the system, and x_p ($p = 1, 2, 3$) the components of the spatial Cartesian coordinates. Equation (4) was first obtained in 1951 by MacDonald [16].

For further treatment, it is more practical to express the free energy in equation (4) as a sum of contributing terms. The effective field then has the more explicit (and transparent) form

$$\boldsymbol{H}_{eff}(\boldsymbol{M}) = \boldsymbol{H}_{ext} + \boldsymbol{h}(t) + \boldsymbol{H}_{dip}(\boldsymbol{M}) + \boldsymbol{H}_{anis}(\boldsymbol{M}) + \boldsymbol{H}_{ex}(\boldsymbol{M}) + \dots \qquad \textbf{(5)}$$

The contributing terms in the effective magnetic field (5) are the static applied external magnetic field, the dynamic external magnetic field, the dipolar field, the crystalline magnetic anisotropy field, the exchange field and other "fields" representing any further possible interactions in the magnetic material. In most experimental situations, the dynamic external magnetic field always fulfills the condition $h(t) \ll H_{ext}$. The explicit **M** dependence shows that the effective magnetic fields are function of the magnetization. The specific forms of the contributing fields in equation (5) will be discussed shortly.

The amplitude of the magnetization $|\mathbf{M}|$ is constant, as can be seen from the equation of motion (3). This constant is called the saturation magnetization M_0. Consequently, the dynamic part of the magnetization has only components perpendicular to the equilibrium direction of the magnetization. Employing the the usual notation, where lower case letters denote the dynamic fields, and upper case letters static fields, the approximate magnetization can be expressed as

$$\boldsymbol{M}(\boldsymbol{r},t) = M_0\, \boldsymbol{e_z} + \boldsymbol{m}_\perp(\boldsymbol{r},t)\,, \tag{6}$$

where $\mathbf{e_z}$ is the unit vector in the static magnetization direction, and $\mathbf{m}_\perp(\mathbf{r},t)$ is the dynamic magnetization component perpendicular to $\mathbf{e_z}$. This notation introduces a coordinate system related to the equilibrium position of the static magnetization, called the *precessional or rotational frame*. Equation (5) can now be written as

$$\boldsymbol{H}_{eff}(\boldsymbol{M}) = \boldsymbol{H}_i + \boldsymbol{h}_{dip}(\boldsymbol{M}) + \boldsymbol{h}_{anis}(\boldsymbol{M}) + \boldsymbol{h}_{ex}(\boldsymbol{M}) + \boldsymbol{h}(t) + \ldots \tag{7}$$

Here the static and dynamic contributions to the effective field are separated. The term $\mathbf{H}_i = \mathrm{H}_i\, \mathbf{e}_z$ represents the resulting static *internal* field, and will be discussed in detail later. It also defines the direction of the equilibrium orientation of the static magnetization.

Inserting expression (6) into the equation of motion (3) and neglecting the higher order terms, the linearized equation of motion is

$$\frac{d\boldsymbol{m}_\perp}{dt} = \gamma\, \boldsymbol{e}_z \times [H_i \boldsymbol{m}_\perp - M_0(\boldsymbol{h}_{dip}(\boldsymbol{m}_\perp) + \boldsymbol{h}_{anis}(\boldsymbol{m}_\perp) + \boldsymbol{h}_{ex}(\boldsymbol{m}_\perp) + \ldots)] - M_0 \boldsymbol{e}_z \times \boldsymbol{h}(t)\,. \tag{8}$$

To solve equation (8), it is necessary to specify the explicit forms of the internal field $\mathbf{H}_i$ and the dynamic fields. The usual approach is to first expand the dynamic part of the magnetization $\mathbf{m}_\perp(\mathbf{r},t)$ into a Fourier series of the form

$$\boldsymbol{m}_\perp(\boldsymbol{r},t) = \boldsymbol{m}_0(t) + \boldsymbol{m}(\boldsymbol{r},t) = \boldsymbol{m}_0(t) + \sum_{\kappa \neq 0} \boldsymbol{m}_\kappa(\boldsymbol{r},t)\,, \tag{9}$$

where $\mathbf{m}_0(t)$ denotes the uniform part of the dynamic magnetization and is usually called the *uniform mode*. The sum over $\kappa \neq 0$ defines the nonuniform part of the dynamic magnetization expanded into the *eigenmodes* of the system. The explicit form of $\mathbf{m}_\kappa(\mathbf{r},t)$ has yet to be established. It is worth mentioning that the uniform mode $\mathbf{m}_0(t)$, as will be discussed later, describes the in-phase and spatially uniform component of the dynamic response. In contrast, the $\mathbf{m}_\kappa(\mathbf{r},t)$ terms represent the spatially nonuniform modes of the system. In the special (and often analyzed) case, when the eigenmodes of the system can be expressed as plane waves, the $\mathbf{m}_\kappa(\mathbf{r},t)$ terms represent nonuniform, propagating excitations called *spin waves*.

Return now to the development of the various contributions to the effective field. Assume that the sample has the form of an ellipsoid with principal axes along the X, Y, and Z axes of the Cartesian coordinate system, and that the principal crystallographic axes are along the XA, YA, and ZA axes. The coordinate system related to the equilibrium magnetization is along the *x*, *y*, and *z* axes. The external static magnetic field is assumed uniform throughout the sample. The various coordinate systems are shown in Fig. 1.

Let us first consider the dipolar field, $\mathbf{H}_{dip}(\mathbf{M})$ for an ellipsoidal sample placed in a uniform external static magnetic field $\mathbf{H}_{ext}$. The magnetization is given by equation (6). It will be necessary to find the equilibrium position of the static magnetization, and its dynamic response to the external field. This problem can be separated into two parts, a static and a dynamic one. Both parts have to fulfill Maxwell's equations and the corresponding boundary conditions. The nominal sample dimension *d* plays a key role in the various approximations used. For $\kappa = k \gg 2\pi/d$ the surface boundary conditions may be neglected, greatly simplifying the problem. Here we have introduced *k* for the wavenumber of short wavelength spin waves to distinguish this special case from the more general eigenvalue κ. When κ is comparable or less than $2\pi/d$, the boundary conditions must be taken into account. Another important consideration is the ratio of the eigenmode wavenumber κ to the wavenumber of ordinary electromagnetic waves, $k_0 = (\omega_\kappa\sqrt{\epsilon})/c$. The value of this ratio determines whether electromagnetic propagation effects can be neglected in the analysis. The neglect of such effects corresponds to the condition $\kappa \gg k_0$. Only this case will be covered in this chapter. For ferrites at 10 GHz with a relative dielectric constant of $\epsilon = 10$, the value of k_0 is 6.6 cm^{-1}. With propagation effects neglected, Maxwell's equations yield

$$\nabla . [\boldsymbol{H}_{dip}(\boldsymbol{r},t) + 4\pi \boldsymbol{M}(\boldsymbol{r},t)] = 0 ,$$

$$\nabla \times \boldsymbol{H}_{dip}(\boldsymbol{r},t) = \boldsymbol{0} . \qquad (10)$$

The static part represents a standard magnetic boundary value problem, the solution of which can be found in any basic electromagnetic theory text (see, e.g., [17]). The solution is a static uniform field inside a uniformly magnetized ellipsoid, given by

$$\boldsymbol{H}_i' = \boldsymbol{H}_{ext} - 4\pi\,[N].\boldsymbol{M}_0\,, \tag{11}$$

where $\mathbf{H}_{ext}$ is the applied uniform external static magnetic field and [N] is the so-called demagnetizing tensor. If the sample is magnetized along one of its principal axes, the demagnetizing tensor is diagonal with components N_X, N_Y, N_Z. Values of these components for different sample geometries can be found, for example, in [18]. As shown in eq. (9), the dynamic component of the magnetization can be separated into uniform and nonuniform terms. The uniform dynamic term can be treated in the same way as the static term, leading to a contribution of the form

$$\boldsymbol{h}_{0\,dip}(t) = -\,4\pi\,[N].\boldsymbol{m}_0(t)\,. \tag{12}$$

Maxwell's equations for the nonuniform part of the dynamic dipole field are

$$\nabla\times\boldsymbol{h}_{dip}(\boldsymbol{r},t) = \boldsymbol{0}\,,$$

$$\nabla.[\boldsymbol{h}_{dip}(\boldsymbol{r},t) + 4\pi\,\boldsymbol{m}(\boldsymbol{r},t)] = 0\,. \tag{13}$$

One now must solve the system of equations (13), the equation of motion (3), and the corresponding boundary conditions. There are several possible approaches to this task. One way is to introduce the magnetic scalar potential

$$\boldsymbol{h}_{dip}(\boldsymbol{r},t) = -\,\nabla\psi(\boldsymbol{r},t)\,. \tag{14}$$

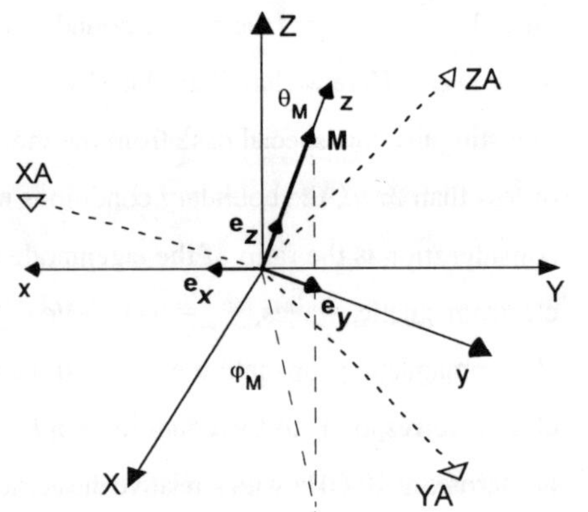

Fig. 1: Coordinate systems. The X, Y, and Z directions are related to the sample principal axes, the XA, YA, and ZA to the magnetocrystalline anisotropy principal axes, and the x, y, and z to the equilibrium position of the magnetization.

Equation (14) automatically satisfies the first of the equations (13). The dynamic dipole field $\mathbf{h}_{dip}(\mathbf{r},t)$ is a result of the response of the magnetic system. The corresponding relation to the eigenmodes of the system is given by the equation of motion. The next step is to include the solution of the equation of motion, and solve the boundary value problem. We will return to this step later when solving these equations for specific geometries. The situation is considerably more simple for spin waves. Then, the $\mathbf{m}_\kappa(\mathbf{r},t)$ term in equation (13) becomes

$$\boldsymbol{m}_\kappa(\boldsymbol{r},t) \rightarrow \boldsymbol{m}_k(t)\exp[-i(sign k)\boldsymbol{k}.\boldsymbol{r}]\,. \tag{15}$$

The eigenwavevector $\boldsymbol{\kappa}$, as mentioned before, is replaced by $\mathbf{k}$, the wavevector of the plane wave component (spatial Fourier component) of the dynamic magnetization. Equation (15) represents one contributing term in equation (9). To keep the terms in the equation real, the summation in (9) has to be performed over all $\pm k \neq 0$ terms. The factor 1/2 is included in the spinwave amplitude $\mathbf{m}_k(t)$, and the relation

$$\boldsymbol{m}_k^*(t) = \boldsymbol{m}_{-k}(t)\,, \tag{16}$$

holds for all complex amplitudes in the expansion. Relation (16) will be important in the development of the spinwave instability theory. Inserting (13) and (15) into Maxwell's equations, for $k \gg k_0$, the k-th spatial Fourier component of the dynamic dipolar field becomes [19]

$$\boldsymbol{h}_{kdip}(t) = -4\pi\frac{\boldsymbol{k}}{k^2}[\boldsymbol{k}.\boldsymbol{m}_k(t)] = -4\pi[N_k].\boldsymbol{m}_k(t)\,, \tag{17}$$

where we have introduced the dipole tensor, $[N_k]$, following [20].

Another approach is based on the solution of a system of integro-differential equations. We start with the second of the equations (13). After some elementary algebra, the equation can be rewritten into Poisson equation [17]. The "magnetic charge" entering the Poisson equation is given by the expression $-4\pi\nabla.\mathbf{m}(\mathbf{r},t)$. The solution of the Poisson equation has a standard form of [17]

$$\boldsymbol{h}_{dip}(\boldsymbol{r},t) = 4\pi\int_{-\infty}^{\infty}[G(\boldsymbol{r},\boldsymbol{r}')].\boldsymbol{m}(\boldsymbol{r}',t)dV'\,, \tag{18}$$

where [**G**(**r**,**r'**)] is the tensorial Green function for the boundary value problem. The system of coupled equations (18) and (8) is then solved. A detailed description of this procedure is discussed in chapter 6, and in [21]. We will return to the specific description of the dipolar fields later. It should be stressed however, that the form of the dipolar fields can be specified only for a given sample geometry. As mentioned above, for spin waves the solution is more simple.

The next contribution to the effective field is the magnetocrystalline anisotropy term. It too can be split into a sum of a static and dynamic components. Both terms can be derived from equation (4) by using the magnetocrystalline anisotropy free energy $F_A(\mathbf{M}(\mathbf{r},t))$. The free energy term F_A can be expressed as an exponential series with respect to the directional cosines of the magnetization vector,relative to the crystal's principal axes. This was shown for the first time by Akulov [22]. Then, one has $F_A(\mathbf{M}(\mathbf{r},t)) = F_A(\alpha_{XA},\alpha_{YA},\alpha_{ZA})$ with α_i the directional cosines relative to the XA, YA, ZA coordinate system (See Fig. 1). From equation (4), the static magnetocrystalline anisotropy term is

$$\boldsymbol{H}_A = -\frac{1}{M_0}\nabla_\alpha\left\{F_A(\alpha_{XA},\alpha_{YA},\alpha_{ZA})\right\}\Big|_{\alpha_{eq}} = (H_A^{XA},H_A^{YA},H_A^{ZA})\,. \tag{19}$$

The dynamic anisotropy contribution to the effective field is

$$\boldsymbol{h}_A(\boldsymbol{r},t) = \sum_{i=XA}^{ZA}\frac{\partial}{\partial\alpha_i}\left\{-\nabla_\alpha F_A(\alpha_{XA},\alpha_{YA},\alpha_{ZA})\right\}\Big|_{\alpha_{eq}}\cdot\frac{\boldsymbol{m}_{\perp i}(\boldsymbol{r},t)}{M_0}$$

$$= [\boldsymbol{A}].\left\{\boldsymbol{m}_0(t) + \boldsymbol{m}(\boldsymbol{r},t)\right\} = \boldsymbol{h}_{A0}(t) + \boldsymbol{h}_A[\boldsymbol{m}(\boldsymbol{r},t)]\,, \tag{20}$$

where ∇_α denotes the gradient with respect to direction cosines, $\boldsymbol{\alpha}_{eq} = \mathbf{M}_0/M_0 = \mathbf{e}_z$ is the equilibrium position of the static magnetization, $m_{\perp i}$ the i-th component of the total dynamic magnetization, and [**A**] the general 3 × 3 anisotropy tensor. Equation (20) was obtained in a linear approximation. The effective field $\mathbf{h}_{A0}(t)$ in equation (20) represents the dynamic anisotropy contribution induced by the uniform precession. The effective field $\mathbf{h}_A(\mathbf{m}(\mathbf{r},t))$ represents the dynamic anisotropy contribution due to nonuniform modes. It is worth mentioning, that both dynamic field contributions introduced in equation (20) give only the first order approximation to the fields. Nevertheless, this first order approximation is used almost exclusively, because (a), it simplifies the solution, and (b), the resulting theoretical predictions are in good agreement with experimental observations. The components of the matrix [**A**] are

$$A_{ij} = \frac{1}{M_0}\frac{\partial}{\partial \alpha_j}\left\{ - \nabla_\alpha F_A(\alpha_{XA}, \alpha_{YA}, \alpha_{ZA})\right\}\Big|_{\alpha_{eq}} = - \frac{1}{M_0^2}\frac{\partial^2}{\partial \alpha_i \partial \alpha_j} F_A(\alpha_{XA}, \alpha_{YA}, \alpha_{ZA})\Big|_{\alpha_{eq}}, \quad \textbf{(21)}$$

$$A_{ij} = A_{ji}.$$

To keep the expressions general, no specific form of the magnetocrystalline free energy is taken at this point. Later, we will discuss some specific examples. The static equilibrium condition for the magnetization direction cannot be specified yet. To do so, all of the terms contributing to the effective field have to be considered. The resulting magnetocrystalline anisotropy contribution to the effective field is then

$$\boldsymbol{H}_A[\boldsymbol{M}(\boldsymbol{r},t)] = \boldsymbol{H}_A + \boldsymbol{h}_{A0}(t) + \boldsymbol{h}_A[\boldsymbol{m}(\boldsymbol{r},t)], \quad \textbf{(22)}$$

with contributing terms given by eqs. (19) and (20).

The last explicitly addressed term in the effective magnetic field is the exchange field. In the classical spin wave theory,this term models the effect of the quantum mechanical exchange interaction between the spins of the system. The standard form used to describe this interaction is [23]

$$\boldsymbol{h}_{ex}[\boldsymbol{M}(\boldsymbol{r},t)] = \frac{D}{M_0}\nabla^2 \boldsymbol{M}(\boldsymbol{r},t). \quad \textbf{(23)}$$

Considering expressions (6) and (9), (23) can be written in the form

$$\boldsymbol{h}_{ex}(\boldsymbol{r},t) = \frac{D}{M_0}\nabla^2 \boldsymbol{m}(\boldsymbol{r},t) = \frac{D}{M_0}\sum_{\kappa \neq 0} \nabla^2 \boldsymbol{m}_\kappa(\boldsymbol{r},t). \quad \textbf{(24)}$$

For spin waves (plane wave spatial Fourier expansions), the dynamic exchange field is

$$\boldsymbol{h}_{ex}(\boldsymbol{r},t) = - \frac{D}{M_0}\sum_{\kappa \neq 0} \boldsymbol{m}_k e^{i(\mathrm{sign}\, k)\left[\omega_k t - \boldsymbol{k}.\boldsymbol{r}\right]}, \quad \textbf{(25)}$$

where D is the so-called *exchange parameter*. The value of the exchange parameter for yttrium iron garnet (YIG) is 5.4×10^{-9} Oe.cm^2. The effect of exchange will be discussed later in more detail.

This concludes the description of the most important contributions to the effective field. As discussed above, we will specify these fields in more detail as we discuss the solution of the equation of motion. A short comment on the equilibrium position of the static part of magnetization should be made first, however. As mentioned before, the direction of the internal magnetic field specifies the equilibrium position of the magnetization, since the magnetization has to be aligned with the internal magnetic field. This statement can be concisely expressed as

$$\boldsymbol{H}_i \times \boldsymbol{e}_z = \boldsymbol{0}. \tag{26}$$

4. Linearized Equation of Motion

The results from the previous section allow one to write the total effective magnetic field as

$$\begin{aligned}\boldsymbol{H}_{eff}(\boldsymbol{r},t) &= \boldsymbol{H}_0 + \boldsymbol{H}_A - 4\pi\,[N].\boldsymbol{M}_0 - 4\pi\,[N].\boldsymbol{m}_0(t) + [A].\boldsymbol{m}_0(t) + \\ &+ \sum_{\kappa\neq0}[A].\boldsymbol{m}_\kappa(\boldsymbol{r},t) + \frac{D}{M_0}\sum_{\kappa\neq0}\nabla^2\boldsymbol{m}_\kappa(\boldsymbol{r},t) + \boldsymbol{h}_{dip}[\boldsymbol{m}(\boldsymbol{r},t)] + \dots\,.\end{aligned} \tag{27}$$

The static term $\mathbf{H}_i = \mathbf{H}_0 + \mathbf{H}_A - 4\pi[\mathrm{N}].\mathbf{M}_s$ represents the static internal field. Inserting (27) with (9) into (8), one obtains

$$\begin{aligned}\frac{d\boldsymbol{m}_0(t)}{dt} + \sum_{\kappa\neq0}\frac{d\boldsymbol{m}_\kappa(\boldsymbol{r},t)}{dt} &= \gamma\,\boldsymbol{e}_z\times\{[(\boldsymbol{H}_0 + \boldsymbol{H}_A - 4\pi\,[N].\boldsymbol{M}_0).\boldsymbol{e}_z]\boldsymbol{m}_0(t)] - \\ &- M_0[-4\pi\,[N].\boldsymbol{m}_0(t) + [A].\boldsymbol{m}_0(t)] - \\ &+ [(\boldsymbol{H}_0 + \boldsymbol{H}_A - 4\pi\,[N].\boldsymbol{M}_0).\boldsymbol{e}_z]\sum_{\kappa\neq0}\boldsymbol{m}_\kappa(\boldsymbol{r},t) - \\ &- M_0[\sum_{\kappa\neq0}[A].\boldsymbol{m}_\kappa(\boldsymbol{r},t) + \frac{D}{M_0}\sum_{\kappa\neq0}\nabla^2\boldsymbol{m}_\kappa(\boldsymbol{r},t) + \\ &+ \boldsymbol{h}_{dip}[\boldsymbol{m}(\boldsymbol{r},t)] + \dots]\} - M_0\,\boldsymbol{e}_z\times\boldsymbol{h}(t)\,.\end{aligned} \tag{28}$$

Equation (28) can be separated into two independent linear equations. One is for the uniform precession,

$$\frac{d\boldsymbol{m}_0(t)}{dt} = \gamma\, \boldsymbol{e}_z \times \{[(\boldsymbol{H}_0 + \boldsymbol{H}_A - 4\pi\, [N].\boldsymbol{M}_0).\boldsymbol{e}_z]\boldsymbol{m}_0(t)] - \\ - M_0[-4\pi\, [N].\boldsymbol{m}_0(t) + [A].\boldsymbol{m}_0(t)]\} - M_0 \boldsymbol{e}_z \times \boldsymbol{h}(t)\,, \tag{29}$$

and the second is for the nonuniform modes of the system

$$\sum_{\kappa\neq 0} \frac{d\boldsymbol{m}_\kappa(\boldsymbol{r},t)}{dt} = \gamma\, \boldsymbol{e}_z \times \{[(\boldsymbol{H}_0 + \boldsymbol{H}_A - 4\pi\, [N].\boldsymbol{M}_0).\boldsymbol{e}_z] \sum_{\kappa\neq 0} \boldsymbol{m}_\kappa(\boldsymbol{r},t) - \\ - M_0\, [\sum_{\kappa\neq 0} [A].\boldsymbol{m}_\kappa(\boldsymbol{r},t) + \frac{D}{M_0} \sum_{\kappa\neq 0} \nabla^2 \boldsymbol{m}_\kappa(\boldsymbol{r},t) + \boldsymbol{h}_{dip}(\boldsymbol{m}_\kappa(\boldsymbol{r},t)) + ...]\}. \tag{30}$$

Equations (29) and (30) represent the general, linear, uniform and nonuniform eigenmode responses of the magnetic system,respectively. Depending on the range of **κ**'s involved, equation (30) describes different types of modes excited in the material. If only small κ's are considered (e.g., in thick films, or when only the in-plane dimensions of the film are important) the exchange contribution in equation (30) can be neglected. Equation (30) then describes the so-called *magnetostatic waves*. For moderate κ's, with exchange effects considered, equation (30) describes the so-called *dipole-exchange modes*. For short wavelengths, or large κ's, equation (30) describes *spin waves*. In the next section, we will address all of the possible solutions of the linearized equation of motion. The approach used here is quite general. Equation (3) also includes nonlinear or so-called *instability effects* if the higher order terms (of second and higher order in **m** and **h**) are not neglected. Some of these effects are described in chapters 8 and 9 of this book, and elsewhere, e.g., [20].

Let us return to the solution of the linearized equation of motion. We will start with the most basic excitation - uniform precession, as described by the solution of the equation (29).

4.1. Solution of Linearized Equation of Motion. Uniform Precession and Ferromagnetic Resonance

Consider first the solution of equation (29) without the external dynamic magnetic field (the so-called *pumping field*) term. We are looking for the lossless, steady state response of the system. We start with the effective field (27) which includes only the external static magnetic field $\mathbf{H}_{ext}$. Later, also the other terms of the effective field will be included. For simplicity let us assume that the external static field $\mathbf{H}_{ext} = H_0\, \mathbf{e}_Z$ is applied in the Z direction of the X, Y, Z coordinate system. Equation (29) then reduces to

$$\frac{d\boldsymbol{m}_0}{dt} = \gamma H_0 \boldsymbol{e}_Z \times \boldsymbol{m}_0(t). \tag{31}$$

Equation (31) can be expressed as two differential equations for the X and Y components of the uniform mode magnetization vector $\mathbf{m}_0$. The solution of equation (31) is

$$\boldsymbol{m}_0(t) = m_0 \cos(\omega_0 t + \varphi_{M0})\boldsymbol{e}_X + m_0 \sin(\omega_0 t + \varphi_{M0})\boldsymbol{e}_Y. \tag{32}$$

The resulting motion of the uniform mode $\mathbf{m}_0(t)$, as described by equation (32), is a circular precession around the direction of the static external field. This precession is in the so-called *Larmor sense* with frequency $\omega_0 = \gamma H_0$. In practical applications, the external pumping field usually has harmonic time dependence, and thus it is convenient to seek the solution of equation (31) in complex form. The solution can then be expressed as $m_{0i}(t) = m_{0i} \cdot \exp(\pm i\omega_0 t)$. Defining complex amplitudes $\mathbf{m}^{\pm} = (m_{0x}\, \mathbf{e}_x \pm m_{0y}\, \mathbf{e}_y)$, equation (31) gives

$$\boldsymbol{m}_0^{\pm}(t) = m_0(\boldsymbol{e}_X \mp i\boldsymbol{e}_Y)e^{\pm i\omega_0 t}, \tag{33}$$

where one can introduce the complex amplitudes

$$\begin{aligned} \boldsymbol{m}_0^{\pm} &= m_0(\boldsymbol{e}_X \mp i\boldsymbol{e}_Y) \\ \boldsymbol{m}_0^{+} &= (\boldsymbol{m}_0^{-})^{*} \end{aligned} \tag{34}$$

The ± signs correspond to the possible choice of the time dependence in the complex variables and * the complex conjugation. Obviously, the relation between solutions (32) and (33) is

$$m_0(t) = Re\left\{m_0^{\pm}.e^{\pm i\omega_0 t}\right\}. \quad (35)$$

A different approach has been used in [24] to calculate the transient response of the uniform precession to the external pumping field, while a similar method was used in [25] for the description of nonlinear ferromagnetic phenomena and in [26] to describe the ferromagnetic resonance effects at large precession angles. The direction of the magnetization vector in relation to the coordinate system can be defined by the polar and azimuthal angles θ_M and φ_M (Fig. 1) respectively. Components of the magnetization vector are then $M_X = M_0 \sin\theta_M \cos\varphi_M$, $M_Y = M_0 \sin\theta_M \sin\varphi_M$, and $M_Z = M_0 \cos\theta_M$. In this representation, $dM_i = \partial M_i/\partial\theta_M)d\theta_M + (\partial M_i/\partial\varphi_M)d\varphi_M$ holds for each component. When the effective field is expressed in its Cartesian components, the equation of motion can be expressed as

$$\frac{d\theta_M}{dt} = -\gamma H_{effx} \sin\varphi_M + \gamma H_{effy} \cos\varphi_M ,$$

$$\frac{d\varphi_M}{dt} = \gamma H_{effz} - \left[\gamma H_{effx}\left(\frac{\cos\theta_M}{\sin\theta_M}\right)\cos\varphi_M + \gamma H_{effy}\left(\frac{\cos\theta_M}{\sin\theta_M}\right)\sin\varphi_M\right]. \quad (36)$$

In the present case, the only nonzero component is the Z component of the effective field. The solution of equation (36) is then

$$\theta_M(t) = const. ,$$

$$\varphi_M(t) = \omega_0 t + \varphi_{M0}. \quad (37)$$

This result is, of course, the same as before. It describes the circular, Larmor precession of the magnetization around the direction of the static field. In both cases, φ_{M0}, denotes the initial phase of the magnetization.

Next, let us investigate the influence of the sample shape. This means that the terms $4\pi[\mathbf{N}].(\mathbf{M}_0 + \mathbf{m}_0(t))$ are kept in equation (29). If the sample is magnetized along one of its principal axes, say the Z axis, of the Cartesian coordinate system, then the demagnetizing tensor will be diagonal.

Turn now to equations (36). It will become clear that it is not possible to obtain closed form solutions, except for special cases. Numerical integration of equations (36) is necessary in general. Although this makes it more difficult to give a simple and illustrative interpretation of the solutions, this approach gives an alternate way to solve for the dynamic magnetic response. It is especially suitable for finding the transient response when there is a sudden change in some parameter of the system. Keeping the demagnetizing term in equation (29), equation (36) yields, after some elementary algebra,

$$\frac{d\theta_M}{dt} = -\gamma 2\pi M_0 [N_X + N_Y] \sin\theta_M \sin 2\varphi_M ,$$
$$\frac{d\varphi_M}{dt} = \gamma (H_0 - 4\pi M_0 N_Z) + \gamma 4\pi M_0 (N_X \cos\varphi_M{}^2 + N_Y \sin\varphi_M{}^2) \cos\theta_M . \tag{38}$$

This equation can only be solved numerically. It simplifies in the special case of $N_X = N_Y = N_\perp$, and under the assumption that the angle θ_M is small. One can use then the approximations $\sin\theta_M \approx \theta_M$ and $\cos\theta_M \approx 1$, and the solution of the second equation (38) is then

$$\varphi_M(t) = \omega_0 t + \varphi_{M0} ,$$
$$\omega_0 = \gamma [H_0 + 4\pi M_0 (N_\perp - N_Z)] . \tag{39}$$

The solution of the first equation is

$$\theta_M(t) = \exp\left[\frac{\gamma 4\pi M_0 N_\perp}{2\omega_0} \cos(2\omega_0 t + \varphi_{M0}') \right] . \tag{40}$$

Equations (39) and (40) describe the elliptical precessional motion of the magnetization vector with frequency ω_0.

Let us now investigate the solution of equation (29), without the anisotropy term, in the precessional frame. This coordinate system is represented in Fig. 1 through its unit vectors $\mathbf{e}_x$, $\mathbf{e}_y$, $\mathbf{e}_z$. The equilibrium position of the magnetization is described by polar and azimuthal angles $\theta_M{}^d$, and $\varphi_M{}^d$. We use the superscript d to stress the connection to the X, Y, Z coordinate system. The transformation matrix from the X, Y, Z to the x, y, z system is

$$[T] = \left\| \begin{matrix} \cos\theta_M^d \cos\varphi_M^d & \cos\theta_M^d \sin\varphi_M^d & -\sin\theta_M^d \\ -\sin\varphi_M^d & \cos\varphi_M^d & 0 \\ \sin\theta_M^d \cos\varphi_M^d & \sin\theta_M^d \sin\varphi_M^d & \cos\theta_M^d \end{matrix} \right\| . \quad (41)$$

Let the external static magnetic field be applied in an arbitrary direction described in the X, Y, and Z coordinate system by polar and azimuthal angles θ_H and φ_H. We will return to the previous example with $\mathbf{H}_{ext}$ along the Z axis later. In the precessional frame, the applied field can be expressed as

$$\boldsymbol{H}_{ext}^{[x,y,z]} = [\boldsymbol{T}].\boldsymbol{H}_{ext}^{[X,Y,Z]} = H_0\left\{\sin(\theta_H - \theta_M^d) + \sin\theta_H\cos\theta_M^d[\cos(\varphi_H - \varphi_M^d) - 1]\right\}\boldsymbol{e}_x +$$
$$+ H_0\sin\theta_H\sin(\varphi_H - \varphi_M^d)\boldsymbol{e}_y + H_0\left\{\cos(\theta_H - \theta_M^d) + \sin\theta_H\sin\theta_M^d[\cos(\varphi_H - \varphi_M^d) - 1]\right\}\boldsymbol{e}_z. \quad (42)$$

In a similar way, the static demagnetizing field contribution is

$$\boldsymbol{H}_{dip}^{[x,y,z]} = -4\pi M_0\left\{\frac{1}{2}(N_X\cos^2\varphi_M^d + N_Y\sin^2\varphi_M^d - N_Z)\sin 2\theta_M^d\,\boldsymbol{e}_x + \right.$$
$$+ \frac{1}{2}(N_X - N_Y)\sin\theta_M^d\,\sin 2\varphi_M^d\,\boldsymbol{e}_y + \quad (43)$$
$$\left. + \left[(N_X\cos^2\varphi_M^d + N_Y\sin^2\varphi_M^d)\sin^2\theta_M^d + N_Z\cos^2\theta_M^d\right]\boldsymbol{e}_z\right\},$$

and the dynamic dipolar field contribution is

$$\boldsymbol{h}_{dip}^{[x,y,z]} = -4\pi[\boldsymbol{T}].[\boldsymbol{N}].[\boldsymbol{T}]^{-1}.\boldsymbol{m}_0^{[x,y,z]} = -4\pi[\mathrm{N}].\boldsymbol{m}_0^{[x,y,z]}, \quad (44)$$

where $[\mathbf{T}]^{-1}$ is the inverse $[\mathbf{T}]$ matrix, and $[\mathbf{N}]$ is the demagnetizing tensor, transformed into the precessional frame, given by

$$[\mathrm{N}] = \begin{Vmatrix} N_{xx} & N_{xy} & N_{xz} \\ N_{yx} & N_{yy} & N_{yz} \\ N_{zx} & N_{zy} & N_{zz} \end{Vmatrix}, \tag{45}$$

where

$$\begin{aligned}
N_{xx} &= (N_X\cos^2\varphi_M^d + N_Y\sin^2\varphi_M^d)\cos^2\theta_M^d + N_Z\sin^2\theta_M^d, \\
N_{xy} &= \frac{1}{2}(N_Y - N_X)\cos\theta_M^d\ \sin2\varphi_M^d = N_{yx}, \\
N_{xz} &= \frac{1}{2}(N_X\cos^2\varphi_M^d + N_Y\sin^2\varphi_M^d - N_Z)\sin2\theta_M^d = N_{zx}, \\
N_{yy} &= N_X\sin^2\varphi_M^d + N_Y\cos^2\varphi_M^d, \\
N_{yz} &= \frac{1}{2}(N_Y - N_X)\sin\theta_M^d\ \sin2\varphi_M^d = N_{zy}, \\
N_{zz} &= (N_X\cos^2\varphi_M^d + N_Y\sin^2\varphi_M^d)\sin^2\theta_M^d + N_Z\cos^2\theta_M^d.
\end{aligned} \tag{46}$$

Inserting the effective field components expressed in the precessional coordinate system into the equation of motion (29), one obtains equations for the complex amplitude components of the dynamic magnetization,

$$\begin{aligned}
(i\omega_0 + \gamma 4\pi M_0 N_{yx})m_x + (\gamma H_{iz} + \gamma 4\pi M_0 N_{yy})m_y &= 0\ , \\
-(\gamma H_{iz} + \gamma 4\pi M_0 N_{xx})m_x + (i\omega_0 - \gamma 4\pi M_0 N_{xy})m_y &= 0\ .
\end{aligned} \tag{47}$$

The $\exp(i\omega_0 t)$ time dependence was used in derivation of equation (47). Solving (47) for the eigenfrequency ω_0 gives

$$\omega_0{}^2 = (\gamma H_{iz} + \gamma 4\pi M_0 N_{xx})(\gamma H_{iz} + \gamma 4\pi M_0 N_{yy}) - (\gamma 4\pi M_0 N_{xy})^2, \tag{48}$$

where H_{iz} is the z component of the internal static field as obtained from eqs. (42) and (43). We define the x and y *stiffness fields* as

$$H_x = (H_{iz} + 4\pi M_0 N_{xx})$$
$$H_y = (H_{iz} + 4\pi M_0 N_{yy})\,. \tag{49}$$

The resonance frequency is then

$$\omega_0 = \gamma \sqrt{H_x . H_y - (4\pi M_0 N_{xy})^2}\,. \tag{50}$$

Return now to the specific case of the static external field $\mathbf{H}_{ext}$ applied in the Z direction. From equation (26), the equilibrium angles are $\theta_M{}^d = \varphi_M{}^d = 0$, $N_{xy} = 0$, and (50) reduces to the well-known Kittel formula for the uniform mode resonance frequency [27, 28]. Equation (47), using the same definition as in (33), gives the complex eigenmode amplitude $\mathbf{m}_0$ as

$$m_0^+ = \frac{m_0}{\sqrt{\varepsilon_0}}(e_X - i\varepsilon_0 e_Y),$$
$$m_0^- = (m_0^+)^*, \tag{51}$$

where we have introduced the uniform precession *ellipticity factor* ε_0, given by

$$\varepsilon_0 = \frac{\omega_0}{\gamma\,[H_0 + 4\pi M_0 (N_Y - N_Z)]} = \frac{\omega_0}{\gamma H_y}\,. \tag{52}$$

The complex amplitude in (51) is normalized, by a procedure that will be explained later. The resulting motion of the magnetization as described by equations (51) and (52) is again elliptical precession in the Larmor sense. The main axes of the ellipse are along the unit vectors $\mathbf{e}_x$ and $\mathbf{e}_y$, and their amplitudes are 1 and ε_0, respectively.

The resulting influence of the sample shape on the eigenfrequency and eigenmode amplitude is described by equations (50), (51) and (52). The main result is that the circular precession, as characterized by the complex amplitude (34), changes due to the influence of the sample shape to an elliptical precession. The elliptical precession is described by the complex amplitude (51) and ellipticity factor (52). Since detailed calculations for specific sample shapes have been carried out and are described elsewhere, we will not go into more detail. All special cases are included in formula (50) for the uniform precession frequency.

The introduced ellipticity factor plays an important role in the understanding of nonlinear processes in finite size samples. We will return to this point later when investigating the forced excitations of the system.

The next important term in the effective field is due to the anisotropy. Let us investigate its influence on the solution of the equation of motion. To do so, we must first be more specific about the exact form of the magnetocrystalline anisotropy free energy. The coordinate system related to the crystallographic axes of the magnetic material is denoted by XA, YA, and ZA. This system need not coincide with the principal axes of the sample ellipsoid, although for simplicity here, we will assume that both coordinate systems coincide. Let us also assume that X = XA = [100], Y = YA = [010], and Z = ZA = [001] are the axes of the cubic magnetocrystalline anisotropy characterizing the material. The corresponding set of polar and azimuthal angles define the magnetization and effective field directions. The magnetocrystalline anisotropy free energy and related effective fields could be in the following analysis kept general. For better physical insight, however, it is more reasonable to work with a specific system. The restriction is made to cubic and uniaxial anisotropy to avoid excessive algebraic complications. The extension to other possible crystallographic forms is straightforward. For simplicity, the uniaxial anisotropy axis is along the Z axis of the coordinate system. The relevant free energy terms are

$$\begin{aligned} &F_A(\alpha_X,\alpha_Y,\alpha_Z) = F_{Ac}(\alpha_X,\alpha_Y,\alpha_Z) + F_{AU}(\alpha_X,\alpha_Y,\alpha_Z) \ , \\ &F_{Ac}(\alpha_X,\alpha_Y,\alpha_Z) = K_1(\alpha_X^2\alpha_Y^2 + \alpha_X^2\alpha_Z^2 + \alpha_Y^2\alpha_Z^2) \ , \\ &F_{AU}(\alpha_X,\alpha_Y,\alpha_Z) = K_U(1 - \alpha_Z^2) \ , \end{aligned} \tag{53}$$

where K_1 is the first cubic anisotropy constant and K_U is the uniaxial anisotropy constant. From (53) and (21), keeping in mind that $\alpha_X{}^2 + \alpha_Y{}^2 + \alpha_Z{}^2 = 1$, one obtains

$$\begin{aligned} \boldsymbol{H}_A &= H_{AX}\boldsymbol{e}_X + H_{AY}\boldsymbol{e}_Y + H_{AZ}\boldsymbol{e}_Z = \\ &= -\frac{2K_1}{M_0}\alpha_X(1-\alpha_X^2)\boldsymbol{e}_X - \frac{2K_1}{M_0}\alpha_Y(1-\alpha_X^2)\boldsymbol{e}_Y + \left[-\frac{2K_1}{M_0}\alpha_Z(1-\alpha_Z^2) + \frac{2K_U}{M_0}\alpha_Z\right]\boldsymbol{e}_Z, \end{aligned} \tag{54}$$

and for the components of the anisotropy tensor [**A**], the relations

$$A_{XX} = -\frac{2K_1}{M_0^2}(1 - 3\alpha_X^2); \; A_{YY} = -\frac{2K_1}{M_0^2}(1 - 3\alpha_Y^2);$$

$$A_{ZZ} = -\frac{2K_1}{M_0^2}(1 - 3\alpha_Z^2) + \frac{2K_U}{M_0}, \tag{55}$$

$$A_{ij} = A_{ji} = 0, \qquad \textit{for } i \neq j.$$

Applying the transformation matrix (44) to the anisotropy field, the components in the precessional frame can be expressed as

$$\boldsymbol{H}_A^{[x,y,z]}(t) = \boldsymbol{H}_A^{[x,y,z]} + [\mathrm{A}].\boldsymbol{m}_0^{[x,y,z]}(t), \tag{56}$$

where [A] is the anisotropy matrix transformed into the precessional frame. Substitution of (54) and (56) into (29) implies that the dynamic anisotropy effect on the uniform precession can be included as an "effective" demagnetizing field. This field is defined through an effective demagnetizing tensor with components

$$\mathrm{N}_{ij} = N_{ij} - \frac{A_{ij}}{4\pi}. \tag{57}$$

This derivation assumed uniform precession, so the results represented by equations (47 - 52) are still valid. However, the demagnetizing tensor components must be replaced by (57), and the static component of the anisotropy field (54) must be added to $\mathbf{H}_i$. Consequently, the magnetic moment still undergoes elliptical precession in the Larmor sense, and only formula (50) for the resonance frequency ω_0 must be modified.

Up to now, only the unforced eigenoscillations of the uniform precession mode of the magnetic system have been investigated. Next, we consider the *driven* or *pumped* response of the system. The pumping is provided by an external, time dependent field $\mathbf{h}(t)$. Harmonic time dependence $h(t) = h_0\, e^{i\omega t}$ is assumed. In a typical experimental situation, the pumping frequency is usually in the microwave region. In this chapter we will assume that the amplitude of the pumping field is small, relative to the static field, so that the response of the system is linear. The amplitudes and phases of the complex field components determine its polarization, which in general is elliptical. In the standard experimental situation, however, the microwave pumping field is linearly polarized. We will treat this problem in general and show some examples of the most common experimental situations. Solution of the torque equation for the driven uniform mode leads to the well-known relation

$$\left\| \begin{matrix} m_{x0} \\ m_{y0} \end{matrix} \right\| = \left\| \begin{matrix} \chi_e^{xx} & -i\kappa_e^{xy} \\ i\kappa_e^{yx} & \chi_e^{yy} \end{matrix} \right\| \left\| \begin{matrix} h_{0x} \\ h_{0y} \end{matrix} \right\| . \tag{58}$$

The tensor in equation (58) is called the *susceptibility tensor*. In the lossless case its components are given by:

$$\chi_e^{xx} = \frac{\gamma M_0 \omega_y}{\omega_0{}^2 - \omega^2} \qquad \chi_e^{yy} = \frac{\gamma M_0 \omega_x}{\omega_0{}^2 - \omega^2},$$
$$\kappa_e^{xy} = \frac{-\gamma M_0 (\omega + i\omega_{xy})}{\omega_0{}^2 - \omega^2} \qquad \kappa_e^{yx} = \frac{-\gamma M_0 (\omega - i\omega_{yx})}{\omega_0{}^2 - \omega^2}. \tag{59}$$

Here, ω_x, ω_y are the so-called stiffness frequencies obtained from the stiffness fields, and $\omega_{xy} = \gamma(4\pi M_0 N_{xy} - M_0 A_{xy})$. The resulting complex amplitude of the uniform precession is

$$m_0 = m_{0x} e_x + m_{0y} e_y =$$
$$= \gamma M_0 \left\{ \left[\frac{\omega_y}{\omega_0^2 - \omega^2} h_{0x} + i \frac{\omega + i\omega_{xy}}{\omega_0^2 - \omega^2} h_{0y} \right] e_x + \right. \tag{60}$$
$$\left. + \left[-i \frac{\omega - i\omega_{xy}}{\omega_0^2 - \omega^2} h_{0x} + \frac{\omega_x}{\omega_0^2 - \omega^2} h_{0y} \right] e_y \right\}.$$

The pumping field clearly has a large influence on the uniform precession response. Following equation (60), the amplitude of the uniform mode and the shape of the precession cone is determined by the frequency of the driving field. The amplitude becomes infinite at $\omega = \omega_0$ because the losses have been neglected in the system. We will address this topic later. Ferromagnetic resonance (FMR) results when the uniform precession mode is driven by the external time dependent magnetic field at its maximum amplitude. The corresponding eigenfrequency is called the *ferromagnetic resonance frequency*. Figure 2 shows a typical ferromagnetic resonance field H_{res} dependence for a YIG sphere magnetized in the (110) plane and plotted as a function of the angle between the applied field and the [100] direction. The measured resonance fields are shown as full circles. The pumping frequency was held constant at 9.11 GHz. The calculated field dependence, shown by a solid line, was obtained from (50) by setting $\varphi_M = \pi/4$.

Returning to the general disussion of the pumping field, equation (61) can be rewritten in the form

$$\boldsymbol{m}_0 = \frac{\gamma M_0}{2}[(\varepsilon_{xx}h_{0x} - i\varepsilon_{xy}h_{0y})\boldsymbol{e}_x + (i\varepsilon_{yx}h_{0x} - \varepsilon_{yy}h_{0y})\boldsymbol{e}_y]. \qquad (61)$$

For $\omega_{xy} = 0$, the ellipticity factor coefficients can be expressed as

$$\varepsilon_{xx} = \frac{1}{\varepsilon_0}\left(\frac{1}{\omega_0 - \omega} + \frac{1}{\omega_0 + \omega}\right) = \frac{1}{\varepsilon_0}(X_A - X_L),$$

$$\varepsilon_{yy} = \varepsilon_0\left(\frac{1}{\omega_0 - \omega} + \frac{1}{\omega_0 + \omega}\right) = \varepsilon_0(X_A - X_L) \quad , \qquad (62)$$

$$\varepsilon_{xy} = \varepsilon_{yx} = \left(\frac{1}{\omega - \omega_0} + \frac{1}{\omega_0 + \omega}\right) = (X_L + X_A),$$

where $X_L = 1/(\omega - \omega_0)$ and $X_A = 1/(\omega + \omega_0)$ represent the Larmor and anti-Larmor terms in the uniform, off-resonance response. From (60) and (62), it follows that for a driven system, the ellipticity of the uniform precession depends on the pumping field orientation and frequency. Generally, the pumping field can be expressed as

$$\boldsymbol{h}_0(t) = h_{0x}(t)\boldsymbol{e}_x + h_{0y}(t)\boldsymbol{e}_y + h_{0z}(t)\boldsymbol{e}_z,$$

where

$$h_{0j} = \frac{h_0}{2}a_j\left[e^{i(\omega t + \delta_j)} + e^{-i(\omega t + \delta_j)}\right] \qquad (63)$$

$$j = x, y, z.$$

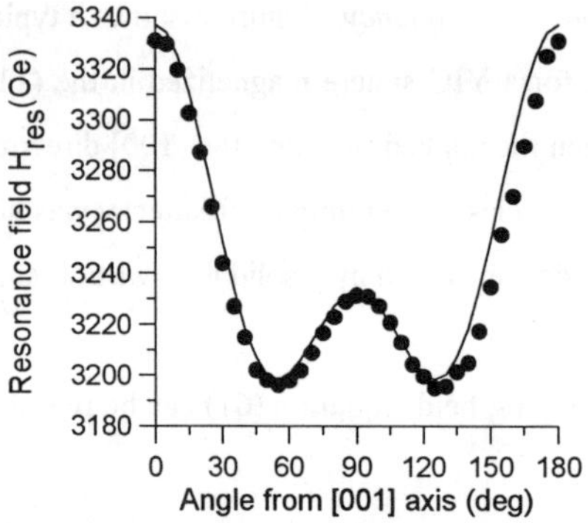

Fig. 2: Ferromagnetic resonance field dependence as a function of the field direction with respect to the [100] cubic anisotropy axis in the (110) plane. The solid symbols represent the data for single crystal yttrium iron garnet. The solid line are calculations using formula (49) The frequency was 9.1 GHz.

The coefficients a_j and individual phases δ_j in equations (63) define the linear, circular or elliptical pumping in Larmor or anti-Larmor sense and cover all the possible experimental configurations.

Let us now consider a specific situation: Consider a sample having the form of a thin, in-plane magnetized film, with the pumping field linearly polarized and applied perpendicular to the static field, as shown in Figure 3. For simplicity we will neglect the anisotropy. This is usually a good approximation for pure, liquid phase epitaxy (LPE) grown YIG films. The uniform precession cone is an ellipse with the major axis along the y direction. Let the linearly polarized pumping field be applied along the x axis. From (61) and (62), for $\gamma H_0 = \omega$, the ellipticity factor has a value of 1, meaning that the precession cone is circular. For the given geometry, this occurs only at field values well above the ferromagnetic resonance. For linear pumping along the y axis, the same condition is fulfilled for $\gamma H_0 = \omega - 4\gamma\pi M_0$. This field value is below the ferromagnetic resonance field for a fixed pumping frequency. Above this field,the ellipticity factor is different from one, and the precession cone is again elliptical. The major axis of the precession cone ellipse in the later case is along the x axis. This change in the orientation of the major axis of the uniform precession cone is important, as it helps to explain some of the effects observed in the nonlinear subsidiary absorption experiments [29, 30]. The analysis is more complex when anisotropy is also taken into account. The main physical conclusions remain the same, though, as in the present case.

4.2 Solution of Linearized Equation of Motion - Nonuniform Modes

4.2.1. Dipole Modes

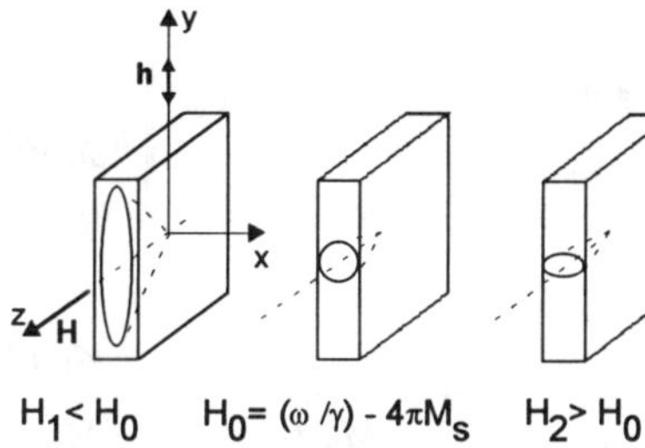

Fig. 3: Schematic illustration of the uniform precession cone for different external field values, as discussed in text.

Let us return now to the solution of equation (30). We first consider the simple case when the exchange field can be neglected. This approximation is called the *non-exchange* or *magnetostatic* limit. In this approximation,only the dipole interactions are considered. Because the difference between the angles of neighboring moments is small, the approximation is called *the long wavelength* approximation. The main interaction involves dipole forces, and the corresponding eigenmodes of the system are called *dipole* or *magnetostatic modes*. To obtain the spatial dependence of the dipole field, the static Maxwell's equations, subject to the electromagnetic boundary conditions are solved. To do so, the particular shape of the sample has to be considered. For a spherical sample, the problem was first solved by Walker [31]. For an in-plane magnetized slab, the problem was solved first by Damon and Eshbach [32]. The use of the static Maxwell's equations is justified only when the electromagnetic wave propagation time within the sample fulfills the condition $\Delta t \ll T_\kappa = 2\pi/\omega_\kappa$, so that the electromagnetic propagation effects can be neglected. If this condition is not fulfilled, the full system of Maxwell's equations must be used. The magnetostatic problem has been treated in many articles and books, mainly in the context of the development of magnetostatic wave devices. Therefore, only the most important results will be summarized here. In the following calculations, the time dependence of the nonuniform eigenmodes is assumed to be of the form $\exp(i\omega_\kappa t)$. The nonuniform magnetization component in (30) is then

$$\boldsymbol{m}_\kappa(\boldsymbol{r},t) = \boldsymbol{m}_\kappa(\boldsymbol{r})e^{i\omega_\kappa t}\,. \tag{64}$$

For spin waves, the result is the well-known dipole field contribution given by equation (17).

From (30), for an ellipsoidal sample, the relation between $\mathbf{h}_{dip}(\mathbf{m}_\kappa)$ and the κ -th mode complex amplitude is

$$\begin{Vmatrix} m_{\kappa x} \\ m_{\kappa y} \end{Vmatrix} = \begin{Vmatrix} \chi_\kappa^{xx} & -i\chi_{a\kappa}^{xy} \\ i\chi_{a\kappa}^{yx} & \chi_\kappa^{yy} \end{Vmatrix} . \boldsymbol{h}_{dip}(\boldsymbol{m}_\kappa), \tag{65}$$

where

$$\chi_\kappa^{xx} = \frac{\gamma M_0 \gamma (H_y - M_0 A_{yy})}{\gamma^2 (H_x - M_0 A_{xx})(H_y - M_0 A_{yy}) - \omega_\kappa^2 - [\gamma M_0 (4\pi N_{xy} - A_{xy})]^2},$$

$$\chi_\kappa^{yy} = \frac{\gamma M_0 \gamma (H_x - M_0 A_{xx})}{\gamma^2 (H_x - M_0 A_{xx})(H_y - M_0 A_{yy}) - \omega_\kappa^2 - [\gamma M_0 (4\pi N_{xy} - A_{xy})]^2}, \tag{66}$$

$$\chi_{a\kappa}^{xy} = \frac{\gamma M_0 [\omega_\kappa - i\gamma M_0 (A_{xy} - 4\pi N_{xy})]}{\gamma^2 (H_x - M_0 A_{xx})(H_y - M_0 A_{yy}) - \omega_\kappa^2 - [\gamma M_0 (4\pi N_{xy} - A_{xy})]^2} = (\chi_{a\kappa}^{yx})^*,$$

are the components of the magnetic susceptibility tensor. Equation (66) is similar to equation (59), with uniform mode resonance frequency ω_0 replaced by the magnetostatic mode eigenfrequency ω_κ, where the anisotropy contribution is expressed explicitly. To solve Maxwell's equations, one can again introduce the scalar potential

$$\boldsymbol{h}_{dip}[\boldsymbol{m}_\kappa(\boldsymbol{r})] = \nabla\psi_\kappa(\boldsymbol{r}). \tag{67}$$

Substituting (67) and (65) into the second of equations (13), one obtains the equation for the magnetic potential inside the sample. For the region outside the sample, Laplace's differential equation for the magnetic potential results. By applying the electromagnetic boundary conditions at the sample boundary, the determinant of the resulting system of equations yields the dispersion relation for the given geometry. The specific form of the solution for different geometries can be found in the cited literature. For the isotropic case of $\chi_\kappa^{xx} = \chi_\kappa^{yy} = \chi_\kappa$ and $\chi_\kappa^{xy} = \chi_\kappa^{yx} = \nu_\kappa$, equation (65) can be rewritten into the form

$$\boldsymbol{m}_\kappa = \chi_\kappa(\nabla\psi_\kappa + i\varepsilon_\kappa \boldsymbol{e}_z \times \nabla\psi_\kappa), \tag{68}$$

where the dipole mode ellipticity factor $\varepsilon_\kappa = \nu_\kappa/\chi_\kappa = \omega_\kappa/(\gamma H_{iz})$ has been introduced. Equation (68) shows that the precession cone of the magnetostatic mode is elliptical. This also applies for the case when the anisotropy is included. The main difference is that for each particular case, the major axis of the elliptical precession cone has to be found, and the vectors defining the plane of the ellipse need not to be perpendicular to each other.

It is necessary also to note that when the anisotropy terms are included, the off-diagonal elements in equation (66) are not purely imaginary, as treated in most of the literature. Consider for example, a thin slab with the x axis perpendicular to the slab plane. Let the expected complex off-diagonal susceptibility tensor component χ_κ^{xy} has the form χ_κ^{xy} = $\mathrm{Re}\chi_\kappa^{xy} + i\mathrm{Im}\chi_\kappa^{xy}$. Then the magnetic potential introduced in [32] must be modified into the form [33,34]

$$\psi^i = \cos(\kappa_z z)(Ae^{i\kappa_{x1}z} + Be^{i\kappa_{x2}z})e^{-i\kappa_y y}. \tag{69}$$

The ansatz (69) reflects the change in the mode profile throughout the film thickness. Here the κ_{xi} is the x component of the excitation wavevector inside the slab, κ_x^e is the x component of the wavevector outside the slab, and κ_y is the in-plane wavevector component. The relation between these components is obtained from the boundary conditions. Equation (69) can be written in the form

$$\psi^i = \cos(\kappa_z z)e^{i\xi x}(Ae^{\eta x} + Be^{-\eta x})e^{-i\kappa_y y} \tag{70}$$

where

$$\xi = \frac{-Re\left(\chi_\kappa^{xy}\right)\kappa_y}{1 + \chi^{xx}}, \quad \eta = \frac{\left\{[\kappa_z^2 + \kappa_y^2(1 + \chi^{yy})](1 + \chi^{xx}) - \left[Re\left(\chi_\kappa^{xy}\right)\right]^2\kappa_y^2\right\}^{1/2}}{(1 + \chi^{xx})}. \tag{71}$$

The dispersion relation is

$$\coth(\eta d) = \frac{\lambda\mu - \kappa^2}{\kappa(\lambda - \mu)} \tag{72}$$

where $\lambda = \eta(1 + \chi^{xx}) - \mathrm{Im}\chi_\kappa^{xy}\,\kappa_y$, $\mu = -\eta(1 + \chi^{xx}) - \mathrm{Im}\chi_\kappa^{xy}\,\kappa_y$, and $\kappa = (\kappa_z^2 + \kappa_y^2)^{1/2}$. For $\mathrm{Re}\chi_\kappa^{xy} = 0$, equations (69) and (72) reduce to the well-known ansatz and dispersion relation, respectively, from the Damon - Eshbach paper [32].

As already mentioned, the case of magnetostatic modes has been treated in many publications (see, e.g., [35]). Much attention has been paid to basic magnetostatic mode properties, which have proven to be important for the understanding of nonlinear phenomena in thin films. One of these basic properties is the scalar product of two eigenmodes. This scalar product was introduced for exchange modes in [36], and for the dipole modes in [37]. This product between the mode $\mathbf{m}_\kappa(\mathbf{r})$ and mode $\mathbf{m}_{\kappa'}(\mathbf{r})$ is defined as [38]

$$\langle \boldsymbol{m}_\kappa | \boldsymbol{m}_{\kappa'} \rangle = \int_{V_S} (i\boldsymbol{e}_z \times \boldsymbol{m}_\kappa)^* . \boldsymbol{m}_{\kappa'} \, dV = 2 V_S \delta_{\kappa\kappa'}, \tag{73}$$

where the integration is over the volume of the sample V_S, and $\delta_{\kappa\kappa'}$ is the Kronecker symbol. The definition (73) was used to normalize the complex amplitude of the uniform mode in equation (51). This definition was also used to explain some size dependent nonlinear processes (see chapter 9 in this book [38, 39]). It is valid also when the exchange interaction is included. Applying this normalization procedure to spin waves, (61) and (73) give

$$\boldsymbol{m}_k(\boldsymbol{r}) = \frac{1}{\sqrt{\varepsilon_k}} (\boldsymbol{e}_{k\|} - i\varepsilon_k \boldsymbol{e}_{k\perp}) \, e^{-i\boldsymbol{k}.\boldsymbol{r}}, \tag{74}$$

where $\varepsilon_k = \omega_k/(\gamma H_{iz})$ or $= \omega_k/(\gamma H_{iz} + Dk^2)$ (with exchange included) is the ellipticity factor of the spin wave. $\mathbf{e}_\|$ and $\mathbf{e}_\perp$ are unit vectors in the direction of the projection of the spinwave wavevector $\mathbf{k}$ on the xy plane and perpendicular to it, respectively as shown in Fig. 4. If the anisotropy is also taken into account, the ellipticity factor becomes complex, and the real value under the square root in (74) has to be taken. Because of anisotropy, the major axis of the spin wave precession cone is not perpendicular to the direction of propagation anymore. This spin wave ellipticity has been successfully used to explain the experimentally observed fine structure in the subsidiary absorption in thin YIG films [40].

4.2.2 Dipole - exchange Modes

From experimental data on thin films, it became evident that sometimes even in the long

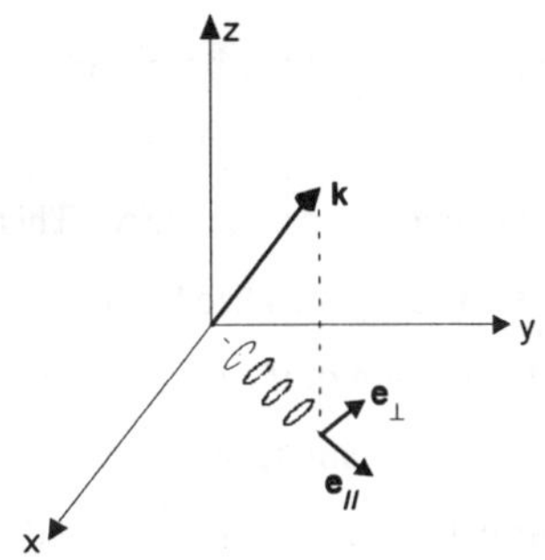

Fig. 4: Schematic illustration of the introduced coordinate system related to the spin-wave wavevector **k**. The ellipses represent the spin-wave precession cones for spin waves propagating along $\mathbf{e}_{//}$.

wavelength limit the exchange interaction cannot be neglected [41-44]. Therefore, the exchange term in eq. (61) has to be included. This case has been extensively studied by Kalinikos and his group, and a beautiful review can be found in [21]. More details can also be found in chapter 6 of this book. The solutions of the equation of motion with exchange included are denoted *dipole - exchange modes*. The presence of the exchange in the equation of motion leads to two major differences from the case of pure dipole modes. First, the exchange field is represented by a differential operator and increases the order of the differential or integro - differential equation describing the problem. The second difference is that the electromagnetic boundary conditions alone are not sufficient, as it is necessary to impose additional conditions on the magnetic moments at the boundaries of the sample. The most straightforward, if oversimplified condition, follows from the nature of the exchange interaction and its behavior at the boundary. Assuming a linear chain of exchange-coupled magnetic moments, the angle between neighboring moments in the long wavelength approximation, varies linearly. A consequence is that the phase difference between the adjacent moments is constant. If a boundary is present, the symmetry is broken leading to a modification of the torque acting on magnetic moments at the boundary. The resulting modified torque acts to keep the phase difference of adjacent moments constant, or the perpendicular components of the magnetic moment relative to the boundary constant. Mathematically, this can be expressed by condition

$$\frac{\partial \boldsymbol{m}_{\kappa}}{\partial \boldsymbol{n}} = 0, \qquad (75)$$

where **n** is a unit vector normal to the boundary. Condition (75) can also be obtained rigorously from the equation of motion when the behavior of the boundary, including the exchange term, is considered. The condition (75) was obtained under the assumption that there are no other torques acting on the magnetic moment at the boundary. This is not true in all practical cases and therefore condition (75) has to be generalized. This generalization has to include additional torques that can influence the motion of the magnetic moment at the boundary. These additional torques can be caused, for example, by "surface anisotropy." This case has been investigated by Rado and Weertman [45], and is discussed in detail in [46]. The more general form of the condition (75) is

$$a \frac{\partial \boldsymbol{m}_\kappa}{\partial \boldsymbol{n}} + b\ \boldsymbol{m}_\kappa = \boldsymbol{0}\,, \tag{76}$$

where the variables a and b are functions of the material parameters involved. The relation b = 0, which gives (75), is often called the *unpinned surface spin condition*. The relation (76) for a = 0 is called the *pinned surface spin* condition. Equation (76) gives the necessary additional exchange condition completing the mathematical description of the dipole - exchange mode behavior. There are two basic theoretical approaches for the treatment of this problem.

The first is based on the solution of the integro-differential equation for magnetization. This includes the solution of Maxwell's equations (18) with the corresponding electromagnetic and exchange boundary conditions. To solve the integro - differential equation, the magnetization is expanded into the infinite series of complete vector eigenfunctions, which satisfy the exchange boundary conditions. This approach is reviewed in [21].

The second approach was developed by Wolfram and deWames [47]. It is a direct extension of the Damon and Eshbach theory [32] for the dipole - exchange case. In this approach, the equation of motion and Maxwell's equations are solved subject to the electromagnetic and exchange boundary conditions. The dynamic dipole field in (13) is expressed as a gradient of the scalar potential $\mathbf{h}_{dip} = \nabla\psi_\kappa$. Neglecting the anisotropy contribution, equation (30) can be converted to a sixth-order differential equation for the scalar magnetic potential of the form

$$\left\{\left[O^2 + O - \Omega_\kappa^2\right]\nabla^2 - O\nabla_z^2\right\}\psi_\kappa = 0, \tag{77}$$

where the differential operator O is given by

$$O = \Omega_H - \frac{D}{4\pi}\nabla^2, \tag{78}$$

and $\Omega_\kappa = \omega_\kappa/(\gamma 4\pi M_0)$ and $\Omega_H = H_{iz}/4\pi M_0$ are the normalized frequencies. The complete solution of equation (77) for a thin film geometry is of the form [48]

$$\begin{aligned} \psi(r) &= C_1 \exp(iQ.\rho)\exp(|Q|z) && (z \le -d/2) \\ &= \sum \psi(Q,k_{zl}) && (-d/2 \le z \le d/2) \\ &= C_2 \exp(iQ.\rho)\exp(-|Q|z), && (z \ge d/2) \end{aligned} \tag{79}$$

where z is the coordinate normal to the surface, and

$$\psi(Q,k_{zl}) = \exp(iQ.\rho)(A_l \exp[ik_{zl}(z+d/2)] + B_l \exp[ik_{lz}(z-d/2)]). \tag{80}$$

The coefficients C_1, C_2, A_l, and B_l, are obtained from the boundary conditions. **Q** and **ρ** are the wavevector and position vector, respectively, in the plane of the film. To go into more detail, one would have to specify the position of the external biasing static magnetic field with respect to the film surface. Usually, parallel and perpendicular (to the surface of the film) cases are treated explicitly [e.g. 48]. One can approximately determine the limits of the magnetostatic and dipole - exchange theories by comparing the magnetostatic and exchange energies. We will follow the classification presented in [48]. With respect to the magnetostatic wavevector amplitude κ, the dipole and exchange energies are equal in YIG for κ values of about 10^6 cm^{-1}. This classification is not as simple, however, when considered in terms of the in-plane wavevector amplitude Q. Here, the comparison can be divided into three regions. For YIG these are: region (I),with $Q < 10^4 \text{cm}^{-1}$, region (II), with $10^4\text{cm}^{-1} \le Q \le 10^7\text{cm}^{-1}$, and region (III), with $10^7\text{cm}^{-1} < Q < Q_{max}$, the value corresponding to the reciprocal lattice vector. The magnetostatic theory is valid for $\kappa < 10^4\text{cm}^{-1}$, and for the magnetostatic branch (the branch with lowest value of κ_{zl}) in the region (I). For exchange branches, the magnetostatic results can be modified to a good approximation in region (I) by simply adding the exchange energy associated with κ_{zl}. This modification is nontrivial, as the quantized values of κ_{zl} depend on Q and in general are not simple multiples of π/d. In the transition region (II), the eigenvectors describing the spin wave states are admixtures of the exchange and magnetostatic branches. The existence and properties of dipole - exchange modes described by these theories have been proven in several recent experimental papers [49, 50, 51].

It is important to note that the above estimates of the magnetostatic theory limits are not absolute. Sometimes the exchange becomes important even for the lowest order magnetostatic branch for Q values that seem to fulfill the above criteria. Such situations happen for example, when the properties of the magnetostatic surface waves are investigated near the critical angle of propagation [52]. The observed "anomaly" [52], was mainly due to the neglect of the exchange effects in the calculations. A similar situation occurs when the excitations under the oblique bias in a thin film are investigated. Even for films where the magnetostatic limit is quite appropriate, it is necessary to consider the exchange term for certain angles of the bias field.

Another special problem related to the measurement of ferromagnetic resonance excitations in magnetic films is due to the finite size of the sample. Some complications may arise if the finite in-plane size of the sample is not considered. It should be stressed that the question of the existence of modes with quantized propagation constants in the film plane is entirely different from the question of the explicit surface pinning. The situation is schematically depicted in Fig. 5. Here, the discrete resonance frequencies corresponding to the quantized in-plane propagation vector **Q** are shown as solid circles. The various discrete frequencies can be labeled by a set of integer numbers (n_x, n_y)

$$\boldsymbol{Q} = (n_x \pi / L_x, n_y \pi / L_y) \,, \tag{81}$$

where L_x and L_y represent the in-plane dimensions of the film. The lowest order mode (1,1) for the in plane magnetized film lies on the magnetostatic volume or surface branch depending upon the sample dimensions and frequency. The angle ϕ_s in Fig. 5 represents the critical angle of propagation for the magnetostatic surface wave. The above discussion does not consider the particular pinning condition of the spins at the surface and is based on the magnetostatic approximation. If the spins are pinned at the film surfaces, then the usual magnetostatic volume and surface modes do not exist. The full dipole - exchange theory then has to be applied, and the lowest order mode (1,1) is strongly influenced by mixing with exchange modes. This was shown experimentally in [50] for a perpendicularly magnetized film.

In concluding this section, it is worth mentioning that the dipole - exchange modes represent an important part of the ferromagnetic resonance spectra. This statement is valid in general not only for very thin films (more often investigated at microwave frequencies), but also in special situations for thick films where the magnetostatic limit would be appropriate. The decision as to whether dipole - exchange mode analysis or the more simple dipole mode analysis has to be used depends on the experimental situation.

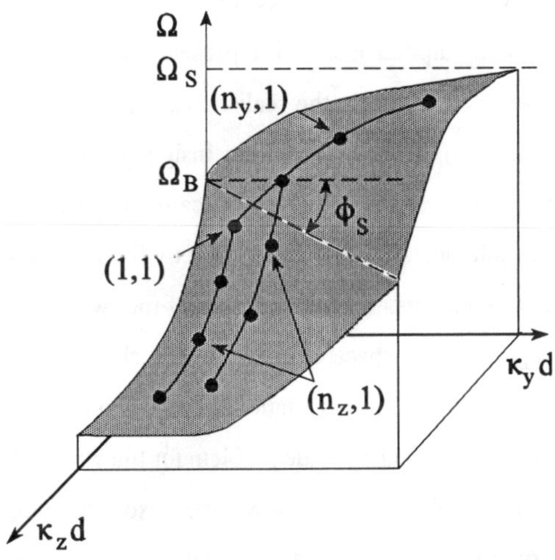

Fig. 5: Schematic illustration of the dispersion surfaces and dispersion curves according to the magnetostatic theory for finite in-plane size slab. κ_y and κ_z are in-plane wavevector componenets. d is the slab thickness, and $\Omega_B^2 = \Omega_H(\Omega_H + 1)$, with $\Omega_H = H/(4\pi M_s)$, and $\Omega_S = \Omega_H + 1/2$.

5. Magnetic excitations in multilayered systems

This section presents a brief

description of ferromagnetic resonance in multisublattice systems. Although this topic has received attention for decades (see, e.g., [2] chapter 9), it deserves further consideration because of the recent developments in the preparation of artificial multilayered structures. In these structures, the coupling between the neighboring layers can be changed from layer to layer. This outline of the ferromagnetic resonance response in these structures is based on the classical approach, a detailed treatment of which may be found in chapter 4 of this book. We introduce only the basic ideas, as illustrated by the example of two magnetic layers separated by a nonmagnetic spacer. The dynamic response of this structure is calculated by first finding the equilibrium orientation of the magnetization in the coupled layers. The equation of motion is then solved for each layer, taking into account all possible interactions and torques at the surfaces and interfaces.

Consider the simple case of two layers ferromagnetically coupled through the spacer, with only the dipolar interaction present. This case of was initially solved in [53]. Although it does not represent the realistic case (the exchange interaction cannot be neglected in this situation), it gives a simple insight into the mode behavior. It can serve also as an introduction to the more complicated problem where all the possible interactions have to be considered. The calculation strictly follows the Damon - Eshbach approach [32], modified for two layers. The thicknesses of the two layers are given by d_1 and d_2, and the layers are separated by a spacer of thickness s as shown in Fig. 6. For this configuration one writes the equation for a magnetostatic potential similar to that given by (79) for each layer. Then, the complete boundary value problem for the whole structure is solved. The solutions consist of two modes. Fig. 6 shows the absolute values of the profiles of the out-of-plane, m_x components of these modes. The calculations are for two different spacer thicknesses and for iron parameters ($4\pi M_0 = 22$ kG, $\gamma = 2.926$) with a fixed in-plane wavevector of 1.82×10^5 cm^{-1}. The profile denoted m^+_x corresponds to the in-phase mode, and the profile denoted m^-_x to the out-of-phase mode. If the spacer thickness decreases to zero, the in-phase mode becomes the regular DE surface mode with frequency corresponding to a mode for a single slab of thickness $d = d_1 + d_2$ and in-plane wavenumber Q. The out-of-phase mode has the profile of the "volume" mode, and for decreasing spacer thickness becomes the first order

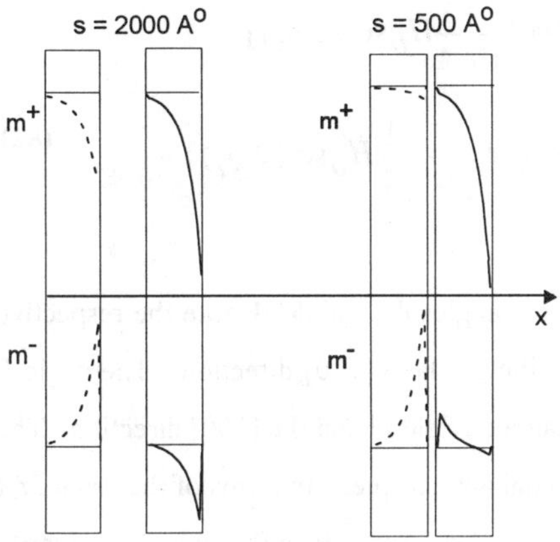

Fig. 6: Mode patterns through the film thickness for a double layer sample, for two spacer thicknesses. The + and - symbols correspond to the in-phase and out-of-phase modes of the system. The calculation is for iron parameters, $4\pi M_s$ = 22 kG, γ = 2.926, and an in-plane wavevector amplitude of 1.82×10^5 cm^{-1}.

volume mode. The frequency of the out-of-phase mode in this approximation approaches the frequency of the upper limit of the spin wave manifold, which is the degenerate frequency of all volume modes for this configuration. This is because the exchange has been neglected. The effect of the exchange is to lift the degeneracy, and the frequency of the out-of-phase mode will approach the frequency of the first exchange volume mode for a single layer of thickness $d_1 + d_2$.

This example is only illustrative, but describes the behavior of the possible resonance modes of a multilayer structure. Let us now briefly address the case when the magnetizations of the two layers have different positions compared with a specified direction. This happens when a small bias field is applied to two anisotropic, anti-ferromagnetically coupled layers. The situation is depicted schematically in Fig. 7. Such films may be characterized, for example, in terms of the usual [100] easy direction of iron, a small uniaxial anisotropy along one in-plane hard axis, and ferromagnetic or antiferromagnetic interlayer exchange coupling. For thin, single domain, in-plane magnetized, double layer sandwich films of thickness d_1 and d_2, and magnetization vectors $\mathbf{M}_1$ and $\mathbf{M}_2$, respectively, one may write the film energy density U per unit area as [54]

$$\frac{U}{M_0} = d_1[H\cos(\theta - \phi_1) + \frac{1}{8}H_A\sin^2(2\phi_1) - \frac{1}{4}H_U\sin(2\phi_1)] +$$

$$+ d_2[H\cos(\theta - \phi_2) + \frac{1}{8}H_A\sin^2(2\phi_2) - \frac{1}{4}H_U\sin(2\phi_2)] + \tag{82}$$

$$+ (d_1 + d_2)\, H_J \cos(\phi_1 - \phi_2)\,.$$

Here, M_0 is the saturation magnetization, and the angles ϕ_1 and ϕ_2 denote the respective angles between $\mathbf{M}_1$ and $\mathbf{M}_2$ vectors and a specific in-plane [100] direction. The angle θ denotes the angle between the applied in-plane magnetic field H and the [100] direction. The fields H_A and H_U characterize the cubic and uniaxial anisotropies. In terms of the usual first order anisotropy energy density parameters, K_1 and K_u, these fields may be written as $H_A = 2K_1/M_0$ and $H_U = 2K_u/M_0$, respectively. The coupling between the layers is specified in terms of the exchange field H_J. Positive H_J gives the minimum energy, with $\mathbf{M}_1$ and $\mathbf{M}_2$ anti-parallel. This interlayer exchange coupling is often expressed in terms of a volume exchange energy density parameter $J = M_0 H_J$, or a surface exchange energy density parameter, $A_{12} = J(d_1 + d_2)/2$. Equation (79) can be used to evaluate the static equilibrium values ϕ_1 and ϕ_2 of a layer's magnetization as a function of H or other parameters.

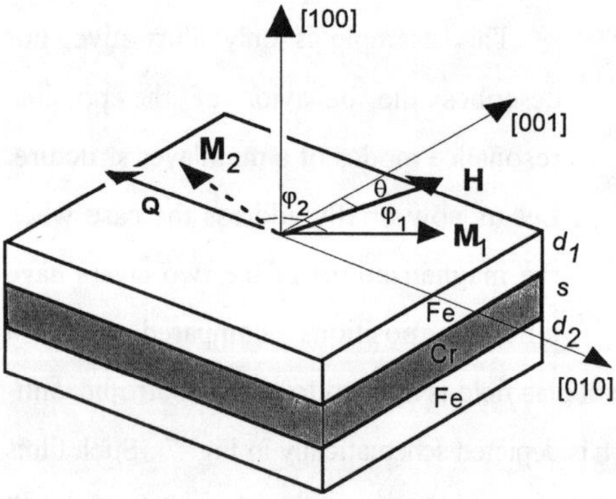

Fig. 7: Geometry of the sandwich Fe/Cr/Fe film. The [100] anistropy axis is perpendicular to the film plane. The in-plane layer magnetizations $\mathbf{M}_1$ and $\mathbf{M}_2$ lie at angles φ_1 and φ_2 with respect to the [001] axis, respectively. The in-plane external magnetic field **H** is at an angle θ with respect to the same axis. **Q** represents the in-plane propagation vector.

The dynamic theory follows the basic micromagnetic approach [55 - 57]. This combines the torque equation for magnetizations $\mathbf{M}_1$ and $\mathbf{M}_2$ for two thin magnetic layers, effective fields $\mathbf{H}_{\text{eff1}}$ and $\mathbf{H}_{\text{eff2}}$ that act on $\mathbf{M}_1$ and $\mathbf{M}_2$, and the appropriate boundary conditions to find the normal modes of the system. The effective fields include terms due to the applied field, exchange, cubic and uniaxial anisotropy, and dipolar interactions. The dipole field Fourier components, $\mathbf{h}_{dip}(\mathbf{Q},X)$, where X denotes the coordinate perpendicular to film plane, are related to the dynamic magnetization Fourier components $\mathbf{m}_1(\mathbf{Q},X)$ and $\mathbf{m}_2(\mathbf{Q},X)$ for the two layers through Maxwell's equations [58], [59] or through Maxwell's equations and the Green's function formalism [43]. This procedure yields two sets of coupled equations involving $\mathbf{m}_1$ and $\mathbf{m}_2$.

The solutions of these coupled equations, which are quite complicated for thick magnetic films, are simplified considerably for the lowest frequency modes and very thin magnetic layers. This simplification results because for very thin magnetic layers, the lowest frequency mode for a single layer is a near-uniform mode. For this mode, $\mathbf{m}_1$ or $\mathbf{m}_2$ is essentially constant across the film thickness. This, in turn, allows one to average the effective fields $\mathbf{H}_{\text{eff1}}$ and $\mathbf{H}_{\text{eff2}}$ over the thickness of the films, simplifying the calculation considerably [57], [59]. Through this averaging process, one invokes the so-called Hoffmann boundary conditions at the magnetic film surfaces [61]. The result is a set of six coupled equations of the form

$$[\boldsymbol{L}_{op} - \omega/\gamma]\begin{bmatrix}\boldsymbol{m}_1\\ \boldsymbol{m}_2\end{bmatrix} = 0, \tag{83}$$

where $\mathbf{L}_{\text{op}}$ represents a matrix of terms containing the external applied field H and various magnetic parameters for the sandwich listed in equation (79). ω is the spinwave frequency, and γ denotes the gyromagnetic ratio for the material. The system of six equations represented by equation (83) can be solved for the normal mode spin wave frequencies and the normal mode magnetization distributions.

Finally, by properly choosing coordinate systems,for example such that the local z_1- and z_2-axes are in the directions of the static components of the $\mathbf{M}_1$ and $\mathbf{M}_2$ vectors in the two magnetic layers, respectively, the number of coupled equations can be reduced from six to four. The solutions again correspond, in certain limits,to in-phase and out-of-phase precessions for the $\mathbf{M}_1$ and $\mathbf{M}_2$ vectors. In the high field limit with the static components $\mathbf{M}_1$ and $\mathbf{M}_2$ parallel to each other, the in-phase mode reduces to the ferromagnetic resonance mode with $\mathbf{M}_1$ and $\mathbf{M}_2$ precessions in-phase. For the out-of-phase mode, the precessional motions for $\mathbf{M}_1$ and $\mathbf{M}_2$ in this limit are 180° out-of-phase. As in bulk, two sublattice, antiferromagnets and ferrimagnets, the in-phase and out-of-phase frequencies can be different. In such systems, the out-of-phase frequency is usually much higher than the in-phase frequency, depending on the applied field $\mathbf{H}$ and the static configurations for $\mathbf{M}_1$ and $\mathbf{M}_2$. The mode frequencies can also be different for opposite spin wave propagation directions. This difference in propagation properties represents an extension of the Damon-Eshbach theory for surface modes. More details of the theory outlined above are found in the references.

6. *Relaxation*

Up to now we have treated the magnetic medium as lossless. The damping term was introduced in equation (3), but was not discussed in detail. In this section we include the damping in the equations of motion phenomenologically, and briefly discuss the implications for the ferromagnetic resonance linewidths observed in experiment. The experimental and theoretical study of the relaxation processes in ferromagnetic resonance yields important information about microscopic properties of magnetic materials. Therefore, knowledge of the losses in magnetic systems is extremely important for their applications.

The damping term in equation (4) can be obtained from simple, nonequilibrium thermodynamics arguments. We will follow the approach of [62] for insulators by considering a region in the sample Ω where energy conservation can be described by the equation

$$\frac{d}{dt}\int_{\Omega} e\,dV = -\oint \boldsymbol{J}_q . d\boldsymbol{S} - \oint (\boldsymbol{E} \times \boldsymbol{H}) . d\boldsymbol{S}\,, \tag{84}$$

where e denotes the energy density, $\mathbf{J}_q$ the density of the heat flow, $\mathbf{E}$ the electric field intensity, $\mathbf{H}$ the magnetic field intensity and $(\mathbf{E} \times \mathbf{H})$ the Poynting vector. dS is a surface element oriented out of the region Ω. After simple manipulation this integral equation leads to a local differential equation of the form

$$\frac{\partial e}{\partial t} + \nabla . \boldsymbol{J}_q + \nabla . (\boldsymbol{E} \times \boldsymbol{H}) = 0 . \qquad \textbf{(85)}$$

Using Maxwell's equations, the last term in equation (85) can be rewritten as

$$\frac{\partial e}{\partial t} = - \nabla . \boldsymbol{J}_q + \boldsymbol{E} . \frac{\partial \boldsymbol{D}}{\partial t} + \boldsymbol{H} . \frac{\partial \boldsymbol{B}}{\partial t} + \boldsymbol{J} . \boldsymbol{E} , \qquad \textbf{(86)}$$

where **J** is a current density. For a magnetic insulator, this equation is

$$\frac{\partial e}{\partial t} = - \nabla . \boldsymbol{J}_q + \boldsymbol{H} . \frac{\partial \boldsymbol{B}}{\partial t} . \qquad \textbf{(87)}$$

If the energy density *e* is expressed through the internal energy density *u* of a unit sample volume, one obtains

$$\frac{\partial u}{\partial t} = - \nabla . \boldsymbol{J}_q + \boldsymbol{H}_{eff} . \frac{\partial \boldsymbol{M}}{\partial t} . \qquad \textbf{(88)}$$

For a reversible process, equation (88) becomes

$$\frac{\partial u}{\partial t} = \boldsymbol{H}_{eq} . \frac{\partial \boldsymbol{M}}{\partial t} + T \frac{\partial s}{\partial t} , \qquad \textbf{(89)}$$

where $\mathbf{H}_{eq}$ is the equilibrium position of the effective magnetic field, T is the temperature in degrees Kelvin, and *s* is the entropy per unit volume of the insulator. Subtracting (89) from (88), we obtain an equation of continuity for the entropy (in this special case) of the form

$$\frac{\partial s}{\partial t} = - \nabla . \boldsymbol{J}_s - \frac{\left(\boldsymbol{H}_{eq} - \boldsymbol{H}_{eff} \right)}{T} . \frac{\partial \boldsymbol{M}}{\partial t} , \qquad \textbf{(90)}$$

where $\mathbf{J}_s = \mathbf{J}_q / T$. Equation (90) represents the "production" of entropy. The first factor in the second term represents a generalized thermodynamic force,responsible for the nonequilibrium processes. In the linear theory the $\partial \mathbf{M} / \partial t$ term can be expressed as a linear combination of generalized thermodynamic forces. If the nonequilibrium process is due only to the relaxation of the magnetization, then this relation is

$$\frac{d\boldsymbol{M}}{dt} = L\frac{\boldsymbol{H}_{eff} - \boldsymbol{H}_{eq}}{T}, \tag{91}$$

where the proportionality constant L depends on the local state parameters. Let us further assume that **M** is in the $\mathbf{H}_{eq}$ direction,and that the magnetization during the relaxation changes its direction, but not its amplitude. Then, equation (91) can be rewritten as

$$M\frac{d\boldsymbol{e}_M}{dt} = \frac{L}{T}\boldsymbol{H}_{eff\perp} = \frac{L}{T}[\boldsymbol{H}_{eff} - (\boldsymbol{H}_{eff}.\boldsymbol{e}_M)\,\boldsymbol{e}_M] = -\frac{L}{TM^2}\boldsymbol{M}\times(\boldsymbol{M}\times\boldsymbol{H}_{eff}). \tag{92}$$

The last term in equation (92) was obtained under the assumption that the amplitude of the magnetization vector is constant. This equation also describes the motion of the magnetization vector due to dissipation. The full equation of motion, including the dissipation process, is given by a sum of the right-hand sides of equation (3) and (92), yielding

$$\frac{d\boldsymbol{M}}{dt} = -\gamma\boldsymbol{M}\times\boldsymbol{H}_{eff} + \frac{\alpha}{M}\boldsymbol{M}\times(\boldsymbol{M}\times\boldsymbol{H}_{eff}). \tag{93}$$

where $\alpha = -L/(TM^2)$. This equation was introduced in 1935 by Landau and Lifshitz [15]. The damping term can also be expressed in other forms, which have been extensively discussed in the literature. A summary can be found, for example, in [6].

To include the damping term in any of the above solutions, it would be necessary to resolve the equation of motion (93) and the corresponding boundary conditions. A simpler approach involves the *complex frequency method* [14], where the damping is taken into account by substituting a complex frequency for the natural resonance frequency. The natural resonance frequency $\omega_0=(\omega_x\omega_y)^{1/2}$ is replaced by a complex frequency

$$\omega_0 \to \omega_0 + i\eta_0, \tag{94}$$

where η_0 is a phenomenological damping rate. Further, the stiffness frequencies must be replaced by

$$\begin{aligned} \omega_x &\to \omega_x + iq_0, \\ \omega_y &\to \omega_y + iq_0. \end{aligned} \tag{95}$$

In these expressions, q_0 is a phenomenological damping rate,assumed to be the same in the x- and y- directions. The necessary relation between relaxation rates η_0 and q_0 is simply obtained from their definition

$$(\omega_x \omega_y)^{1/2} + i\eta_{i0} = [(\omega_x + iq_0)(\omega_y + iq_0)]^{1/2}. \tag{96}$$

After some simple algebra, one obtains the relation

$$q_0 = \frac{2(\omega_x \omega_y)^{1/2}}{(\omega_x + \omega_y)} \eta_0. \tag{97}$$

The proportionality factor reduces to unity for spheres. With this relation established, it is possible to substitute the modified expressions for the stiffness frequencies into the susceptibility equations that become complex. Introduction of the dissipation term into the equation of motion limits the power absorbed by the specimen at ferromagnetic resonance. This allows the experimental evaluation of the dissipation processes by measuring the so-called *resonance* or *effective linewidth*, as discussed in much of the literature. Recent interest in new preparation techniques for magnetic thin films has led to careful evaluation of experimental results. Because of the sample geometry and the low magnetic losses, many modes are excited in a standard FMR experiment. The mode frequencies (at constant external field) or resonance fields (at constant pumping frequency) can be easily obtained using the procedures described above. The situation is not as simple for the resonance linewidth. If the number of excited modes is large, or the separation of the modes is not large enough, the modes overlap and the curves are deformed. This complicates the extraction of the linewidths from the measurements, and fitting procedures are therefore required. In the simplest approach, the power absorbed by the sample is given by [63]

$$P_{abs} = \int_V \mathbf{h} . \frac{d\mathbf{M}}{dt} dV, \tag{98}$$

For multimode excitations one can follow, for example, the approach used in [64]. There the DE boundary value problem is solved and the integral (98) is calculated for the thin film geometry using the proper criteria for amplitudes and excitation conditions. The losses can be then introduced into the calculation using a Lorenzian absorption curve for each excited magnetostatic mode. The absorption curve has the form

$$L(H) = \frac{2I}{\pi \Delta H \left[1 + \left(\frac{H - H_r}{\Delta H/2} \right)^2 \right]} . \tag{99}$$

is the intensity of the excited mode, $\Delta H = 2\eta_0 / \gamma$ the resonance linewidth, and H_r the resonance field at a given frequency. To calculate the mode intensities, the coupling with the real excitation structure (waveguide, resonator, microstrip line etc.) also has to be considered. The above approach allows one to calculate the absorption spectra and compare them with the experimental results, providing one method of investigating the losses in the film. If the intensities of the excited modes are correctly recognized, information about the pinning conditions at the film surfaces can be also obtained. The full dipole-echange mode analysis must be carried out, but the approach is essentially the same as described above.

We next address the influence of the low or moderate conductivity of the sample on the measured losses. Spin wave resonance in metals is reviewed in [65]. The linewidth in metals has been extensively studied for several decades and a comprehensive review of earlier results is given in [3]. The contribution to the linewidth in metals can be separated into two parts. One part is connected with the exchange effects, and the other is related to the relaxation processes as influenced by the conductivity of the sample. Calculations for thin films can be found in [66]. Here, we only address the influence of conductivity for magnetic fields far from resonance, so that the effects are associated with the measured FMR tail response. The details of the calculations for medium thick slabs may be found in [67] and [68]. The geometry of the problem is shown in Fig. 8. The uniform microwave field $\mathbf{h}(t) = \mathbf{h}e^{i\omega t}$ is assumed to drive the uniform mode magnetization $\mathbf{m}_\perp(t) = \mathbf{m}_\perp e^{i\omega t}$. $\mathbf{h}(t)$ and $\mathbf{m}_\perp(t)$ combine to produce the dynamic magnetic induction $\mathbf{b}(t)$ which induces a microwave electric field $\mathbf{e}(\mathbf{r},t)$ inside the sample. This electric field in turn, generates eddy currents that result in ohmic losses. The critical assumption in the model is that the additional microwave fields produced

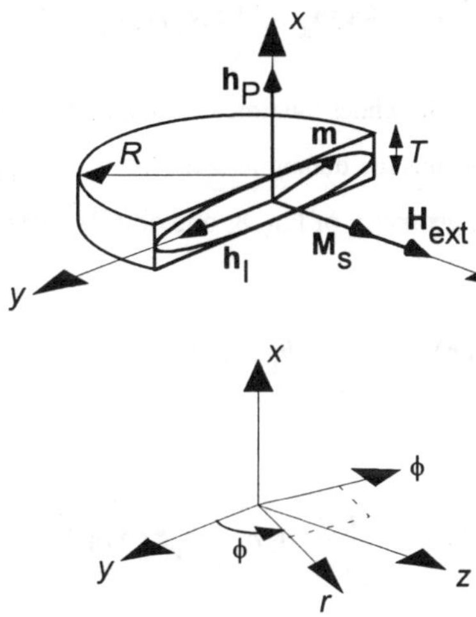

Fig. 8: Diagrams of the (x,y,z) coordinate system for a ferrite disk of thickness T and radius R, and the (r, ϕ, x) coordinate system for the eddy current problem. The disk is centered at $x = 0$ and is perpendicular to the x axis. The in-plane magnetic field $\mathbf{H}_{ext}$ and the static magnetization $\mathbf{M}_s$ are along the z-axis. The (r, ϕ, x) coordinate system has the r- axis in the disk plane, at an angle ϕ relative to the positive y- axis.

by eddy currents are small compared with the driving field $\mathbf{h}(t)$ and do not appreciably modify the uniform mode response. It is also important to emphasize that this model is valid only in the far-from-resonance limit and is useful for materials with moderate conductivity. For high conductivity materials, the eddy current fields are large, and a full self consistent solution of the boundary value response problem is needed (see, e.g., [68]).

The electric fields and corresponding eddy current distribution for the in-plane magnetized disk geometry are best calculating using the cylindrical coordinate system shown in Fig. 8. Disk surfaces are at $X = \pm\, T/2$, the r- and ϕ- axes are in-plane, with the r- axis at an angle ϕ relative to the positive Y- direction, and the ϕ axis at ϕ relative to the z-direction. It is assumed that the X directed currents are suppressed by the thin disk geometry, i.e., the X components of the induced electric field and current are neglected.

One then starts with the uniform in-plane microwave field $\mathbf{h}(t) = he^{i\omega t}\mathbf{y}$, and the corresponding dynamic magnetization $\mathbf{m}_\perp(t)$. The electric field is written as $\mathbf{e}(\mathbf{r},t) = \mathbf{e}(\mathbf{r})e^{i\omega t}$. The function $\mathbf{e}(\mathbf{r})$ may be obtained from Maxwell's equations according to

$$\nabla \times \boldsymbol{e}(\boldsymbol{r}) = -\frac{i\,\omega}{c}\boldsymbol{b}\,, \tag{100}$$

where c is the speed of the light and $\mathbf{b}$ is given by

$$b = h + 4\pi m_\perp - 4\pi N_\perp m_x e_x - 4\pi N_\parallel m_y e_y. \qquad (101)$$

Note that **b** defines the time dependent magnetic induction $\mathbf{b}(t) = \mathbf{b}e^{i\omega t}$ inside the sample. It also contains the demagnetizing terms, the applied microwave field **h**, and the $\mathbf{m}_\perp$ response. For the geometry of Fig. 8, one obtains a complex electric field amplitude component given by

$$e_r(\boldsymbol{r}) = \frac{i\omega}{2c}[2b_y x \sin\phi], \qquad (102)$$

and

$$e_\phi(\boldsymbol{r}) = -\frac{i\omega}{2c}[b_x r - 2b_y x \cos\phi]. \qquad (103)$$

Note that the electric field has both r and ϕ components, and that $\mathbf{e}(\mathbf{r})$ depends on r, ϕ, and x. The electric field given by equations (95) and (96) also satisfies the $\nabla.\mathbf{e}(\mathbf{r}) = 0$ condition. From Ohm's law, the microwave electric field generates a current density distribution $\mathbf{j}(\mathbf{r})e^{i\omega t} = \sigma\, \mathbf{e}(\mathbf{r})e^{i\omega t}$, where σ denotes the microwave conductivity at frequency ω.

Let the time averaged eddy current power dissipation for the disk be denoted by P_E. This power dissipation may be obtained by integrating the time averaged eddy current power dissipation density $1/2\sigma\,|\mathbf{e}(\mathbf{r})|^2$ over the entire volume of the disk. This gives

$$P_E = \frac{1}{2}\int_V \sigma\left(|e_r(\boldsymbol{r})|^2 + |e_\phi(\boldsymbol{r})|^2\right)dV = V\frac{\sigma\omega^2}{8c^2}\left[\frac{R^2}{2}|b_X|^2 + \frac{T^2}{3}|b_Y|^2\right], \qquad (104)$$

where the sample volume, $V = \pi R^2 T$ has been explicitly factored out. As far as the $|b_X|^2$ and $|b_Y|^2$ factors in equation (104) derive from uniform microwave field and dynamic magnetization response throughout the disk volume, these factors are constant. This means that the average eddy current dissipation density has two terms, which scale with the square of the disk radius R, and the square of the disk thickness T.

The remainder of the formal analysis is straightforward and involves writing $|b_X|^2$ and $|b_Y|^2$ in terms of the microwave field and magnetic response functions. For the problem here, the field is linearly polarized along the Y- axis. This means that $|b_X|^2$ and $|b_Y|^2$ contain common h^2 factors that may be factored out. The result is a working equation for the time averaged eddy current power dissipation in the disk

$$P_E = V\frac{\sigma\,\omega^2 h^2}{8c^2}\left[\frac{R^2}{2}U_X + \frac{T^2}{3}U_Y\right], \tag{105}$$

where U_X and U_Y are given by

$$\begin{aligned} U_X &= (1-N_\perp)^2\,[(4\pi\kappa')^2 + (4\pi\kappa'')^2], \\ U_Y &= 1 + 2.(1-N_\|).4\pi\chi_Y' + (1-N_\|)^2[(4\pi\chi_Y')^2 + (4\pi\chi_Y'')^2]. \end{aligned} \tag{106}$$

for the pumping microwave field in-plane of the disk, and

$$\begin{aligned} U_X &= 1 + 2(1-N_\perp)4\pi\chi_X' + (1-N_\perp)^2[(4\pi\chi_X')^2 + (4\pi\chi_X'')^2], \\ U_Y &= (1-N_\|)^2[(4\pi\kappa')^2 + (4\pi\kappa'')^2]. \end{aligned} \tag{107}$$

for the pumping microwave field perpendicular to the disk plane. The above expressions simplify considerably in the high field limit, which is pertinent here. In this limit, the $4\pi\chi_{X,Y}$ and $4\pi\kappa'$ terms are much bigger than the $4\pi\chi_{X,Y}''$ and $4\pi\kappa''$ terms, showing that the high field FMR tail is not dependent on the magnetic losses. This changes the analysis of the measured high field results in these materials considerably.

We would like to make a final comment about the analysis presented in this section. It should be realized that the sample thickness plays an extremely important role. The presented results are based on the experimental results obtained on relatively thick samples (several hundred microns or more). If the thickness of the sample is reduced to below about 100 μm, the in-plane microwave pumping FMR tail is dominated again by the magnetic properties of the sample, and the classical effective linewidth technique can be used even for samples with moderate conductivity. The thickness limit varies depending upon the conductivity and the pumping frequency used. The analysis presented in the previous paragraph allows one to estimate the necessary limits for some given sample parameters. It also allows one to choose the strategy of the experiment, depending on the desired information.

7. New FMR experimental techniques

Since the uniform precession resonance frequency for available magnetic materials is in the microwave and millimeter frequency band, most experimental techniques carry out microwave investigation of ferromagnetic resonance phenomena. Recently, several new experimental techniques have emerged that combine microwave pumping with additional experimental tools. Each of them has certain advantages and drawbacks compared to the classical microwave approach. These techniques make it possible to directly observe the spatial distribution of the modes excited in the magnetic system. The first of these, developed by Pelzl *et al.*, known as the *photoacoustic and photothermal FMR detection technique*, is reviewed in [69], and in chapter XIII of this book. The second, pioneered by Wettling and Jantz [70] and further developed in the past ten years for spatially resolved FMR measurements is called the *Brillouin light scattering technique*. We will briefly introduce their basic ideas. Details can be found in the chapter 13 and the referred literature.

7.1 Photothermal methods

In principle, the photothermally based FMR techniques use two approaches. The first involves the thermal modulation of the FMR response by application of a focused light beam on the microwave field pumped sample. Consequently, the FMR is locally influenced due to the temperature variation. If the beam is modulated, the FMR response follows the modulation cycle, allowing the spatial mapping of the FMR absorption. This approach is called the *photothermally modulated* or *PM-FMR* technique [71].

The second approach is based on the mirage effect, where the beam passing over the sample is deflected due to the modulation in the permitivity of the air. This modulation is caused by the heat generated near the sample by the FMR absorption. By scanning the surface of the sample with the beam, a lateral image of the FMR absorption in the sample can be obtained. This technique is called *laser beam deflection,* or *PD-FMR* technique.

In the following discussion, we briefly describe only the PM-FMR technique. The experimental setup after [71] is shown in the Fig. 9. The microwave pumping field is supplied by a coaxial coil, and the ferromagnetic resonance is recorded via a pickup coil as shown.

The planes of the coils are kept perpendicular to each other to reduce the off-resonance coupling. The registered FMR signal is modulated due to the change in the intensity of the laser beam focused on the sample. The intensity of the beam is modulated by means of mechanical chopper. The perturbation is caused by the periodic laser induced heating of the affected spot. This local heating causes a periodic shift in the resonance frequency. The sign of the frequency shift depends on the position of the heated spot with respect to a node of the excitation within the sample. To increase the sensitivity of the system,the frequency shift is detected at the side of the FMR line using lock-in detection. The position of the focus in the experimental configuration of Fig. 9 is controlled by a movable mirror, allowing for the

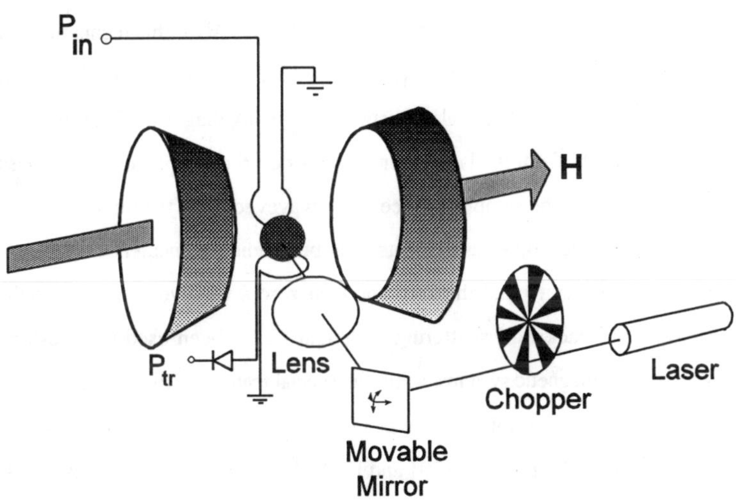

Fig. 9: Experimental setup for probing of the dynamic magnetization on the surface of a YIG sphere by PM - FMR technique. P_{in} represents the input microwave pumping signal. P_{tr} stands for the detected photothermally modulated FMR signal. H is the externally applied static magnetic field.

scanning of the sample surface. Throughout the experiment, the amplitude of the periodic frequency shift is recorded as a function of the focus position. The result of this procedure is shown in Fig. 18 of the chapter XIII. It corresponds to the transverse magnetization distribution for the (3,1,0) magnetostatic mode excited in a 2 mm diameterYIG sphere. Similar results for (2,1,0) and (4,3,0) magnetostatic modes can be found in [71]. The results clearly show the potential capabilities of the PM-FMR technique for the investigation of spinwave excitations in magnetic structures. The further details of this and related techniques can be found in chapter XIII of this book.

7.2 Brillouin Light Scattering Measurements of FMR

The second experimental technique is Brillouin light scattering (BLS), which allows one to determine the number of excited uniform precession magnons per unit volume of the sample. Because the measurements of dissipative losses by BLS are extremely difficult, other techniques are more appropriate for this purpose. Also, this technique is more appropriate for samples with well-prepared surfaces, and usually requires an optically transparent sample, which represents an additional restriction. The block diagram of the system is depicted in the Fig. 10a. Light from the laser source incides on the sample and either passes through it, or is reflected from the sample surface. The wavevector of the uniform precession is zero and therefore, in reflection measurements, near perpendicular incidence on the sample is required. This is necessary to find the zero and near zero k wavevectors in the reflected light. Both forward and backward scattering techniques have been used to investigate ferromagnetic resonance in magnetic systems, using the special feature of the multipass, tandem Sandercock design Fabry-Perrot interferometer [72]. Further details of this technique applied to FMR excitations can be found in [70] and [74]. Fig. 10b illustrates the process that leads to the shift in the frequency of the scattered light. The frequency shift occurs via the interaction of light with the magnetic excitations in the sample. The polarization of the light passing through the sample is changed by the precessing magnetization via magneto-optic effects. The transverse precessing magnetization $\mathbf{m}_{\perp}$ leads to a small rotation of the polarization ϕ_0 oscillating at the frequency ω_0. A crossed polarizer generates the frequency modulated signal, with the amplitude [75]

$$E^t = \frac{e^{(-\alpha_o L/2)} E^i \phi_0}{2},$$

$$\phi_0 = \frac{F d m_\perp}{M_0}, \tag{108}$$

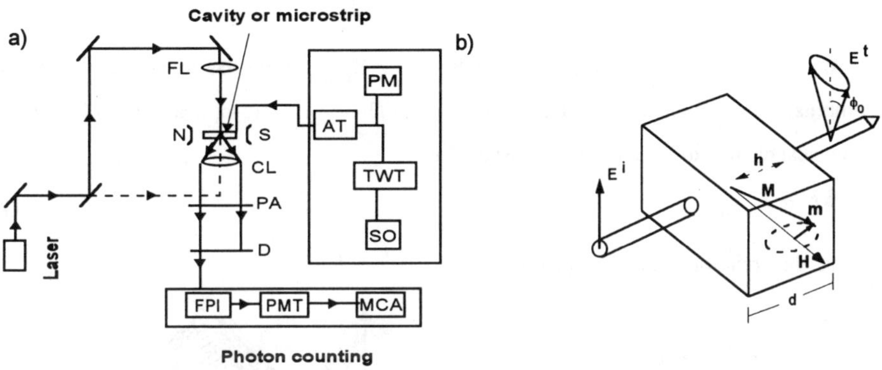

Fig. 10: Experimetal setup for combined FMR and inelastic light scattering measurements. (a) Schematic diagram showing the combined FMR and Brillouin light scattering specrometer. CL - collecting lens, FL - focusing lens, PA - polarization analyser, D - diaphragm and apperture. The photon counting block includes the Fabry-Perrot interferometer, photomultiplayer and multichanel analyser. The FMR microwave block includes the source oscillator, TWT amplifier, power meter and attenuator. (b) schematic description of the variation of the light polarization produced by the interaction of the light with the precessing magnetization.

where E^i is the incident light electric field intensity, α_0 the optical absorption constant, F the Faraday constant, *d* the sample thickness and M_0 the saturation magnetization. F is related to magneto-optical effects. If only magnetic circular birefringence is considered, F is related to the circular birefringence constant K_{MCB} [75] by

$$F = \frac{\pi M_0 K_{MCB}}{\lambda n}, \tag{109}$$

where λ is the light wavelength and n is the refractive index. The transmitted intensity is then given by

$$\frac{I^t}{I^i} = \left|\frac{E^t}{E^i}\right|^2 = e^{-\alpha_o d}\left|\frac{\phi_0}{2}\right|^2 = \frac{\pi^2 K_{MCB}^2 d^2 m_\perp^2}{4\lambda^2 n^2}. \tag{110}$$

If other magneto-optical effects are involved, the K_{MCB}^2 term has to be replaced by a combination of magneto-optical constants [75]. These magneto-optical effects lead to a dynamical change in the dielectric constant $\delta\varepsilon$

$$\delta\varepsilon = K_{MCB}\, m_\perp. \tag{111}$$

Using this relation, equation (109) gives

$$\frac{I^t}{I^i} = \frac{|k_L|^2\, \delta\varepsilon^2 d^2}{16\,\varepsilon^2}, \tag{112}$$

where $\varepsilon = n^2$ and $k_L = 2\pi n/\lambda$ have been used. Equation (112) is a well-known equation for the intensity of the scattered light from spin-wave modes [76]. It allows one to estimate the scattering cross section of the light scattered from magnons. This was done experimentally in [74], where the number of uniform precession magnons excited at FMR was varied by changing the input microwave power on the sample. As

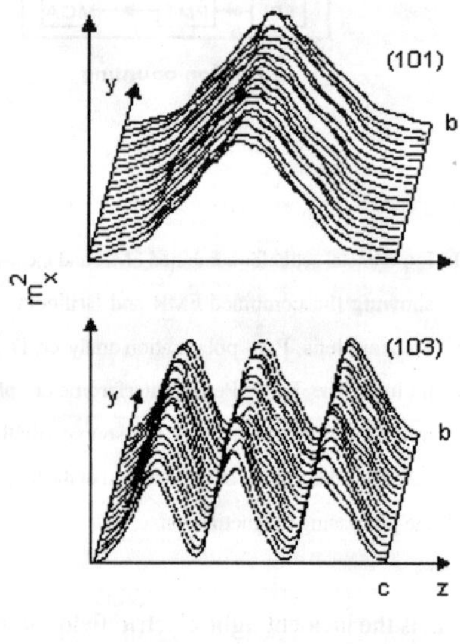

Fig. 11: Spatial variations of the squared magnetization amplitude for (101) and (103) modes, as measured by Brillouin light scattering. The applied external field is uniform.

expected, linear dependence of the scattering intensity as a function of the number of uniform mode magnons was observed. The measurements gave a cross section of approximately 3 $\times 10^{-9}$ counts/s mW at incident laser power of 4 mW. The scattering process can also be regarded as inelastic light scattering from spin waves with wavevector $\kappa_m = 0$. A series of combined BLS-FMR measurements were made on YIG in [73], on permalloy films and YIG in [74], and on layered systems in [77].

Another demonstration of this combined technique was presented in [78] where the authors could visualize the spatial distribution of the dynamic magnetization distributions. These results are shown in Fig. 11. These plots represent the measured spatial distribution of magnetostatic resonance modes (101) and (103) excited by the applied microwave field. c and b represent the in-plane dimensions of the sample. The external magnetic field is applied along the z axis. Fig. 11 shows the pinning of the excited modes at the sample edges,perpendicular to the applied static magnetic field. The amplitude distributions are well described by a sinusoidal function with the wavevector amplitude $n\pi/c$.

Acknowledgments

The author is indebted to M. Hurben, Department of Physics, Colorado State University, USA, for his stimulating discussion, careful reading, and editing of the manuscript. The helpful discussion with Drs. G. Wiese and M. Mendik during the preparation of the present work is also greatly appreciated.

References

[1] J. H. E. Griffiths, Nature **158**, 670 (1946).

[2] C. W. Haas and H. B. Callen, in Magnetism, edited by G. T. Rado and H. Suhl, (Academic Press New York 1963), pp.450 -549.

[3] Ferromagnetic Resonance, edited by S. V. Vonsovskii, Pergamon Press Ltd. 1966.

[4] A. G. Gurewich, Magnetic Resonance in ferrites and antiferromagnetics, Science Moscow 1973 (In Russian).

[5] P. E. Wigen, T. S. Stakelon, H. Puszkarski, and P. Yen, AIP Conf. Proc. **29**, 670 (1976).

[6] C. E. Patton, Magnetic Oxides, Part 2, edited by D. J. Craik (John Willey and Sons, London 1975).

[7] B. Heinrich, S. T. Purcell, J. R. Dutcher, K. B. Urquhart, J. F. Cochran, and A. S. Arrot, Phys. Rev. B **38**, 12879 (1988).

[8] J. J. Krebs, P. Lubitz, A. Chaiken, and G. A. Prinz, J. Appl. Phys. **67**, 5920 (1990).

[9] H. Hundenquit, J. Mag. Magn. Mat. **93**, 336 (1991).

[10] R. Kordecki, R. Mackenstock, J. Pelzl, E. Becker, G. Dumpich, and G. Suran, J. Mag. Magn. Mat. **93**, 281 (1991).

[11] P. C. Dorsey, S. E. Bushnell, R. G. Seed, and C. Vittoria, J. Appl. Phys. **74**, 1242 (1993).

[12] M. Abe, T. Itoh, and Y. Tamaura, Thin Solid Films **216**, 155 (1992).

[13] Q. H. F. Vrehen, IEEE Trans. Mag. **MAG - 4**, 479 (1968).

[14] C. E. Patton, and T. Kohane, Rev. Sci. Instrum. **43**, 76 (1971).

[15] L. D. Landau, and E. M. Lifshitz, Phys. Zs. Soviet Union **8**, 153 (1935).

[16] J. R. MacDonald, Proc. Phys. Soc. **A64**, 968 (1951).

[17] J. D. Jackson, Classical Electrodynamics, John Wille and Sons 1975.

[18] J. A. Osborn, Phys. Rev. **67**, 351 (1945).

[19] C. Herring, and C. Kittel, Phys. Rev. **81**, 869 (1951).

[20] M. Chen, and C. E. Patton, Chapter 3 in Nonlinear Phenomena and Chaos, Edited by P. E. Wigen, (World Scientific 1994).

[21] B. A. Kalinikos, Chapter 2 in Linear and Nonlinear Spin Waves in Magnetic Films and Superlattices, Edited by M. G. Cottam, World Scientific, (1994).

[22] N. S. Akulov, Ferromagnetism, ONTI Moscow and Leningrad 1939.

[23] B. Lax, and K. J. Button, Microwave ferrites and ferrimagnetics, (McGraw Hill, New York 1962), p.175.

[24] J. Jasenek, Phys. Stat. Sol. (b) **95**, K131 (1979).

[25] F. Waldner, J. Phys. C. **21**, 1243 (1988).

[26] K. Gnatzing, H. Dötsch, M. Ye, and A. Broekmeyer, J. Appl. Phys. **62**, 4839 (1987).

[27] C. Kittel, Phys. Rev. **71**, 270 (1947).

[28] C. Kittel, Phys. Rev. **73**, 155 (1948).

[29] W. D. Wilber, J. G. Booth, C. E. Patton, G. Shrinivasan, and R. W. Cross, J. Appl. Phys. **64**, 5477 (1988).

[30] G. Wiese, Z. Phys. B**91**, 57 (1993).

[31] L. R. Walker, Phys. Rev. **105**, 390 (1957).

[32] R. W. Damon, and J. R. Eshbach, J. Phys. Chem. Solids **19**, 308 (1960).

[33] R. Gieniusz and L. Smoczyński, J. Magn. Magn. Mat. **66**, 366 (1987).

[34] R. A. Lemons, and B. A. Auld, J. Appl. Phys. **52**, 7360 (1981).

[35] P. Kabos, V. S. Stalmachov, Magnetostatic waves and their applications, (Chapmann and Hall, London 1994).

[36] E. Schlömann, R. I. Joseph, J. Appl. Phys. **32**, 1006 (1961).

[37] G. Wiese, Nichtlineare Dynamic und Instabilitäten in ferromagnetischen Resonanzexperimenten, PhD. thesis, Technische Hochschule Darmstadt 1991. G. Wiese, Z. Phys. B**82**, 453 (1991).

[38] G. Wiese, H. Benner, Z. Phys. B**79**, 119 (1990).

[39] P. H. Bryant, Phys. Rev. B**39**, 4363 (1989).

[40] G. Wiese, L. Buxman, P. Kabos, and C. E. Patton, J. Appl. Phys. **75**, 1041 (1994).

[41] J. D. Adam, T. W. O'Keefe, and R. W. Patterson, J. Appl. Phys. **50**, 2446 (1979).

[42] B. A. Kalinikos, N. G. Kovshikov, and A. N. Slavin, Sov. Phys. JEPT Lett. **38**, 413 (1983).

[43] B. A. Kalinikos, and A. N. Slavin, J. Phys. C**19**, 7013 (1986).

[44] B. A. Kalinikos, N. G. Kovshikov, and A. N. Slavin, Sov. Phys. JEPT **94**, 303 (1988).

[45] G. T. Rado, and J. R. Weertman, J. Phys. Chem Solids **4**, 315 (1959).

[46] C. E. Patton, Phys. Reports **103**, 251 (1984).

[47] T. Wolfram, and R. E. DeWames, Progr. Surf. Sci. **2**, 233 (1972).

[48] R. E. deWames, and T. Wolfram, J. Appl. Phys. **41**, 987 (1970).

[49] R. Henry, S. D. Brown, P. E. Wigen, and P. J. Besser, Phys. Rev. Lett. **28**, 1272 (1972).

[50] A. V. Pomyalov, and P. E. Zil'berman, Sov. Phys. Solid State **26**, 2111 (1984).

[51] J. Barak, S. M. Bhagat, and C. Vittoria, J. Appl. Phys. **59**, 2521 (1986).

[52] P. Kabos, C. E. Patton, and W. D. Wilber, Phys. Rev. Lett. **53**, 1962 (1984); erratum in **54**, 851 (1985).

[53] P. Grünberg, J. Appl. Phys. **51**, 4338 (1980).

[54] P. Kabos, C. E. Patton, M. O. Dima, and D. B. Church, J. Appl. Phys. **75**, 3553 (1994).

[55] J. Barnas, and P. Grünberg, J. Mag. Magn. Mat. **82**, 186 (1989).

[56] B. Hillebrands, Phys. Rev. B**41**, 530 (1990).

[57] R. L. Stamps, Phys. Rev. B**49**, 339 (1994).

[58] J. F. Cochran, J. Rudd, W. B. Muir, B. Heinrich, and Z. Celinski, Phys. Rev. B**42**, 508 (1990).

[59] J. F. Cochran, B. Heinrich, and A. S. Arrot, Phys. Rev. B**34**, 7788 (1986).

[60] R. L. Stamps, and B. Hillebrands, Phys. Rev. B**44**, 12 417 (1991).

[61] F. Hoffmann, Phys. Status Solidi **41**, 807 (1970),
F. Hoffmann, A. Stankoff, and H. Pascard, J. Appl. Phys. **41**, 1022 (1970).

[62] S. Barta, Solid State Physics II, (Slovak Technical University 1991) (In Slovak).

[63] M. Sparks, Ferromagnetic relaxation theory, (McGraw Hill 1964), p.34.

[64] O. von Geisau, Photermische Untersuchungen der ferromagnetischen Resonanz kollektiver magnetischer Anregungen in YIG, PhD. thesis, Ruhr-Universität, Bochum 1993, p.74.

[65] Z. Frait, and D. Fraitova, Chapter 1 in Spin waves and magnetic excitations, edited by A. S. Borovik-Romanov and S. K. Sinha, (Elsevier Science Publishers 1988).

[66] T. Wolfram, and R. E. DeWames, Phys. Rev. B**4**, 3125 (1970).

[67] J. R. Truedson, K. D. McKinstry, P. Kabos, and C. E. Patton, J. Appl. Phys. **74**, 2705, (1993).

[68] J. R. Truedson. P. Kabos, K. D. McKinstry, and C. E. Patton, J. Appl. Phys. **76**, 432 (1994).

[69] J. Pelzl and U. Netzelmann, in Photoacoustic, Photothermal and Photochemical Processes at surfaces and thin films, Edited by P. Hess, Springer series: Topics in Current Physics, vol. 47, Springer - Verlag Berlin, Heidelberg 1989.

[70] W. Wettling, and W. Jantz, J. Mag. Magn. Mat. **45**, 364 (1984).

[71] F. Rödelsperger, O. von Geisau, H. Benner, and J. Pelzl, in Proceedings of the Congress Ampére, Athens 1992.

[72] Interferometer system manufactured by JRS, Ltd., Zwillikerstrasse 8, 8910 Affoltern a. A., Switzerland.

[73] W. D. Wilber, W. Wettling, P. Kabos, C. E. Patton, and W. Jantz, J. Appl. Phys. **55**, 2533 (1984).

[74] G. Shrinivasan, C. E. Patton, and J. G. Booth, J. Appl. Phys. **63**, 3344 (1988).

[75] W. Wettling, M. G. Cottam, and J. R. Sandercock, J. Phys. C**8**, 211 (1975).

[76] B. A. Auld, and D. A. Wilson, J. Appl. Phys. **38**, 333 (1967).

[77] S. O. Demokritov, J. Mag. Magn. Mat. **126**, 291 (1993).

[78] S. M. Rezende, and A. Azevedo, Phys. Rev. B**44**, 7062 (1991).

SUBMILLIMETER SPECTROSCOPY OF ANTIFERROMAGNETIC DIELECTRICS. RARE-EARTH ORTHOFERRITES

A.M.BALBASHOV*), G.V. KOZLOV, A.A.MUKHIN and A.S.PROKHOROV

Institute of General Physics of the Russian Acad. Sci., 38 Vavilov St., 117942 Moscow, Russia
**) Moscow Power Engineering Institute, 14 Krasnokazarmennaya St., 105835 Moscow, Russia*

1. Introduction

The resonance absorption lines of many antiferromagnets fall directly in the range of submillimeter (SBMM) wavelengths. In this region the measurements of the magnetic spectra of antiferromagnets, i. e. the frequency dependences of the real (μ') and imaginary (μ'') parts of the magnetic permeability, are considered to be a difficult task for the conventional spectroscopic techniques used in the microwave and IR ranges [1]. However, the application of the SBMM spectroscopy technique rapidly improving in recent years [2,3], which is based on the use of tunable monochromatic oscillators like the backward-wave oscillators (BWO) as sources of radiation, makes the situation quite different. In the methodical sense, the BWO spectroscopy is most convenient for measuring the magnetic spectra of antiferromagnets at SBMM wavelengths, combining the most significant achievements of the IR spectroscopy and the microwave technology. Actually, on the one hand, the high «quality» of the operating radiation - its sufficiently high intensity (~10 mW), high monochromaticity ($\Delta\nu/\nu \sim 10^{-5}$), and high degree of polarization (~99.99%) - permits to perform measurements with high resolution in frequency ($\sim 10^{-3}$ cm^{-1}) and high signal-to-noise ratio.

On the other hand, as in IR spectroscopy, measurements are carried out in free space with the help of simple measuring schemes. This makes it possible, firstly, to tune continuously the frequency of the operating radiation over a wide range, and, as a consequence, to record the measurable quantities in the form of the spectra rather than at separate frequency points. Secondly, a simple distribution of the field in the operating wave greatly simplifies the organization of the measurements and all the subsequent calculations. Spectra can be recorded on the real time scale. For instance, it takes about 10 s to record the $T(\nu)$ spectra containing 100 points [3,4].

In this chapter we shall mainly present the results of our studing by means Of the submillimeter BWO spectroscopy the dynamic magnetic properties of the rare-earth orthoferrites $RFeO_3$, where R is a rare-earth ion. The rare-earth orthoferrites have proved to be an interesting class of the weak ferromagnets for investigations of various types of magnetic interactions, in particuliar, anisotropic R-Fe exchange interaction [5-7] determining their rich magnetic properties. The most striking feature of these materials is the existence of various spin reorientation phase transitions (SRPT) in them. This permits to consider these magnets as good model objects for studying different magnetic (exchange) interactions, as well as the phase transitions in the systems with interacting order parameters.

The dynamic properties of orthoferrites and, in particular, the behaviour of modes of magnetic excitations at SRPT, should be rather peculiar [9-11]. The fact is that two types of magnetic excitations (modes) exist in orthoferrites, pertaining to the Fe subsystem (antiferromagnetic resonance (AFMR) modes) and to the R subsystem (R modes). These modes interact with each other, what is particularly strongly manifested at spin-reorientation phase transitions and can essentially change their dynamics. The study of the dynamic properties of rare-earth orthoferrites in the SRPT region is one of the key aspects of this chapter. Another important problem to be discussed is a role of the R subsystem in the formation of the dielectric properties of orthoferrites.

2. General Characteristics of Rare-Earth Orthoferrites

2.1 Crystal and magnetic symmetry

Rare-earth orthoferrites $RFeO_3$, where R is a rare-earth ion or yttrium ion, refer to a large class of materials of the ABO_3 type, having a distorted structure of the perovskite $CaTiO_3$. The crystall structure of orthoferrites is described by orthorhombic space group D_{2h}^{16}-Pbmn [12,13]. Their unit cell contains four $RFeO_3$ molecules. The Fe^{3+} ions are located in centrosymmetrical positions having distorted octahedral oxygen environment [13]. The local symmetry of the environment of rare-earth ions is described by the point group C_s , containing only one non-trivial symmetry element - the mirror plane parallel to the ab-plane of the crystal.

The magnetic structure of orthoferrites is closely associated with their crystalline symmetry. First of all, it is determined by a strong antiferromegnetic superexchange Fe^{3+}-O^{2-}-Fe^{3+}, which leads to antiferromagnetic spin ordering of Fe^{3+} ions of the G-type (chess order) [14]. The corresponding Neel temperatures (T_{N1}) for various orthoferrites amount to 620-740 K [15]. So far as the crystal and the magnetic unit cells coincide in orthoferrites, it is enough to use four magnetic sublattices of Fe^{3+} ions (M_i, i=1,2,3,4) and four ones of R^{3+} ions (M_i, i=5,6,7,8) for characterizing their magnetic structures.

Under the symmetry operations of the space group D_{2h}^{16} the vector components of magnetic moments M_i are transformed via one another, forming the basis of the reducible representation of this group. However, the symmetric combinations of these vectors [14,16], numbering to eight,

$$\begin{aligned}
\mathbf{F} &= \mathbf{M}_1 + \mathbf{M}_2 + \mathbf{M}_3 + \mathbf{M}_4, & \mathbf{f} &= \mathbf{M}_5 + \mathbf{M}_6 + \mathbf{M}_7 + \mathbf{M}_8,\\
\mathbf{G} &= \mathbf{M}_1 - \mathbf{M}_2 + \mathbf{M}_3 - \mathbf{M}_4, & \mathbf{g} &= \mathbf{M}_5 - \mathbf{M}_6 + \mathbf{M}_7 - \mathbf{M}_8,\\
\mathbf{A} &= \mathbf{M}_1 - \mathbf{M}_2 - \mathbf{M}_3 + \mathbf{M}_4, & \mathbf{a} &= \mathbf{M}_5 - \mathbf{M}_6 - \mathbf{M}_7 + \mathbf{M}_8,\\
\mathbf{C} &= \mathbf{M}_1 + \mathbf{M}_2 - \mathbf{M}_3 - \mathbf{M}_4, & \mathbf{c} &= \mathbf{M}_5 + \mathbf{M}_6 - \mathbf{M}_7 - \mathbf{M}_8,
\end{aligned} \tag{2.1}$$

Table 1. Irreducible representation of the D_{2h}^{16} space group and magnetic group, corresponding to various magnetic configurations in $RFeO_3$

Γ_i	I	$\bar{2}_x$	$\bar{2}_y$	Components of the basis vectors						Magnetic group		
				Fe^{3+} - ions			R^{3+} - ions					
Γ_1	1	1	1	A_x	G_y	C_z			c_z	m	m	m
Γ_2	1	1	-1	F_x	C_y	G_z	f_x	c_y		$\underline{m}$	$\underline{m}$	m
Γ_3	1	-1	1	C_x	F_y	A_z	c_x	f_y		$\underline{m}$	$\underline{m}$	m
Γ_4	1	-1	-1	G_x	A_y	F_z			f_z	$\underline{m}$	$\underline{m}$	m
Γ_5	-1	1	1				g_x	a_y		$\underline{m}$	$\underline{m}$	$\underline{m}$
Γ_6	-1	1	-1						a_z	m	m	$\underline{m}$
Γ_7	-1	-1	1						g_z	m	m	$\underline{m}$
Γ_8	-1	-1	-1				a_x	g_y		m	m	$\underline{m}$

can already be transformed according to the irreducible representations Γ_i of the space group D_{2h}^{16}, moreover all of them being one-dimensional (Table 1).

It is convenient to use the indices of the representations of the D_{2h}^{16} group to characterize the possible magnetic configurations, determined by the nonzero components of the basis vectors **F**, **G**, etc. For example, the magnetic configuration characterized by one nonzero z-component of the **G** vector, (G_z), as well as more complex magnetic configurations G_zF_x, $G_zF_xC_y$, $G_zF_xC_yf_xc_y$, refers to representation Γ_2. The angular magnetic configuration, i.e. $F_xG_zF_zG_x$, in which the ferromagnetism and antiferromagnetism vectors (**F** and **G**) lie in the ac-plane of the crystal, corresponds to the reducible representation Γ_{24}.

2.2. Magnetic interactions, thermodynamic potential and spin reorientation origin

A particular type of magnetic configuration in orthoferrites is determined by the relation of quantities of various microscopic interactions in crystals. The strongest of them is the antiferromagnetic isotropic Fe-Fe exchange interaction resulting in spin ordering of the G type of the Fe^{3+} ions (G>>F,A,C). More weak (anisotropic) interactions stipulate the deviation of the magnetic moments from the purely antiferromagnetic configuration, and also stabilize their orientation with respect to the crystal axes. The most important of them is the antisymmetric Fe-Fe exchange of the Dzyaloshinskii-Morya type, which causes, in particular, the appearance of a spontaneous weak-ferromagnetic moment. The same interaction, together with single-ion anisotropy of the Fe subsystem and the magnetodipole Fe-Fe interaction , contributes to the energy of magnetocrystallographic anisotropy [6]. A phenomenological approach is usually used to describe the Fe subsystem (see, for instance [6,7,17]), based on the expansion of the thermodynamic potential Φ_{Fe} in powers of the basis vector components. In the two-sublattice approximation (M_1=M_3, M_2=M_4 or A=C=0) [6,7,17]

$$\Phi_{Fe}(F,G) = \Phi_0(G) + {}^1/{}_2AF^2 + {}^1/{}_2D(FG)^2 - d_1F_xG_z - d_3F_zG_x - M_0FH + \Phi_A{}^{Fe}(G) \quad (2.2)$$

where

$$\Phi_A^{Fe}(G) = {}^1/_2 K_{ac}^{Fe} G_z^2 + {}^1/_2 K_{ab}^{Fe} G_y^2 + {}^1/_4 K_2 G_z^4 + {}^1/_4 K_2' G_y^4 + {}^1/_2 K_2'' G_y^2 G_z^2 \quad (2.3)$$

is the anisotropy energy, M_0 is the saturation magnetic moment of Fe sublattices ($M_0=5\mu_B$ per Fe^{3+} ion, μ_B is the Bohr magneton); A,D are the isotropic exchange constants; d_1 and d_3 are the Dzyaloshinskii constants. We use the dimensionless vectors **F** and **G** ($\mathbf{F}=(\mathbf{M}_1+\mathbf{M}_2)/2M_0$, $\mathbf{G}=(\mathbf{M}_1-\mathbf{M}_2)/2M_0$).

However, in spite of a great importance of the indicated interactions in the Fe subsystem, it is impossible to explain a wide variety of the magnetic properties of orthoferrites only by these interactions. Their properties, particularly in the low-temperature region, are determined by the type of the rare-earth ion. This is strongly manifested in various temperature dependences of the weak-ferromegnetic moment, in the presence of the compensation points, and mainly in the existence of different spin-reorientation phase transitions either spontaneous or induced by the external magnetic field [6,8]. It is found at present that such a variety of properties of orthoferrites is associated first of all with the individual magnetic features of the rare-earth ion in the crystal and its interaction with the Fe subsystem [5,8].

At temperatures T>5 K the rare-earth ions in orthoferrites are in a paramagnetic state, and the R-Fe interaction leads to their polarization (magnetization). The anisotropic R-Fe exchange makes the main contribution to this interaction. This is associated with the fact, that the isotropic R-Fe exchange of the Fe^{3+}-R^{3+} pair is actually completely compensated in the crystal because of antiferromagnetic ordering of the Fe subsystem. As for the R-R interaction, it is manifested only at low temperatures (T<5 K), leading in some cases to antiferromagnetic ordering of the R subsystem.

At room temperatures the influence of the rare-earth ions on the magnetic properties of orthoferrites is negligible. This is confirmed particularly by the fact, that all orthoferrites with either magnetic or nonmagnetic rare-earth ions have at high temperatures the same magnetic configuration Γ_4, being stabilized by the anisotropy energy of the Fe subsystem. On lowering temperature, when the polarization contribution of the rare-earth ions to the thermodynamic potential induced by the R-Fe interaction begins to increase ($\sim 1/T$), a substantial difference appears in the behaviour of orthoferrites with various rare-earth ions. An important part is thereby played by either the anisotropic properties (susceptibility, g-tensor) of the rare-earth ion itself or the anisotropy of the R-Fe interaction, determining all together the additional contribution to the magnetic anisotropy energy, the weak-ferromagnetic moment, etc, [5-8].

Note thereby, that the usual purely phenomenological approach is not effective for describing the properties of the R subsystem. The quantum-phenomenological approach [7] is more effective. According to it, the R subsystem is described by virtue of microscopic Hamiltonians (that of the crystalline field, the exchange ones, etc.) obtained by using the symmetry properties of the system, as well as the already known data on the ion spectrum in crystal. The parameters of these Hamiltonians are determined from the experiment.

At high temperatures (T>5K), when the R subsystem is paramagnetic, its state is uniquely determined by the effective fields caused by the Fe subsystem (provided that both subsystems are in thermodynamic equilibrium with respect to each other). Therefore, when describing the static properties of the system, it is logical to successively exclude the variables of the R subsystem expressing them via the basis vectors of the Fe subsystem. The effective (renormalized) thermodynamic potential thus obtained depending only on the variables of the Fe subsystem (F,G) has in the high-temperature limit the form [6,7]

$$\Phi(F,G) = \Phi_0(G) + {}^1/_2AF^2 + {}^1/_2D(FG)^2 - \bar{d}_1F_xG_z - \bar{d}_3F_zG_x - \tau_1H_xG_z - \tau_3H_zG_x - M_0\Sigma_k(1+\eta_k)F_kH_k - {}^1/_2\Sigma_k\,\chi_R{}^kH_k{}^2 + \Phi_A(G), \tag{2.4}$$

where $\Phi_A(G)$ is given by Eq.(2.3) with the renormalized anisotropy constants $K_{ac,ab} = K_{ac,ab}{}^{Fe} + K_{ac,ab}{}^{R}$. The presence of the paramagnetic R subsystem results for $\Phi(F,G)$ in Eq.(2.4) not only in renormalizing the coefficients of the initial thermodynamic potential of the Fe subsystem, but also in the appearance of new terms. Thus the coefficients η_k (k=x,y,z), $\tau_{1,3}$ in Eq.(2.4) are related exclusively to the R-Fe interaction. The concrete temperature dependence of the coefficients of the thermodynamic potential is determined by the spectrum of the rare-earth ion in the crystalline field [7]. Minimizing the thermodynamic potential $\Phi(F,G)$ with respect to F (F<G), let us reduce it to

$$\Phi(G) = -{}^1/_2\Sigma_k\chi_\perp{}^kH_k{}^2 - {}^1/_2\Sigma_{kl}\Delta\chi_\perp{}^{kl}G_kG_lH_kH_l - m_xG_xH_x - m_zG_xH_z - {}^1/_2\Sigma_k\,\chi_R{}^kH_k{}^2 + \bar{\Phi}_A(G) \tag{2.5}$$

where the anisotropy energy $\bar{\Phi}_A(G)$ has the form Eq.(2.3) with the replacement $K_{ab} \rightarrow K_{ab} + d^2/A$,

$$m_x = \tau_1 + M_0(1+\eta_x)d/A, \quad m_z = \tau_3 - M_0(1+\eta_z)d/A \tag{2.6}$$

- are the weak-ferromagnetic moments along the axes a(x) and c(z), respectively;

$$\chi_\perp{}^k = \chi_\perp{}^{Fe}(1+\eta_k)^2, \quad \Delta\chi_\perp{}^{kl} = \chi_\perp{}^{Fe}(1+\eta_k)(1+\eta_l)/(1+A/DG_0{}^2),$$
$$\chi_\perp{}^{Fe} = M_0{}^2/A = M_0/2H_E \tag{2.7}$$

We have neglected in Eq. (2.5) small renormalization Dzyaloshinskii constants and assumed $d_1 = -d_3 = d$.

The thermodynamic potential in the form Eq.(2.5) is the basis for the analysis of the magnetic properties of the rare-earth orthoferrites, particularly spin-reorientation phase transitions as well as spin excitations (AFMR) in the Fe subsystem.

2.3. Dynamic properties and spin excitation in RFeO₃: general features

Dynamic magnetic properties of orthoferrite are determined both by the antiferromagnetically ordered Fe subsystem and the paramagnetic rare-earth subsystem. There are four branches of spin waves and consequently four types of homogeneous oscillations in the Fe subsystem [18]. Two of them are exchange modes lying at $\nu \approx 10^3$ cm^{-1}, and other two are AFMR modes lying at the SBMM wavelengths. The AFMR modes were studied by conventional antiferromagnetic resonance methods [19-23], Far-infrared spectroscopy [24], inelastic neutron scattering [25], Raman scattering [9,10,26], and in the recent time - by the methods of submillimeter BWO spectroscopy [1,11,27-34]. The experimental data have shown a diversity and strong dependence of AFMR mode behaviour on the R ion type. That indicates an important contribution of the R subsystem to the formation of orthoferrite dynamic properties determined by the electronic transitions inside the ground R ion multiplet split by the crystal and exchange fields. In particular, due to R-Fe interaction, R modes play an important role in the soft modes' behaviour of AFMR modes at SRPTs [11].

In this section we shall confine ourselves to the dynamic properties of the Fe subsystem. We shall describe the dynamics of the Fe subsystem on the basis of the Landau-Lifshitz equations

$$(M_o/\gamma)\dot{\mathbf{F}} = [\mathbf{F}\Phi_F] + [\mathbf{G}\Phi_G] + \mathbf{R}_F\,, \quad (M_o/\gamma)\dot{\mathbf{G}} = [\mathbf{F}\Phi_G] + [\mathbf{G}\Phi_F] + \mathbf{R}_G, \tag{2.8}$$

where $\gamma=2\mu_B/h$ is the gyromagnetic ratio, $\Phi_F= \partial\Phi/\partial\mathbf{F}$, $\Phi_G= \partial\Phi/\partial\mathbf{G}$, $\mathbf{R}_F$, $\mathbf{R}_G$ are the relaxational terms, which we have not specified for the present. The influence of the rare-earth ion on the dynamics of the Fe subsystem is taken into account in this case only through the thermodynamic potential $\Phi=\Phi(\mathbf{F},\mathbf{G})$ renormalized by the R-Fe interaction and determined by Eq.(2.4). This approximation is valid when the frequencies of the R modes are higher than the AFMR ones. In this case we can describe the R subsystem in a quasi-static approximation, i.e. as a subsystem, following instantaneously the Fe subsystem. Such an approach allows to directly relate the basic dynamic properties of the system to the static parameters determined according to Eqs. (2.4) and (2.5). We shall discuss the dynamic effects connected with the R subsystem beyond of this approximation below in section 8.

Let us deduce the resonance frequencies and tensors of the high-frequency magnetic susceptibility obtained on the basis of Eqs. (2.4), (2.5), (2.8) at H=0 for various magnetic phases.

$\Gamma_4(G_xF_z)$ phase. M_1 mode of the $\Gamma_{23}(G_zF_xF_y)$ symmetry:

$$\varpi_1 = \gamma\,(2H_EK_{ac}/M_0)^{1/2}\,, \tag{2.9}$$

M_2 mode of the Γ_{14} $(G_yG_xF_z)$ symmetry:

$$\varpi_2 = \gamma\,(2H_EK_{ab}/M_0)^{1/2}\,, \tag{2.10}$$

$$\hat{\chi}(\omega)=\begin{vmatrix} \chi_{xx} & i\chi_{xy} & 0 \\ -i\chi_{xy} & \chi_{yy} & 0 \\ 0 & 0 & \chi_{zz} \end{vmatrix}\,, \tag{2.11}$$

where $\chi_{xx}(\varpi) = \chi^x{}_{rot}N_1(\varpi)$, $\chi_{yy}(\varpi) = \chi^y{}_{\perp}N_1(\varpi)$, $\chi_{xy}(\varpi) = -(\varpi/\varpi_1)(\chi_{xx}\chi_{yy})^{1/2}$,

$$\chi_{zz}(\varpi) = \chi^z{}_{\perp}N_2(\varpi),\ \chi^x{}_{rot} = m_x{}^2/K_{ac},\ N_{1,2}(\varpi) = \varpi_{1,2}{}^2/(\varpi_{1,2}{}^2 - \varpi^2 + i\varpi\Delta\varpi_{1,2}) \tag{2.12}$$

$\Delta\varpi_{1,2}$ is the linewidth ($\Delta\varpi_{1,2} << \varpi_{1,2}$). The mode of the $\Gamma_{23}(G_zF_xF_y)$ symmetry corresponds to the oscillations of the quantities G_z,F_x,F_y, which are transformed according to the reducible representation Γ_{23} of the D_{2h}^{16} group, and analogously for other modes [35].

$\Gamma_1(G_y)$ phase. M_1 mode of the $\Gamma_2(G_zF_x)$ symmetry:

$$\varpi_1 = \gamma\,(2H_EK_{bc}/M_0)^{1/2}\ , \tag{2.13a}$$

M_2 mode of the $\Gamma_4(G_xF_z)$ symmetry:

$$\varpi_2 = \gamma\,(2H_EK_{ba}/M_0)^{1/2}\ , \tag{2.13b}$$

$$\hat{\chi}(\omega)=\begin{vmatrix} \chi_{xx} & 0 & 0 \\ 0 & \chi_{yy} & 0 \\ 0 & 0 & \chi_{zz} \end{vmatrix}\ , \tag{2.14}$$

where $\chi_{xx}(\varpi) = (\chi'^{x}_{rot} + \chi^{x}{}_{\perp})N_1(\varpi)$, $\chi_{zz}(\varpi) = (\chi'^{z}_{rot} + \chi^{z}{}_{\perp})N_2(\varpi)$, $\chi_{yy}(\varpi) \approx 0$,

$\chi'^{x}_{rot} = (m_x)^2/K_{bc}$, $\chi'^{z}_{rot} = (m_z)^2/K_{ba}$, $K_{bc} = K_{ac} - K_{ab} + K''_2 - K'_2$,

$$K_{ba} = -K_{ab} - K'_2, \tag{2.15}$$

$\Gamma_2(G_zF_x)$ phase. M_1 mode of the $\Gamma_{34}(F_yF_zG_x)$ symmetry:

$$\varpi_1 = \gamma\,(2H_EK_{ca}/M_0)^{1/2}\ , \tag{2.16a}$$

M_2 mode of the $\Gamma_{12}(G_yF_xG_z)$ symmetry:

$$\varpi_2 = \gamma\,(2H_EK_{cb}/M_0)^{1/2}, \tag{2.16b}$$

$$\hat{\chi}(\omega)=\begin{vmatrix} \chi_{xx} & 0 & 0 \\ 0 & \chi_{yy} & i\chi_{yz} \\ 0 & -i\chi_{yz} & \chi_{zz} \end{vmatrix}\ , \tag{2.17}$$

where $\chi_{xx}(\varpi) = \chi^{x}{}_{\perp}N_2(\varpi)$, $\chi_{yy}(\varpi) = \chi^{y}{}_{\perp}N_1(\varpi)$, $\chi_{zz}(\varpi) = \chi^{z}_{rot}N_1(\varpi)$,

$\chi_{yz}(\varpi) = -(\varpi/\varpi_1)(\chi_{zz}\chi_{yy})^{1/2}$, $\chi^{z}_{rot} = m_z^{\,2}/K_{ca}^{\,eff}$, $K_{ca} = -K_{ac} - K_2$,

$$K_{cb} = K_{ab} - K_{ac} - K_2 + K''_2, \tag{2.18}$$

The quantities $\chi_\perp^{k}$ (k=x,y,z), $K_{ac,ab}$, K_2, K'_2, K''_2, $m_{x,z}$ included in Eqs.(2.10)-(2.18) determine not only the dynamic, but also the static magnetic properties of the crystal. Therefore, in the following sections we shall see that the study of the dynamic susceptibility in the region of resonance frequencies provides direct information about magnetic configurations of the system, phase transitions between them, anisotropy energy and other characteristics.

3. Growing and physical properties of orthoferrite single crystals

Historically, orthoferrite single crystals were grown by melt - flux growing methods in 1956 [36]. As a rule the melt - flux is used on the base of PbO - PbF_2 and the crystallization process is conducted by spontaneous nucleation or on seed crystals fastened on a platinum seed holder. However, in this case the orthoferrite crystals are contaminated by a marked amount of the Pb ions in composition. This contamination deteriorates very much optical and electrodynamical properties of this material and excludes their many technical applications, although for obtaining crystals of multi-component substituted composition, in many cases the flux method is only applicable. On the other hand, there exists a more adequate orthoferrite growing technique. That is the floating zone (FZ) crystal growth [37,38] on equipment with radiation (light) heating of the liquid zone. This technology permits to obtain very high quality, high-purity materials in view of large uniform crystals with dimensions: diameter up to 20, length 80 mm, that opens ways to making large single crystal specimens useful for correct macro scale magnetic, optic, acoustic etc. physical experiments and producing of some technical devices on these materials. Fig.1 gives a view of FZ single crystals of $YFeO_3$ and $DyFeO_3$ orthoferrites. The main problems of FZ orthoferrite single crystals is obtaining materials with a high level of structure perfection in macro and micro scales. For this purpose the next operation procedures are used in the crystal growth process:

1. low speed of FZ crystal growth 5 - 10 mm/h;
2. use of high quality seed crystals at the crystal growing beginning;
3. very smooth development of the initial stage of crystal growth at transition from seed to crystal with a large diameter;
4. high temperature (1500°C) annealing of growing crystal in the growth process with the help of a tube shape electrical furnace, placed immediately below the floating zone;
5. use of gas atmosphere over melt with controlled composition and pressure pure oxygen or inert gas, CO_2, etc under the pressure up to 100 atm.

The crystal structure perfection level can be accurately determined by the X-ray rocking curve value. Table 2 gives the crystal growth condition and structure perfection parameters of some FZ single crystal orthoferrites. According to these rocking curve values, the dislocation density of FZ single crystals is estimated at 10^5 - 10^6 cm^{-2}. In the best cases dislocation density approaches 10^4 cm^{-2} and with an account taken of the absence of such macro defects as cracks, foreign inclusions, voids etc., these crystals can be considered as sufficiently perfect.

Another problem in orthoferrite crystal growing is reaching micro perfection, which is determined by the valence stabilization of metallic ions, mainly Fe^{3+}, and exclusion of impurity influence on the electrical compensation of materials. The best methods of investigation of micro perfection orthoferrite single crystals is research of their optical spectral absorption in visible and near infrared region. Optical properties greatly differ for flux-melt grown and FZ single crystals. Optical absorption of flux-melt grown crystals is considerably higher than FZ

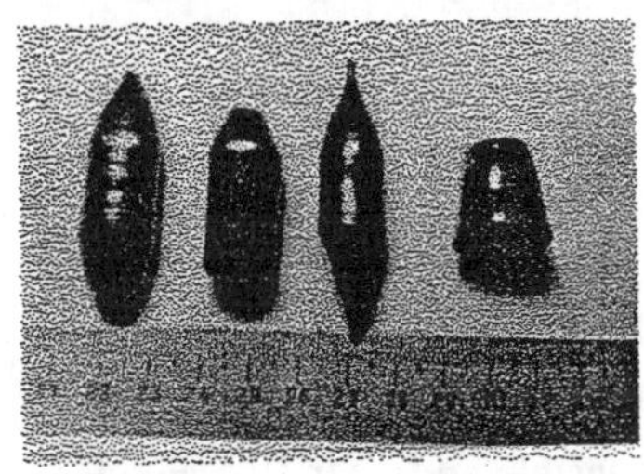

Fig.1. Photograph of the orthoferrite single crystals grown by floating zone melting.

Table 2. Structure perfection parameters of FZ orthoferrites single crystals grown under different conditions.

Gas atmosphere	Composition	Crystal growth speed, mm/h	Domain wall coercitivity, Oe	Peak number on rocking curve	Rocking curve width angle, sec		Blocks disorientation angle, sec
					general	minial	
Air	$YFeO_3$	6	0.5	1-2	40	28	30
Air	$YFeO_3$	10	1	1-2	50	30	40
Air	$YFeO_3$	13	2	2-3	60	30	40-60
N_2	$(SmTb)FeO_3$	8	5	2-3	300	60	60-150

crystals in visible and near IR. The main reason is Pb contamination in flux crystals and the following transition of Fe^{3+} ions into Fe^{4+} . Naturally, impurity ions with 4 valence (Si for example) lead to Fe^{3+} - Fe^{2+} transition. At the same time in the absence of impurities the valence balance depends on the vacancy concentration material and, consequently, on partial pressure of oxygen in high temperature melt growth or thermotreatment. As a result, the whole electrical balance of material is a complex function of its purity and thermal equilibrium conditions in treatments.

The equilibrium partial oxygen pressure at high temperature turns out to depend on a number of rare earth elements. By our estimation this value varies in a wide range of 10^2 -10^7 Pa at a temperature level of 1700-1800°C (melting points). Fig.2 presents results of thermodynamic investigation of equilibrium O_2 pressure at comparatively low temperatures, which show this dependence on the number of rare earth element. Fig.3 illustrates the influence of the gas pressure in the FZ crystal growth of samarium orthoferrite in comparison with the growing process of this orthoferrite from flux-melt. Similar data also exist for other orthoferrites. It follows from these results that each FZ orthoferrite crystal requires strictly determined oxygen pressure in the growing. But experiments indicate that in many cases crystals can be grown in air, but improving of their optical absorption parameters require annealing under gas pressure control condition at temperatures 1400-1500°C for a long enough time. For some orthoferrite compositions equilibrium atmosphere in crystal growing is obligate because air conditions do not permit to get quality materials. For example, mixed $Nd_{0.8}Pr_{0.2}FeO_3$ composition has best parameters when grown in ambient N_2 gas melt.

Crystal growing of mixed ortoferrite composition, for example, $(SmTb)FeO_3$, $(SmEr)FeO_3$, $(NdPr)FeO_3$, $(YLu)FeO_3$, $(SmPr)FeO_3$, etc requires more finer growth conditions, namely, low growth rate 3-5 mm/h, highly uniform heating of the melted zone, and so on.

On substitution of Fe in orthoferrite composition for nonmagnetic elements Ga, Al, crystal growing by FZ gives marked improvement of crystal transparency in visible and near IR - Fig.4, what may be connected with decreasing material optical density.

Of all the spectra of orthoferrite compositions, the best relaxation and optical parameters have materials with diamagnetic R elements, La, Y, Lu, among which the yttrium orthoferrite is most technological in FZ crystal growth and permits to obtain the crystals with extremely high optical transparency in visible and IR, very high mobility of domain wall up to 13000 cm/sOe at 300 K and 50000 cm/s Oe at liquid nitrogen and narrowest antiferromagnetic resonance line width in the submillimeter region. The La orthoferrites have predisposition to twinning in the crystallization process and Lu orthoferrites feature incongruent melting of stoichiometric compound and its crystallization by FZ should be conducted from liquid zone of displaced composition with corresponding increasing of technological process complexity. To obtain mixed orthoferrite crystals, one has to take into account segregation phenomena in the crystallization process. The coefficient of distribution is increased on mixing of the R elements with large ionic radius differences, for example Sm-Lu, Eu-Lu etc. The basic magnetic properties of mixed orthoferrite with substitution of R and Fe elements are quite known [6,8] and FZ crystallization permits to obtain many crystals of mixed composition besides, for example, compounds with Cr, Bi in large concentration, and some others.

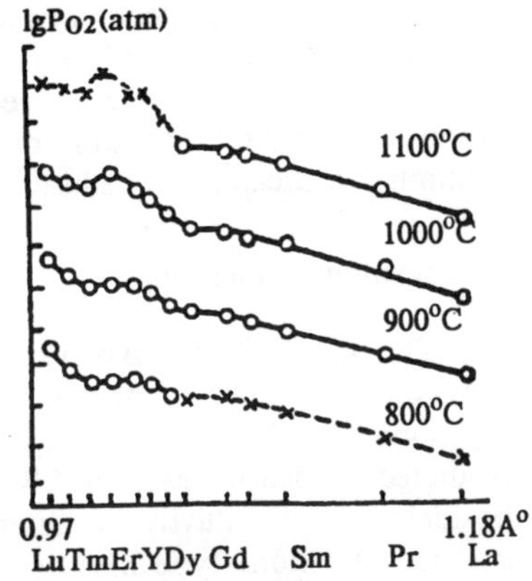

Fig.2. Equilibria isotherm P_{O2} for different orthoferrites.

High frequency relaxation properties of orthoferrite single crystals involve intrinsic mechanisms of relaxation of this type of materials, which depend on the

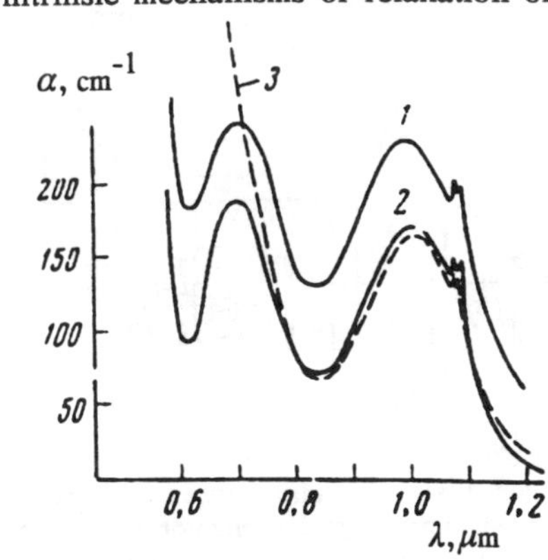

Fig.3.Absorption spectra of $SmFeO_3$: 1 - grown at P = 1.1 10^6 Pa, 2 - grown in air, 3 - grown from PbO flux-melt.

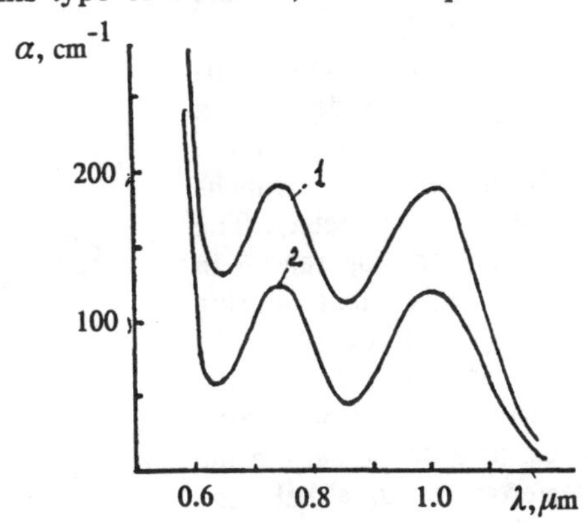

Fig.4. Absorption spectra of $YFeO_3$ (1) and $YFe_{0.8}Ga_{0.2}O_3$ (2) at IR frequency region.

general structure of perfection of crystals. The optical transparency in visible and IR can be used as a criterion of perfection in impurity-free crystals. In doped and mixed crystals nonintrinsic relaxation mechanisms are important, and distinction of effects responsible for the broadening of resonance line is very difficult and requires some additional verification procedures.

4. Experimental method

The possibilities of BWO spectroscopy has already been used to some extent to study the properties of magnets [39-41]. Now the submillimeter BWO spectrometer «Epsilon» has all of the enumerated advantages [3,4]. This device was initially designed for dielectric measurements at SBMM wavelengths, i. e. for recording the spectra of the dielectric permittivity of materials [3,42]. Beginning from 1984, we have widely used the «Epsilon»-type BWO spectrometer for measuring the magnetic spectra of antiferromagnetic dielectrics [1,11].

Figure 5 shows the basic measuring system of «Epsilon» for T and φ measurements. The T vs ν and φ vs ν spectra are recorded separately in two stages. The T vs ν spectrum is recorded with the help of a simple «transmission» scheme, and the sections isolated in Fig.5 by dashed lines are not used. The generator is operated in the frequency scanning mode. The scanning is carried out at different points by varying the BWO supply voltage in steps. The signal at the detector is recorded as a function of the radiation frequency. In order to eliminate the instrumental functions of the spectrometer, the entire procedure is repeated twice, with the sample in the channel and without it.

The second stage involves the recording of the φ vs ν spectrum. All the elements of the channel are now used, and the system represents a Rozhdestvenskii two-beam polarization interferometer (Mach-Zehnder interferometer). The interferometer operates in such a way that during frequency scanning, one of its mirrors is displaced all the time with the help of a servo system in order to sustain the interferometer in a balanced state. The measurable quantity in this case is the displacement $\Delta(\nu)$ of this mobile mirror. In

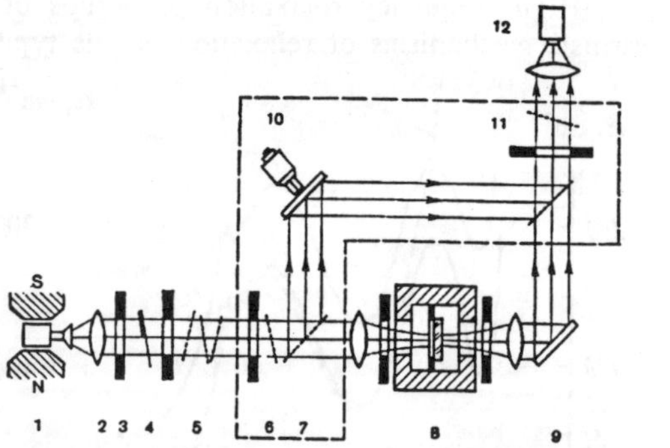

Fig.5. Measuring channel of the «Epsilon» spectrometer. 1 - radiation generator (BWO); 2 - teflon lenses; 3 - absorbing diaphragms; 4 - modulator; 5 - grid attenuator; 6 - polarizer; 7 - dividing grids; 8 - thermostat with sample; 9 - mirror (phase modulator); 10 - phase-shift compensator; 11 - analyzer; 12 - radiation detector. The sections isolated by the dashed lines are not used in T(ν) measurements.

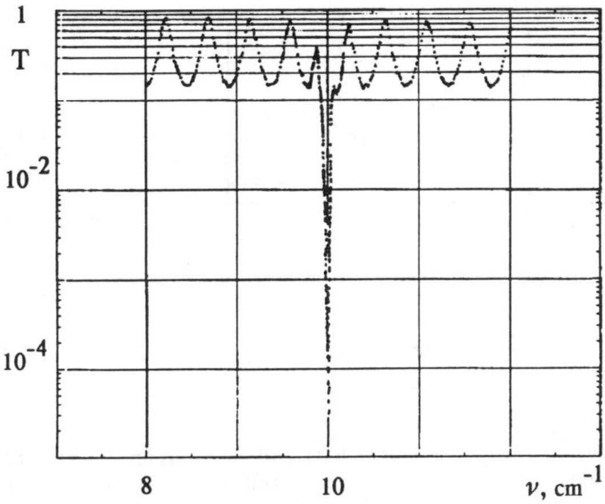

Fig.6 The transmission spectrum of $YFeO_3$ plate (d=2.05mm) at 300K near the F-mode (h||b, e||c-axis).

other words, we practically register a change in the optical thickness of the sample with frequency. As in the case of recording T vs ν spectrum, the basic measurement in this case (with the sample in the channel) is also preceded by gauge measurement, i.e. by the measurement of $\Delta_0(\nu)$ in an empty channel.

The transmission spectrum $T(\nu)$ is calculated by a termwise division of two data files, i.e. the division of the basic file by the gauge file. The phase spectrum $\varphi(\nu)$ of the sample is determined from the difference $\Delta(\nu)$ - $\Delta_0(\nu)$. Clearly, because of the complex instrumental function of the BWO spectrometer, the accuracy of T vs ν and φ vs ν measurements depends strongly on the frequency matching of the basic and the gauge spectra. The reproducibility of results in our spectrometer is 1-0.1% (depending on the scanning rate) for $T(\nu)$ measurements and 1 μm for $\Delta(\nu)$ measurements. These values represent the typical dispersion of the data on T and Δ at an individual point on the frequency scale upon a comparison of several successively recorded spectra.

The speed of measurements on «Epsilon» is as follows: a spectrum of 100 points for a retuning of one BWO is registered in 10 to 30 sec for $T(\nu)$ and 30 sec for $\varphi(\nu)$. The calculation of $\varepsilon'(\nu)$, $\varepsilon''(\nu)$ and $\mu'(\nu)$, $\mu''(\nu)$ takes another few minutes.

The basic parameters of the spectrometer are as follows:
frequency range ν=2-35 cm^{-1}, frequency resolution $\Delta\nu/\nu = 10^{-4}$-10^{-5}, dynamic range 40-50 dB, temperature region 4.2 - 700 K, characteristic transverse dimensions of the samples are 10×10 mm^2.

When measuring the spectra of non-magnetic substances, in order to calculate ε' and ε'' it is necessary to measure at each frequency two parameters characterizing the wave interaction with sample. As experience shows, an optimum pair for transparent samples ($T>10^{-4}$) is T and φ, and for opaque samples - R and ψ. The accuracies, 1-5% over ε' and 5-10% over ε'' may be considered as standard for BWO spectrometer.

In the case of magnetic materials ($\mu \neq 1$) dielectric measurements become considerably complicated. For determining the four unknown values $\varepsilon'(\nu)$, $\varepsilon''(\nu)$ and $\mu'(\nu)$, $\mu''(\nu)$, generally speaking, it is necessary to measure all four parameters of the wave interaction with the sample, i.e. T, φ, R and ψ. However, in many practically

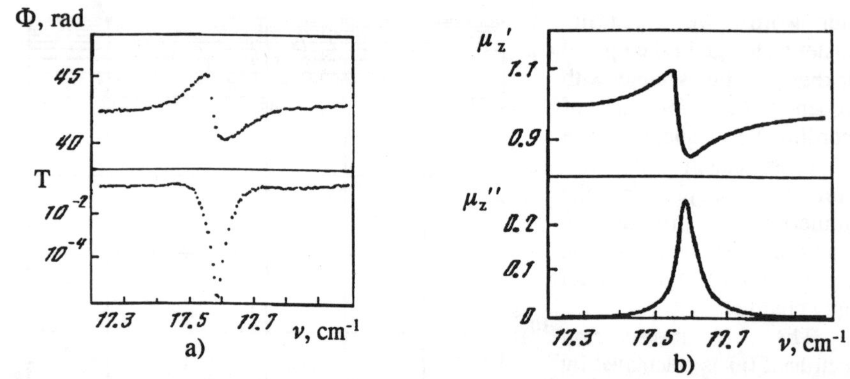

Fig.7. The phase shift $\Phi(\nu)$ and transmission $T(\nu)$ spectra of $YFeO_3$ plate (d=0.818 mm) at 300K near the AF mode (a) and the magnetic permeability spectra $\mu(\nu) = \mu'(\nu) + i\mu''(\nu)$ calculated from $\Phi(\nu)$, $T(\nu)$ data.

important cases one can manage to simplify the problem. The fact is that at submillimeter wavelengths magnetic permeabilities of the antiferromagnets are very small, except for the narrow parts of the spectrum situated near the resonant frequencies of the magnon modes. In this case, dielectric spectra $\varepsilon'(\nu)$ and $\varepsilon''(\nu)$ of the substance outside the magnetic absorption lines can be determined by the usual method, for instantce by $T(\nu)$ and $\varphi(\nu)$, and then the dependences $\mu'(\nu)$ and $\mu''(\nu)$ can be calculated from the data $T(\nu)$ and $\varphi(\nu)$ inside the line using the values of ε' and ε'' detemined outside the lines.

The antiferromagnetic resonance modes (quasiferromagnetic (F) and quasiantiferromagnetic (AF)) were observerd in all orthoferrites investigated. Fig.6 presents the transmission spectrum of the yttrium orthoferrites as an example. The common feature of such spectra is the presence of the periodic oscillations, resulting from the interference of the waves inside the plane-parallel plates. Besides the interference oscillations, the spectra $T(\nu)$ contain sufficiently narrow absorption lines, which correspond to the quasiferromagnetic and quasiantiferromagnetic AFMR modes according to the excitation conditions.

Using experimental dependences $T(\nu)$, $\varphi(\nu)$, the spectra of the permeability $\mu=\mu'+i\mu''$ are calculated according to the well-known formulae for a plane-parallel plate. The dependences $\varepsilon'(\nu)$ and $\varepsilon''(\nu)$ were obtained using the oscillating $T(\nu)$ curves beyond the absorption lines and then were interpolated into the resonance region. The typical results of such measurements are shown in Fig.7.

The magnetic spectra of measured orthoferrites are well described within experimental accuracy by the harmonic oscillator model for a permeability dispersion:

$$\mu_k = 1 + \Delta\mu_k \nu_0^2 / (\nu_0^2 - \nu^2 + i\nu\Delta\nu), \tag{4.1}$$

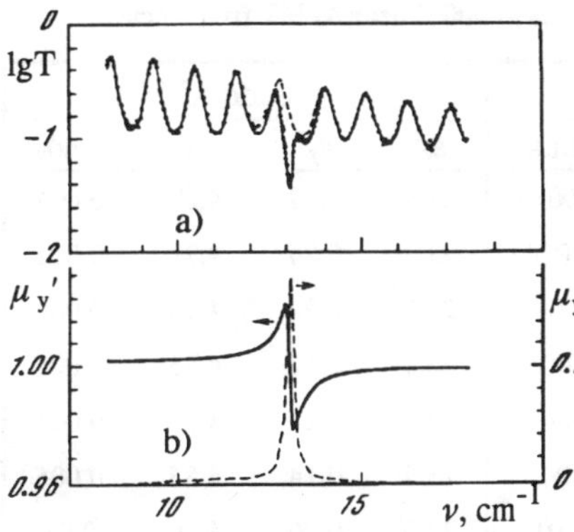

Fig.8 The transmission spectrum of $TmFeO_3$ plate (d=0.808 mm) at 300K near the F-mode (points-experiment, dashed line - calculation for μ=1, solid line - calculation for $\mu \neq 1$) (a) and the magnetic permeability spectra calculated from T(ν) data (b).

where ν_0 is the resonance frequency, $\Delta\nu$ is the linewidth, $\Delta\mu_K$ is the mode contribution to the static permeability, k=x,y,z. As a result, for determining the parameters of the AFMR modes we usually measured the transmission spectra only and then fitted them by least squares, using the simple model of the damped harmonic oscillator Eq.(4.1).

Fig.8 presents the transmission spectra of the thulium orthoferrite as an example. The calculation of the magnetic spectra (parameters of the AFMR modes) from the experimental transmission spectra T(ν) were performed in two stages. The dependences $\varepsilon'(\nu)$ and $\varepsilon''(\nu)$ are fitted at μ=1 using the oscillating T(ν) curves beyond the absorption lines. The dependences given in the form of polynomials were then interpolated into the resonance region. For the orthoferrites in SBMM range $\varepsilon'(\nu) >> \varepsilon''(\nu)$ and therefore it is quite enough to measure only T (ν) spectra without measuring the phase shift itself for determining the dielectric parameters. In the second stage of fitting the dielectric parameters $\varepsilon'(\nu)$ and $\varepsilon''(\nu)$ in the absorption line region were considered to be already given and the matching of the oscillator parameters ν_0, $\Delta\nu$, $\Delta\mu_K$ in Eq.(4.1) was performed in such a way that the calculated T(ν) spectra provided the best fit to the data to within the plotting accuracy.

5. Basic magnetic and dielectric properties of orthoferrites at submillimeter wavelengths

5.1. Dielectric permittivity and magnetic permeability of $RFeO_3$ at room temperatures. Parameters of the AFMR modes

We have investigated the magnetic and dielectric properties of the orthoferrites $RFeO_3$ (R=Y, Tm, Dy, Gd, Ho, Er, Tb) at submillimeter wavelengths ν=100-1000 GHz (3-33 cm^{-1}) [29]. Single $RFeO_3$ crystals were grown by float zoning with radiative heating. The samples for measurements were a-cut plane-paralled plates with the transverse dimensions about 1x1 cm and thickness from 0.2 to 1.5 mm.

For all orthoferrites we have studied $\varepsilon'(\nu) \approx$ const and $\varepsilon''(\nu)$ is a linear function of the frequency at SBMM wavelengths. Table 3 presents the values of the $\varepsilon'(\nu)$ and $\varepsilon''(\nu)$ components along the crystal axes b(y) and c(z), as well as the dielectric loss tangent $\tan\delta=\varepsilon''/\varepsilon'$ obtained by interpolating the dependences ε' and ε'' into the region of the AFMR lines. The values of the refractive index $n \approx (\varepsilon')^{1/2}$ outside the

Table 3. Dielectric parameters of the orthoferrites at 300 K near AFMR frequencies.

Crystals	F - mode AFMR				AF - mode AFMR			
	ε_z'	ε_z''	n_z	$\tan\delta_z$	ε_y'	ε_y''	n_y	$\tan\delta_y$
$YFeO_3$	25.4	0.04	5.04	0.0016	21.9	0.09	4.68	0.004
$TmFeO_3$	27.8	0.50	5.27	0.018	22.2	0.26	4.71	0.012
$DyFeO_3$	26.2	0.26	5.12	0.01	24.0	0.20	4.90	0.008
$GdFeO_3$	24.7	0.13	4.97	0.005	23.5	0.12	4.85	0.005
$HoFeO_3$	26.2	0.23	5.12	0.009	22.8	0.17	4.78	0.007
$ErFeO_3$	26.2	0.14	5.12	0.005	23.3	0.14	4.83	0.005
$TbFeO_3$	26.0	0.26	5.10	0.01	24.0	0.26	4.90	0.01

AFMR lines are also given there. From data of the Table 3 we can conclude that orthoferrites feature a sufficiently small magnitude of $\tan\delta$ ($\approx$0,01-0,001). These values of $\tan\delta$ are quite close to those observed for such high-quality dielectric materials like fused quartz.Note, that together with a high refractive index, the orthoferrites also have a large birefringence at SBMM wavelengths. For example, it follows from Table 3 that the difference between the refractive indices in $TmFeO_3$ amounts to n_z-n_y=0.56, being 10 times as large as the difference between the refractive indices of the ordinary and extraordinary rays in crystalline quartz.

The values of the parameters of both AFMR modes for the orthoferrites we have investigated are listed in Table 4. It is seen that the contributions of both modes to static permeability are virtually the same, however, their linewidths differ greatly.

We have also studied the AFMR-mode behaviour in the magnetic field H| |m| |c-axis. Fig.9 shows a resonance frequency increase in the magnetic field for $TmFeO_3$. A dependence of resonance frequencies on the magnetic field is widely used in the well-known technique for the resonance line recording at a fixed radiation frequency with the scanning by the magnetic field. In such experiments the AFMR linewidths are measured in units of the magnetic field strength ΔH. The relation between ΔH and the linewidth $\Delta\nu$ measured can be found, if the dependence of the AFMR frequency $\nu_0(H)$ on the magnetic field H is known . In the orthoferrites for H| |m| |c ($\Gamma_4(G_xF_z)$ - magnetic phase)

$$\nu_1(H)=[\nu_1^2(0) + (\gamma/2\pi)^2 H(H+H_D^z)]^{1/2},$$

$$\nu_2(H)=[\nu_2^2(0) + (\gamma/2\pi)^2 H_D^z(H+H_D^z)]^{1/2} \quad (5.1)$$

for F- and AF-modes respectively, where $H_D^z=H_E m_z/M_0\approx 10^5$ Oe is the Dzyaloshinskii-like effective field along c-axis. For a small field $\nu_{1,2}(H) = \nu_{1,2}(0) + \alpha_{1,2}H$, where $\alpha_{1,2}=\gamma^2 H_D^z/8\pi^2\nu_{1,2}(0)$.

We have carried out the measurements of the magnetic spectra of orthoferrites in the magnetic field H=3,6 kOe (H| |m| |c) to determine the coefficients $\alpha_{1,2}$. The

Table 4. AFMR mode parameters of the orthoferrites at 300 K.

Crystals	F - mode AFMR				AF - mode AFMR			
	ν_1, GHz	$\Delta\nu_1$, GHz	$\Delta\mu_y \cdot 10^4$	$\mu_y''(\nu_0)$	ν_2, GHz	$\Delta\nu_2$, GHz	$\Delta\mu_z \cdot 10^4$	$\mu_z''(\nu_0)$
$YFeO_3$	299.4	0.9	8	0.27	526.8	2.1	7.5	0.19
$TmFeO_3$	402	7.5	8.5	0.04	698	8.4	9	0.075
$DyFeO_3$	379.5	10.5	9	0.03	510	10.5	7	0.034
$GdFeO_3$	305	12	8	0.02	605.7	13	8	0.037
$HoFeO_3$	378	15	8	0.02	597	9	7.5	0.051
$ErFeO_3$	377	8.0	8	0.04	673	7	8	0.077
$TbFeO_3$	322	20	8	0.013	537	15	6.5	0.023

Table 5. AFMR mode parameters of the orthoferrites for the magnetic field H | | c at 300 K.

Crystals	F - mode AFMR			AF - mode AFMR		
	α_1, GHz/ kOe	H_D^z, kOe	ΔH, kOe	α_2, GHz/ kOe	H_D^z, kOe	ΔH, kOe
$YFeO_3$	1.90	145	0.47	1.10	145	2
$TmFeO_3$	1.85	185	4.0	1.0	175	9
$DyFeO_3$	1.65	165	6.0	1.45	190	7
$GdFeO_3$	1.10	90	10.	0.5	80	25
$HoFeO_3$	1.25	125	12.	0.9	130	10

results of these measurements are presented in Table5. On calculation H_D^z ($H_D^z = 8\pi^2\alpha_{1,2}\nu_{1,2}(0)/\gamma^2$) and ΔH ($\Delta H = \Delta\nu/\alpha$) listed in Table 5, we use the values of $\nu_{1,2}(0)$ and $\Delta\nu$ from Table 4 and take the values of γ for the g-factor equal to two. It is seen that the minimum value of the AFMR linewidth is observed in $YFeO_3$ for the quasiferromagnetic AFMR mode.

Thus, the orthoferrites are the materials possessing very high-quality magnetic and dielectric parameters in SBMM range. As dielectric materials, the orthoferrites weakly absorb at SBMM

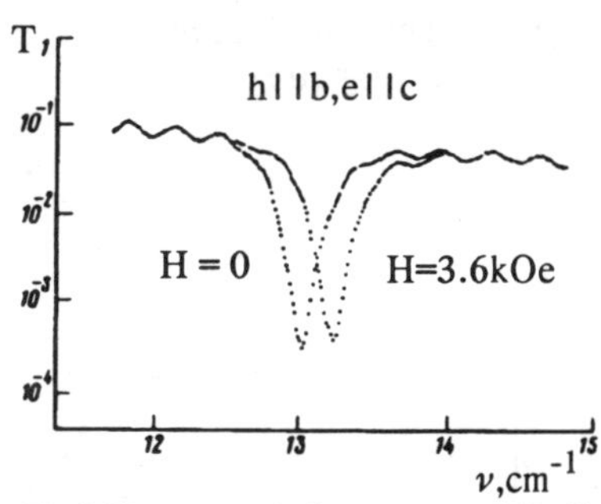

Fig.9 The transmission spectra of $TmFeO_3$ plate (d=2.7 mm) at 300K near the F mode in the magnetic field H | | c–axis.

wavelengths (small angular dielectric loss tangent) and they have high refractive indices and birefrigence. As magnetic materials, rare-earth orthoferrites possess narrow high-quality resonance lines and high linear coupling coefficients between the AFMR frequencies and the external magnetic field.

5.2. Temperatures behaviour of the AFMR modes in YFeO3

We consider now the behaviour of the AFMR modes in $YFeO_3$ orthoferrite in a wide temperature range. In $YFeO_3$ (Y^{3+} is a nonmagnetic ion) at all temperatures in zero magnetic field the magnetic configuration Γ_4 is preserved. The behaviour of the AFMR modes in $YFeO_3$ are determined by the Fe subsystem only without the influence of the R subsystem on it.

We studied single crystals grown by Drs.V.V.Fedotova and A.P.Guess (Institute of Solid State Physics, Minsk) from a solution in the melt of a boron-barium solvent (type I), as well as our single crystals grown by float zoning with radiative heating (type II). The AFMR frequencies (Fig.10) for both modes below 70 K reach saturation, being practically independent of temperature. As the temperature is increased up to T_N the AFMR mode frequencies decreased; that results from the reduction of the lattice magnetization and anisotropy fields of the Fe subsystem at $T \rightarrow T_N$. It is interesting to note the presence of a weak maximum at $T \approx 300$ K in the temperature dependence of $\nu_1(T)$. Maximum results from the competition between the negative single-ion contribution and the positive exchange and magnetic dipole ones into the anisotropy constant K_{ac}.

The contributions of both modes at T_{room} are actually the same ($\Delta\mu_{y,z} = 4\pi\rho\chi_\perp$) and do not vary with temperature. As for the linewidths $\Delta\nu$, their temperature behaviour for M_1 and M_2 modes is substantially different (see Fig. 10b). The linewidth of the M_2 mode increases monotonically with temperature. M_2 mode quality factor reaches the value of 2000 at T=100 K. The linewidth of the M_1 mode remains practically the same with temperature.

At low temperature the behaviour of the M_1 mode is different for I and II samples. We have observerd the single M_1 mode with cooling up to 5 K in samples of the type I. The linewidth of the M_1 mode was practically independent of temperature with cooling (Fig.10b).

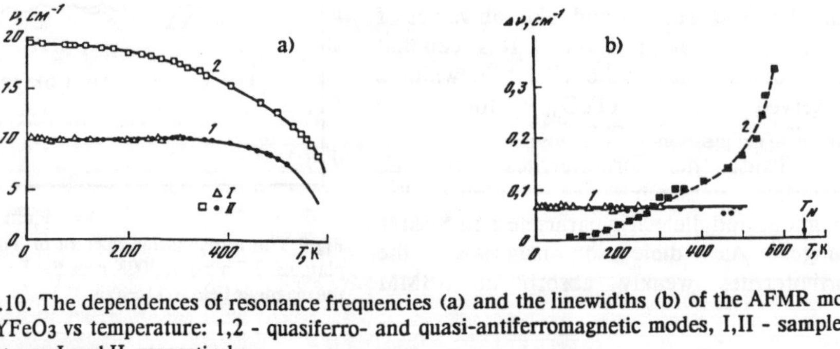

Fig.10. The dependences of resonance frequencies (a) and the linewidths (b) of the AFMR modes in YFeO3 vs temperature: 1,2 - quasiferro- and quasi-antiferromagnetic modes, I,II - samples of the types I and II, respectively.

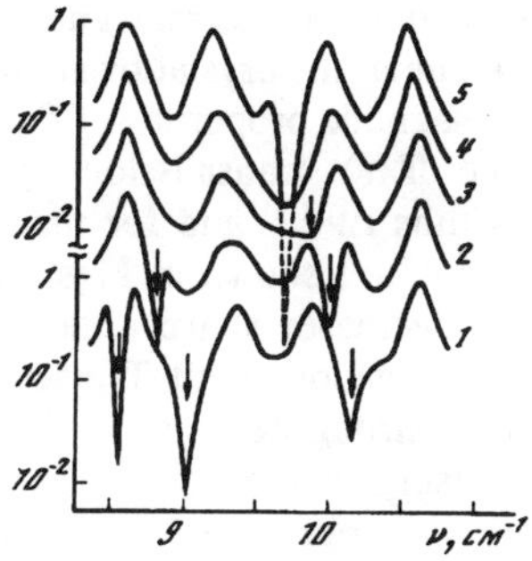

Fig.11. Transmittance spectra for $YFeO_3$ samples of the II type at low temperatures and h | | b: 1 - T= 5K, 2 - 30, 3 - 60, 4 - 80, 5 - 180. The scale of ordinates is given for the upper (5) and the lower (1) curves only. The intermediate spectra are arbitrary shifted with respect to each other. The dashed line indicates the M_1 mode, arrows - the unknown low-temperature absorption lines.

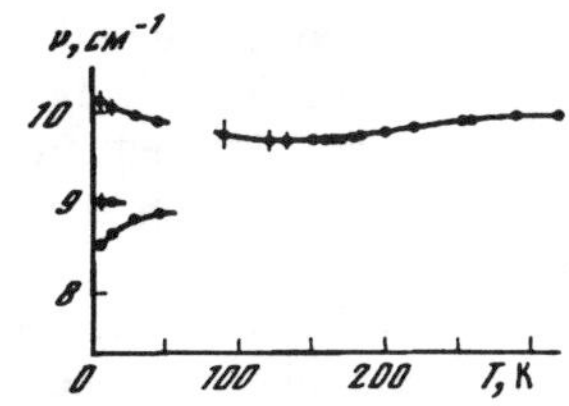

Fig.12. Temperature behaviour of the resonance frequencies of the modes observed in $YFeO_3$ samples of the type II at h | | b [27].

In the II type samples in the vicinity of the M_1 mode novel narrow absorption lines are inflamed in the transmission spectra on cooling, beginning from the temperature 50 K (Fig.11) [27]. At first there are only two of them (see spectrum at 30 K), and then near the liquid helium temperature at ν=9 cm^{-1}, a third line is appeared. Fig.12 shows the temperature behaviour of absorption lines frequencies we have observed in the vicinity of the M_1 mode.

One of possible reasons for such a strong difference in the temperature behaviour of the M_1 mode in samples of the types I and II and appearence of the impurity modes may result from various concentrations of the impurity Fe^{2+} or Fe^{4+} ions. Another reason of the appearence of the impurity modes is due to an occupation of a small part of Y sites by the Fe^{3+} ions [43].

6. Dynamics of the spin reorientational phase transitions

In many orthoferrites the decrease in temperature causes a strong increase of the rare-earth contribution to the anisotropy energy, leading to a change in signs of the effective anisotropy constants, and as a consequence to the spin reorientational phase transitions. These transitions are accompanied by the anomalies of the magnetic, magnetoelastic and other properties of the crystal [5-8]. However, this is most clearly manifested in the dynamic magnetic properties : in soft modes and other excitations. As a rule, the dynamics of the spin-reorientation phase transitions in rare-earth orthoferrites is sufficiently complex due to interaction of various types of excitations [11,22,24,30,44,45]. Below we shall consider the most simple examples of the dynamics of the first- and second-order spin-reorientation transitions, being realized in $DyFeO_3$ and $TmFeO_3$, respectively.

6.1. Dynamics of the first-order phase transition Γ_4-Γ_1 in $DyFeO_3$

The dysprosium orthoferrite is the only unsubstituted orthoferrite in which with lowering temperature a spin reorientational phase transition occurs at T_M=40÷50 K from the weakly ferromagnetic phase $\Gamma_4(G_xF_z)$ into the antiferromagnetic phase $\Gamma_1(G_y)$ (Morin-type spin-reorientation phase transition) [6,46,47]. In the course of this transition the antiferromagnetic vector G of the Fe^{3+} ions is abruptly reoriented in the ab-plane of the crystal from the a- to the b-axis.

Now consider the dynamic magnetic properties of $DyFeO_3$ [28]. The temperature dependences obtained from the $T(\nu)$ transmission spectra for the resonance frequencies $\nu_0^{(1,2)}$, for the linewidths $\Delta\nu^{(1,2)}$ and for the mode contributions $\Delta\mu_y^{(1)}$ and $\Delta\mu_z^{(2)}$ to the static permeability, are shown in Figs. 13,14. It can be seen from Fig. 13 that the frequency $\nu_0^{(2)}$ decreases sharply, becomes lower than $\nu_0^{(1)}$ and reaches a minimum at the phase transition point T_M=50 K. This behaviour of the $\nu_0^{(2)}$ frequency is an evidence of a strong decrease in the anisotropy energy in the ab-plane of the crystal with decreasing temperature. As for the $\nu_0^{(1)}$ frequency of the quasiferromagnetic mode, it has no appreciable anomaly at the phase transition point.

The contributions of the quasiferromagnetic mode $\Delta\mu_y^{(1)}(T)$ and quasi-antiferromagnetic mode $\Delta\mu_z^{(2)}(T)$ (Fig. 14), as well as linewidths $\Delta\nu^{(1,2)}(T)$ (Fig.13), exhibit the strongest anomalies at spin-reorientation phase transition point.

The abrupt changes of $\Delta\mu_y^{(1)}$, $\Delta\mu_z^{(2)}$, $\Delta\nu^{(1,2)}$ observed at $T=T_M$, as well as the fact that the $\nu_0^{(2)}(T)$ frequency of the quasi-antiferromagnetic mode softens only at one temperature, indicate that the spins in $DyFeO_3$ are reoriented via a first-order phase transition. This agrees with the results of static measurements [47].

The observed behaviour of the AFMR mode parameters in $DyFeO_3$ at a Morin-type phase transition is well explained on the whole on the basis of the approach discussed above in section 2 and by the corresponding formulae for the resonance frequencies and the dynamic susceptibility. Now consider this problem in more details. Note, first of all, that the softening of the $\nu_0^{(2)}(T)$ frequency observed on the approach to the phase transition point reflects, according to (2.10b), (2.13b), the temperature dependence of the anisotropy constant $K_{ab}(T)$ which reverses sign near T_M. At the phase transition point, when $K_{ab}(T_M)=-K'_2/2$, the frequency of the soft mode is thereby finite and continuous, but its slope $d\nu_0^{(2)}/dT$ reverses sign. The

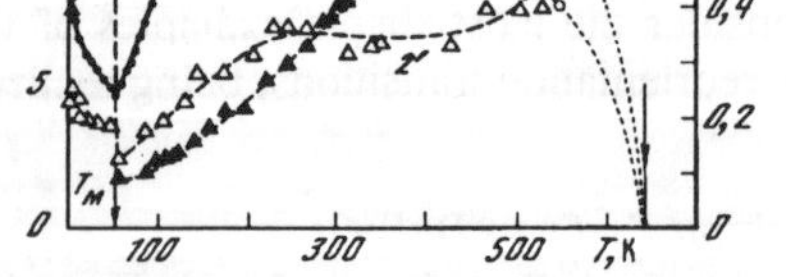

Fig.13. Dependences of the AFMR frequencies and the linewidths for $DyFeO_3$ vs temperature 1 - ν_1, 1' - $\Delta\nu_1$, 2 - ν_2, 2' - $\Delta\nu_2$.

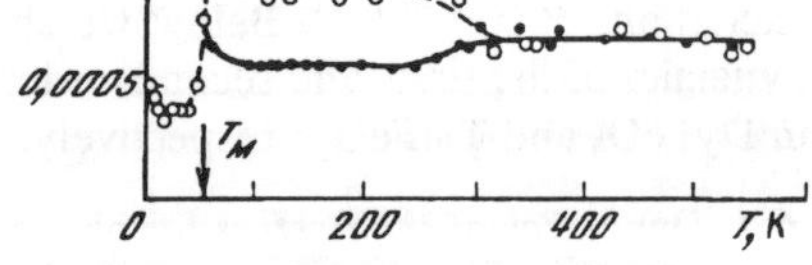

Fig.14. Temperature dependence of contributions of two AFMR modes to static permeability in $DyFeO_3$: 1 - $\Delta\mu_y^{(1)}$, 2 -

frequency at the phase transition point

$$\nu_0^{(2)}(T) = (\gamma/2\pi)(-K'_2/2\chi_\perp)^{1/2}, \qquad (6.1)$$

is determined by a fourth-order effective anisotropy constant, so that K'_2 can be directly estimated, if the quantity $\chi_\perp = M_0/2H_E$ is known. The latter can be easily estimated from the values of $\Delta\mu_x^{(2)}$ measured for the quasi-antiferromagnetic mode at T=300 K, when $\eta_{zz} \ll 1$ and $\chi_\perp^z = \chi_\perp^{Fe}$. As a result, putting $\Delta\mu_z^{(2)} = 7 \cdot 10^{-4}$ (see Fig. 14), $\chi_\perp^{Fe} = \Delta\mu_z^{(2)}/4\pi\rho = 0.7 \cdot 10^{-5}$ cm³/g ($\rho = 8.0$ g/cm³ is the density of $DyFeO_3$), $\gamma = 1.76 \cdot 10^7$ Hz/Oe we get $K'_2 = -3.5 \cdot 10^4$ erg/g, which agrees in order with the data of reference [6] ($K'_2 = -2 \cdot 10^4$ erg/g).

We consider now the behaviour of the permeability tensor at a Morin-type phase transition. The largest discontinuity is observed here for the component $\Delta\mu_z^{(2)}$ of the static contribution to the permeability. According toEq. (2.12), (2.15) we get

$$[\Delta\mu_z^{(2)}(\Gamma_1) - \Delta\mu_z^{(2)}(\Gamma_4)]/4\pi\rho = \chi^z_{rot}(T_M), \qquad (6.2)$$

The quantity χ^z_{rot} is the static differential susceptibility in the antiferromagnetic phase $\Gamma_1(G_y)$ relative to spin rotation in the ab-plane, being inversely proportional to the anisotropy constant in this plane. Since the anisotropy constant decreases at the phase transition point, this causes a large jump of the susceptibility on the transition to the Γ_1 phase. Unlike the antiferromagnetic Γ_1 phase, susceptibility along the z-axis above the phase transition point, i. e. in the Γ_4 phase, is independent on K_{ab} and is determined only by the quantity $\chi_\perp^z$ (see Eq.(2.12)) that makes it small. Knowing the susceptibility jump at the transition point ($\chi^z_{rot}(T_M) = 3.1 \cdot 10^{-4}$ cm³/g) and the value of K'_2, we can estimate a weakly ferromagnetic moment along the c-axis: $m_z = 2.3$ emu/g (i.e. 0.11 μ_B/ion); this agrees with the data of references [46,47].

At the transition point the jump $\Delta\mu_y^{(1)}$ of the quasi-ferromagnetic mode agrees with the model considered. The fact that in the antiferromagnetic Γ_1 phase the observed value of $\Delta\mu_y^{(1)}$ does not vanish, as it should according to Eqs.(2.14), (2.15), can be due to an inaccurate orientation of the sample.

6.2. Dynamics of the second-order phase transitions Γ_4-Γ_{24}-Γ_2 in $TmFeO_3$

The spin reorientation occurred most widely among orthoferrites is realized for the case of $TmFeO_3$: $\Gamma_4(G_xF_z)$-$\Gamma_{24}(G_{xz}F_{zx})$-$\Gamma_2(G_zF_x)$. It is accomplished in the form of two second-order phase transitions (Γ_4-Γ_{24} at T_1 and Γ_{24}-Γ_2 at T_2, $T_1 > T_2$), being accompanied by a smooth reorientation of the weak ferromagnetic moment **F** from the c- to the a-axis and of the antiferromagnetic vector **G** from the a- to the c-axis.

The behaviour of the AFMR frequencies in $TmFeO_3$ was studied by the magnetic resonance technique [19], nonelastic neutron scattering studies [25], Raman scattering of light [26] and the SBMM BWO spectroscopy [33].

Figures 15,16 present temperature dependences of the resonance frequencies and contributions of the AFMR modes to the static permeability obtained from fitting the SBMM transmission spectra. Together with narrow AFMR modes ($\Delta\nu/\nu = 10^{-2}$), the wide absorption lines were observed related to electronic transitions inside the

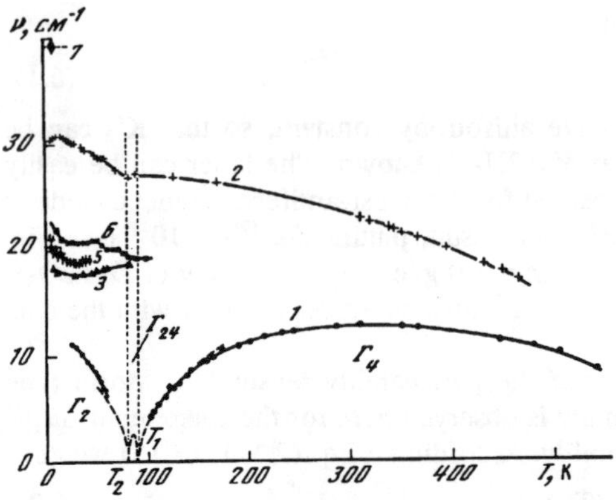

Fig.15. Temperature dependences of the AFMR M_1 (1) и M_2(2) and the rare-earth R_1(3), R_1' (4), R_1'' (5), R_2 (6), R_3 (7) mode frequencies in $TmFeO_3$.

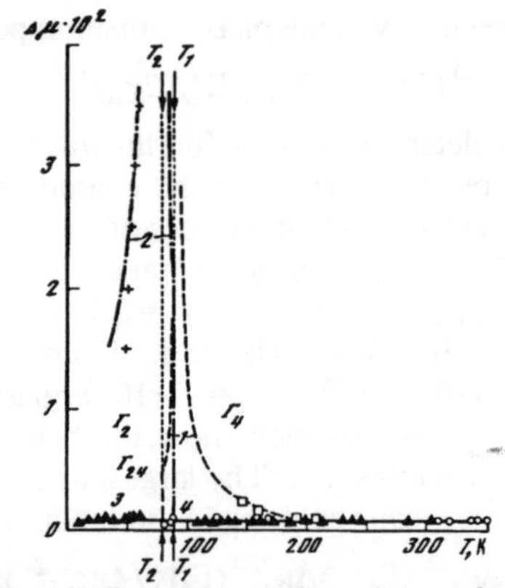

Fig.16. Temperature dependence of the contributions of AFMR modes M_1 и M_2 to the static permeability in $TmFeO_3$: 1 - $\Delta\mu_x^{(1)}$, 2 - $\Delta\mu_z^{(1)}$, 3 - $\Delta\mu_y^{(1)}$, 4 - $\Delta\mu_z^{(2)}$.

Tm^{3+} ion (R mode), which we shall consider in the next section. Let us discuss here the anomalies of the AFMR modes in the spin reorientation region.

The quasiferromagnetic AFMR M_1 mode is a soft mode for this reorientation, whose frequency ν_1 softens on both boundaries (T_1 and T_2) (Fig. 15). This is a direct consequence of the decrease in the anisotropy constant K_{ac} in the ac-plane.

The soft mode contributions $\Delta\mu_x^{(1)}$ and $\Delta\mu_z^{(1)}$ undergo on the upper T_1 and the lower T_2 bounds, respectively, strong anomalies (divergences) in the phase transition region (Fig. 16). The $\Delta\mu_y^{(1)}$ contribution remains virtually constant.

As for the behaviour of the second AFMR mode (quasi-antiferromagnetic M_2 mode), it has no such strongly pronounced anomalies in the spin reorientation region as the M_2 mode does. We should only note a small decrease in its frequency ν_2 in the T_1-T_2 interval, as well as a linear decrease of the contribution of this mode $\Delta\mu_z^{(2)}$ and its vanishing at $T \le T_2$.

Let us analyze the observed behaviour of the AFMR modes on the basis of Eqs. (2.9)-(2.18) for the dynamic susceptibility in the phases Γ_2 and Γ_4, and corresponding expressions for the angular Γ_{24} phase (see also [11,25,48]). In the vicinity of the spin reorientation region, where K_{ac} reverses sign, we can put $K_{ac}(T)=K_0(T-T_1)/T_1$ providing a good agreement with the experiment near the reorientation region. . The quantity K_0 is determined by the slope $d\nu_1^2/dT=1.3\cdot10^3 GHz^2/K$. It is necessary to take an account of the deviation of the dependence $K_{ac}(T)$ from the linear one far away from $T_{1,2}$.

The anomalous behaviour of the soft mode contributions can also be completely explained in the framework of the approach used. Thus, the increase in the contribution $\Delta\mu_x^{(1)}=4\pi\chi^x_{rot}$ observed at $T\rightarrow T_1$ is due to the increase in the rotation susceptibility χ_{rot}^x, which diverges at the point T_1 according to Eq. (2.12), while at the point T_2 on the transition from the angular Γ_{24} to the Γ_2 phase abruptly vanishes (see Fig. 16). The growth of the contribution $\Delta\mu_z^{(1)}=4\pi\chi^z_{rot}$ at $T\to T_2$ can be explained analogously.

Figure 16 shows the temperature dependences of the contributions $\Delta\mu_x^{(1)}$ and $\Delta\mu_z^{(1)}$ (curves 1,2) calculated by formulae similar to Eqs.(2.12), (2.18). The values of the weakly ferromagnetic moments were taken from the results of static measurements [6,49,50] : m_x^0=1.2 Gcm^3/g, m_z^0=3 Gcm^3/g, and the anisotropy constants were determined from the temperature dependences ν_1(T) (see above): $K_0=1.2\cdot10^5$ erg/g, $K_2=1.5\cdot10^4$ erg/g. We thereby put M_0=100 Gcm^3/g, $H_E=6.4\cdot10^6$ kOe, ρ=8 g/cm^3. A satisfactory agreement between theory and experiment is seen from Fig. 16 near the reorientation region.

7. Rare-Earth Modes

In the present and following sections we shall analyze the part played by the rare-earth subsystem in the formation of the dynamic properties of orthoferrites at SBMM wavelengths. At sufficiently high temperatures, when the R subsystem is in a paramagnetic state, its dynamic properties will be mainly determined by single-ion electronic excitations in rare-earth ions. We shall call them rare-earth modes, or R modes. In the first place we shall be interested in the R modes caused by electronic transitions inside the ground multiplet of a rare-earth ion split in the crystal field, external magnetic field and exchange field due to the Fe subsystem. Such rare-earth modes were observed in some orthoferrites ($TmFeO_3$ [33,34,51], $HoFeO_3$ [30,45,51], $YbFeO_3$ [23,24,51], $ErFeO_3$ [52,51]) and orthochromites ($ErCrO_3$ [44]).

7.1. Spectrum and dynamic susceptibility of the rare-earth ions in RFeO3. Classification of R modes.

Spectrum of the rare-earth ions in $RFeO_3$ is determined in primarily by the crystal field, the exchange and dipolar R-Fe interactions and the external magnetic field. The low-symmetry crystal field described by the point group C_s is the strongest interaction for the R-ions.

The ground multiplet of the non-Kramers R-ions (Ho^{3+}, Tb^{3+}, Tm^{3+}...) having an even number of the electrons and integer quantum number J of the total angular momentum, split by the crystal field into a series of singlets characterized by one of the two one-dimensional irreducible representations of the C_s group (A_1 and A_2). For some cases the spectra of the non-Kramers R ions have a quasi-doublet structure (Tb^{3+}, Ho^{3+}) [53,54]. For the Kramers R-ions (Dy^{3+}, Er^{3+}, Yb^{3+}...) characterized by an odd number of 4f-electrons and a half-integer J, the ground multiplet splits by the crystal field into doublets. Each doublet is described by the Kramers-conjugated wave functions.

The exchange R-Fe interaction and the external magnetic field cause the change in the R-ion spectrum: splitting of doublets and quasidoublets, shift of the «center of gravity» or the position of singlet levels [6,7]. Thereby the corresponding wave functions acquire an admixture of other states. Such mixing occurs especially strongly in the case of two nearby states.

Dynamic magnetic susceptibility $\chi_R(\varpi)$ of the R subsystem is determined by the magnetodipolar electron transitions between the energy levels of the R ions. For the case of the isothermal R-subsystem, whose temperature coincides with the temperature of the thermostat (crystal lattice), one can obtain the following expressions for the susceptibility $\chi(\varpi,T)$ of the R ions, noninteracting with each other (see otherwise [55])

$$(\hat{\chi}_R)_{lk}=\sum_{n,m}[\mathrm{Re}(\Delta\chi_{lk}^{nm})-\frac{i\omega}{\omega_{nm}}\mathrm{Im}(\Delta\chi_{lk}^{nm})]\frac{\omega_{nm}^2}{\omega_{nm}^2-\omega^2}+\Delta\chi_{lk}\frac{i\delta}{i\delta+\omega}\ , \tag{7.1}$$

where $\chi_{lk}^{nm}=\frac{1}{4}N\sum_p\frac{\mu_{nm}^l\mu_{nm}^k\exp(-\beta E_m)}{h\omega_{mn}Z}$, $\chi_{lk}=\frac{1}{4}N\sum_p\sum_m\frac{\Delta\mu_{mm}^l\Delta\mu_{mm}^k\exp(-\beta E_m)}{TZ}$,

$\mu_{nm}^l=\langle n|\mu^l|m\rangle$ is the matrix element of the operator for the magnetic R-ion moment μ between the states $|n\rangle$ and $|m\rangle$ with the energies E_n, E_m; $\varpi_{nm}=(E_n-E^m)/h$ is the frequency of the transition $n\rightarrow m$; $\Delta\mu_{nm}^k=\mu_{nm}^k-\langle\mu^k\rangle$, $\langle\mu^k\rangle$ is the k-component of the average R ion moment; $Z=\Sigma_m\exp(-\beta E_m)$; N is the number of R-ions per 1 cm^3; $\beta=1/T$. The states $|n\rangle$, $|m\rangle$ depend on the index p, pertaining to the four crystallographically inequivalent positions of R-ions. In Eq. (7.1) the quantity $\delta\rightarrow+0$ characterizes an infinitely weak interaction of the R ions with the thermostat, ensuring the isothermality of the R-subsystem, and permits a correct limit of the dynamic susceptibility to the static one when $\varpi\rightarrow 0$.

The R-ions occupy the off - centrosymmetric positions (C_s symmetry) though the space symmetry (D_{2h}^{16}) of the orthoferrite crystals remains centrosymmetric. As a result, even within the ground multiplet of the R-ion not only magnetodipolar but also electrodipolar electronic transitions turn out to be allowed. Hence this transitions should determine not only the magnetic but also the dielectric and the magnetoelectric properties of the orthoferrites.

The dynamic dielectric polarizability $\alpha_R(\varpi)$ due to electrodipolar electronic transitions inside the ground multiplet of R-ion can also be written in the form Eq. (7.1), however replacing the operator μ with the operator **d** for the electrodipolar moment. Note that the electrodipolar transitions between various R ion multiplets are distinctly observed in the optical range (see, for instance, [53,56]). As for the observation of such transitions inside the ground R ion multiplet falling in the far IR and SBMM frequency ranges, the studies of $TmFeO_3$ carried out by us provide the positive answer on this question (see below).

Further concretization of the dynamic susceptibilities is determined by the selection rules for the matrix elements of the operators μ, d and by their relation for various inequivalent positions. In a first approximation the matrix elements of the d and μ operators can be calculated on the basis of «pure» wave functions of the R ion in the crystal field. For the non-Kramers ions the matrix elements are non-zero for the following component of μ and d [56]:

Table 6. Symmetry and activity of the electronic excitations in Fe- and R-subsystems of the orthoferrites with the non-Kramers rare-earth ions.

Magnetic phase	Symmetry of the excitations (representations of the group D_{2h}^{16})	AFMR-modes in Fe subsystem	R modes			
			Transition $A_1 \rightarrow A_2$		Transitions $A_1 \rightarrow A_1$, $A_2 \rightarrow A_2$	
		h	h	e	h	e
Γ_4	Γ_{14}	h_z	-	-	h_z	-
$G_xF_zA_y$	Γ_{23}	h_x, h_y	h_x, h_y	-	-	-
	Γ_{58}	-	-	e_z	-	-
	Γ_{67}	-	-	-	-	e_z, e_y
Γ_2	Γ_{12}	h_x	h_x	-	(h_x)	-
$G_zF_xC_y$	Γ_{34}	h_y, h_z	$h_y(h_z)$	-	$h_z(h_y)$	-
	Γ_{56}	-	-	(e_x)	-	e_x
	Γ_{78}	-	-	$e_z(e_y)$	-	$e_y(e_z)$
Γ_1	Γ_1	-	-	-	-	-
$G_yA_xC_z$	Γ_2	h_x	h_x	-	-	-
	Γ_3	-	h_y	-	-	-
	Γ_4	h_z	-	-	h_z	-
	Γ_5	-	-	-	-	-
	Γ_6	-	-	-	-	e_x
	Γ_7	-	-	-	-	e_y
	Γ_8	-	-	e_z	-	-

μ_z, d_x, d_y allowed for $A_1 \leftrightarrow A_1$ or $A_2 \leftrightarrow A_2$,
μ_x, μ_y, d_z allowed for $A_1 \leftrightarrow A_2$, (7.2)

The unit cell of an orthoferrite includes four rare-earth ions, the spectra of which are completely identical to one another. Therefore, the R modes corresponding to these electronic transitions in the R ion are four-fold degenerate and pertain to four different types of vibrations of the variables of the R subsystem. This degeneracy is removed when the exchange and dipole interactions of the R ions with each other as well as with the Fe^{3+} ions are taken into account (the analogue of the Davidov splitting of the exciton modes).

To classify the excitations in a crystal one has to take into account the space crystalline and magnetic symmetry of the system. The symmetry of the electronic excitations can also be easily characterized by the representations of the $D_{2h}^{16\text{ - Pbnm}}$ group (see for instance [35]).The classification of the vibrations and their activity for magnetic configuration Γ_2 (F_xG_z), $\Gamma_4(F_zG_x)$, $\Gamma_1(G_y)$ with an account of different electron transitions between the states of the non-Kramers R ions in the crystal field are presented in Table 6. There are two different types of vibrations , some of them referring to the reducible representations Γ_{14}, Γ_{23} (phase Γ_4), Γ_{12}, Γ_{34} (phase Γ_2), Γ_1 - Γ_4 (phase Γ_1) are centrosymmetric and are excited only by the magnetic field, the other ones - are non-centrosymmetric and excited only by the electric field. The components **e** and **h** in brackets in Table 6 excite the corresponding R-modes only in case of mixing the A_1 and A_2 states through the R-Fe interaction which, as a rule, is noticeably weaker than the splitting in the crystal field (in particular for $TmFeO_3$).

7.2. Magnetodipolar and electrodipolar R modes in $TmFeO_3$

To study the nature of the electronic transitions inside the ground multiplet of the R-ions, we have performed the polarization measurements of the SBMM abrosption and the far IR reflection spectra of the $TmFeO_3$ orthoferrite [34].

Transmission spectra of the a-, b-, c-cut plane-parallel plates of $TmFeO_3$ were measured using the submillimeter «Epsilon» BWO-spectrometer in the frequency range ν=3-33cm^{-1} at T=4.2-500K. As shown in Fig.17, the transmission spectra exhibit intensive absorption lines M_1, R_1, R_1', R_1'', R_2 which bulge out of the background of oscillations caused by interference in the sample. Altogether seven absorption lines were observed in the transmission spectra.Two of them, which are rather narrow ($\Delta\nu/\nu \approx 10^{-2}$) and are observed over the whole temperature interval, are the already known AFMR modes in the Fe-subsystem - M_1 and M_2 (Fig. 15). The remaining broader absorption lines ($\Delta\nu/\nu \approx 1$) R_1(h||c, e||b), R_1'(h||a, e||b), R_1''(h||b, e||a), R_2(h||b, e||c), R_3(h||b, e||c) which are observed only at T< 100 K are assigned to R -modes.

The reflection spectra were measured using the Fourier spectrometer Bruker IFS 113v at temperatures 300, 80, and 20 K in the range of ν = 10-100 cm^{-1} . An important peculiarity of the reflection spectra is that the shape of the lines allows a qualitative conclusion whether the mode has to be assigned to a magnetodipolar or electrodipolar electronic transition (R-mode). For the electrodipolar R-mode the reflection coefficient increaces while approaching its resonance frequency from below and decreases while approaching it from above, wereas for the magnetodipolar R-mode the opposite is true. The examples of the reflection spectra exhibiting a series of lines with different shape are shown in Fig.18. The shapes of the M_1, R_1(h||c, e||a), R_3'(h||b, e||a), R_3''(h||a, e||b) lines indicate the magnetic nature of the corresponding modes and the shape of the R_2,

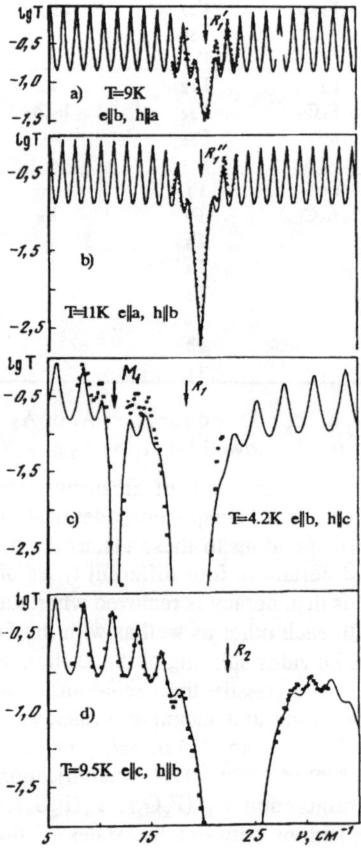

Fig.17. Transmission spectra of the c-cut (a and b) and a-cut (c and d) plane-parallel $TmFeO_3$ samples. In Figs. 17 and 18 points indicate the experimental results, the solid lines the calculations using the model of magnetic and dielectric oscillators for $\mu(\nu)$ and $\varepsilon(\nu)$

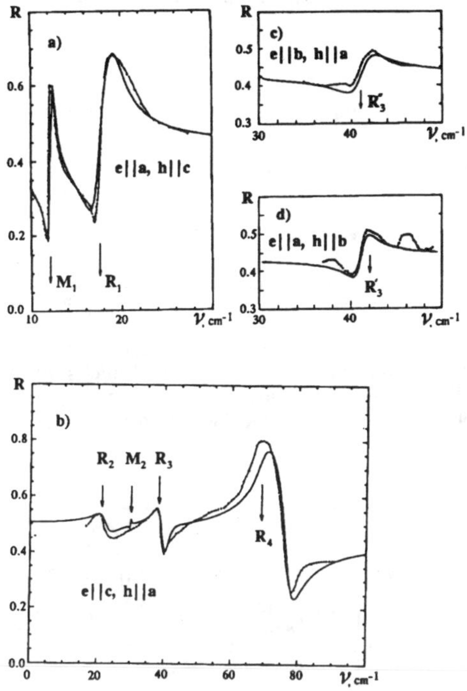

Fig.18. Reflectivity spectra for b-cut (a and b) and c-cut (c and d) plane-parallel $TmFeO_3$ samples [34].

Table 7. The classification of the observed R-modes in $TmFeO_3$.

R-mode	Transition	Symmetry	Excitation condition
R_1	$E_1(A_1) \rightarrow E_2(A_1)$	Γ_{34}	h_z (h_y)
R_1'	$E_1(A_1) \rightarrow E_2(A_1)$	Γ_{78}	e_y (e_z)
R_1''	$E_1(A_1) \rightarrow E_2(A_1)$	Γ_{56}	e_x
R_2	$E_2(A_1) \rightarrow E_3(A_2)$	Γ_{78}	e_z (e_y)
R_3	$E_1(A_1) \rightarrow E_3(A_2)$	Γ_{78}	e_z (e_y)
R_3'	$E_1(A_1) \rightarrow E_3(A_2)$	Γ_{34}	h_y (h_z)
R_3''	$E_1(A_1) \rightarrow E_3(A_2)$	Γ_{12}	h_x
R_4	$E_1(A_1) \rightarrow E_4(A_2)$	Γ_{78}	e_z (e_y)

R_3, R_4(h||a, e||c) lines their dielectric nature.

The identification of the observed R-modes with the certain electronic transitions of the Tm^{3+} ground multiplet as well as their symmetry are presented in Table 7.

To determine the parameters of the observed modes, the transmission and the reflection spectra were fitted, using the harmonic oscillator model for the magnetic permeability and dielectric permittivity

$$\mu(\nu)=1+\Sigma\Delta\mu_\alpha\nu_\alpha^2/(\nu_\alpha^2-\nu^2+i\nu\Delta\nu_\alpha), \qquad (7.3)$$

$$\varepsilon(\nu)=\varepsilon^\infty+\Sigma\Delta\varepsilon_\alpha\nu_\alpha^2/(\nu_\alpha^2-\nu^2+i\nu\Delta\nu_\alpha), \qquad (7.4)$$

where ν_α is the resonance frequency, $\Delta\nu_\alpha$ the linewidth and $\Delta\mu_\alpha$ $\Delta\varepsilon_\alpha$ the contributions of the α-th mode to the static magnetic permeability or dielectric permittivity, respectively. As a result, the temperature dependences of the resonance frequencies (Fig.15,19), linewidths and mode contributions (Fig.20) to the static magnetic permeability and dielectric permittivity were derived.

Note that the revealed R_2, R_3, R_4 electrodipolar modes have rather high strengths which therefore give an appreciable contribution to the dielectric permittivity of $TmFeO_3$ (to our estimates, at T=20 K $\Delta\varepsilon_{23}^z(R_2)=5$, $\Delta\varepsilon_{13}^z(R_3)=2$, $\Delta\varepsilon_{14}^z(R_4)=6$).Thus, R-modes play an important role not only in the magnetic, but also in the dielectric properties of $TmFeO_3$.

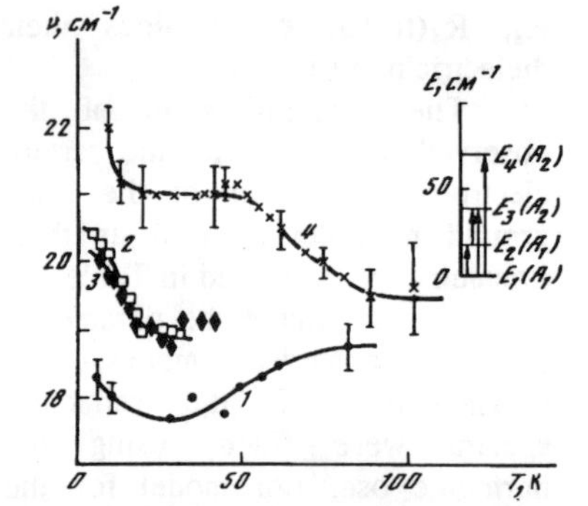

Fig.19. Temperature dependences of the rare-earth mode frequencies: $R_1(1)$, $R_1'(2)$, $R_1''(3)$, $R_2(4)$ in $TmFeO_3$. Insert: the lower part of the spectra of the ground multiplet of the ion Tm^{3+}

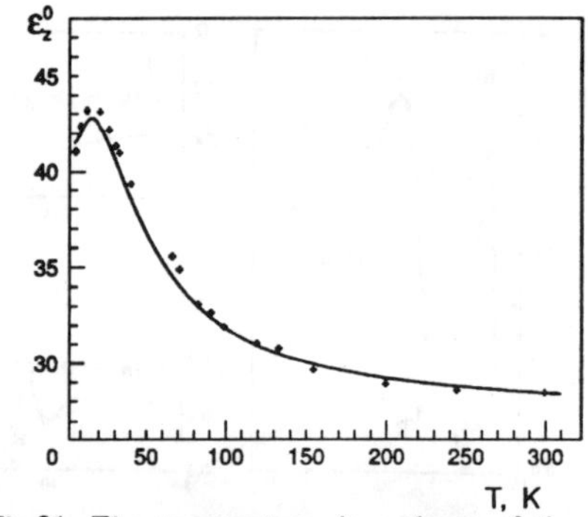

Fig.21. The temperature dependence of the dielectric permittivity of $TmFeO_3$ at the frequency range ν = 5.6 -9 cm^{-1}: points - experiment, solid lines - theory.

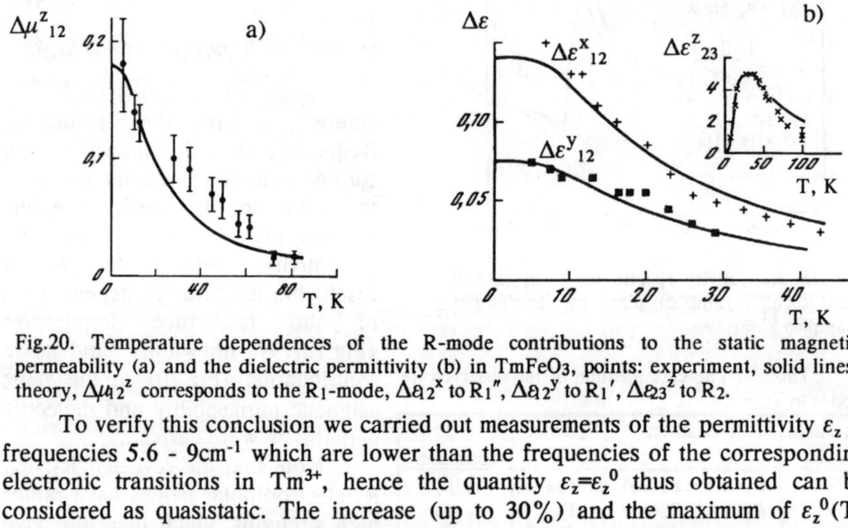

Fig.20. Temperature dependences of the R-mode contributions to the static magnetic permeability (a) and the dielectric permittivity (b) in $TmFeO_3$, points: experiment, solid lines: theory, $\Delta\mu_{12}^z$ corresponds to the R_1-mode, $\Delta\varepsilon_{12}^x$ to R_1'', $\Delta\varepsilon_{12}^y$ to R_1', $\Delta\varepsilon_{23}^z$ to R_2.

To verify this conclusion we carried out measurements of the permittivity ε_z at frequencies 5.6 - 9cm^{-1} which are lower than the frequencies of the corresponding electronic transitions in Tm^{3+}, hence the quantity $\varepsilon_z=\varepsilon_z^0$ thus obtained can be considered as quasistatic. The increase (up to 30%) and the maximum of $\varepsilon_z^0(T)$ observed at low temperature (Fig. 21) confirm the conclusion about the appreciable contribution to ε_z of the electrodipolar electronic transitions in Tm^{3+}.

The temperature dependences of $\Delta\mu_{ij}^k$, $\Delta\varepsilon_{ij}^k$ as well as ε_z^0 (Fig.20) are determined by changing populations of Tm^{3+} energy levels (see Eq.(7.1)). In particular, for $\Delta\varepsilon$, ε_z^0 we have

$$\Delta\varepsilon_{ij}^z(T) = \frac{2N\,(d_{ij}^z)^2}{(E_j-E_i)}\,\frac{\exp(-\beta E_i) - \exp(-\beta E_j)}{Z(T)}, \tag{7.5}$$

$$\varepsilon_z^0(T) = \varepsilon_z^\infty + \Delta\varepsilon_{23}^z + \Delta\varepsilon_{13}^z + \Delta\varepsilon_{14}^z + \Delta\varepsilon_{15}^z + ..., \tag{7.6}$$

$$Z(T) = 1 + e^{-\beta E_2} + e^{-\beta E_3} + e^{-\beta E_4} + pe^{-\beta E_5}, \tag{7.7}$$

where E_i are the crystal field energy levels of Tm^{3+} (in cm^{-1}): $E_1=0$, $E_2=18$, $E_3=39$, $E_4=70$, $E_5=350$ (E_5 is the center of a «gravity» for $p=9$ excited Tm^{3+} ground multiplet levels), $\bar{d}_{ij}^z$ matrix element of the electrodipolar moment, renormalized by the electron-phonon interaction, $\bar{d}_{ij}^z \approx d_{ij}^z(1 + \Lambda_{e-l}(\varepsilon_z-1)/4\pi)$, Λ_{e-l} the electron-phonon interaction constant. As a result we obtained a good agreement between theory and experiment (Fig.20) and determined the corresponding matrix elements: $\mu_{12}^z=4\mu_B$, $\bar{d}_{12}^x=0.033D$, $\bar{d}_{12}^y=0.023D$, $\bar{d}_{13}^z=0.155D$, $\bar{d}_{14}^z=0.356D$, $\bar{d}_{23}^z= 0.5D$. Note that the electron-phonon interaction causes a strong renormalization of d_{ij}^z.

Thus, the study carried out using $TmFeO_3$ as an example has shown that the R-subsystem plays an important role in the formation not only of magnetic, but also of dielectric and magnetoelectric properties of orthoferrites as well as orthochromites and orthoaluminates.

In the present section we discussed in fact only one of the aspects of the problem mentioned above, which includes the dynamics of the R subsystem only, without taking into account of interaction with the Fe subsystem. The latter is not less interesting particularly in the region of spin-reorientation phase transitions.

8. Influence of the rare-earth modes on the AFMR-mode behaviour and dynamics of spin reorientational phase transitions

In the present section we shall consider the influence of the interaction between the rare-earth modes and the AFMR modes on the dynamic properties of orthoferrites. The strongest manifestation of the interaction of these modes should be expected in the region of the spin-reorientational phase transitions, since the latter are in turn due to the exchange (dipole) interaction of the Fe and R subsystems. Just this provides an explaination to the fact that in various orthoferrites the observed behaviour of the AFMR modes is qualitatively different even at spin-reorientational transitions of one and the same type [9,10,23,—26,30,33,44,45,52].

We shall analyze below the dynamics of the interacting Fe and R subsystems for the most widely occurring spin reorientation Γ_4-Γ_{24}-Γ_2 in the ac-plane as an example and then discuss the corresponding experimental data.

8.1. Spin-Hamiltonian of the R-subsystem, thermodynamical potential and equation of motion

Consider for definiteness orthoferrites with the non-Kramers rare-earth ions of the Ho^{3+}, Tb^{3+} type, the ground state of which in the crystal field is a quasidoublet (two close singlets), separated from the excited states by the energy $\Delta E=100$ K. Let us assume that wave functions of the quasi-doublet pertain to two different representations of the point group C_s. Thus, at $T<\Delta E$ the behaviour of the R subsystem can be described in the two-level approximation on the basis of an effective Hamiltonian with spin $S_{eff}={}^1/_2$, which we write in the form, following [7,30]

$$H_{eff} = -\sum_i [\Delta_{cf}\sigma_\zeta{}^i + (\mu_0{}^i(H + aF) + BG_z)\sigma_\xi{}^i] - \tfrac{1}{2}\sum_{i,j}\lambda_{ij}\sigma_\xi{}^i\sigma_\xi{}^j + \sum_i \Delta E_{vv}{}^i, \quad (8.1)$$

where $\boldsymbol{\sigma}^i=(\sigma_\xi{}^i, \sigma_\eta{}^i, \sigma_\zeta{}^i)$ are Pauly matrices of the i-th R-ion; $2\Delta_{cf}$ is the quasidoublet splitting in the crystal field; $\boldsymbol{\mu}_0{}^i=(\mu_x, \pm\mu_y, 0)$ is the quasidoublet magnetic moment; a and B are respectively the isotropic and anisotropic R-Fe exchange constants, and λ_{ij} are the R-R interaction constants. The «±» signs correspond to two crystallographically nonequivalent positions of the R-ions. The quantities $\Delta E^i_{vv}=\Delta E^i_{vv}(H,F,G)$ represent the Van Vleck-like shift of the «center of gravity» of the R-ion quasidoublet on account of admixing of the excited states by the R-Fe interaction and by the external field.

The R subsystem will be described with its nonequilibrium state defined by the average values of the Pauly matrices of the i-th ion, omitting henceforth the averaging symbol ($\langle\sigma^i\rangle=\boldsymbol{\sigma}^i$). The dynamic equations for σ^i are of the form [30,57]

$$^1/_2\hbar\,\dot{\boldsymbol{\sigma}}^i = -[\boldsymbol{\sigma}^i\,\boldsymbol{\Phi}^i] + \mathbf{R}_i\,, \quad (8.2)$$

where $\hbar$ is the Planck's constant, $\boldsymbol{\Phi}_i=\partial\Phi/\partial\boldsymbol{\sigma}^i$ and $\Phi(F,G,\boldsymbol{\sigma}^i)$ is the thermodynamic potential of the nonequilibrium state of the system, $\mathbf{R}_i$ are the relaxation terms, which we shall describe phenomenologically

$$\mathbf{R}_i = -\lambda_0\boldsymbol{\Phi}^i - \lambda_1[(\sigma^i)^2\boldsymbol{\Phi}^i - \boldsymbol{\sigma}^i(\boldsymbol{\sigma}^i\boldsymbol{\Phi}^i)] - \lambda_2\boldsymbol{\sigma}^i(\boldsymbol{\sigma}^i\boldsymbol{\Phi}^i)\,, \quad (8.3)$$

where $\lambda_{0,1,2}$ are the dissipation constants. Changing over to the approximation of two R-sublattices corresponding to two nonequivalent positions of the R-ions ($\boldsymbol{\sigma}_{1,2}$) and introducing the variables $\mathbf{f}=(\boldsymbol{\sigma}_1+\boldsymbol{\sigma}_2)/2$, $\mathbf{c}=(\boldsymbol{\sigma}_1-\boldsymbol{\sigma}_2)/2$, we obtain

$$(M_B/\gamma)\dot{\mathbf{f}} = [\mathbf{f}\boldsymbol{\Phi}_f] + [\mathbf{c}\boldsymbol{\Phi}_c] + \mathbf{R}_f\,, \qquad (M_B/\gamma)\dot{\mathbf{c}} = [\mathbf{f}\boldsymbol{\Phi}_c] + [\mathbf{c}\boldsymbol{\Phi}_f] + \mathbf{R}_c, \quad (8.4)$$

where $\gamma=2\mu_B/\hbar$, $M_B=\mu_B N$, μ_B is the Bohr magneton, N is the number of ions per cm^3.

We describe the dynamics of the Fe subsystem on the basis of the Landau-Lifshitz equations (2.8). We represent the nonequilibrium thermodynamic potential using the molecular field approximation in the form [7]

$$\Phi(F,G,f,c) = \tilde{\Phi}_{Fe}(F,G) - N[f_\xi(\mu_x(H_x+aF_x)+BG_z) + f_\zeta\Delta_{cf} + c_\xi\mu_y(H_y+aF_y) +$$

$$+ \tfrac{1}{2}\lambda_f f_\xi{}^2 + \tfrac{1}{2}\lambda_c c_\xi{}^2 + \tfrac{1}{2}T(S(\sigma_1)) + S(\sigma_2))]\,, \quad (8.5)$$

where $S(\sigma)=\ln 2 - {}^1/_2(1+\sigma)\ln(1+\sigma) - {}^1/_2(1-\sigma)\ln(1-\sigma)$; $\sigma_{1,2} = |\,\mathbf{f}\pm\mathbf{c}\,|$, $\lambda_{f,c}$ are the R-R interaction constants;

$$\tilde{\Phi}_{Fe}(F,G) \equiv \Phi_{Fe}(F,G) + \sum_i \Delta E^i_{vv} = \tfrac{1}{2}AF^2 + \tfrac{1}{2}D\,(FG)^2 - d(F_xG_z - F_zG_x) - M_0FH -$$

$$\tau_1^{vv}H_xG_z - \tau_3^{vv}H_zG_x - \tfrac{1}{2}\sum_k \chi_k^{vv}H_k{}^2 + \Phi_A(G)\,, \quad (8.6)$$

$$\Phi_A(G) = {}^1/_2K_{ac}{}^0G_z^2 + {}^1/_2K_{ab}{}^0G_y^2 + {}^1/_4K_2G_z^4 + {}^1/_4K_2'G_y^4 + {}^1/_2K_2''G_y^2G_z^2 \qquad (8.7)$$

Φ_{Fe} is the thermodynamic potential of the Fe subsystem, renormalized by the Van Vleck contribution to the quasidoublet energy. The most significant contribution is made by these corrections to the anisotropy constants ($K_{ac}{}^0 = K_{ac}{}^{Fe} + K_{ac}{}^{vv}$, $K_{ab}{}^0 = K_{ab}{}^{Fe} + K_{ab}{}^{vv}$).

8.2. Stability of the magnetic phases and spin reorientation

We calculate the equilibrium values of the order parameters in the phases Γ_2, Γ_4, Γ_{24} and analyze their stability. Since the temperature is low enough and the Fe sublattices are saturated i.e.$D \rightarrow \infty$, we can assume $FG=0$. Minimizing $\Phi(F,G,f,c)$ with respect to F, f, c ($F<<1$), we find for the phases Γ_2, Γ_{24}, Γ_4 that $f_\eta = F_y = c = 0$ and

$$f_\xi = f_0 \sin\psi \equiv \Delta_{ex}{}^0 G_z/(\overline{T} - \overline{\lambda}_f),\ f_\zeta = f_0\cos\psi,\ F_x = F^0G_z,\ F_z = F^0G_x, \qquad (8.8)$$

where $\overline{T} = \Delta_R/\mathrm{th}(\Delta_R/T)$, $f_0 = \mathrm{th}(\Delta_R/T)$, $F_0 = (d + Na\mu_x f_\xi G_z)/A$, $\overline{\lambda}_f = \lambda_f + \Delta\lambda_f G_z^2$

$$\Delta\lambda_f = N(a\mu_x)^2/A,\ \sin\psi = \Delta_{ex}/\Delta_R,\ \Delta_R{}^2 = \Delta_{cf}{}^2 + \Delta_{ex}{}^2,\ \Delta_{ex} = \Delta_{ex}{}^0 G_z + \overline{\lambda}_f f_\xi$$
$$\Delta_{ex}{}^0 = B + a\mu_x d/A. \qquad (8.9)$$

Substituting Eq.(8.8) in Eq.(8.5) we obtain the thermodynamic potential that depends only on G. In collinear phases we have $G_x = \pm 1$, $G_z = 0$ (Γ_4) and $G_x = 0$, $G_z = \pm 1$, (Γ_2). The stability of these phases is determined from the conditions of positive definiteness of the increment of $\Phi(G)$ when G deviates from the equilibrium state in Γ_4 and Γ_2 phases:

$$\Delta\Phi_{\Gamma_4} = \tfrac{1}{2}K_{ac}\Delta G_z{}^2 + \tfrac{1}{2}K_{ab}\Delta G_y{}^2,\quad \Delta\Phi_{\Gamma_2} = \tfrac{1}{2}K_{ca}\Delta G_x{}^2 + \tfrac{1}{2}K_{cb}\Delta G_y{}^2, \qquad (8.10)$$

where $K_{ac} = K_{ac}{}^o - N(\Delta_{ex}{}^0)^2/(\overline{T} - \lambda_f)$, $K_{ab} = K_{ab}{}^0 + d^2/A \equiv \overline{K}_{ab}{}^{Fe} + K_{ab}{}^{vv}$,

$$K_{ca} = N(\Delta_{ex}{}^0)^2(1+\varepsilon)/(\overline{T} - \overline{\lambda}_f) - K_{ac}{}^o - K_2,\quad \overline{K}_{ab}{}^{Fe} = K_{ab}{}^{Fe} + d^2/A,$$
$$K_{cb} = K_{ab} + K_2'' - K_{ac}{}^o - K_2 + N(\Delta_{ex}{}^0)^2/(\overline{T} - \overline{\lambda}_f),\ \varepsilon = \Delta\lambda_f/(\overline{T} - \overline{\lambda}_f),\ \overline{\lambda}_f = \lambda_f + \Delta\lambda_f, \qquad (8.11)$$

The quantities K_{ac}, K_{ab}, K_{ca}, K_{cb} determine the AFMR frequencies and characterize the rigidity of the system when G deviates from equilibrium, provided that F, f, c follow G in equilibrium fashion. For $T>>\Delta_R$ when $\overline{T} \rightarrow T$ they turn into the usual effective anisotropy constants in the corresponding planes. The thermodynamic potential of the system can be represented in this case in the form (2.5).

At temperatures high enough, when $K_{ab}{}^0 > 0$ and $K_{ac}{}^0 > 0$, the phase Γ_4 is stable. As temperature is lowered, the negative R-Fe exchange contribution is increased and at $K_{ac}(T_1) = 0$ the phase Γ_4 loses the stability (the first spin reorientation phase transition) and the system goes into the angular phase Γ_{42}. The reorientation is finished by the second spin reorientation phase transition when $K_{ca}(T_2) = 0$ and $G_z = \pm 1$ (phase Γ_2).

We note that in the course of spin reorientation the splitting of the ground quasidoublet $2\Delta_R$ increases on account of the increase in exchange contribution, attaining its maximum in the phase Γ_2. This is the reason for the advantage in the energy of the rare-earth subsystem over the anisotropy energy of the Fe subsystem, stabilizing the phase Γ_4. These two competition factors equalize each other in the course of reorientation. In view of that the given spin reorientation (in $TbFeO_3$, as an example) can be regarded as the magnetic phase transition of the Jan-Teller-type[7].

8.3. Dynamics of interacting Fe and R subsystems

There are four resonance frequencies in the system, two of them corresponding to the AFMR modes (quasiferromagnetic M_1 and quasi-antiferromagnetic M_2), and other two corresponding to resonance rare-earth modes $R_{3,4}$. Since the oscillations of the Fe and R subsystems are usually coupled, the relation of one or another mode to the Fe or R subsystem is to some extent conventional, expecially in the vicinity of the frequency crossing. Nevertheless, far away from the region of the frequency crossing , when the interaction of oscillations of the Fe and R subsystems is not strong, such a classification is justified, since the corresponding modes differ strongly in their parameters (linewidths, intensities). This enables us to treat them either as the AFMR modes or as the rare-earth ones. We adduce here the expressions for the resonance frequencies in different phases [30].

Γ_2 phase.

Modes M_2 and R_4 of symmetry Γ_{12}:

$$\varpi_\pm^2 = {}^1/_2(\varpi_E\varpi_{cb} + \varpi_R\varpi_R') \pm [{}^1/_4(\varpi_E\varpi_{cb} - \varpi_R\varpi_R')^2 + \varpi_R^2\varpi_{cb}\varpi_{int}^x]^{1/2}, \qquad (8.12)$$

Modes M_1 and R_3 of symmetry Γ_{34}:

$$\varpi_\pm^2 = {}^1/_2(\varpi_E\varpi_{ca} + \varpi_R\varpi_R'') \pm [{}^1/_4(\varpi_E\varpi_{ca} - \varpi_R\varpi_R'')^2 + \varpi_R^2\varpi_{ca}\varpi_{int}^y]^{1/2}, \qquad (8.13)$$

where $\varpi_{cb,ca} = \gamma K_{cb,ca}/M_0$, $\varpi_R = 2\Delta_R/h$, $\varpi_R', \varpi_R'' = \varpi_R - 2\lambda_{f,c}f_0\cos^2\psi/h$,

$$\varpi_{int}^{x,y} = \gamma(a\mu_{x,y}\cos\psi)^2 f_0/\mu_{Fe}\Delta_R. \qquad (8.14)$$

The quantities $K_{ca,cb}$ and the equilibrium values of f_0 , ψ are determined respectively by Eqs. (8.11), (8.9).

As is seen from the formulae adduced, the treated modes of the symmetry Γ_{12} represent the quasi-antiferromagnetic AFMR mode M_2 of the Fe subsystem and the rare-earth mode R_4 interacting with one another. Neglecting the interaction between AFMR and R modes which is weak enough, i.e. assuming $\varpi_{int}^{x,y}=0$ (a=0), we can obtained for the frequencies

$$\varpi_+ = \varpi_2(Fe) = (\varpi_E\varpi_{cb})^{1/2}, \ \varpi_- = \varpi_4(R) = (\varpi_R\varpi_R')^{1/2}, \qquad (8.15)$$

$$\varpi_+ = \varpi_1(Fe) = (\varpi_E\varpi_{ca})^{1/2}, \ \varpi_- = \varpi_3(R) = (\varpi_R\varpi_R'')^{1/2}. \qquad (8.16)$$

Γ_{24} phase.

In this phase all four modes are generally speaking interacting (the oscillation symmetry corresponds to the reducible Γ_{1234} representation of the group D_{2h}^{16}). They differ, however, in the degree of interaction. The analysis shows, that the strongest interaction is that of the modes M_1 and R_4; it is determined principally by an anisotropic R-Fe interaction exchange-enhanced on account of ϖ_E ($\approx\varpi_E\varpi_R^2\varpi_{int}$).The interaction between the M_1 and R_4 modes exceeds their interaction with the M_2 and R_3 modes, as well as the interaction of the latter with one another. Therefore, neglecting the interaction of the M_1 and R_4 modes with the M_2 and R_3 ones, as well as of the latter with one another, we get at a=0 for the corresponding resonance frequencies

$$\varpi_2(Fe) = (\varpi_E \overline{\varpi}_{ab})^{1/2}, \quad \varpi_3(R) = (\varpi_R \varpi_R'')^{1/2}, \tag{8.17}$$

$$\varpi_\pm^2 = {}^1/_2(\varpi_E \overline{\varpi}_{ac}^0 + \varpi_R \varpi_R') \pm [{}^1/_4(\varpi_E \overline{\varpi}_{ac}^0 - \varpi_R \varpi_R')^2 + \varpi_E \varpi_R^2 \varpi_{int}]^{1/2}, \tag{8.18}$$

where $\overline{\varpi}_{ab} = \gamma(K_{ab} + K_2'' G_z^2)/M_0$, $\overline{\varpi}_{ac}^0 = \gamma(K_{ac}^0 + 3K_2 G_z^2) G_x^2/M_0$,

$$\varpi_{int} = \gamma(\Delta_{ex}^0 \cos\psi)^2 f_0 G_x^2/\mu_{Fe}\Delta_R, \tag{8.19}$$

while the rest quantities f_0, ψ, ϖ_R, ϖ_R', ϖ_R'' are defined by Eqs. (8.11), (8.14).

Γ_4 phase.

Mode of symmetry Γ_{14}. There is only one resonance mode M_2, representing quasi-antiferromagnetic AFMR mode of the Fe subsystem. Its frequency is determined by Eq. (8.17) with G_x=1, G_z=0.

Modes of symmetry Γ_{24}. There are three resonance modes for the oscillations of this symmetry. Two of them can be reffered to the rare-earth $R_{3,4}$ modes excited mainly by the fields oriented respectively along the b- and a-axis, while the other one pertains to the quasiferromagnetic AFMR mode M_1. Just as in the angular Γ_{42} phase, the strongest interaction takes place between the M_1 and R_4 modes, which, however interact weakly with the R_3 modes. The corresponding resonance frequencies are therefore determined by the Eqs. (8.17), (8.18) with G_x=1, G_z=0, $\cos\psi$=1.

Let us consider the temperature dependence of the resonance frequencies in the course of spin reorientation Γ_4-Γ_{42}-Γ_2. Note first of all that the nature of the frequency behaviour of the AFMR and the rare-earth modes is qualitatively different, depending on the relation between the resonance frequency of the R subsystem $\varpi_R \approx \varpi_R' \approx \varpi_R''$ and the frequency of the quasiferromagnetic mode without contribution of the exchange quasidoublet splitting: $\varpi_{10}^{Fe} = (\varpi_E \varpi_{ac}^0)^{1/2}$, where $\varpi_{ac}^0 = \gamma K_{ac}^0/M_0$.

Figure 22 shows the temperature dependences of the resonance frequencies for two cases: $\varpi_R > \varpi_{10}^{Fe}$ and $\varpi_R < \varpi_{10}^{Fe}$. In the former case, when the quasidoublet splitting is large enough, the softening of the quasiferromagnetic AFMR mode $\varpi_- = \varpi_1(Fe)$ occurs at the phase transition points T_1 and T_2 (Fig.22a).

In the latter case (Fig.22b), when a quasidoublet splitting is not large and the frequencies of the R subsystem lie lower than the AFMR ones, the situation turns out to be different. The frequency of the quasiferromagnetic AFMR mode $\varpi_1(Fe) = \varpi_+$ undergoes no softening on approaching to the upper spin-reorientation bound (T_1), while the frequency $\varpi_4(R) = \varpi_-$ of one of the rare-earth modes really softens. At the other end of the spin reorientation the quasiferromagnetic mode gets softened in the Γ_2 phase. These and other more complicated types of the resonance mode behaviour are possible in rare earth orthoferrites.

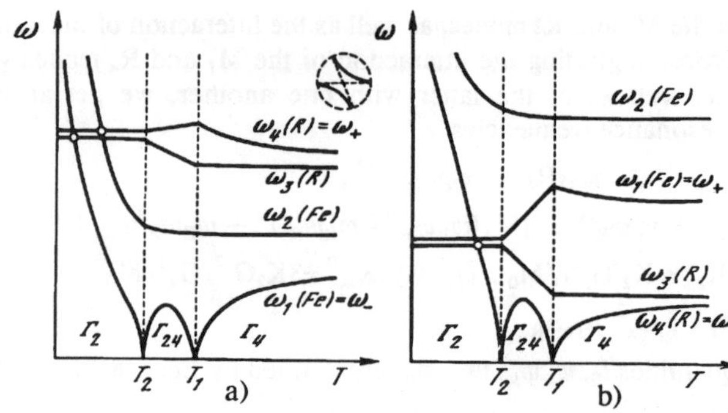

Fig.22. Temperature dependences of the resonance frequencies in RFeO3 in the course of spin reorientation Γ_4-Γ_{42}-Γ_2: a - $\varpi_R > \varpi_{10}^{Fe}$, b - $\varpi_R < \varpi_{10}^{Fe}$.

8.4 The analysis of the resonance frequency behaviour in the course of spin reorientation in RFeO3 (R=Tm,Tb,Yb,Er,Dy)

We now discuss the behaviour of the resonance frequencies (of the AFMR and R modes) at SRPT Γ_4-Γ_{24}-Γ_2 in $TmFeO_3$, $TbFeO_3$, $YbFeO_3$, $ErFeO_3$ from the standpoint of the mentioned- above approach . Despite of the fact, that in all these orthoferrites one and the same spin reorientation occurs, the resonance frequency behaviour in them differ qualitatively. The reason for that consists evidently in peculiarities of an electronic spectra of the corresponding R-ions and in different dynamics of the interacting Fe and R subsystems.

$TmFeO_3$. In this orthoferrite the R mode frequencies exceed the frequency of a quasiferromagnetic AFMR mode M_1 and do not prevent from its softening (see Fig. 15). The case, corresponding to Fig.22a ($\varpi_R > \varpi_{10}^{Fe}$), is therefore realized for $TmFeO_3$, justifying the approach used by us (see sections 2 and 6), in which the R subsystem is taken into account in a quasistatic approximation by renormalizing the thermodynamic potential.

$TbFeO_3$. According to the Raman data [26], while approaching the upper spin-reorientation bound (T_1=6.5 K), any softening of a quasiferromagnetic AFMR mode M_1 is completely absent. As follows from section 8.3, this can be related to the presence of a low-lying R mode interacting with M_1. Actually, the splitting of the ground quasidoublet of the Tb^{3+} ion in the Γ_4 phase is extremely small ($2\Delta_{cf} < 1cm^{-1}$ [7]), and the spin-reorientation mechanism Γ_4-Γ_{24}-Γ_2 is due to an additional exchange quasidoublet splitting in the Γ_2 phase. Therefore, the frequency behaviour in $TbFeO_3$ corresponds to the case depicted in Fig. 22b ($\varpi_R < \varpi_{10}^{Fe}$). As $T \to T_1$ the frequency of the M_1 mode is finite , while the R mode $\varpi_4(R)$ is the soft mode.

$YbFeO_3$. Fig.23a presents temperature dependences of the resonance frequencies for $YbFeO_3$. Experimental data (points) are taken from [24]. Together with the two AFMR modes, the R modes were also observed in $YbFeO_3$ related to electronic transitions inside the ground Kramers doublet of the Yb^{3+} ion. One of

these modes ($\varpi_3(R)$) retains its frequency value in the spin-reorientation region, while the other mode ($\varpi_4(R)$), according to the experimental data [23], softens either at T_1 or at T_2. It is seen from Fig. 23a that in $YbFeO_3$, as well as in $TbFeO_3$, the softening of a quasiferromagnetic AFMR mode M_1 is absent. Moreover, its frequency increases as $T \to T_1$. It is obviously due to its interaction with the lower lying R modes. Thus, the resonance frequency behaviour in $YbFeO_3$ corresponds qualitatively to the case $\varpi_R < \varpi_{10}^{Fe}$ (see Fig. 22b). But as compared to the model considered in section 8.3, there are some distinctions for the non-Kramers ions. We therefore adduce below the simplified formulae for the resonance frequencies in orthoferrites with the non-Kramers R-ions (see also [35,44]).

$\Gamma_4(F_zG_x)$ phase. Mode Γ_{14}:

$$\varpi_2(Fe) = (\varpi_E \varpi_{ab})^{1/2}, \tag{8.20}$$

modes Γ_{23} ($\varpi_1(Fe)$, $\varpi_{3,4}(R)$):

$$\varpi_\pm^2 \approx (\varpi_E \varpi_{ac} + \varpi_R^2)/2 \pm [(\varpi_E \varpi_{ac} - \varpi_R^2)^2 + 4\varpi_R^2 \varpi_E \varpi_{int}^z]^{1/2}, \quad \varpi_3(R) \approx \varpi_R, \tag{8.21}$$

$\Gamma_2(F_xG_z)$ phase. Modes Γ_{34}($\varpi_1(Fe)$, $\varpi_3(R)$):

$$\varpi_\pm^2 \approx {}^1/_2(\varpi_E \varpi_{ca} + \varpi_R^2) \pm {}^1/_2[(\varpi_E \varpi_{ca} - \varpi_R^2)^2 + 4\varpi_R^2 \varpi_E \varpi_{int}^x]^{1/2}, \tag{8.22}$$

modes Γ_{12}($\varpi_2(Fe)$, $\varpi_4(R)$):

$$\varpi_\pm^2 \approx {}^1/_2(\varpi_E \varpi_{cb} + \varpi_R^2) \pm {}^1/_2[(\varpi_E \varpi_{cb} - \varpi_R^2)^2 + 4\varpi_R^2 \varpi_E \varpi_{int}^y]^{1/2}, \tag{8.23}$$

where $\varpi_R h = 2(\Delta_x^2 G_x^2 + \Delta_y^2 G_y^2 + \Delta_z^2 G_z^2)^{1/2}$, $\varpi_{int}^k = \Delta_k^2 / h k_B T$ (k=x,y,z), $2\Delta_{x,y,z}$ is the exchange splitting of the ground doublet of the Yb^{3+} ion in the phases $\Gamma_4(G_x)$, $\Gamma_1(G_y)$ and $\Gamma_2(G_z)$, respectively; $\varpi_{ac} = \gamma K_{ac}^0/M_0 + \varpi_{int}^x$, $\varpi_{ab} = \gamma K_{ab}^0/M_0 + \varpi_{int}^x$, $\varpi_{ca} = -\gamma(K_{ac}^0 + K_2)/M_0 + \varpi_{int}^z$, $\varpi_{cb} = \gamma(K_{ab}^0 + K_2'')/M_0 + \varpi_{ca}$, K_{ac}^0, K_{ab}^0 are the anisotropy constants in the corresponding crystal planes without of the

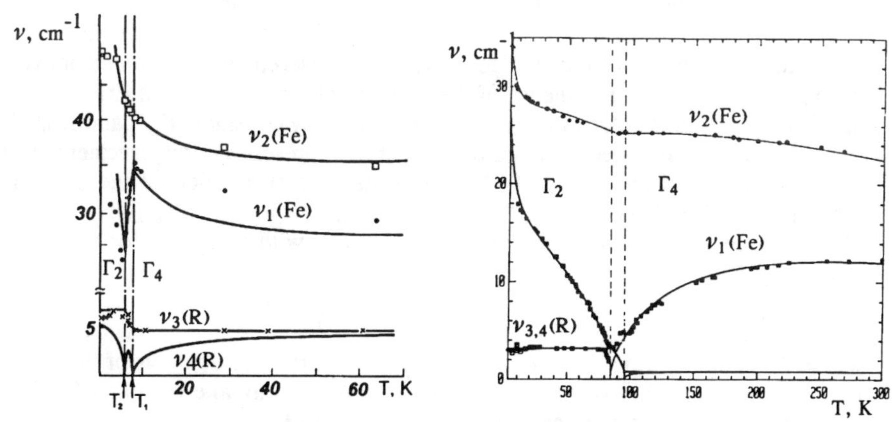

Fig.23. Temperature dependences of the resonance frequencies in $YbFeO_3$ (a) and $ErFeO_3$ (b) in the course of orientational phase transition Γ_4-Γ_{24}-Γ_2. Points - experiment, solid lines - theory; 1 - $\nu_1(Fe)$, 2 - $\nu_2(Fe)$, 3 - $\nu_3(R)$, 4 - $\nu_4(R)$.

contribution from the exchange splitting of the ground Yb^{3+} doublet. Taking the latter into account, these constants are equal to

$$K_{ac} = K_{ac}^0 + (\Delta_x^2 - \Delta_z^2)/T,\ K_{ab} = K_{ab}^0 + (\Delta_x^2 - \Delta_y^2)/T, \qquad (8.24)$$

We have omitted the formulae for frequencies in the canted phase because of their awkwardness. We used the high-temperature approximation ($\Delta_k << T$) while obtaining Eqs. (8.20)-(8.23), as well as neglected the R-R interaction and some small corrections related to the isotropic R-Fe exchange.Using Eqs. (8.20)-(8.23) we were capable to describe rather well the experimental temperature dependences of the resonance frequencies in $YbFeO_3$ (Fig. 23a) with the following values of the constants: $\varpi_E=2.5\cdot10^5$ GHz, $\Delta_x=3.35$ K, $\Delta_z=5.2$ K, $K_{ac}^0=1.86$ K, $K_{ab}^0=3.2$ K, $K_2=0.47$ K.

$ErFeO_3$. In this orthoferrite the AFMR (M_1 and M_2) mode behaviour was studied by nonelastic neutron [25] and Raman[10] scattering studies, and the BWO spectroscopy technique. Together with the familiar AFMR modes we have revealed the R modes identified with the electronic transitions in the Kramers doublet of the Er^{3+} ion split in the exchange field of the Fe subsystem. Figure 23b shows the temperature dependences of the corresponding resonance frequencies in $ErFeO_3$, the main peculiarity of which is the softening of the quasiferromagnetic mode frequency ϖ_1(Fe) in the region of spin-reorientation transitions Γ_2-Γ_{24}-Γ_4. However, this softening occurs only on the lower spin-reorientation bound T=87 K. The value of the gap thereby observed on the temperature dependence of the M_1 mode frequency is close (at T_2) to the value of the ground Er^{3+} doublet splitting in the Γ_2 phase (≈ 3 cm^{-1}) obtained in the optical studies [58]. An interaction (repulsion) of the AFMR and R mode frequencies was observed near T_2.

Thus, $ErFeO_3$ has a lower lying R mode, preventing from the full softening of the quasiferromagnetic mode frequency ϖ_1(Fe) and determining for it the gap value in the SRPT region. One should point out that the behaviour of the ϖ_1(Fe) mode in $ErFeO_3$ qualitatively differs from the behaviour of the similar mode in $YbFeO_3$, where the lower lying R mode is also present. The discrepancy mentioned above can be explained by a strong influence of the excited states of Er^{3+} ions in $ErFeO_3$, which determine the Van Vleck contribution to the anisotropy constants K_{ac}^{vv} and K_{ab}^{vv}.

Consider this aspect in more detail. Let us represent the displacements of the center of gravity of the ground (i=1) and excited (i≥2) doublets of the Er^{3+} ion in the form $\Delta E_{vv}^{(i)}=const+\Delta E_{ac}^{(i)}G_z^2+\Delta E_{ab}^{(i)}G_y^2$, where $\Delta E_{ac,ab}^{(i)}$ are the phenomenological constants. In this case by analogy with [7] we have for the dependences $K_{ab,ac}^{vv}(T)$

$$K^{vv}_{ab,ac}(T)=2\sum_i \Delta E^{(i)}_{ab,ac} exp(-\beta E_i^0) / \sum_i exp(-\beta E_i^0), \qquad (8.25)$$

where E_i^0 is the energy of centers of gravity of unperturbed doublets of the Er^{3+} ion in the crystal field: $E_1^0=0$, $E_2^0=68$ K, while it is enough to take the other levels into account by introducing their center of gravity $E_3^0=170$ K.

An important feature for describing the $ErFeO_3$ properties is the fact, that the Van Vleck contribution of Er^{3+} ions to the anisotropy energy is determined by the matrix elements of an R-Fe exchange operator between those states of the Er^{3+} ion in a crystal field, between which the transition frequencies exceed the AFMR ones significantly. Therefore, at the AFMR frequencies the corresponding contribution of the Er subsystem to the anisotropy energy can be regarded in a quasistatic approximation, as it is due to the very part of the variables of the Er subsystem, which follows the variables of the Fe subsystem instantaneously. Thus, it is natural to connect the strong softening of the quasiferromagnetic M_1 mode at the SRPT in $ErFeO_3$ with the growth of the negative Van Vleck contribution to the anisotropy constant K_{ac} in the ac-plane, as it is not frozen unlike the contribution related to the exchange splitting of the ground Er^{3+} doublet.

The softening of the M_1 mode can also be interpreted as a result of its interaction with the higher lying R modes determined by the electronic transitions from the ground to the excited doublets of the Er^{3+} ion.

Temperature dependences of the resonance frequencies of $ErFeO_3$ calculated on the basis of the effects mentioned above permitted to describe the experiment well (see Fig. 23,b), using the following values of the system parameters: $\varpi_E=2.25\cdot10^5$ GHz, $K_{ac}^{(0)} = K_{ac}^{Fe} + \Delta E_{ac}^{(1)} = -0.72$ K, $\Delta E_{ac}^{(2)} = 0.85$ K, $\Delta E_{ac}^{(3)} = 1.55$ K, $K_{ab}^{(0)} = K_{ab}^{Fe} + \Delta E_{ab}^{(1)} = 3.1$ K, $\Delta E_{ab}^{(2)} = -1.05$ K, $\Delta E_{ab}^{(3)} = -1.1$ K, $\Delta_x = 0.6$ K, $\Delta_z = 2.4$ K.

Comparison between the three different terms in the expression for the anisotropy constant $K_{ac}(T)=K_{ac}^{Fe}+K_{ac}^{vv}(T)+(\Delta_x^2-\Delta_z^2)/T$ shows that the decrease and change in sign of $K_{ac}(T)$ at $T=T_1$ occurs mainly due to a competition between the first and the second terms (in $ErFeO_3$) and the first and the third terms (in $YbFeO_3$ and $TbFeO_3$).

Note in conclusion that to some extent the same situation as in $ErFeO_3$ takes also place in $DyFeO_3$. The mechanism of the SRPT of the Morin type (Γ_4-Γ_1) in $DyFeO_3$, according to 6, is due to the competition between the anisotropy energy in the ab-plane of the Fe subsystem and the corresponding Van Vleck contribution from the Dy^{3+} ions, which is not frozen at the AFMR frequencies. The exchange splitting of the ground quasidoublet of the Dy^{3+} ion, being too small in the Γ_4 and Γ_1 phases, does not thereby play an appreciable part in the mechanism of the given spin reorientation. This explains the possibility of the quasistatic description of the Dy subsystem at the AFMR frequencies (section 6.1).

9. High-frequency magnetic properties and new spin reorientational transitions in HoFeO3

Peculiarities of the dynamics of spin reorientational transitions considered above due to interaction between the AFMR and R modes are particularly strongly manifested in $HoFeO_3$. According to optical data [53,54,59] and our results [45], the energy of the splitting of the ground quasidoublet of the Ho^{3+} ion falls for $HoFeO_3$ directly in the region of AFMR frequencies. A comparatively high sensitivity of the dynamic magnetic properties of $HoFeO_3$ to variations in its magnetic structure enabled us to reveal fine structure of spin reorientation in this orthoferrite which occur in the form of the three SRPTs $\Gamma_4(G_x) \rightarrow \Gamma_{24}(G_xG_z) \rightarrow \Gamma_{12}(G_yG_z) \rightarrow \Gamma_2(G_z)$ [30] rather than two $\Gamma_4(G_x) \rightarrow \Gamma_{24}(G_xG_z) \rightarrow \Gamma_2(G_z)$ hitherto assumed [6].

We observed altogether four absorption lines in the transmission spectra. Two of them, quite narrow ($\Delta\nu/\nu \approx 10^{-2}$) and observable in the entire investigated temperature range, represent the quasiferromagnetic (ν_1(Fe)) and quasi-antiferromagnetic (ν_2(Fe)) AFMR modes. The two other lines, excited respectively by fields h||a and h||b, are broader ($\Delta\nu/\nu \approx 1$) and are observed only at low temperatures (T<120 K), when their intensities are high enough. In the Γ_2 phase at T<40K their frequencies are practically the same, while on going into the Γ_4 phase (T>60K) only one mode (ν_3(R)) in the h||b polarization is observed in the investigated frequency band, whose frequency decreases from 7 to 3 cm^{-1}. We have identified them with the rare-earth modes ($R_{3,4}$) related to electronic transitions between the energy levels of the ground quasidoublet of the Ho^{3+} ion. Their frequencies $\nu_{3,4}$(R) in the Γ_2 phase correspond to Ho^{3+} quasidoublet splitting obtained from the optical investigations ($2\Delta_R$=6-7.5 cm^{-1}) [53,54,59], and the observed decrease of ν_3(R) on going into the Γ_4 phase, due to vanishing of the exchange contribution Δ_{ex}, agrees well with the quasidoublet splitting by the crystal field ($2\Delta_R=2\Delta_{cf}$=3,3 cm^{-1}) [59].

A characteristic feature of the behaviour of the resonance frequencies (Fig. 24) is the softening of both AFMR frequencies in the region T=35-65 K, indicating a decrease of the anisotropy energy in different planes of the crystal. At $T_1=58\pm2$K, $T_2=51\pm1$ K, and $T_3=39\pm2$ K the temperature dependences of the resonance frequencies, as well as of the linewidths and the mode contributions [30] reveal anomalies that point to the existence of spin reorientational phase transitions at these temperatures.

Analysis of these date allowed us to conclude that spin reorientation in $HoFeO_3$ proceeds in the form of three phase transitions (Γ_4-Γ_{24}-Γ_{12}-Γ_2), in the course of which the vector G is deflected from the a-axis (Γ_4) at the upper boundary T_1 and rotates smoothly in the ac-plane (Γ_{24}) and then, without reaching the c-axis (10-20°), it is jumpwise reoriented at T_2 into the bc plane (Γ_{12}) in which it already rotates smoothly to the c-axis

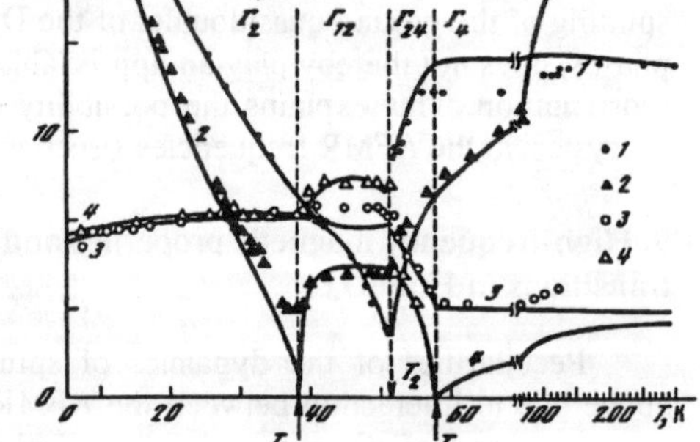

Fig. 24. The temperature dependences of the resonance frequencies in $HoFeO_3$: points – experiment, solid lines – theory:1-ν_1(Fe), 2-ν_2(Fe), 3-ν_3(R), 4-ν_4(R) . Insert — the SRPT scheme.

at $T_3(\Gamma_2)$. Further detailed studies of the static [60] and acoustic [61] properties of $HoFeO_3$ have confirmed these phase transformations in $HoFeO_3$.

A feature of the spin reorientation in $HoFeO_3$ is that a both anisotropy constants K_{ac} and K_{ab} reverse sign near the SPRT region. The mechanism of the temperature dependence $K_{ac}(T)$ is determined mainly by the Zeeman contribution due to the anisotropy of the exchange splitting of the ground quasidoublet of the Ho^{3+} ion (see Eq.(8.11)). As for $K_{ab}(T)$, according to Eq(8.11), there is no such contribution there. Therefore the observed decrease of K_{ab} with lowering T can be naturally attributed to the temperature dependence of the Van Vleck contribution $K_{ab}^{vv}(T)$ determined by formula, similar to Eq.(8.25).

Our numerical calculations [30] of the resonance frequencies (Fig.24) as well as linewidths and mode contributions have accounted well for the experimental data. As a result, parameters of the principal magnetic interactions in $HoFeO_3$ have been obtained.

10. Possible technical applications of orthoferrite single crystals

Single crystals of rare-earth orthoferrites have numerous wonderful physical properties, in that number: very high (gigantic) magnetic anisotropy, large values of mobility and low coercivity of magnetic domain wall, good transparency in infrared light and high value of magnetooptical Faraday constant, very intensive resonant phenomena in submillimeter region and pronounced magneto- acoustical properties, etc. Therefore, they can find various uses in numerous technical applications. Such using has been limited by the absence of effective technology of the growing of large and perfect orthoferrite single crystals. This technology now exists as a crystal growth by Floating Zone method with radiation heating [37,38].

In this brief review possible technical use of orthoferrite single crystals includes the following:

1.The modulation of visible and IR light by moving the domain wall in thin orthoferrite slice cut perpendicularly to optical axis. In such $YFeO_3$ modulator the maximum modulation frequency is equal to 1GHz while using highly mobile wedge-like head of stripe domain. The domain wall velocity can reach the value of 20 km/s at the domain wall mobility higher 1.5×10^4 cm/s Oe [62,63].

2.Magnetooptical sensors of magnetic fields and currents on orthoferrite single crystals can possess high sensitivity at the level of 10^{-3} - 10^{-4} Oe and very large dynamical range - maximum field up to 10^5 Oe [64].

3.With the help of orthoferrite slice with mobile stripe domain structure one can realize a contactless reading head for magnetic audio or video types. Magnetooptical reading of information with the help of slice domain structure has already been used [65].

4.Light matrix-space modulators of 128x128 bit capacity were realized using the orthoferrite slices cut perpendicularly to optical axis with unlimited time memory. At induced surface coercivity of about 30 Oe the information input velocity can reach

10^{10} b/s by current matrix circuit at magnetooptical reading of written information (at wavelength 0,63 mkm) [66].

5.Optical isolator for IR region (at 1.3 and 1.5 mkm) can be designed using yttrium orthoferrite single crystals slices cut perpendicularly to the optical axis. Such isolator for light beam divergence about 0.4 degrees exhibits insertion loss about 1dB and backward loss about 21 dB (at 1.3 mkm). Wonderful peculiarity of such isolator is the possibility to avoid using of permanent magnets for orthoferrite saturation, as the latter possesses the property of holding the saturated state [67].

6. TE-TM unidirectional mode converter, utilizing single crystalline yttrium orthoferrite substrate at wavelength 1.55 mkm without external magnetic field, exhibits power loss about 0.18 dB at 100% conversion, being much smaller than the loss in case of garnet film using [68].

7. Use of orthferrite crystal substrates is of great interest for the production of high temperature superconducting thin films [69]. Such substrates possess adequate high-frequency properties and good crystallographic match with HTSC film. The existence of weak ferromagnetism in the substrate can give the possibility for designing new functional devices, using, for example, the interaction between magnetostatic waves in the substrate and HTSC film.

8. Pronounced magnetoacoustical properties of orthoferrite crystals, $YFeO_3$ and $TmFeO_3$ in particular, and the existence of large nonlinear magnetoacoustical effects open new possibility for using these materials in technical applications [70,71].

9.Orthoferrites are the materials possessing very high-quality magnetic and dielectric parameters in SBMM range. As dielectric materials, the orthoferrites weakly absorb at SBMM wavelengths (small angular dielectric loss tangent) and they have high refractive indices and birefrigence. As magnetic materials, rare-earth orthoferrites possess narrow high-quality resonance lines and high linear coupling coefficients between the AFMR frequencies and the external magnetic field.

Owing to such a combination of high-frequency magnetic and dielectric properties the rare-earth orthoferrites would be successfully used in millimeter and submillimeter devices. Modern zone floating technology allows to receive mixed orthoferrite compositions (for instance, $Y_{1-x}Lu_xFeO_3$ [72]), in which the desired frequency values of resonant AFMR modes could be chosen.

11. Conclusion

Resuming the principal results of the present chapter, we note that the magnetic and dielectric properties of orthoferrites in the submillimeter wavelength region have, on the one hand, a series of common features (similar dielectric parameters ε' and ε'', the presence of the AFMR modes of the Fe subsystem with the close parameter values at $T \approx T_{room}$), while on the other hand, a strong dependence of their properties on the R ion type, in particular, on its spectrum in the crystal and exchange fields, is observed as a rule on lowering temperature. This is manifested, in the first place, by the presence of diverse R modes determined by the electronic transitions inside the ground multiplet of the R ion. These modes, interacting with the AFMR ones, make the behaviour of the latter qualitatively different, especially in the spin-reorientation region. As a result, even for the same type of the spin-reorientation phase transitions the soft mode could be represented either by the AFMR mode ($TmFeO_3$), or by the R mode ($YbFeO_3$). More complicated situations are also possible, when on one spin-reorientation boundary the soft mode is the AFMR mode, while on the other it is the R mode ($HoFeO_3$).

Realization of one or another type of the mode behaviour depends on the relation between the eigenfrequencies of the AFMR and R modes far away from the spin-reorientation phase transitions, the exchange splitting anisotropy of the R ion ground quasidoublet and other factors. This permits to obtain direct information about the energy of magnetic anisotropy, the parameters of anisotropic R-Fe interaction and some other quantities.

The important role of the R subsystem is manifested by its influence not only on the magnetic, but also, in some cases, on the dielectric properties of orthoferrites. This is due to an ambiguous nature of electronic transitions in the R ion, i. e. the magneodipolar and electrodipolar ones, the reason for which is the absence of the symmetry center for the local R ion positions in orthoferrites. The ambiguous character of electronic transitions in the R ions was distinctly manifested in $TmFeO_3$, where we observed the magnetodipolar R modes together with the electrodipolar R modes in the Tm subsystem, determining an appreciable contribution to the permittivity of the crystal along the c-axis.

We cannot exclude the fact, that in other orthoferrites similar dielectric anomalies are also possible. Many of the effects and regularities studied in the present paper can also be revealed in the rare-earth orthochromites, orthoaluminates and other related materials.

Acknowledgment

This work was supported by the Russian Foundation for Fundamental Investigations, grant 93-02/17123.

References

1. A.A.Volkov , Yu.G.Goncharov, G.V.Kozlov, K.N.Kocharyan, S.P.Lebedev, A.S.Prokhorov , A.M.Prokhorov, Pisma v Zh. Eksp. Teor. Fiz. **39** (1984) 140.
2. Volkov A.A., Goncharov Yu.G., Kozlov G.V., LebedevS.P., Prokhorov A.M., Infrared Phys. **25**, (1985) 369.
3. Volkov A.A., Goncharov Yu.G., Kozlov G.V., LebedevS.P., in *Submillimeter dielectric spectroscopy of solids,* ed. G.V.Kozlov (Nauka, Moscow, 1990).
4. A.A.Volkov, Yu.G.Goncharov, G.V.Kozlov, S.P.Lebedev, V.I.Maltsev, Prib.& Tekh. Eks. (1984) 236.
5. R. L. White, J. Appl. Phys. **40** (1968) 1061.
6. K.P.Belov, A.K.Zvezdin, A.M.Kadomtseva, R.Z.Levitin, *Orientational transitions in rare-earth magnetics* (Nauka, Moscow, 1979), p. 317.

7. A.K.Zvezdin, V.M.Matveev, A.A.Mukhin, A.I.Popov, *Rare-earth ions in magnetically ordered crystals* (Nauka, Moscow, 1985), p. 296.
8. K.P.Belov, A.K.Zvezdin, A.M.Kadomtseva, Sov.Scince Rev., sec.A, Phys. Rev. **9** (1987) 118.
9. R.M.White , R.J.Nemanich , C.Herring, Phys. Rev. **25** (1982) 1822.
10. N.Koshisuka , K.Hayashi, J.Magn. and Magn. Mater. **31/34** (1983) pt.2 569.
11. A.A.Mukhin, A.S.Prokhorov , in *Submillimeter dielectric spectroscopy of solids,* ed. G.V.Kozlov (Nauka, Moscow, 1990).
12. S.Geller , E.A.Wood, Acta crystallogr. **10** (1956) 563.
13. M.Marezio, J.P.Remeika, P.D.Dernier, Acta crystallogr. **B 26** (1970) 2008.
14. W.C.Koehler, E.D.Wollan, W.K.Wilkinson, Phys. Rev. **118** (1960) 58.
15. D.Treves, J. Appl. Phys. **36** (1965) 1033.
16. W.C.Koehler, E.O.Wollan, J. Phys. Chem. Solids **2** (1957) 100.
17. E.A.Turov, *Physical properties of magnetically ordered crystals*, ed. by S.V.Vonsovskii (Publishing House of the Acad. of Sci. of the USSR, Moscow, 1963), p. 224.
18. G.F.Herrmann, Phys. Rev. **A133** (1964) 1334.
19. R.C.Le Craw, R.Wolf, E.M.Georgy , F.B.Hagedorn, J.C.Hensel, J.P.Remeika J. Appl. Phys. **39** (1968) 1019.
20. E.M.Gyorgy, J.P.Remeika, F.B.Hagedorn, J. Appl. Phys. **39** (1968) 1369.
21. V.I.Ozogin, V.G.Shapiro, K.G.Gurtovoi , E.A.Galst'yn, A.Yu.Chervonenkis, Zh. Eksp. Teor. Fiz. **62** (1972) 2221.
22. A.M.Balbashov, Yu.M.Gufan, P.Yu.Marchjukov, E.G.Rudashevsky, Zh. Eks. Teor. Fiz. **94** (1988) 305.
23. N.K.Dan'shin, G.G.Kramarchjuk, M.A.Sdvizkov, Pisma v Zh. Eks. Teor. Fiz. **44** (1986) 85.
24. K.B.Aring, A.J.Sievers, J.Appl.Phys. **41** (1970) 1197.
25. V.G.Shapiro, J.D.Axe, J.P.Remeika, Phys. Rev. **B10** (1974) 2014.
26. S.Venugopalan, M.Dutta, A.K.Ramdas, J.K.Remeika, Phys. Rev. **B27** (1983) 3115.
27. A.M.Balbashov, A.A.Volkov, G.V.Kozlov, S.P.Lebedev, A.S.Prokhorov, Fiz. Tverd. Tela **27** (1985) 270.
28. A.M.Balbashov, A.A.Volkov, S.P.Lebedev, A.A.Mukhin, A.S.Prokhorov, Zh. Eksp. Teor. Fiz. **88** (1985) 974.
29. A.M.Balbashov, A.A.Volkov, G.V.Kozlov, S.P.Lebedev, A.S.Prokhorov, Zh. Tekh. Fiz. **56** (1986) 421.
30. A.M.Balbashov, G.V.Kozlov, S.P.Lebedev, A.A.Mukhin, A.Yu.Pronin, A.S.Prokhorov, Zh. Eksp. Teor. Fiz. **95** (1989) 1092.
31. I.A.Zorin, A.M.Kadomtseva, G.V.Kozlov, S.P.Lebedev, M.M.Lukina, A.A.Mukhin, A.S.Prokhorov, Fiz. Tverd. Tela **31** (1989) 184.
32. A.M.Balbashov, Vo Han Fuk, S.P.Lebedev, G.V.Kozlov, A.A.Mukhin, A.Yu.Pronin, A.S.Prokhorov, Zh. Eksp. Teor. Fiz. **90** (1986) 289.

33.A.M.Balbashov, A.A.Volkov, G.V.Kozlov, S.P.Lebedev, A.A.Mukhin, A.Yu.Pronin, A.S.Prokhorov,A.M.Prokhorov, Pisma v Zh. Eks. Teor. Fiz. 42 (1985) 456.
34. A.A.Mukhin, A.Yu.Pronin, A.S.Prokhorov, G.V.Kozlov, V.Zelezny, J.Petzelt, Physics Letters A 153 (1991) 499 .
35.V.G.Barjakhtar, I.M.Vitebsky, D.A.Yablonsky, Zh. Eksp. Teor. Fiz. 76 (1979) 1381.
36.Remeika J.P. , J.Amer. Chem.Soc.78 (1956) 4259.
37.Okada T., Matsumi K., Makino H., Nec.Res.Devel. (1970) N19 102.
38.A.M.Balbashov, S.K.Egorov., J.Cryst.Growth, 52 (1981) 498.
39.E.A.Vinogradov, N.A.Irisova, T.S.Mandelstam, A.M.Prokhorov, T.A.Shmaonov, Prib.& Tekh. Eks. (1967) N5 192.
40.A.S.Prokhorov, E.G.Rudashevsky, L.V.Velikov, IEEE Trans. Microwave Theory and Techn. 22 (1974) 1064.
41.V.M.Naumenko, V.N.Fomin, V.V.Eremenko, Prib.& Tekh. Eks. (1967) N5 223.
42.G.V.Kozlov, A.A.Volkov, S.P.Lebedev, Uspekh. Fiz. Nauk 135 (1981) 515.
43.A.M.Balbashov, A.G.Berezin, Yu.V.Bobryshev, P.Yu.Marchukov, I.N.Nikolaev, Y.Pacses, L.Pust, E.G.Rudashevsky, V.V.Shushpanov, Zh. Eks. Teor. Fiz. 101 (1992) 1397.
44.I.E.Golovenchits, V.A.Sanina, Zh. Eks. Teor. Fiz. 69 (1975) N4 1301.
45.A.M.Balbashov, G.V.Kozlov, S.P.Lebedev et al., Pisma v Zh. Eks. Teor. Fiz. 43 (1986) 33.
46.G.Gorodetsky, B.Sharon , S.Shtrikman, J. Appl. Phys. 39 (1968) 1371.
47.K.P.Belov , A.K.Zvezdin , A.M.Kadomtseva , I.B.Krinetsky, Zh. Eks. Teor. Fiz. 67 (1974) 1974.
48.J.R.Shine, Phys. Rev. Lett. 20 (1968) 728.
49.G.Gorodetsky , D.Treves, Phys. Rev. A 135 (1964) 97.
50.K.P.Belov, V.N.Derkachenko, A.M.Kadomtseva, T.L.Ovchinnikova, V.A.Timofeeva, V.A.Khokhlov, Fiz. Tverd. Tela 17 (1975) 3328.
51.B.T.Smith , J.Yamamoto , E.E.Bell, J. Opt. Soc. Amer. 65 (1975) N5 605.
52.N.K.Dan'shin, N.M.Kovtune, M.A.Sdvizkov, Zh. Eks. Teor. Fiz. 89 (1985) 203.
53.H.Schuchert , S.Hufher, R.Faulhaber, Ztschr. Phys. 220 (1969) 280.
54.A.P.Malozemoff , R.L.White, Solid State Commun. 8 (1970) 665.
55.P.Fulder, *Handbook on the Physics and Chemistry of Rare- Earths*, Ed. K.A.Gsechneidner, Tr. Eyring, L.Eyring (Amsterdam, North-Holland, 1979).
56.A.P.Malozemoff, J. Phys. and Chem. Solids 32 (1971) 1669.
57.A.K.Zvezdin, A.A.Mukhin, Prepr. IOFAN , Moscow, 1986) N225, p. 29.
58.D.L.Wood, L.M.Homes, J.P.Remeika, Phys. Rev. 185 (1969) N2 689.
59.J.C.Walling, R.L.White, Phys. Rev. B 10 (1974) 4748.

60.G.N.Vorob'ev, A.M.Kadomtseva, I.B.Krynetsky, A.A.Mukhin, Zh. Eksp. Teor. Fiz. 95 (1989) N3 1049.
61.N.K.Danshin, S.V.Zherlitsin, S.S.Zvada, A.A.Mukhin, M.A.Sdvizkov, V.D.Fill, Fiz. Tverd. Tela, 31 (1989) 198.
62.A.Ya.Chervonenkis "Magnetooptic properties and possible device application of bubble domain materials with high magnetooptic figure of merit", Proced.of Winter School on New magnetic materials Poland, (1975) 45.
63.M.V.Chetkin, A.I.Akhutkina, A.P.Kuzmenko and S.N.Gadestky, J.Appl. Phys., 53 (1982) 7865.
64.A.M.Balbashov,A.Ya.Chervonenkis , *Magnetic materials for microelectronics* (Energia, Moscow, 1979) p.216.
65.A.M.Balbashov,A.Ya.Chervonenkis , in *Magnetic components of automatic and computer technique* (Control Science Institute, Moscow , 1976), p.15 .
66.A.A.Komlev, A.K.Stolyarov, A.L.Mikaelyan, A.Ya.Chervonenkis, A.M.Balbashov, Kvantovaya electronika, 4 (1977) 1933.
67.A.M.Balbashov,I.Y.Parsegov , F.V.Lisovskii, A.V.Nikolaev, *About the possibility of orthoferrite using for optical infrared isolator*, presented at Conf. MMM-Intermag (1994).
68.M.Abe,M.Gomi and S.Nomura , Proceed.of the Intern. Conf., Japan, Sept.-Oct. (1980) 782.
69.R.Ramesh, A.Inam,W.A.Bonner, R.England, B.J.Wilkons, B.J.Meagher, L.Nazar, X.D.Wu, M.S.Hegde, C.C.Chang, T.Venkatesan, H.Padamsee, Appl.Phys.Lett., 55 (1989) 1138.
70.A.M.Balbashov,N.K.Dan'shin , A.I.Izotov, M.A.Sdvizhkov, L.T.Tsymbal, Fiz. Tverd. tela, (St. Petersburg), (1989) N7 .
71.A.Yu.Lebedev, V.I.Ozhogin, A.Yu.Yakubovskii, Intermag. Conf., Digest, NH-11, Grenoble (1981); Zh. Eksp. Teor. Fiz., 85 (1983) 1059.
72.G.V.Kozlov, S.P.Lebedev, A.A. Mukhin, I.V.Fedorov, A.S.Prokhorov, A.M.Balbashov,I.Y.Parsegov , IEEE Trans. on Magnetics, 29 (1993), 3443.

RECENT ADVANCES IN STUDIES ON THE MAGNETIC STRUCTURE OF NONCRYSTALLINE OXIDES

Gopalan Srinivasan
Department of Physics, Oakland University
Rochester, Michigan 48309-4401, USA

1. Introduction

This review is concerned with magnetically ordered amorphous compounds which contain at least one 3-d transition metal ion and oxygen. *We do not discuss here transition metal ions in glass forming materials such as borates, silicates or phosphates,* metallic alloy glasses, or glassy metals. A detailed discussion on the magnetic order in such materials is provided in Refs. 1 - 7.

Most amorphous magnetic materials of current interests have been metallic alloys with long range ferromagnetic and/or antiferromagnetic exchange interactions. In noncrystalline insulators, one would expect the magnetic structure to be less complicated than that of metallic alloys since superexchange interactions couple only nearest neighbor cations, according to their bond distances and angles. However, there has been a relatively few studies on the magnetic order in amorphous oxides. This lack of research activity may be attributed to failures in the past to achieve ordering at ambient temperature in several noncrystalline oxides.[8-19] Since the materials lack the periodic crystalline order, there is an inherent frustration in the exchange coupling between the neighboring spins. Therefore, the antiferromagnetic superexchange interaction between transition metal ions which leads to a long range order in crystalline materials does not seem to result in an ordered structure in their amorphous counterparts.

There has been a renewed interest in magnetism in amorphous (a-) oxides in recent years mainly due to the observation of a ferrimagnet-like behavior in several compounds. A long range magnetic order in amorphous oxides was first reported for $FeGaO_3$.[20] Sugimoto and Hiratsuka succeeded in preparing amorphous ferrites in phosphate glasses.[21,22] A ferrimagnetic character is also observed in a- $Bi_2O_3 . Fe_2O_3$ substituted with magnetic and nonmagnetic compounds. $BiFeO_3$ is an antiferromagnet in the crystalline state and is a paramagnet at room temperature in the amorphous state. However, a long range order with Curie temperatures $T_c = 400 - 800$ K is observed in rapidly quenched and radio frequency (rf) sputtered samples of $Bi_2O_3 . Fe_2O_3$ substituted with ZnO, CaO, CuO or ABO_3 (AB = BaTi, PbTi, PbZr)-type compounds.[23-28] In samples prepared by rapidly quenching, one observes a spontaneous moment at room temperature. The moment is attributed to magnetically ordered clusters in the samples.[23-25] Important observations in films prepared by rf sputtering are as follows. As-deposited samples are paramagnetic. A spontaneous moment is observed only in films annealed at high temperatures. A loss of spontaneous moment is observed when crystallization commences. Mossbauer studies show evidence for a long range magnetic order. Thus high temperature annealing appears to be a key ingredient for achieving a long range magnetic order in rf sputtered samples.[26-28]

A research program at Oakland University on magnetic ordering in amorphous oxides was initiated in 1990. The basic objective of the research was to investigate the magnetic structure of noncrystalline oxides prepared by rf sputtering techniques. Compounds of $Bi_2O_3 . Fe_2O_3$ substituted with nonmagnetic perovskites ABO_3, ZnO, CuO, or CaO have been studied so far. Our efforts have resulted in the following key accomplishments: (i) samples with nonmagnetic perovskites show a record room temperature magnetization, (ii) films containing zinc, copper, or calcium oxide are anisotropic amorphous magnets with a large uniaxial or planar anisotropy, and (iii) the nature and composition of ordered clusters in some samples have been inferred from magnetic characterization of the films.[29-40] Further details on our studies are provided in subsequent sections.

In this review, first we discuss ordering in noncrystalline transition metal oxides. The magnetic structure of mixed oxides including amorphous garnets is considered in Section 3. Studies on ternary oxides with specific emphasis on research performed at Oakland University are described in Section 4. Finally, results of recent theoretical studies on this topic are summarized in Section 5.

2. Amorphous 3d-Oxides

The emphasis of this section is on the magnetic structure in pure transition metal oxides. There have been a very few such studies. Systems investigated so far include chromium dioxide and ferric oxide. a-CrO_2 films remain paramagnetic whereas a-Fe_2O_3 shows a random antiferromagnetic (speromagnetic) order. Table 1 summarizes results of studies on these oxides.

Magnetic measurements on a-CrO_2 films prepared by sputtering techniques show that the films are paramagnetic down to the liquid helium temperature.[8] Films of a-Fe_2O_3 prepared by thermal decomposition or by sputtering are reported to order antiferromagnetically or speromagnetically at temperatures ranging from 48 to 220 K.[8-12] The sample obtained by thermal decomposition of an aerosol on heated substrates show a magnetic transition at a Néel temperature T_N of 80 K. The paramagnetic moment is 2.5 μ_B per iron ion, indicating an antiferromagnetic coupling between Fe^{3+} ions. Mössbauer studies indicate a distorted octahedral environment for iron ions.[9,10] The speromagnetic transition occurs at a slightly higher temperature in films prepared by diode sputtering from ferric oxide targets.[8] However, for films prepared from a target of Fe_3O_4, a much higher T_N of 215 - 220 K and a complex temperature dependence for the susceptibility are reported.[11]

Table 1: Magnetic order and ordering temperature for amorphous 3d oxides.

Compounds	Magnetic order	T_N (K)	Ref.
CrO_2	Paramagnetic	–	8
Fe_2O_3	Speromagnetic	100	8
	Speromagnetic	80	9, 10
	Antiferromagnetic	215 - 200	11
	Antiferromagnetic	48 - 62	12

The most interesting features of the results in Tables 1 are (i) a relatively low ordering temperature and (ii) the absence of a net magnetic moment in the amorphous state. For example, crystalline Fe_2O_3 orders either antiferromagnetically or ferrimagnetically below 950 K or 1020 K; however, the amorphous analog orders at very low temperatures. In the crystalline oxide, the long range order results from antiferromagnetic superexchange interactions between cations in two or more unequal sublattices. Such a special topology seems to be nonexistent in the amorphous state. Further, magnetic measurements do not show any evidence for a sharp transition at T_N in the amorphous state.[9,10] Such a behavior may be attributed to a broad distribution in the strength of antiferromagnetic exchange between iron ions.

3. Mixed Oxides

We consider next magnetic properties of mixed oxides containing Fe_2O_3 and one of the following compounds: (i) R_2O_3 (R = Y, Gd, Dy, Er, Tm, Yb, Lu and Eu) with the composition corresponding to amorphous counterparts of crystalline rare-earth iron garnets (RIG) $R_3Fe_5O_{12}$ and (ii) gallium oxide, zinc oxide, strontium oxide, bismuth oxide, cobalt oxide or copper oxide. Several of these compounds order magnetically. With the exception of $Fe_2O_3 \cdot CuO$ which orders below 725 K, the ordering temperature for the noncrystalline oxides is at least an order of magnitude smaller than for the crystalline oxides.

3.1 *Amorphous Garnets*

Samples of noncrystalline $R_3Fe_5O_{12}$ were prepared by several techniques including flash-evaporation, rapid quenching, pyrolysis, and sputtering. Details on the nature of magnetic structure for the samples are given in Table 2. Simpson and Lucas prepared a-YIG by flash-evaporation of 3Y:5Fe alloys in an oxygen atmosphere.[18] Susceptibility data for the sample indicated antiferromagnetic interactions between iron ions. Gyorgy and others prepared a-YIG by twin roller-quenching techniques.[13] X-ray diffraction (XRD) studies on these red colored flakes revealed a crystalline contamination of less than 3%. Figure 1 shows the temperature dependence of susceptibility χ for the flakes. A broad maximum in χ centered at 40 K is observed. Theoretical fit to the high temperature χ data yielded a paramagnetic Curie temperature of -500 K and an effective magnetic moment of 4.4 Bohr magneton (μ_B) per iron moment. Mossbauer data at 4.2 K showed a well resolved six-line pattern, indicative of an ordered magnetic structure.[13] Samples of YIG prepared by pyrolysis, however, were not purely amorphous and contained microcrystals of YIG which behaved superparamagnetically at room temperature.[15] a-RIG prepared by coprecipitation orders magnetically with an ordering temperature ranging from 14 K to 25 K.[14] Finally, Mössbauer studies provided evidence for ferrimagnetic order in a-DyIG and EuIG.[16]

Table 2: Magnetically ordered amorphous garnets.

Compound	Magnetic order	T_N (K)	Ref.
YIG	Antiferromagnetic	40	13
	Antiferromagnetic	14 - 25	14
DyIG	Antiferromagnetic	14 - 25	14
	Ferrimagnetic	70	16
RIG (R=Gd, Dy, Er, Gd, Lu, Tm, Yb)	Antiferromagnetic	14-25	14
EuIG	Ferrimagnetic	62	16

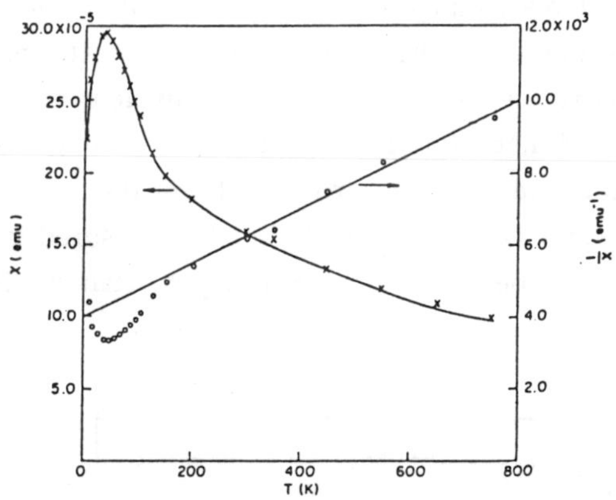

Fig. 1. Data on the variation of susceptibility with temperature for a-YIG synthesized by double roller quenching techniques. A magnetic transition at 40 K is evident from the data. [after Ref. 13]

3.2 *Other Mixed Oxides*

The nature of magnetic order and the ordering temperature for some mixed oxides are provided in Table 3. A spin-glass type character is observed for a-$ZnFe_2O_4$ and $CoFe_2O_4$.[19] A superparamagnetic character is reported for a-Fe_2O_3 containing SrO or Bi_2O_3.[41-44] A long range ferromagnetic or ferrimagnetic order is evident in oxides containing Fe and Ga or Fe and Cu.[20,33] Details of these studies are considered next.

Table 3 Magnetic order in mixed oxides containing Fe.

Compound	Magnetic order	Ordering temp. (K)	Ref.
$Fe_2O_3.ZnO$	spin glass	60	19
$Fe_{0.69}Ga_{1.37}O_3$	ferromagnetic	155	20
$(Fe_2O_3)_{0.65}.(SrO)_{0.35}$	Superparamagnetic	–	41-44
$Fe_2O_3.Bi_2O_3$	Superparamagnetic	–	41-44
$Fe_2O_3.CoO$	spin glass?	284, 86	19
$Fe_2O_3.CuO$	Ferrimagnetic	720	33

3.2.1 a-$Fe_{0.69}Ga_{1.37}O_3$

A ferromagnetic order in noncrystalline oxides was first reported for nonstoichiometric a-$FeGaO_3$.[20] Figure 2 shows the measured magnetization M versus applied static field H at 4.2 K. The amorphous oxide is a hard magnet showing hysteresis and remanence. The important results of this study are: (i) the oxide orders ferromagnetically at 155 K; (ii) the ferromagnetism is attributed to the presence of divalent and trivalent iron ions; (iii) it is suggested that the ferromagnetic superexchange associated with $Fe^{2+} \Leftrightarrow Fe^{3+}$ electron transfer is stronger than the antiferromagnetic superexchange between trivalent iron ions; (iv) the measured isomer shifts at room temperature are in the range corresponding to octahedrally coordinated Fe^{2+} and Fe^{3+} ions.

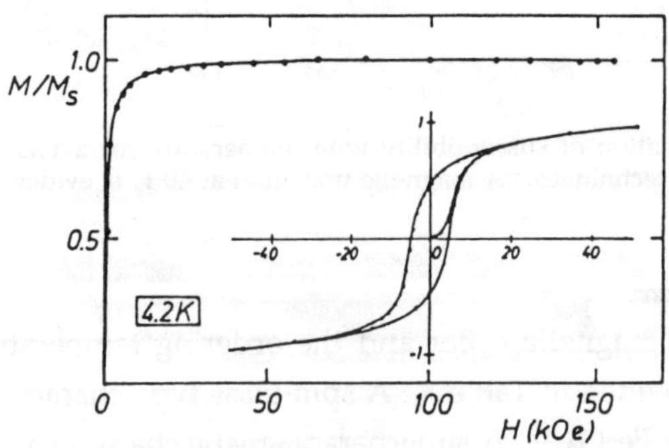

Fig. 2: Data on magnetization versus static field for amorphous $FeGaO_3$. [after Ref. 20]

3.2.2 *a-*$CuFe_2O_4$

A long range magnetic order with a planar anisotropy is reported in as-deposited films of amorphous copper ferrite prepared by rf sputtering techniques.[33] Data on the temperature dependence of the saturation magnetization, in-plane anisotropy field, and gyromagnetic ratio were obtained on 1 μm thick films sputtered in a 50% oxygen and 50% argon atmosphere. The films show a large magnetization and anisotropy field at room temperature. The magnetization, Curie temperature, and gyromagnetic ratio are in agreement with the values expected for crystalline copper ferrite. A paramagnetic to antiferromagnetic transition at 50 K is evident in the magnetization data. The film is proposed to contain magnetically ordered clusters of copper ferrite and Fe_2O_3.[33]

Figure 3 shows the temperature dependence of $4\pi M$ measured with a static field of 5 kOe applied parallel to the sample plane. The low temperature data are shown separately on an expanded scale. The key features of the data are: (i) one observes a rapid drop in $4\pi M$ with increasing T at very low temperatures; (ii) a peak in the magnetization is observed at 50 K ; (iii) for T above 50 K, the variation of $4\pi M$ with T is somewhat similar to the behavior expected for ferrimagnetic materials; and (iv) the Curie temperature T_C is 720±10 K. Similar features have been reported for $4\pi M$ versus T data for amorphous ferrites in phosphate glasses.[21,22] However, the magnetization achieved in rf sputtered samples is larger at least by an order of magnitude. The magnetic moment in the sample is attributed to ordered clusters of copper ferrite for the following reasons. First, the magnetization at 4.2 K corresponds to a magnetic moment of 0.9 μ_B per formula unit which is about 20% smaller than the moment for crystalline $CuFe_2O_4$.[45] Second, the T_C and gyromagnetic ratio for the amorphous compound are in excellent agreement with the reported values for $CuFe_2O_4$.[45] One also observes other important features in Fig.3, namely, the rapid decrease in $4\pi M$ at low temperatures and the peak in $4\pi M$ at 50 K. The initial decrease in $4\pi M$ with increasing T arises when weakly coupled spins become paramagnetic due to thermal agitation. Finally, the maximum in the magnetization at 50 K could be attributed to an ordering of clusters of Fe_2O_3. A similar phase transition at 48 - 100 K is observed in oxides containing trivalent Fe ions.[8,12]

The planar anisotropy character for the amorphous film was evident from ferromagnetic resonance (FMR) studies at x-band frequencies. The anisotropy field was determined from FMR data on the effective magnetization $4\pi M_{eff} = 4\pi M - H_A$, where H_A is the anisotropy field, and $4\pi M$ values in Fig.3. Positive (negative) values of H_A indicate anisotropy fields

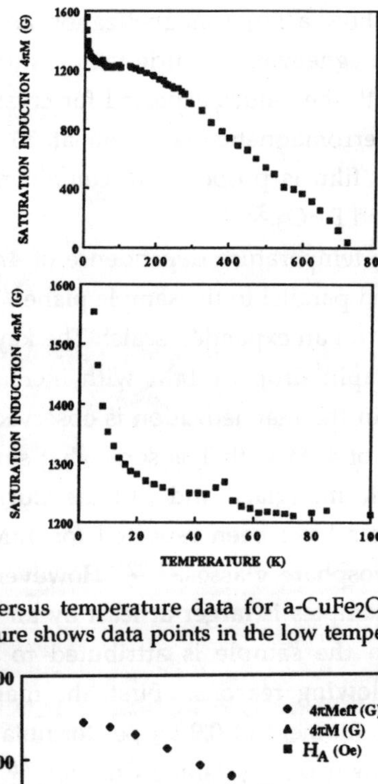

Fig. 3. Magnetization versus temperature data for a-$CuFe_2O_4$ prepared by rf sputtering techniques. The lower figure shows data points in the low temperature region. [after Ref. 33]

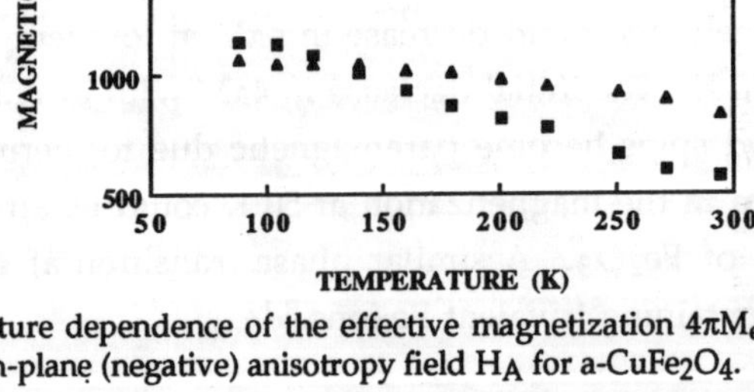

Fig. 4. Temperature dependence of the effective magnetization $4\pi M_{eff}$ obtained from FMR studies and the in-plane (negative) anisotropy field H_A for a-$CuFe_2O_4$. [after Ref. 33]

perpendicular (Parallel) to the film plane. Figure 4 shows the variation in H_A with temperature. The anisotropy in the amorphous film could be due to intrinsic mechanisms such as single-ion anisotropy and possible contributions from strains at the glass-film interface.

The temperature dependence of the effective gyromagnetic ratio is shown in Fig.5. A substantial increase in γ as T decreases is observed. Such a pronounced temperature dependence for γ is usually observed near the compensation point in ferrimagnetic crystalline oxides where the sublattice magnetizations cancel each other.[47] However, the data in Fig.4 do not indicate the presence of a compensation point in the amorphous magnet. Investigations on the exact nature of magnetic ordering in the film by Mössbauer and neutron diffraction techniques are necessary for an understanding of the data in Fig.5.

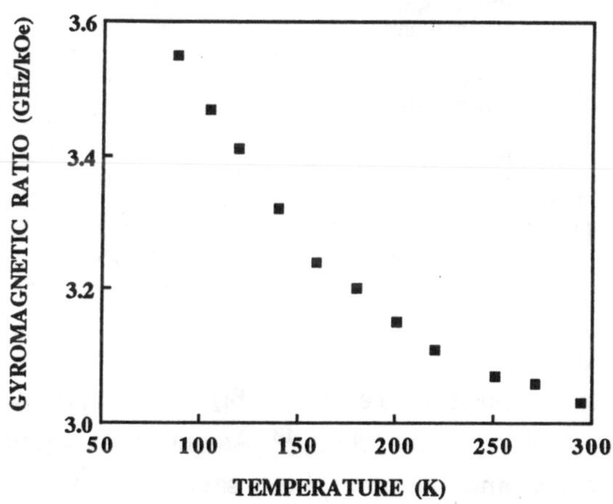

Fig. 5. The effective gyromagnetic ratio γ as a function of temperature for a-$CuFe_2O_4$ obtained from FMR at 9.1 GHz. [after Ref. 33]

4. Ternary Oxides

A long range magnetic order is observed in a family of ternary amorphous oxides containing Fe_2O_3, Bi_2O_3 and one of the following compounds: Li_2O, K_2O, CaO, ZnO, CuO, and ABO_3 (AB = BaTi, PbTi, PbZr).

The oxides show a ferromagnetic or ferrimagnetic character with a relatively high Curie temperature, on the order of 300 - 900 K. The samples are prepared by either rapid quenching or rf sputtering. A spontaneous moment is observed in as-quenched samples. However, as-sputtered samples are paramagnetic at room temperature and films annealed in air at high temperatures show a spontaneous moment. Table 4 provides details on the nature of magnetic order and ordering temperatures in ternary oxides.

Table 4. Magnetic order and ordering temperatures for substituted amorphous compounds of $Fe_2O_3 . Bi_2O_3$.

Substituent	Magnetic order	Ordering temp.(K)	Ref.
Li_2O	ferromagnetic	800	48
K_2O	ferromagnetic	673	49
CaO	ferromagnetic	760	50
	ferrimagnetic	600 - 750	34, 37
ZnO	speromagnetic	460 - 520	23, 24
	ferrimagnetic	600	31, 35
CuO	ferrimagnetic	650	25
	ferrimagnetic	700 - 750	30, 36
ABO_3 (AB = BaTi, PbTi, PbZr)			
	ferrimagnetic	473 - 773	26, 32

4.1 $Fe_2O_3 . Bi_2O_3 . Li_2O$

Oxides of the composition $(Fe_2O_3)_{0.5}$ $(Bi_2O_3)_{0.5(1-x)}$ $(Li_2O)_{0.5x}$ were prepared for $x = 0 - 1$ by rapid quenching.[48] As-quenched samples are x-ray amorphous for $x < 0.55$ and the amorphous samples show a spontaneous moment at room temperature. Figure 6 shows the variation of the room temperature M with x. The largest M of 4.5 emu/g is obtained for $x = 0.5$. Mössbauer absorption spectra shown in Fig. 7 provide evidence for the presence of two magnetic phases in the sample, a ferromagnetic phase (β-) with a Curie temperature of 800 K and a short range antiferromagnetic phase (α-) with a Néel temperature of 200 K.[48] High resolution electron microscopy investigations reveal the presence of a dense nanometer sized less disordered phase identified as the ferromagnetic component in the samples.[51]

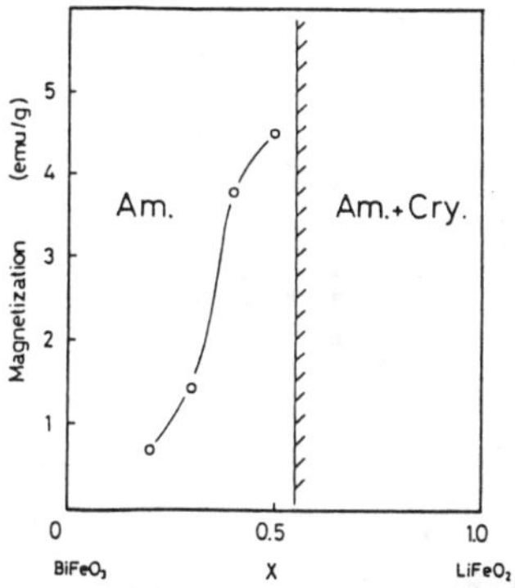

Fig. 6. Room temperature magnetization in an applied static field of 15 kOe for a- $(Fe_2O_3)_{0.5}$ $(Li_2O)_{0.5x}$ $(Bi_2O_3)_{0.5(1-x)}$ samples prepared by rapid quenching. [after Ref. 48]

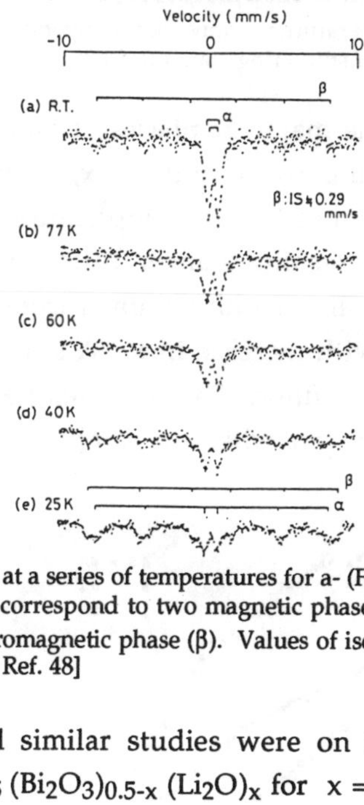

Fig. 7. Mössbauer spectra at a series of temperatures for a- $(Fe_2O_3)_{0.5}$ $(Li_2O)_{0.25}$ $(Bi_2O_3)_{0.25}$. The spectral components correspond to two magnetic phases, a short-range antiferromagnetic component (α-) and a ferromagnetic phase (β). Values of isomer shifts indicate the absence of divalent iron ions. [after Ref. 48]

We performed similar studies were on the magnetic properties of samples of $(Fe_2O_3)_{0.5}$ $(Bi_2O_3)_{0.5-x}$ $(Li_2O)_x$ for $x = 0 - 0.4$ prepared in oxygen atmospheres by rf sputtering techniques.[52] As-sputtered films are x-ray amorphous and paramagnetic. The films develop a spontaneous moment M when annealed at $T_a \geq 675$ K as shown in Fig.8. The data show a maximum in M for $T_a = 775 - 875$ K, even though x-ray diffraction data do not show the

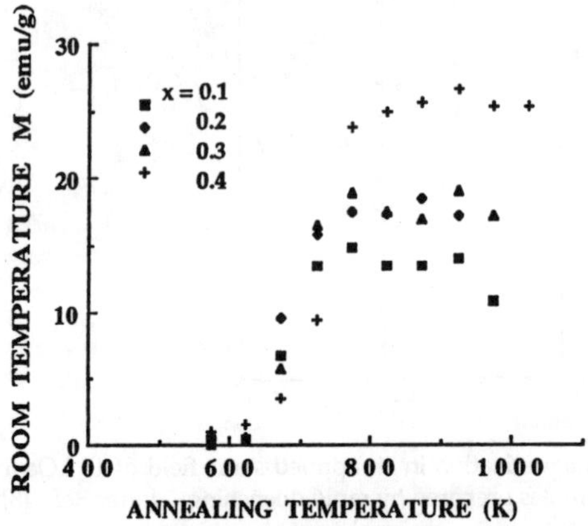

Fig. 8. Magnetization at room temperature vs. annealing temperature T_a data for a- $(Fe_2O_3)_{0.5}$ $(Bi_2O_3)_{0.5-x}$ $(Li_2O)_x$ synthesized by rf sputtering. [after Ref. 52]

precipitation of any crystalline magnetic phases. One also observes a general increase in M with increasing concentration, x, of *nonmagnetic* Li in the samples. Data on the variation of M with temperature shown in Fig. 9 reveal an increase in M with increasing concentration x of Li_2O. Both the gyromagnetic ratio γ and the line-width obtained from x-band ferromagnetic resonance studies decrease with increase in x. A strong dependence of γ on temperature is observed for the films. For example, the data in Fig. 10 for x = 0.1 indicate a decrease in γ by 15% when the sample is warmed up from 77 K to 300 K.

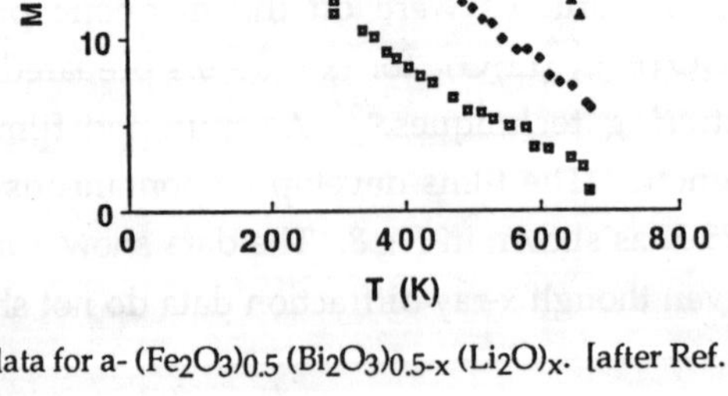

Fig. 9. M vs. T data for a- $(Fe_2O_3)_{0.5}$ $(Bi_2O_3)_{0.5-x}$ $(Li_2O)_x$. [after Ref. 52]

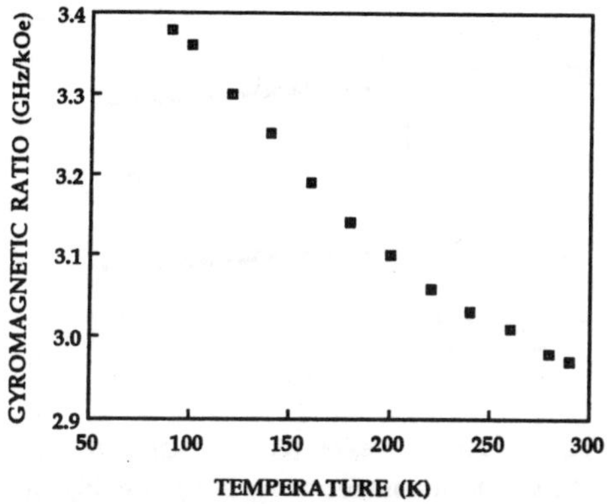

Fig. 10. Variation of gyromagnetic ratio γ with temperature for a- $(Fe_2O_3)_{0.5}$ $(Bi_2O_3)_{0.4}$ $(Li_2O)_{0.1}$. [after Ref. 52]

4.2 $Fe_2O_3 \, . \, Bi_2O_3 \, . \, K_2O$

Films of $(Fe_2O_3)_{0.734}$ $(Bi_2O_3)_{0.183}$ $(K_2O)_{0.083}$ were prepared by spin coating a gel of nitrates.[49] Figure 11 shows the dependence of the sample magnetic moment at room temperature as a function of annealing temperature T_a. Samples annealed in air at T_a below 500°C are paramagnetic. A spontaneous moment is measured in films annealed at higher T_a, and the maximum in M occurs for T_a = 600°C. X-ray diffraction data in Fig. 12 for annealed samples show the precipitation of antiferromagnetic phases such as

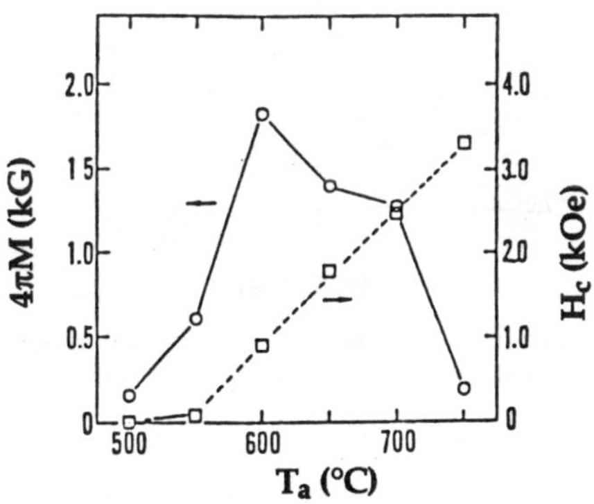

Fig. 11. The dependence of room temperature saturation induction $4\pi M$ and coercivity H_c on the annealing temperature T_a for films of $(Fe_2O_3)_{0.734}$ $(Bi_2O_3)_{0.183}$ $(K_2O)_{0.083}$ prepared by sol-gel techniques. [after Ref. 49]

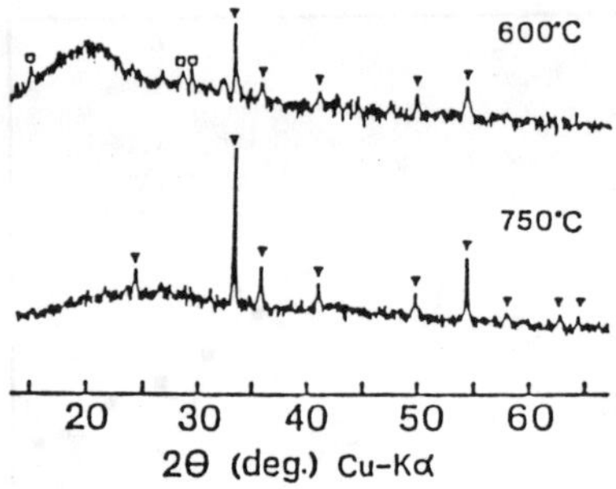

Fig. 12. X-ray diffraction patterns for samples of $(Fe_2O_3)_{0.825}$ $(K_2O)_{0.083}$ $(Bi_2O_3)_{0.092}$ annealed at 600°C and 750°C. Crystalline antiferromagnetic phases $Bi_2Fe_4O_9$ (□) and α-Fe_2O_3 (▼) were identified in the films. [after Ref. 49]

α-Fe_2O_3 and $Bi_2Fe_4O_9$, but ferromagnetic or ferrimagnetic crystalline phases are absent. Hysteresis loop data, shown in Fig. 13, indicate a hard magnetic character for the films. The origin of ferromagnetism in the system is unknown.[49]

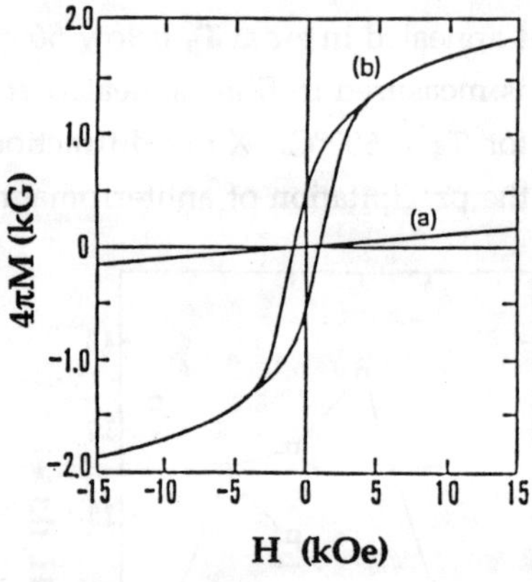

Fig. 13. Magnetization versus static field data at room temperature for films of $(Fe_2O_3)_{0.734}$ $(K_2O)_{0.083}$ $(Bi_2O_3)_{0.183}$ annealed at 500 °C (a) and 600°C (b). The samples were prepared by sol-gel techniques. [after Ref. 49]

4.3 $Fe_2O_3 . Bi_2O_3 . CaO$

A ferromagnetic character in the oxide was reported for ribbons prepared by rapid quenching.[50] The composition over which amorphous samples could be obtained is shown in the phase diagram in Fig. 14. Curves of constant room temperature M are also sketched. The largest M is achieved for $(Fe_2O_3)_{0.65}$ $(Bi_2O_3)_{0.15}$ $(CaO)_{0.2}$ and, as shown in Fig. 15, a substantial reduction in the magnetization occurs when the sample is heated above the

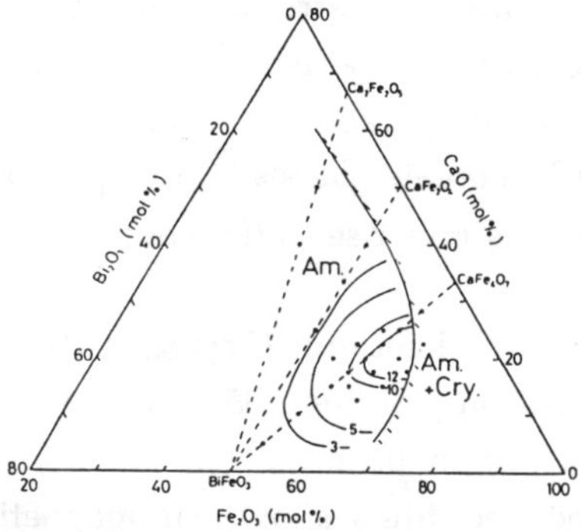

Fig. 14. The phase diagram showing the composition range for amorphous (Am) and crystalline (Cry) mixed oxides of $Fe_2O_3 . CaO . Bi_2O_3$. The hatched side corresponds to the mixed phase region. Curves of constant room temperature magnetization for the amorphous compounds are also shown. [after Ref. 50]

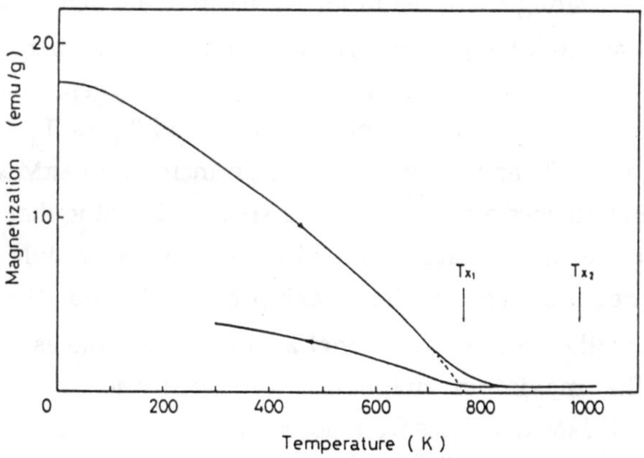

Fig. 15. M vs. T data for increasing and decreasing temperatures for a- $(Fe_2O_3)_{0.65}$ $(CaO)_{0.2}$ $(Bi_2O_3)_{0.15}$ for an applied static magnetic field of 10 kOe. M is 17.8 emu/g extrapolated to 0 K. Crystallization temperatures are indicated in the figure. The approximate Curie temperature is 760 K. [after Ref. 50]

crystallization temperature of 766 K and cooled back to room temperature. A Curie temperature of about 760 K is determined for the ribbon. Mössbauer spectra show that the glass is composed two amorphous components, a ferromagnetic phase and an antiferromagnetic phase with a T_N of 80 K.

In a- Fe_2O_3 . Bi_2O_3 . CaO thin films prepared by reactive rf sputtering in a pure oxygen or oxygen-argon atmosphere, we observe paramagnetic character in as-deposited samples and a ferrimagnetic behavior in films annealed at high temperatures. Ferromagnetic resonance studies show a uniaxial or planar anisotropy in annealed films. The magnetic parameters for the sputtered films are comparable to values reported for samples of amorphous Fe_2O_3 . CaO in borate glasses.[53] It is proposed that magnetically ordered amorphous clusters give rise to the observed ferrimagnetic character in the annealed films.[34]

Films with the chemical formula $(Fe_2O_3)_{0.5}$ $(Bi_2O_3)_{0.5-x}$ $(CaO)_x$ were deposited for x = 0 - 0.5 on Corning 7059 glass or silicon substrates. The sputtering was carried out in pure oxygen or 50% oxygen + 50% argon atmospheres. As-deposited films show paramagnetic behavior at room temperature for all x-values. Films containing CaO develop a spontaneous moment when annealed in air at temperatures T_a. Figure 16 shows the variation of $4\pi M$ with T_a for samples with x = 0.1 - 0.5. The data were obtained by first heating the films in air for 60 min. at each T_a and then by measuring $4\pi M$ at room temperature with an in-plane static magnetic field of 5 kOe. The results are for films sputtered in a pure oxygen atmosphere. In Fig.16, $4\pi M$ remains small and independent of x and T_a for $T_a \leq 450$°C. For films annealed at 500°C and above, one sees an increase in $4\pi M$ with T_a and the magnetization reaches a maximum for a specific T_a value that depends on x. Further increase in T_a is accompanied by a decrease in $4\pi M$. Figure 17 shows the x-dependence of the maximum $4\pi M$ and the T_a-value corresponding to the maximum magnetization. One notices a substantial variation in the maximum magnetization with the concentration of nonmagnetic CaO; $4\pi M$ decreases from about 970 G for x = 0.1 to 350 G for x = 0.5. The T_a-value corresponding to the maximum in $4\pi M$ increases with x. Since the amount of Fe in the films is independent of x, data in Fig. 16 and 17 show the variation in $4\pi M$ with the amount of nonmagnetic CaO.

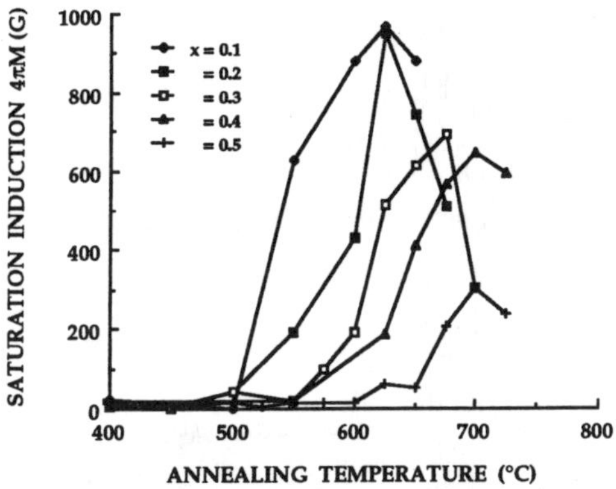

Fig. 16. Room temperature saturation induction vs. annealing temperature data for rf sputtered films of $(Fe_2O_3)_{0.5}(Bi_2O_3)_{0.5-x}(CaO)_x$ for x =0.1 - 0.5. The films were sputtered in a pure oxygen atmosphere and the annealing was done for 60 min. in air. [after Ref. 34]

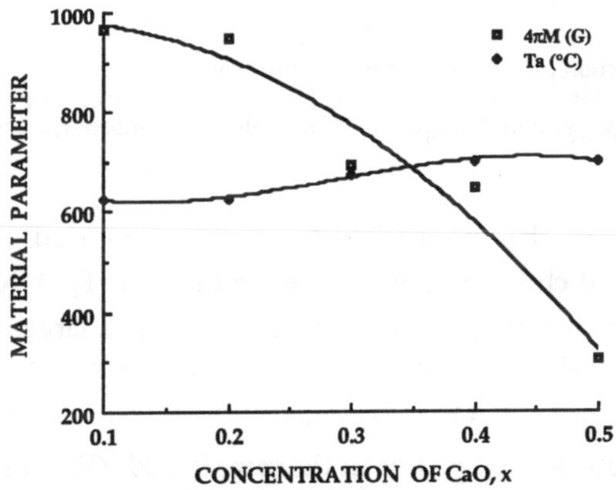

Fig. 17. The variation with concentration of CaO, x, of the maximum room temperature saturation induction 4πM and the annealing temperature for the maximum 4πM, for samples of $(Fe_2O_3)_{0.5}(Bi_2O_3)_{0.5-x}(CaO)_x$. [after Ref. 34]

X-ray diffraction measurements were made to investigate the effects of high temperature annealing on the crystal structure of the films. Such data in Fig.18 are for unannealed and annealed films with x = 0.1. It is obvious from the diffraction pattern that the as-sputtered film is amorphous. The data for T_a = 550°C show the presence of crystalline antiferromagnetic $BiFeO_3$ and nonmagnetic $Bi_{3.11}Ca_{0.89}O_{5.56}$. When T_a is increased to 650°C, crystalline antiferromagnetic phases $BiFeO_3$ and $Bi_2Fe_4O_9$ are present in the film.

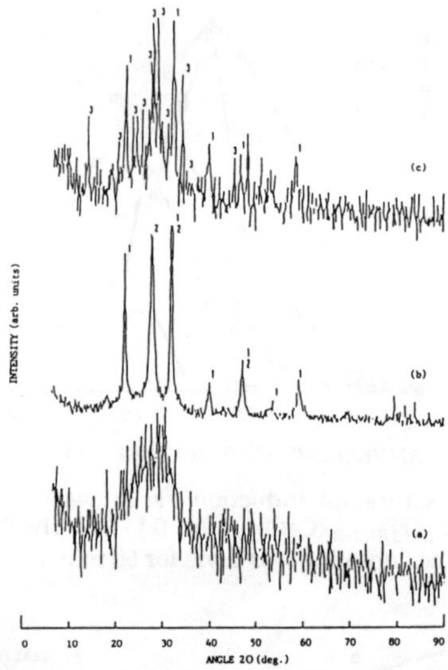

Fig. 18. X-ray diffraction pattern at room temperature for $(Fe_2O_3)_{0.5}$ $(Bi_2O_3)_{0.4}$ $(CaO)_{0.1}$ for (a) as-deposited and for films annealed at (b) 550°C and (c) 650°C. The crystalline peaks of $BiFeO_3$, $Bi_{3.11}Ca_{0.89}O_{5.56}$ and $Bi_2Fe_4O_9$ are represented by numbers 1, 2 and 3, respectively. [after Ref. 34]

We attribute the magnetization in the annealed samples to the presence of ordered clusters for the following reasons. First, XRD data for the annealing temperature at which the sample develops a large moment, i.e., T_a = 550°C, do not show the presence of crystalline ferrimagnetic phases such as γ-Fe_2O_3, Fe_3O_4 or $Bi_3Fe_5O_{12}$. The magnetization corresponds to a volume fraction of 10 - 50% of such phases in the sample and XRD is sensitive enough for the identification of such a large amount of crystalline phases. Second, the Curie temperature for the films do not agree with the values expected for the above crystalline ferrimagnets (results of magnetization studies are discussed later). Finally, it is clear from the data in Figs. 16 and 18 that when Fe is lost from the amorphous structure at high T_a -values in the form of crystalline antiferromagnetic phases, a decrease in $4\pi M$ is observed. Since the sample with x = 0, i.e., pure $BiFeO_3$, remains paramagnetic even when subjected to annealing at high temperatures, the presence of CaO appears to be essential for the formation ordered amorphous clusters.[34]

In order to investigate the nature of amorphous clusters in our films, data on T dependence of $4\pi M$ were obtained for an inplane static field of 5 kG. Figure 19 shows such data for annealed samples with x = 0.2 and 0.5. One observes a linear decrease in $4\pi M$ with increasing T for large x-values. The Curie temperature was determined by linear extrapolation of high temperature $4\pi M$ data, and T_c varies from a minimum of 600±10 K for x = 0.2 to a maximum of 750±10 K for x = 0.5. The Curie temperatures are either higher or lower than the ordering temperatures for the crystalline ferrimagnets γ-Fe_2O_3, Fe_3O_4 and $Bi_3Fe_5O_{12}$. This observation is in agreement with the XRD data in Fig.18 which do not show the presence of ferrimagnetic crystalline phases. One can obtain information regarding the nature of magnetic ordering in the clusters from $4\pi M$-values at low temperatures. For example, the magnetization at 77 K for the sample with x = 0.2 corresponds to a moment of 1 μ_B per formula unit. The observed moment is much smaller than the expected value of 5 μ_B for a parallel alignment of trivalent iron. Thus the magnetic ordering in the amorphous clusters appears to be ferrimagnetic. The magnetic parameters for the present system are in general agreement with the results reported for similar amorphous compounds in borate glasses, $(B_2O_3)_{0.45-x}(Fe_2O_3)_{0.55}(CaO)_x$.[53] The compounds in Ref. 53, prepared by rapid quenching, show a maximum room temperature magnetization of 19 emu/g and a T_c of 750 K for samples with x = 0.15. A long range magnetic order in the sample was evident from Mössbauer studies.

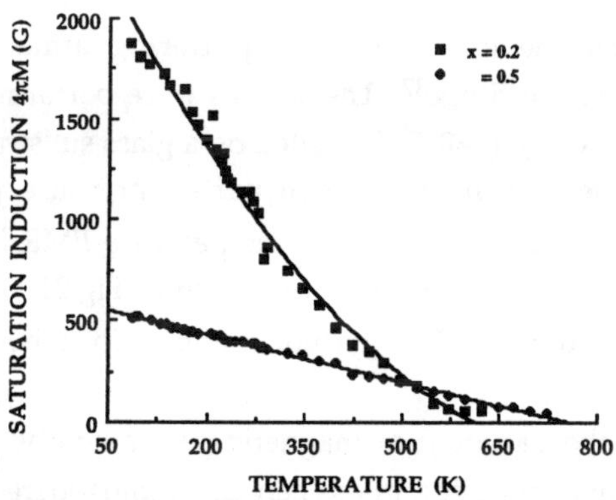

Fig. 19. The dependence of the saturation induction $4\pi M$ on temperature for annealed films of $(Fe_2O_3)_{0.5}(Bi_2O_3)_{0.5-x}(CaO)_x$ for x = 0.2 and 0.5. [after Ref. 34]

In addition to a large magnetic moment, our annealed samples were found to show a uniaxial or planar anisotropy. The anisotropy field H_A was estimated from FMR data at 9 GHz on $4\pi M_{eff} = 4\pi M - H_A$ and $4\pi M$ values measured with the Faraday technique. Figure 20 shows the temperature dependence of H_A for annealed films with x = 0.2 - 0.5. Films with x = 0.1 - 0.4 show a uniaxial anisotropy perpendicular to the film plane. A planar anisotropy character is seen in the film with x = 0.5. As the temperature is increased, the magnitude of anisotropy field decreases.

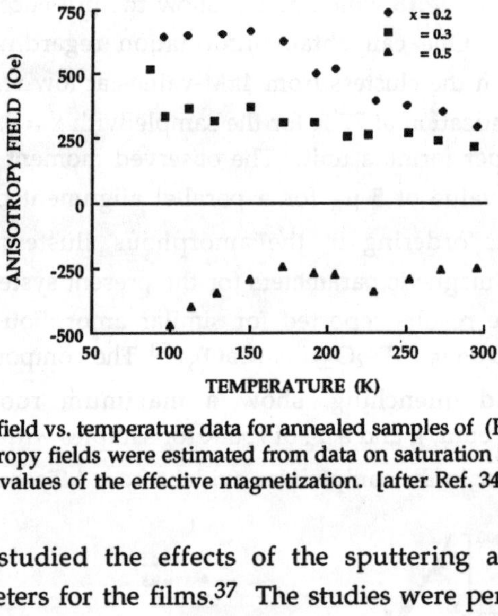

Fig. 20. Anisotropy field vs. temperature data for annealed samples of $(Fe_2O_3)_{0.5}(Bi_2O_3)_{0.5-x}(CaO)_x$. The anisotropy fields were estimated from data on saturation induction and ferromagnetic resonance values of the effective magnetization. [after Ref. 34]

We also studied the effects of the sputtering atmosphere on the magnetic parameters for the films.[37] The studies were performed on films of $(Fe_2O_3)_{0.5}(Bi_2O_3)_{0.1}(CaO)_{0.4}$ (x=0.4) deposited on a glass substrate in a mixed argon-oxygen atmosphere with the oxygen partial pressure $p(O_2)$ varying from 0 to 100%. The variation of the room temperature FMR line-width and gyromagnetic ratio with T_a for the films are shown in Fig.21. The line-width data show a broad minimum. The gyromagnetic ratio generally increases with increase in T_a.[37]

Consider now the variation of magnetic parameters with the oxygen partial pressure.[37] Films sputtered in oxygen rich atmospheres show a large magnetization, and γ for the samples are smaller than for films sputtered in argon rich atmospheres. The anticipated presence of divalent iron ions in samples prepared in argon rich atmospheres could account for the data in Fig. 21. The divalent iron has a magnetization smaller than for trivalent iron and a relatively large γ-value. Also the presence of Fe^{2+} is expected to result in a large ΔH due to conductivity assisted line-broadening effects.

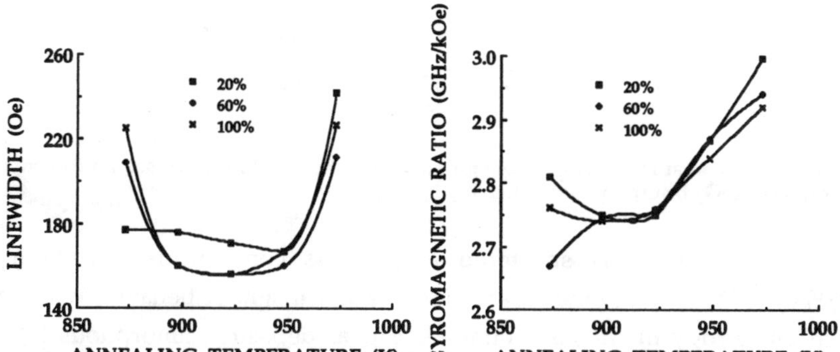

Fig. 21: The annealing temperature dependence of the room temperature FMR line-width ΔH and the gyromagnetic ratio γ for films of a - $(Fe_2O_3)_{0.5}$ $(CaO)_{0.4}$ $(Bi_2O_3)_{0.1}$ sputtered at 20, 60 and 100% oxygen partial pressures. [after Ref. 37]

4.4 Fe_2O_3 . Bi_2O_3 . ZnO

A ferromagnet-like character in Fe_2O_3 . Bi_2O_3 . ZnO is observed by Suzuki and co-workers in samples prepared by melt-quenching and ionized cluster beam deposition techniques.[24] Figure 22 shows some of their results on the variation of room temperature magnetization vs. Zn content. The magnetization increases with increase in Zn concentration and reaches a maximum for 20 mol.% ZnO. Further increase in Zn content results in a partial crystallization and a decrease in magnetization. The Curie temperature for the oxides is in the range 460 to 515 K, depending on the Zn content. A drastic reduction in the magnetization at room temperature is found to occur in samples heated above the crystallization temperature. Mössbauer measurements on the oxides do not show any evidence for long range ferromagnetic ordering in the material. The authors concluded that the amorphous oxide is a speromagnet consisting of iron ions tetrahedrally surrounded by oxygen.

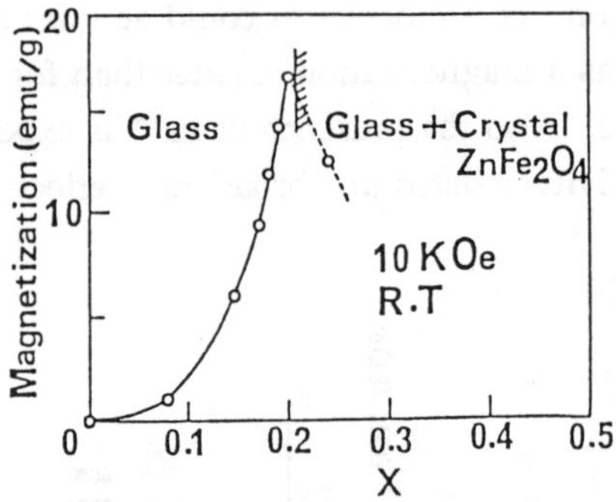

Fig. 22: Magnetization at room temperature for an applied field of 10 kOe versus Zn content for amorphous Fe_2O_3 Bi_2O_3 ZnO. [after Ref. 24]

Our studies on rf sputtered amorphous compounds $(Fe_2O_3)_{0.5}$ $(Bi_2O_3)_{0.5-x}$ $(ZnO)_x$ led to the observation of a ferrimagnetic behavior with an uniaxial anisotropy in the compound.[31] The as-deposited amorphous films are paramagnets for all x-values. The samples develop a magnetic moment and an uniaxial anisotropy on annealing in air at temperatures T_a = 400-775°C. As shown in Fig. 23, the room temperature magnetization shows a maximum for T_a = 500 - 700°C, depending on the x-value. Further increase in T_a results in the crystallization of phases such as 24 Bi_2O_3·ZnO, $BiFeO_3$ and α-Fe_2O_3 in the sample and is accompanied by a decrease in the magnetization. A maximum 4πM of 1900 G was obtained for the sample with x = 0.3.

Measurements on field dependent M shows a soft-magnetic character for the films. Such data at room temperature in Fig.24 are for H perpendicular to the plane of a film with x = 0.125 and annealed at 500°C. The coercive force varies from 90 Oe for H parallel to the film plane to 180 Oe for the perpendicular orientation. The linear increase in M for H above 1 kOe may be attributed to coexisting paramagnetic clusters in the sample.

Figure 25 shows the temperature dependence of 4πM for a series of samples annealed at 650 - 700°C. Since the amount of Fe in the samples is independent of x, it is clear form the data that the magnetization depends

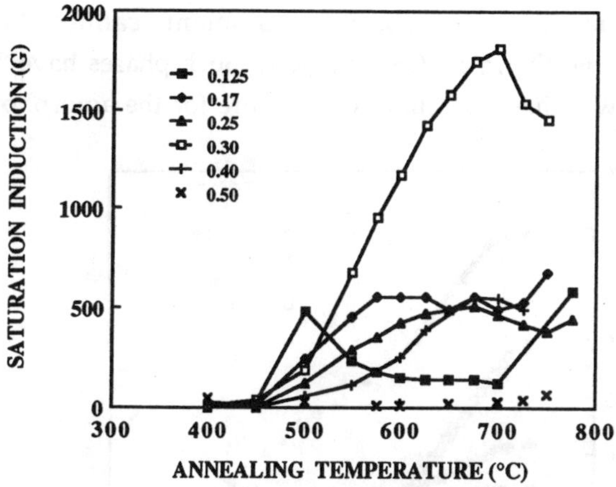

Fig. 23. Room-temperature saturation induction $4\pi M$ as a function of annealing temperature T_a for samples of a - $(Fe_2O_3)_{0.5}$ $(ZnO)_x$ $(Bi_2O_3)_{0.5-x}$ prepared by rf sputtering in a 50% O_2 + 50% Ar atmosphere. The annealing was done in air at T_a for 60 min. [after Ref. 31]

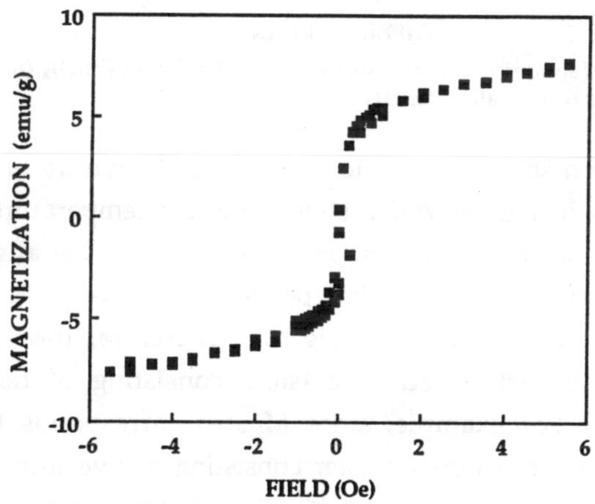

Fig.24. Data on the static field H dependence of room temperature magnetization for a-$(Fe_2O_3)_{0.5}$ $(Bi_2O_3)_{0.375}$ $(ZnO)_{0.125}$ annealed at 500°C for H perpendicular to the film plane.

sensitively on the amount of nonmagnetic Zn in the sample. The measured T_c is about 600±10 K and is lower than the crystallization temperature. The absence of any systematic variation in T_c with x indicates that high temperature annealing results essentially in the formation of a specific type of magnetically ordered clusters in all Zn substituted samples. It is obvious from the T_C-value that the observed moment cannot be due to microcrystalline $Bi_3Fe_5O_{12}$, γ-Fe_2O_3, or Fe_3O_4. Such phases have T_C-values either higher or lower than the Curie temperature for the amorphous films.

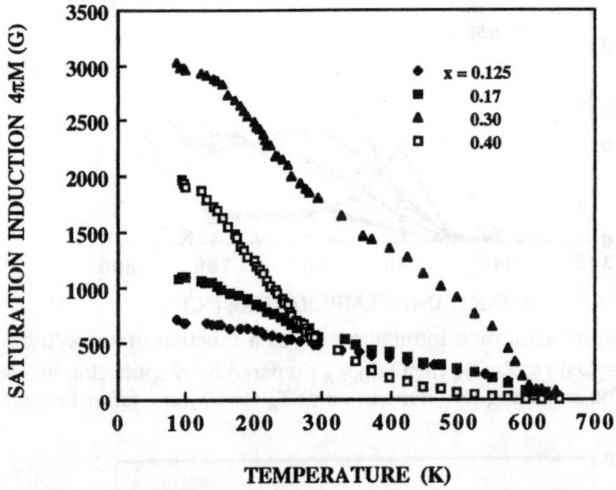

Fig. 25 M vs. T data for rf sputtered annealed samples of a - $(Fe_2O_3)_{0.5}$ $(ZnO)_x$ $(Bi_2O_3)_{0.5-x}$. The Curie temperature is 600 K. [after Ref. 31]

One can obtain some information regarding the nature of clusters in the films from the data in Fig. 26 which shows the low temperature moment as a function of x. The expected magnetic moment for a parallel alignment of Fe^{3+} ions is 5 μ_B per formula unit at 0 K, independent of the x-values, and is a factor of 5-10 larger than the moment in Fig.26. Therefore, the relatively small magnetization can be attributed to clusters consisting of ferrimagnetically coupled iron ions. For example, since M also corresponds to the average magnetic moment per iron ion, a cluster consisting of five iron ions with a 2:3 spin structure will have a M-value of 1 μ_B which is equal to the moment for the sample with x= 0.3. The fact that the Curie temperature is independent of x indicates that the dominant superexchange between magnetic ions within clusters are not significantly altered by the concentration of Zn.

The presence of an uniaxial anisotropy in the amorphous samples was clearly evident from ferromagnetic resonance studies at 9.2 GHz. Figure 27 shows the variation of H_A with T for samples with x = 0.4. The perpendicular anisotropy fields are small so that the easy axis of magnetization is parallel to the sample plane. The data show an overall decrease in H_A as T increases. We also noticed a pronounced dependence of the gyromagnetic ratio γ on T. As shown in Fig. 28 the γ-value increased from 2.75 GHz/kOe at 293 K to 3.02 GHz/kOe at 80 K.

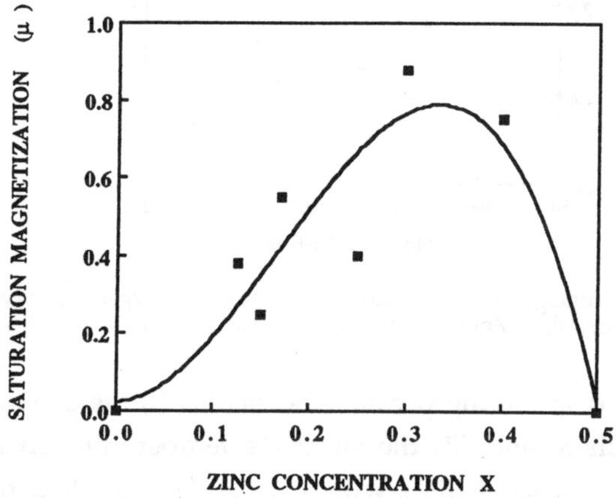

Fig. 26. Saturation magnetization M (in Bohr magneton μ_B) per formula unit at 80 K as a function of concentration of ZnO in a - $(Fe_2O_3)_{0.5}$ $(ZnO)_x$ $(Bi_2O_3)_{0.5-x}$ prepared by rf sputtering. [after Ref. 31]

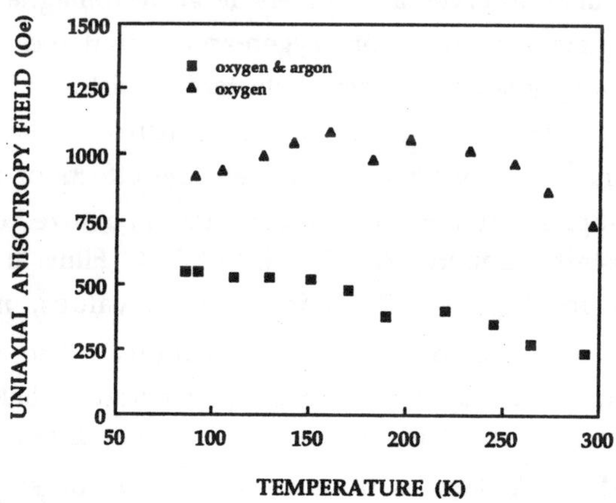

Fig. 27. Uniaxial anisotropy field obtained from FMR and static magnetization measurements for annealed (T_a = 650°C) films of a- $(Fe_2O_3)_{0.5}$ $(ZnO)_{0.4}$ $(Bi_2O_3)_{0.1}$ sputtered in oxygen and oxygen-argon. [after Ref. 35]

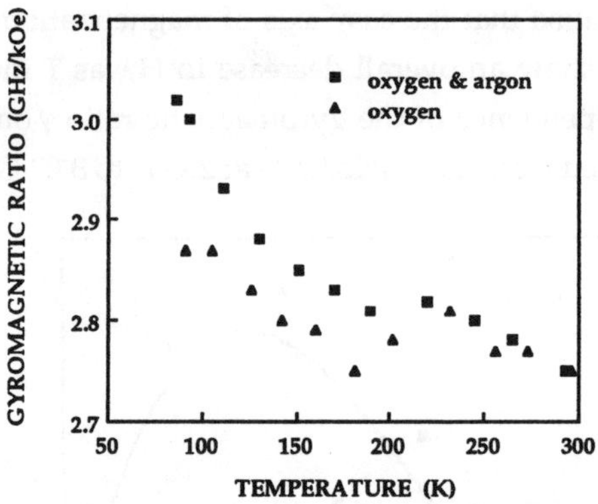

Fig. 28. Variation of gyromagnetic ratio γ with temperature for oxygen and oxygen + argon sputtered samples of $(Fe_2O_3)_{0.5}$ $(ZnO)_{0.4}$ $(Bi_2O_3)_{0.1}$ annealed at 650°C. [after Ref. 35]

The effects of sputtering parameters such as (i) the composition of the sputtering atmosphere and (ii) the substrate temperature on the magnetic properties of such annealed films with x = 0.4 were studied through static magnetization and ferromagnetic resonance measurements.[35,38] For samples sputtered in a mixed oxygen-argon atmosphere, the magnetization and the uniaxial anisotropy field are smaller than for films deposited in a pure oxygen atmosphere. One also observes a relatively large ferromagnetic resonance line-width and gyromagnetic ratio for oxygen-argon sputtered samples. The possible presence of magnetically ordered clusters consisting of divalent iron ions in samples deposited in oxygen-argon atmospheres may account for the above observations.[35] In order to investigate the effects of the substrate temperature T_s, samples with x = 0.4 were deposited in a mixed oxygen-argon atmosphere at substrate temperatures T_s = 450 - 875 K. Films are found to be x-ray amorphous for $T_s \leq 675$ K. For higher T_s-values, *nonmagnetic* crystalline phases are precipitated in the sample. The samples are paramagnetic at room temperature for substrate temperatures below 475 K. A spontaneous magnetic moment is observed in films deposited at higher T_s. Figure 29 shows M vs. T_s data for films on glass or silicon substrates. The saturation induction $4\pi M$ increases with increase in T_s. Figure 30 shows M vs. T data for a sample sputtered at 525 K. The Curie temperature is found to decrease with increasing T_S. A uniaxial anisotropy character in the films is evident from ferromagnetic resonance studies. Thus data on structural and magnetic parameters indicate the possible presence of ordered clusters in films sputtered at high temperatures.[38]

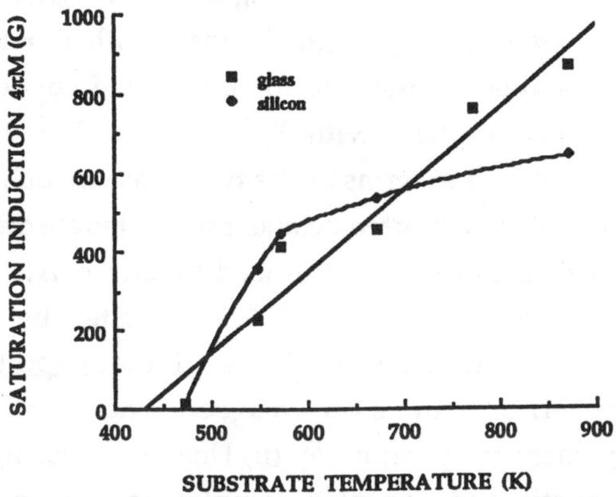

Fig. 29. Data on room temperature $4\pi M$ as a function of substrate temperature T_S for $(Fe_2O_3)_{0.5}$ $(ZnO)_{0.4}$ $(Bi_2O_3)_{0.1}$ sputtered on glass or silicon substrates in oxygen - argon atmospheres. The films retain the amorphous structure for $T_S \leq 575$ K. At higher T_S, nonmagnetic crystalline phases are formed. The spontaneous moment was suggested to arise from ordered clusters formed *in situ* in the film. [after Ref. 38]

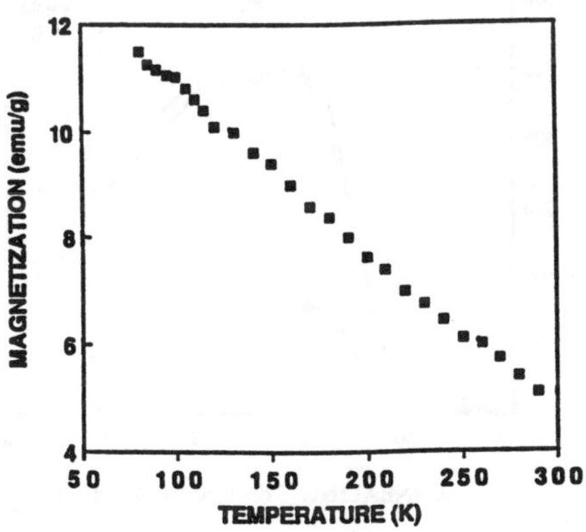

Fig. 30. Magnetization as a function of temperature for $(Fe_2O_3)_{0.5}$ $(Bi_2O_3)_{0.1}$ $(ZnO)_{0.4}$ films sputtered in a mixed oxygen-argon atmosphere at a substrate temperature of 575 K.

4.5 $Fe_2O_3 . Bi_2O_3 . CuO$

Soeya and others[25] prepared rapidly quenched amorphous samples of the composition $(Fe_2O_3)_{0.5}$ $(Bi_2O_3)_{0.5-x}$ $(CuO)_x$ (x = 0 to 0.3). Small angle X-ray scattering studies reveal the presence of 10 nm clusters. The crystallization temperature obtained from differential thermal analysis is 633 K. Amorphous samples with $x > 0.1$ show a spontaneous moment at room temperature which increases with increasing x. The magnetization originates from two magnetic phases, a ferromagnetic phase with $T_c = 650$ K and with a magnetic structure similar to the one for crystalline copper ferrite, and a random antiferromagnetic phase with $T_N = 200$ K . Mössbauer absorption spectra could be understood in terms of the two-phase model, and the spectral parameters indicate Fe^{3+} in distorted octahedral and tetrahedral sites.[25]

A ferrimagnetic behavior characterized by an uniaxial anisotropy was observed in similar films with x = 0 - 0.5 deposited by rf sputtering.[36] Important results of our studies are as follows. (i) the as-sputtered film with x = 0.5, pure copper ferrite, show a ferromagnetic character with an in-plane anisotropy field at room temperature.[33] (ii) Unannealed samples with x = 0.1 - 0.4 are paramagnetic and annealed samples are ferrimagnetic with an uniaxial anisotropy. As shown in Fig. 31, a maximum in the magnetization occurs for T_a = 500-600°C. Phases such as CuO and $Bi_2Fe_4O_9$ are formed when

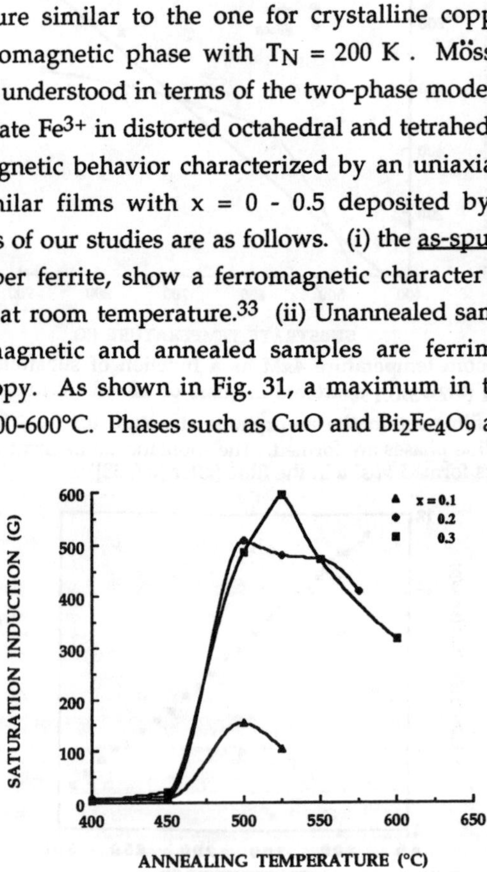

Fig. 31. Annealing temperature T_a versus room temperature saturation induction $4\pi M$ data for rf sputtered films of $(Fe_2O_3)_{0.5}$ $(CuO)_x$ $(Bi_2O_3)_{0.5-x}$. The films were sputtered in a 50%O_2 + 50%Ar atmosphere and annealed in air for 60 min. The magnetization was measured with an in-plane static field of 5 kOe. [after Ref. 36].

the films are crystallized.[36] (iii) The largest $4\pi M$- value of 3 kG at room temperature is observed for the sample with x = 0.4. A soft-ferromagnetic nature with a coercive force less than 10 Oe is evident from M vs. H data. (iv) The Curie temperature determined from M vs. T data in Fig. 32 ranges from 460 to 480°C.[36] The magnetization in μ_B at 80 K is plotted as a function of x in Fig. 33. One observes a linear increase in M with x. A positive identification of the nature and composition of magnetically ordered clusters in annealed samples was possible from the data on the Curie temperature, low

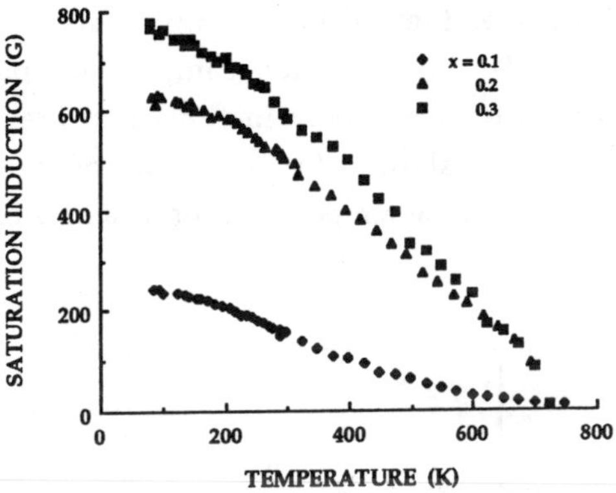

Fig. 32. $4\pi M$ measured at 5 kOe vs. T for samples of $(Fe_2O_3)_{0.5}$ $(CuO)_x$ $(Bi_2O_3)_{0.5-x}$ sputtered in 50%O_2 + 50% Ar and annealed at 500 - 600°C. The Curie temperatures for the films, 460 - 480°C, are in agreement with T_c for crystalline copper ferrite. [after Ref. 36]

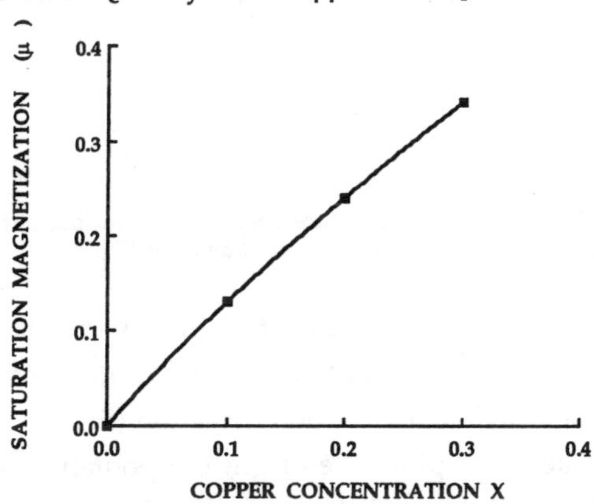

Fig. 33. Variation of the saturation magnetization per formula unit M at 80 K as a function of the concentration x of CuO in annealed rf sputtered films of $(Fe_2O_3)_{0.5}$ $(CuO)_x$ $(Bi_2O_3)_{0.5-x}$. The data indicate a linear increase in M with x, and the magnetization is proposed to arise from copper ferrite like clusters. [after Ref. 36]

temperature magnetization, and g-value. The T_C-value of 460-480°C, the magnetization of 1.1-1.3 μ_B per copper ion, and the g-value of 2.1-2.3 are in excellent agreement with the values expected for crystalline copper ferrite. Thus the amorphous clusters have a composition and spin structure similar to $CuFe_2O_4$.

4.6 $Fe_2O_3 . Bi_2O_3 . ABO_3$

Fujii, Kajima and others reported coexisting ferromagnetism and ferroelectricity in rf sputtered films of a-Fe_2O_3 Bi_2O_3- ABO_3 (AB = PbTi, PbZr, BaTi) are as follows.[26-28] As-deposited samples are paramagnetic. A ferrimagnetic moment is observed only in films that are *annealed* at high temperatures. Figure 34 shows the variation of room temperature magnetization with the annealing temperature for a sample of with $PbTiO_3$.[27]

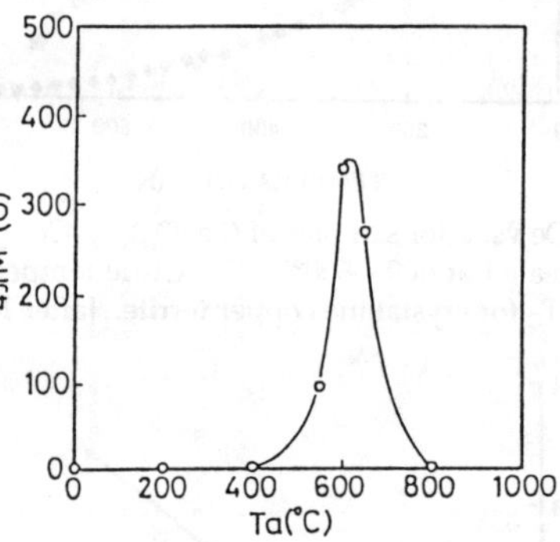

Fig. 34: Magnetization 4πM at room temperature versus the annealing temperature data for $(Fe_2O_3)_{0.62}$ $(Bi_2O_3)_{0.13}$ $(PbTiO_3)_{0.25}$ amorphous films. [after Ref .27].

A loss of spontaneous magnetic moment is observed when crystallization commences. The authors performed high resolution transmission electron microscopy (TEM), selected area electron diffraction (SAED), x-ray diffraction, and Mossbauer studies to explore the origin of spontaneous moment in the annealed films. The studies reveal the presence of a magnetically "dead" 150 nm thick crystalline surface layer. As shown in Figs. 35 and 36, when the surface layer is etched away, the sample magnetization increases, and XRD data did not indicate any crystalline phases in the film interior. Further, the absence of any microcrystals in the film was confirmed by TEM and SAED studies. Kajima, et al, therefore, concluded that the spontaneous moment originated from *purely amorphous interior layers* of the films. A long range order was clearly evident from Mossbauer studies.[28]

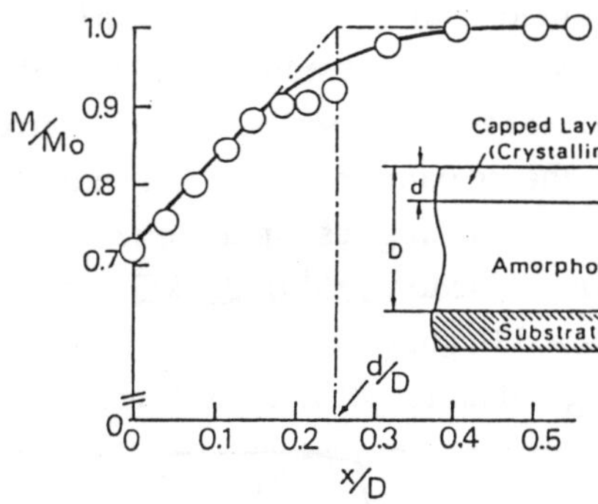

Fig. 35. Variation of the room temperature magnetization with the film thickness for annealed films of $(Fe_2O_3)_{0.74}(Bi_2O_3)_{0.16}(PbTiO_3)_{0.1}$. X-represents the thickness of the surface layer removed by etching the film. M increases with X and levels off when X = d = 145 nm, i.e., when the magnetically dead crystalline top layer of the film is removed. [after Ref. 28]

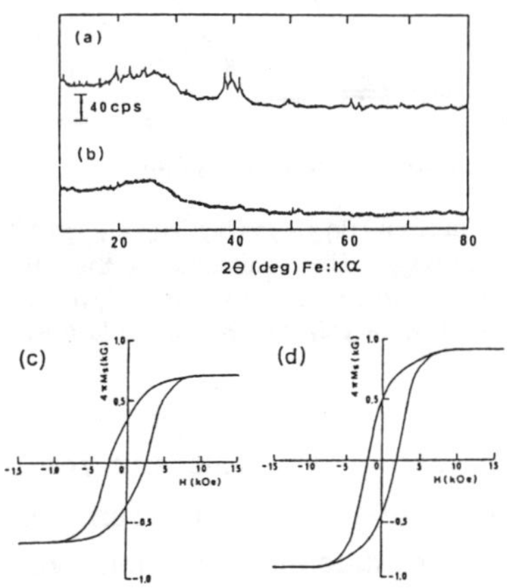

Fig. 36 Changes in the X-ray diffraction pattern, (a) and (b), and in the magnetic hysteresis loop, (c) and (d) when a 145 nm thick magnetically dead surface layer is removed from a sample of $(Fe_2O_3)_{0.74}(Bi_2O_3)_{0.16}(PbTiO_3)_{0.9}$. It is evident from the data that the magnetization arises from the amorphous interior of the film. [after Ref. 28].

Our studies on similar compounds show a ferrimagnetic character for the oxides. Measurements were done on 1 μm thick films sputtered in 50% oxygen and 50% argon atmospheres.[29,32] Results of our investigation are summarized below. (i) As-deposited films are paramagnets at room temperature for all x-values. The films develop a ferrimagnetic moment when annealed at temperatures T_a = 100 - 750°C. Figure 37 shows data on the magnetization M versus T_a for samples substituted with 0.2 $BaTiO_3$, 0.1 $PbTiO_3$ and 0.5 $PbZrO_3$. The significance of these three specific substitutions is that the M value is the largest for these samples in their respective series. It is clear from Fig.37 that the annealing temperature for the maximum in M depends on the nature of nonmagnetic substitution. The highest magnetization of 40 emu/g, corresponding to a saturation induction 4πM of 3.8 kG is achieved for the film containing 0.5 $PbZrO_3$. The 4πM-values are comparable to that for spinel and hexagonal ferrites and are larger than for yttrium- and rare-earth iron garnets.

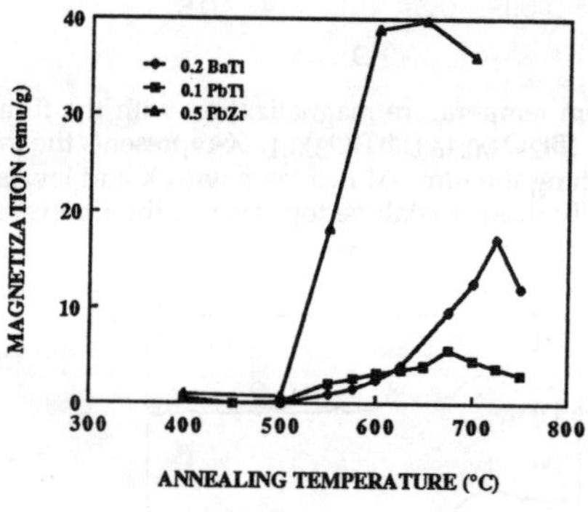

Fig. 37 4πM vs. T_a data for samples of $(Fe_2O_3)_{0.5-x/2}$ $(Bi_2O_3)_{0.5-x/2}$ $(ABO_3)_x$ for $(ABO_3)_x$ = $(BaTiO_3)_{0.2}$, $(PbTiO_3)_{0.1}$ and $(PbZrO_3)_{0.5}$. The significance of these three specific samples is that the room temperature magnetization is the highest for the films in their respective series. The film with $(PbZrO_3)_{0.5}$ shows a 4πM of 3.8 kG, the largest value reported for any amorphous oxide. [after Ref. 29]

The amorphous films show hysteresis and remanence. Figure 38 shows M vs. H data at room temperature for samples with $BaTiO_3$ and $PbZrO_3$ annealed at 650°C. The static magnetic field was applied parallel to the sample plane. Data on the magnetization with the static field applied perpendicular to the sample plane showed similar features. We did not find any evidence for magnetocrystalline anisotropy in the films.

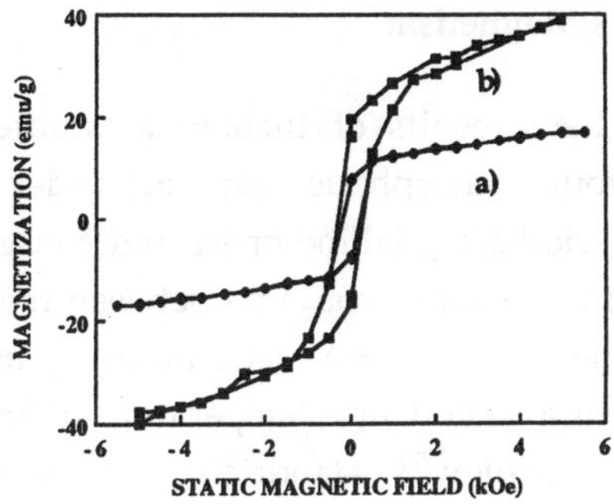

Fig. 38. In-plane static field dependence of the room temperature magnetization for annealed samples of (a) $(Fe_2O_3)_{0.4}(Bi_2O_3)_{0.4}(BaTiO_3)_{0.2}$ and (b) $(Fe_2O_3)_{0.25}(Bi_2O_3)_{0.25}(PbZrO_3)_{0.5}$. The coercive field for these 50% O_2 + 50% Ar sputtered samples is about 200 - 300 Oe. [after Ref. 32]

X-ray diffraction studies do not show the precipitation of any crystalline ferrimagnetic phases in the samples. The magnetic parameters obtained for our films samples are quantitatively different from the results of Fujii, et al on a similar system.[26] Their samples which were sputtered in argon rich atmospheres show a relatively small M and T_c, and a large coercive field. Our studies also show a similar reduction in the magnetization for samples sputtered in pure argon or argon rich atmospheres. We recognized that samples could develop an oxygen deficiency when sputtered under such conditions, resulting in the formation of divalent iron ions. The presence of divalent iron is expected to lead to a reduction in the magnetization and changes in the nature and strength of superexchange interactions. This could possibly account for the enhanced magnetization for our samples, about a factor of two to eight larger than the values reported by Fujii.[26]

Further studies are essential for an understanding of (i) the magnetically ordered microcrystalline or the amorphous phase, (ii) the nature of magnetic ordering, and (iii) the superexchange mechanism that gives rise to the long range order. For example, the sample with 0.5 $PbZrO_3$ show a low temperature moment of 4.4 μ_B per iron ion. Thus the clusters are likely to contain <u>ferromagnetically</u> coupled trivalent iron ions.

5 Theory of Amorphous Magnetism

We summarize here results of theoretical studies that address an important question: would amorphous materials order magnetically even though they lack the periodic crystalline order and consequently there is an inherent frustration in the exchange coupling between the neighboring spins. Equally important is the nature of magnetic ordering in such systems. A ferromagnetic order with a finite Curie temperature in amorphous materials was first predicted by Gubanov.[54] Handrich[55] estimated the spontaneous moment for such structures with a specific type of magnetic atoms and showed that the magnetization would depend sensitively on the randomness (or fluctuations) in the structure and that the magnetization would vanish for a certain critical value of the structural fluctuations. Based on an effective field model, Simpson showed that an antiferromagnetic order would exist in amorphous oxides with a T_N proportional to the square root of the average number of interacting nearest neighbors.[56] Results of this model were in general agreement with magnetic parameters for amorphous yttrium and gadolinium iron garnets.[18]

Kaneyoshi in a series of theoretical works in recent years showed that ferrimagnetism could exist in a two dimensional amorphous structure.[46,57-59] In this model, magnetic parameters were estimated for a two dimensional network consisting of spin-1/2 and spin-1 ions. The calculated susceptibility shows features expected for a ferrimagnet. The theory predicts a collinear or canted ferrimagnetic structure, with a random distribution for the canting angle, depending on the strength and fluctuation in the exchange constants. Important features such as a compensation point and hysteresis are also expected in the temperature dependence of the magnetization in such systems. Some of these predicted features have been observed in the temperature dependence of M in amorphous ferrites and other oxides.[33,40]

The current theoretical understanding of the nature of magnetism and magnetic order in amorphous oxides is quite inadequate. Further, it is not clear whether speromagnetism and other types of short range order can be established by neutron scattering experiments. It needs to be established whether the observed magnetism is truly long order or due to ordering of some microclusters. Additional theoretical and experimental studies are essential for an understanding of magnetism in in amorphous oxides.

Acknowledgments

The author is grateful to Professor M. S. Seehra, West Virginia University, Dr. B. S. Chao, Energy Conversion Devices, Troy, Michgan, Dr. B. U. M. Rao, J. Chen, D. Hagerty, R. Hasse, S. Cheney, and S. Hunter for assistance with the preparation and characterization of oxide films. Financial support for the research was provided by the Office of Naval Research and the Petroleum Research Fund, administered by the American Chemical Society.

REFERENCES

1. J. M. D. Coey, J. Appl. Phys. **49** (1978) 1646.

2. J. M. D. Coey and D. H. Ryan, IEEE Trans. Magn. **MAG-20** (1984) 1278.

3. *Amorphous Magnetism*, Eds. H. O. Hooper and A. M. de Graaf (Plenum Press, New York, 1973).

4. *Amorphous Magnetism II*, Eds. R. A. Levy and R. Hasegawa (Plenum Press, New York, 1977).

5. K. Moorjani and J. M. D. Coey, *Magnetic Glasses* (Elsevier, Amsterdam, 1984).

6. K. Inomata, S. Hashimoto, and S. Nakamura, J. Appl. Phys. **27** (1988) L883.

7. K. Inomata, S. Hashimoto, S. Okuno, E. Matsubara, and Y. Waseda, J. Appl. Phys. **67** (1990) 5796.

8. N. Heiman and N. S. Kazuma, J. Appl. Phys. **50** (1979) 7633.

9. A. M. van Diepen and Th. J. A. Popma, J. de Physique Coll. **C6** (1979) 755.

10. A. M. van Diepen and Th. J. A. Popma, Solid State Commu. **27** (1978) 121.

11. L. A. Marusak and R. Messier, J. Appl. Phys. **50** (1979) 7630.

12. T. Shigematsu, Y. Bando, and T. Takada, J. de Physique Coll. **C2** (1979) 153.

13. E. M. Gyorgy, K. Nassau, M. Eibschutz, J. V. Waszczak, C. A. Wang, and J. C. Shelton, J. Appl. Phys. **50** (1979) 2883.

14. W. Girnus, H. Beuthien, R. Priess, and W. Gunsser, J. Magn. Magn. Mater. **54-57** (1986) 225.

15. Th. J. A. Popma and van Diepen, A. M., Mat. Res. Bull. **9** (1974) 1119.

16. F. J. Litterst, J. Tejada, and G. M. Kalvius, J. Appl. Phys. **50** (1979) 7636.

17. T. Matsuzawa, K. Okamura, T. Shishido, and S. Yajima, J. de Physique Coll. **C2** (1979) 149.

18. A. W. Simpson and J. M. Lucas, J. Appl. Phys. **42** (1971) 2181.

19. S. N. Okuno, S. Hashimoto, K. Inomata, S. Morimoto, and A. Ito, J. Appl. Phys. **69** (1991) 5072.

20. J. M. D. Coey, E. Devlin, and R. J. Gambino, J. Appl. Phys. **53** (1982) 7810.

21. M. Sugimoto and N. Hiratsuka, Jpn. J. Appl. Phys. **21** (1982) 197.

22. M. Sugimoto and N. Hiratsuka, J. Magn. Magn. Mater. **31-34** (1983) 1533.

23. N. Ota, M. Okubo, S. Masuda, and K. Suzuki, J. Magn. Magn. Mater. **54-57** (1986) 293.

24. S. Suzuki, H. Onodera, M. Sakurai, S. Masuda, A. Matsumoto, and H. Sadamura, IEEE Trans. Magn. **MAG-22** (1986) 1090.

25. S. Soeya, S. Nakamura, and N. Ichinose, J. Appl. Phys. **68** (1990) 2875.

26. T. Fujii, S. Jinzenji, Y. Asahara, and A. Kajima, J. Appl. Phys. **64** (1988) 5434.

27. A. Kajima, T. Kaneda, H. Ito, T. Fujii, I. Okamoto, T. Kimura, and K. Ohdan, J. Appl. Phys. **69** (1991) 3663.

28. A. Kajima, T. Kaneda, H. Ito, T. Fujii, and I. Okamoto, J. Appl. Phys. **70** (1991) 3760.

29. B. U. M. Rao, G. Srinivasan, V. S. Babu, and M. S. Seehra, J. Appl. Phys. **69** (1991) 5463.

30. B. U. M. Rao and G. Srinivasan, J. Appl. Phys. **70** (1991) 6317.

31. B. U. M. Rao and G. Srinivasan, Phys. Rev. B **44** (1991) 395.

32. B. U. M. Rao and G. Srinivasan, Appl. Phys. Lett. **58** (1991) 2441.

33. G. Srinivasan, B. U. M. Rao, J. Zhao, and M. S. Seehra, Appl. Phys. Lett. **59** (1991) 372.

34. B. U. M. Rao and G. Srinivasan, IEEE Trans. Magn. **28** (1992) 2444.

35. B. U. M. Rao and G. Srinivasan, J. Magn. Magn. Mater. **103** (1992) L228.

36. B. U. M. Rao and G. Srinivasan, J. Magn. Magn. Mater. **111** (1992) 249.

37. G. Srinivasan, B. U. M. Rao, and U. Saligram, Appl. Phys. A **44** (1992) 549.

38. G. Srinivasan and B. U. M. Rao, J. Appl. Phys. **73** (1993) 5713.

39. G. Srinivasan and B. U. M. Rao, J. Magn. Magn. Mater. **123** (1993) 304.

40. G. Srinivasan, B. U. M. Rao, and J. Chen, IEEE Trans. Magn. MAG- (1993)

41. K. Tanaka, K. Hirao, and N. Soga, J. Appl. Phys. **69** (1991) 7752.

42. K. Tanaka, K. Kajima, and T. Yoko, J. non-cryst. solids **109** (1989) 289.

43. K. Tanaka and N. Soga, J. non-cryst. solids **95-96** (1987) 255.

44. K. Tanaka, H. Hirao, and N. Soga, in*The Physics of non-crystalline solids* (Taylor and Francis, London, 1992), p. 296.

45. J. smit and H. P. J. wijn, *Ferrites* (John Wiley, New York, 1959).

46. T. Kaneyoshi, Solid State Commun. **69** (1989) 91.

47. W. H. von Aulock, *Handbook of Microwave Ferrites* (Academic Press, New York, 1965).

48. S. Nakamura and N. Ichinose, Jpn. J. Appl. Phys. **28** (1989) 984.

49. K. Yamaguchi, K. Matsumoto, and T. Fujii, IEEE Trans. Magn. **28** (1992) 2447.

50. S. Nakamura and N. Ichinose, J. Non-cryst. Solids **95-96** (1987) 849.

51. S. Nakamura, V. Hirotsu, and N. Ichinose, Jpn. J. Appl. Phys. **30** (1991) L844.

52. J. chen, S. Cheney, and G. Srinivasan, *in press*, J. Appl. Phys. (1994).

53. Y. Nagata, T. Inoh, and K. Ohta, IEEE Trans. Magn. **MAG-23** (1987) 2317.

54. G. Gubanov, Fiz. tverd. Tela. **2** (1960) 502

55. K. Handrich, Phys. Stat. Sol. **32** (1969) K55.

56. A. W. Simpson, Phys. Stat. Sol. **40** (1970) 207.

57. T. Kaneyoshi, Phil. Mag. Lett. **55** (1987) 69.

58. T. Kaneyoshi and M. Jascur, J. Phys.: Condens. Matter **4** (1992) L281.

59. T. Kaneyoshi and M. Jascur, J. Magn. Magn. Mater. **115** (1992) 217.

SPECTRUM OF MICROWAVE SPIN WAVES IN MAGNETIC MULTILAYERS

B.A.Kalinikos and P.A.Kolodin
St.Petersburg Electrotechnical University
197373 St.Petersburg, Russia

1 Introduction

In recent years the layered structures containing magnetic and nonmagnetic films have attracted strong interest. This interest is stimulated by their unique spin-wave properties. For example, it is possible to tailor the dispersion properties of spin waves in layered structures by stacking several magnetic and nonmagnetic films (see e.g.[1-4] and references therein). In this meaning multilayered structures become a convenient medium for both the study of spin-wave processes themselves and for modeling linear and nonlinear processes in dispersive media in general.

Such structures are also important for the design of the novel family of magnetostatic wave and magnetostatic wave - optic devices for analog signal processing[1,5].

The description of spin-wave processes in multilayered structures is based, in general, on consideration of two aspects: interactions in spin-system of each film formed the structure and interfilm coupling. Each of these aspects includes two main contributions. One of them is the long-range dipole-dipole interaction and the other one is the short-range exchange interaction.

The first calculations of spin-wave properties of layered structures were restricted to the case of magnetostatic (pure dipolar) waves[1-14]. In these investigations the exchange interaction in spin systems of ferromagnetic films and interlayer exchange coupling were neglected. Other investigations were devoted to the case of pure exchange modes in multilayers[15-20]. Such modes are easily studied experimentally by ferromagnetic resonance technique. Experimental investigations have shown that exchange interlayer coupling is significant only for the distances of several nanometers[21-23].

Further investigations in this field have brought to the simultaneous consideration of both dipole and exchange interactions[24,25]. But since this work was done for the purposes of Brillouin light scattering experiments the behavior of full spin-wave spectrum for the wide range of wave vectors has not been investigated. In other words the shape of spin-wave dispersion branches for the vide wave vector region has not been studied.

In this Chapter theoretical investigations of spin-wave spectrum applicable to microwave experiments are presented. Exact and approximate spin wave dispersion equations which are valid for the wide range of values of in-plane wave vectors are derived and discussed. We consider the multilayered structure consisting of an arbitrary number of ferromagnetic films separated from each other by dielectric layers. The direction of magnetization is also arbitrary. Our calculations are based on tenzorial the Green's function method initially elaborated for single ferromagnetic film[26–28] and later extended for multilayered structure[29–31]. The model includes both dipole and exchange interactions as well as electrodynamic and exchange boundary conditions conditions, but the interlayer exchange coupling is neglected. Such model is applicable when the distances between ferromagnetic films (the thicknesses of dielectric spacers) are large compared with the length of effective exchange coupling. This approximation is valid, for example, for microwave experiments concerned with spin wave propagation in YIG-GGG-YIG... structures.

The chapter is organized as follows. Section 2 includes the main theoretical calculations and qualitative discussion. In the Section 3 the detailed analytics of the perpendicularly magnetized double-layered ferromagnetic film structure is given. It includes the exact and approximate dispersion relation and numerical analysis of spin-wave spectrum. Section 4 presents the numerical analysis of the double-layered theoretical film structure magnetized at arbitrary direction. In Section 5 the results of theoretical investigations of spin-wave spectrum of tangentially magnetized screened ferromagnetic film structure are presented. The extension of the theory to the case of infinite film stack is given in the Section 6.

2 Dipole-exchange spin waves in multilayered ferrite-dielectric structure

2.1 Formulation of the problem

Let us consider a structure consisting of N isotropic ferromagnetic films separated by dielectric (nonmagnetic) layers. The structure is assumed to be unbounded in its plane and magnetized to saturation by a static bias magnetic field $\mathbf{H}$ of an arbitrary direction which is determined by angles θ_0 and φ (see Fig.1). The thicknesses of ferromagnetic films L_i (i=1,2,...,N) and dielectric layers a_i as well as the saturation magnetizations of ferromagnetic films $\mathbf{M_{0i}}$ are arbitrary values. The ζ-axis of the coordinat system $\xi\eta\zeta$ is always assumed to be oriented along the direction of propagation of spin waves in the structure. For convenience we introduce for each ferromagnetic film the coordinate system $x_i y_i z_i$ in which the axis z_i is parallel to the direction of saturation magnetization $\mathbf{M_{0i}}$ of the film. The transition from the coordinate system $\xi\eta\zeta$ to the coordinate system $x_i y_i z_i$ can be done by means of orthogonal transformations of rotation through angles φ and θ_{0i}. The matrices of these transformation are of the form:

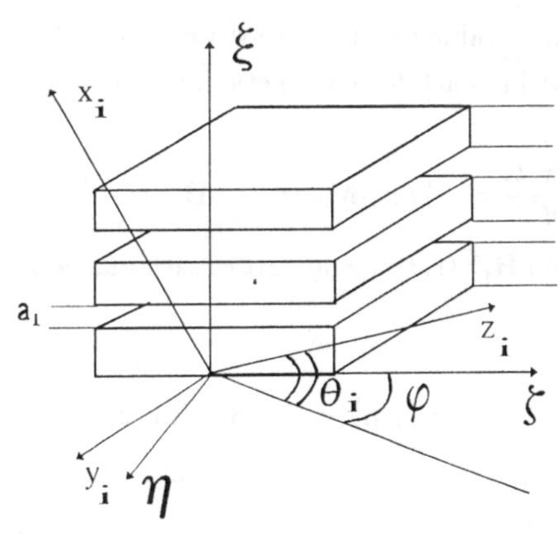

Figure 1: Layered ferrite-dielectric structure and coordinate system.

$$\hat{\mathbf{T}}_{\varphi} = \begin{pmatrix} 1 & 0 & 0 \\ 0 & \cos\varphi & \sin\varphi \\ 0 & -\sin\varphi & \cos\varphi \end{pmatrix} , \tag{1}$$

$$\hat{\mathbf{T}}_{\theta_{\mathbf{i}}} = \begin{pmatrix} \cos\theta_i & 0 & \sin\theta_i \\ 0 & 1 & 0 \\ -\sin\theta_i & 0 & \cos\theta_i \end{pmatrix} . \tag{2}$$

The values of the internal magnetic fields $\mathbf{H_i}$ and their directions θ_i for each film can be found using the solution of the magnetostatic problem when the value $\mathbf{H_0}$ and the direction θ_0 of the external magnetic field are known

$$\begin{cases} H_i \sin\theta_i = & H_0 \sin\theta_0 - M_{0i}\sin\theta_i \quad , \\ M_i \sin 2\theta_i = & 2H\sin(\theta_0 - \theta_i) \ . \end{cases} \tag{3}$$

We stress that in general case $\mathbf{H_i}$ and θ_i are different for each ferromagnetic film formed the structure.

To solve the boundary-value problem, we write the equations of motion for the magnetizations $\mathbf{M_i}(\mathbf{r}, t)$ within each ferromagnetic film (number $i = 1, 2, 3, ..., N$)

$$\frac{\mathbf{M_i}(\mathbf{r}, t)}{\partial t} = -|g|\mu_o\left[\mathbf{M_{0i}}(\mathbf{r}, t), \ \mathbf{H_i^{eff}}(\mathbf{r}, t)\right] \ . \tag{4}$$

The effective magnetic field $\mathbf{H_i^{eff}}(\mathbf{r}, t)$, being part of each magnetic-moment equation, is considered to be equal to

$$\mathbf{H_i^{eff}}(\mathbf{r}, t) = \mathbf{H_i} + \mathbf{h}_d(\mathbf{r}, t) + \alpha_i \nabla^2 \mathbf{m}_i(\mathbf{r}, t) \quad , \tag{5}$$

where $\mathbf{h_d}$ is the variable dipole field, α_i is the exchange constant, $\mathbf{m_i}(\mathbf{r}, t) = \mathbf{M_i}(\mathbf{r}, t) - \mathbf{M}_{0i}$ is the variable magnetization.

The solution of the system consisting of the linearized equations of motion for magnetizations followed from Eq.(4) and the magnetostatic equations are written in the form of plane waves

$$\mathbf{m}_i(\mathbf{r}, t) = \mathbf{m}_{ik}(\xi)\ e^{-(ik_\zeta \zeta - i\omega t)} \quad , \tag{6}$$

$$\mathbf{h}_d(\mathbf{r},t) = \mathbf{h}_{dk}(\xi)\ e^{-(ik_\zeta\zeta - i\omega t)} \quad . \tag{7}$$

Here and below we assume that the in-plane wave vector k_ζ is always positive. Due to the linearity of the problem the Fourier-amplitude $\mathbf{h}_{dk}(\xi)$ of variable dipole field Eq.(7) can be represented as the superposition of all Fourier-components of the dipole fields of all ferromagnetic films formed the layered structure

$$\mathbf{h}_{dk}(\xi) = \sum_{i=1}^{N} \mathbf{h}_{dki}(\xi) \quad . \tag{8}$$

This is the one of central ideas of our approach. Physically it reflects dipole-dipole interaction in a spin-system of a film as well as coupling between ferromagnetic films in a structure through their "individual" dipole fields.

To define the dipole field of each ferromagnetic film, we make use of the tensorial Green function $\hat{\mathbf{G}}$ (see e.g[28]):

$$\mathbf{h}_{dki}(\xi, k_\zeta) = \int_{-\infty}^{\infty} \hat{\mathbf{G}}_{\xi\eta\zeta}(\xi,\xi';k_\zeta)\mathbf{m}_{ik}(\xi',k_\zeta)d\xi' \quad . \tag{9}$$

The Green function in Eq.(9), being defined in the coordinate system $\xi\eta\zeta$, is equal to

$$\hat{\mathbf{G}}_{\xi\eta\zeta}(\xi,\xi';k_\zeta) = \begin{pmatrix} G^P - \delta(\xi-\xi') & 0 & iG^Q \\ 0 & 0 & 0 \\ iG^Q & 0 & -G^P \end{pmatrix} \quad , \tag{10}$$

where the components are

$$G^P = \frac{k_\zeta}{2}\ e^{-k_\zeta|\xi-\xi'|} \quad , G^Q = G^P\ \mathrm{sgn}(\xi-\xi') \quad . \tag{11}$$

As the second step, integral form for the dipole field (9) is substituted in the system of linearized Landau-Lifshitz equations of motion for magnetizations followed from (4). It is convenient to write each equation of motion in the coordinate system $x_iy_iz_i$ where due to the linearization the vector of variable magnetization can be considered two-dimensional

$$\mathbf{m}_{ki}(\xi) = \mathbf{e}_{xi}m_x(\xi) + \mathbf{e}_{yi}m_y(\xi) = \begin{pmatrix} m_x \\ m_y \end{pmatrix} \quad . \tag{12}$$

These calculations lead to the system of N coupled homogeneous integro-differential equations for the Fourier amplitudes of variable magnetizations

$$\hat{\mathbf{F}}_i \mathbf{m}_{ik}(\xi, k_\zeta) = \int_{-\infty}^{\infty} \hat{\mathbf{G}}_{\xi\eta\zeta}(\xi, \xi'; k_\zeta)\mathbf{m}_{ik}(\xi', k_\zeta)d\xi' \quad , \tag{13}$$

where $\hat{\mathbf{F}}_i$ is the linear matrix-differential operator

$$\hat{\mathbf{F}}_i = \begin{pmatrix} \frac{\omega_{Hi}}{\omega_{Mi}} - \alpha_i(\nabla_\xi^2 - k_\zeta^2) & -i\frac{\omega}{\omega_{Mi}} \\ i\frac{\omega}{\omega_{Mi}} & \frac{\omega_{Hi}}{\omega_{Mi}} - \alpha_i(\nabla_\xi^2 - k_\zeta^2) \end{pmatrix} \tag{14}$$

with $\omega_{Hi} = |g|\mu_0 H_i$, $\omega_{Mi} = |g|\mu_0 M_{0i}$.

2.2 General dispersion relation

To solve the set of integro-differential equations (13) we apply the expansion of the Fourier amplitudes of variable magnetizations $\mathbf{m}_{ik}$ in an infinite series of complete orthogonal vector-functions - spin-wave modes (SWM) $\mathbf{S}_{in}^p(\xi)$:

$$\mathbf{m}_{ik}(\xi) = M_{0i} \sum_n^{\infty} \sum_{p=1}^{2} m_{in}^p \mathbf{S}_{in}^p(\xi) \quad , \tag{15}$$

where the m_{in}^p are spin-wave mode amplitudes, and $\mathbf{S}_{in}^1(\xi) = \mathbf{S}_{in}^2(\xi)^* = (\mathbf{e}_{xi} + i\mathbf{e}_{yi})\Phi_{in}(\xi)$. The spin-wave modes $\mathbf{S}_{in}^p(\xi)$ are found as the eigenfunctions of the differential-matrix operator $\hat{\mathbf{F}}_i$ with appropriate exchange boundary conditions (surface spin pinning conditions). The form of the functions $\Phi_{in}(\xi)$ is determined by the exchange boundary conditions. These functions are equal to

$$\Phi_{in}(\xi) = \cos \kappa_{in}(\xi - \sum_{j=1}^{i-1}(L_j + a_{j-1}))(1 + \delta_{0n})^{-\frac{1}{2}} \tag{16}$$

in the case of unpinned surface spins ($n = 0, 1, 2, \ldots$), and

$$\Phi_{in}(\xi) = \sin \kappa_{in}(\xi - \sum_{j=1}^{i-1}(L_j + a_{j-1})) \tag{17}$$

in the case of pinned surface spins ($n = 1, 2, 3 \ldots$), where $\kappa_{in} = \pi n / L_i$, $a_0 \equiv 0$. It is worth noting that in the case of perpendicularly magnetized ferromagnetic film the functions $\Phi_{in}(\xi)$ give transverse distributions of variable magnetization of standing spin waves. In other words, these functions describe the distributitions of the magnetization across a film thickness in the regime of spin wave resonance. Substituting expansion Eq.(15) into Eq.(13) and using the orthogonality condition for SWM we obtain an infinite system of coupled linear algebraic equations for SWM amplitudes m^p_{in}

$$\frac{\Omega_{in'}+\omega}{\omega_{Mi}} m^1_{in'} + (m^1_{in'} + m^2_{in'})\frac{\cos^2\theta_i}{2} = \frac{1}{2}\sum_{j=1}^{N}\frac{\omega_{Mj}}{\omega_{Mi}}\sum_n\Big\{m^1_{jn}(A_j P^{ij}_{nn'} + B_j Q^{ij}_{nn'}) + m^2_{jn}((E_j - C_j)P^{ij}_{nn'} + (B_j - D_j)Q^{ij}_{nn'})\Big\} \quad , \tag{18}$$

$$\frac{\Omega_{in'}-\omega}{\omega_{Mi}} m^2_{in'} + (m^1_{in'} + m^2_{in'})\frac{\cos^2\theta_i}{2} = \frac{1}{2}\sum_{j=1}^{N}\frac{\omega_{Mj}}{\omega_{Mi}}\sum_n\Big\{m^1_{jn}((E_j + C_j)P^{ij}_{nn'} + (B_j + D_j)Q^{ij}_{nn'}) + m^2_{jn}(A_j P^{ij}_{nn'} + B_j Q^{ij}_{nn'})\Big\} \quad , \tag{19}$$

where $\Omega_{in} = \omega_{Hi} + \alpha_i\omega_{Mi}(\kappa^2_{in} + k^2_\zeta)$,

$$A_i = \cos^2\theta_i - \sin^2\theta_i\cos^2\varphi - \sin^2\varphi \quad ,$$

$$B_i = i\sin 2\theta_i\cos\varphi \quad ,$$

$$C_i = i2\sin\varphi\cos\varphi\sin\theta_i \quad ,$$

$$D_i = 2\cos\theta_i\sin\varphi \quad ,$$

$$E_i = \cos^2\theta_i - \sin^2\theta_i\cos^2\varphi + \sin^2\varphi \quad .$$

Matrix elements $P^{ij}_{nn'}$ and $Q^{ij}_{nn'}$ representing dipole interaction between SWM are given by:

$$P^{ij}_{nn'} = \frac{2}{L_i}\int_{-\infty}^{\infty}\int_{-\infty}^{\infty} G^P(\xi,\xi')\Phi_{in}(\xi)\Phi_{jn'}(\xi')d\xi d\xi' \quad , \tag{20}$$

$$Q^{ij}_{nn'} = \frac{2}{L_i}\int_{-\infty}^{\infty}\int_{-\infty}^{\infty} G^Q(\xi,\xi')\Phi_{in}(\xi)\Phi_{jn'}(\xi')d\xi d\xi' \quad . \tag{21}$$

Coupled algebraic systems of equations for SWM amplitudes (18),(19) describe exactly dipole-exchange wave process in the layered structure since they were derived with simultaneous regard for both the dipole-dipole interaction (inside each film and between all films) and exchange interaction inside each film as well as electrodynamic and exchange boundary conditions.

It should be noted that when the films formed the structure are moved apart (in this case $P^{ij}_{nn'}$, and $Q^{ij}_{nn'}$, tend to zero for $i \neq j$), the system of Eqs.(18),(19) splits into N uncoupled systems of equations describing the wave proceses in the single films. Such systems for a free ferromagnetic film were obtained in[27].

We now come to the determination of the dispersion characteristics for spin waves in the layered ferrite-dielectric film structure. The condition of the vanishing of the determinant of the infinite systems Eqs.(18),(19) yields the exact dispersion equation for propagating dipole-exchange spin waves. So the problem of determination of the exact dispersion law $\omega(k_\zeta)$ for SW in the layered structure is reduced to the problem of calculating eigenvalues of the block matrix of the infinite system Eqs.(18),(19). The calculations of eigenvalues of such a matrix could be done, for example, by means of the well known QR-algorithm (see, e.g[32]).

A discussion is now given of some features of spin-wave spectrum formation. We consider first the situation of double film structure. The process of formation of the spin-wave spectrum for such structure can be described as follows: the rapprochement of two ferromagnetic films leads to the dipole interaction (coupling) between them and, as a result, to the splitting of initial spin-wave branches corresponding to the separate films. To illustrate this process let us first qualitatively consider interaction of two identical ferromagnetic films with the multimode spectrum. Two typical situations are possible: first, when the free ferromagnetic films have spectra described by noncrossing curves (see Fig.2 (a)); second, when hybridization of the initial dispersion branches of free ferromagnetic films occurs (see Fig.2 (c)). The dipole interaction between ferromagnetic films with any type of initial spectrum results in the reconstruction of the spectrum: each of the dispersion branches splits into two branches, one of which corresponds to the symmetric distribution of the variable magnetization over the thickness of the layered structure $\mathbf{m}(\xi)$ and another corresponds to the antisymmetric distribution $\mathbf{m}(\xi)$ (see Fig.3).

Also it should be noted that the dipole interaction between ferromagnetic films depends strongly on the spin-wave wavevector value. Namely, when k_ζ tends to zero the dipole interaction vanishes. It means that the origins of spin-wave dispersion

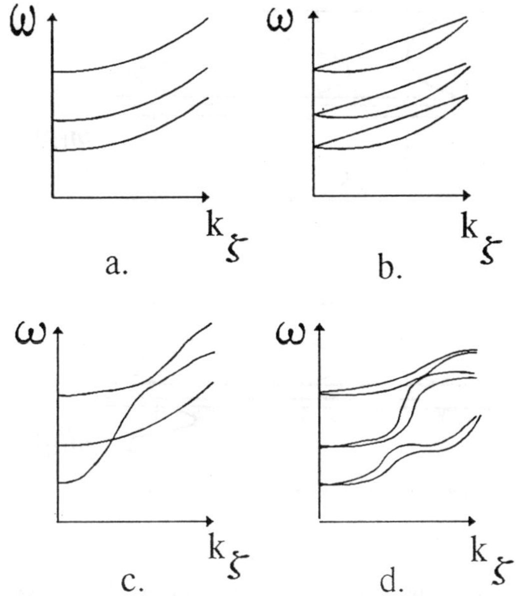

Figure 2: Spin-wave spectrum formation process for double-layered ferromagnetic film structure.

branches of the layered film structure are spin-wave resonance frequencies of the separate films formed the structure [1]. For the case of large k_ζ ($k_\zeta \geq 10^5$ cm^{-1}) the dipole field is localized near the film surfaces and, therefore, has small effect on the neighboring films. The region of the strong dipole interaction for the case of identical ferromagnetic films lies in the range of wavevectors where $k_\zeta L_i = 0.5$.

[1]The frequencies of spin-wave resonances of layered structure are also determined by the exchange coupling of ferromagnetic films. The theoretical treatment of the effect of the exchange coupling on spin-wave spectrum formation which could be easeally incorporated in the present theory is given in[29]. In particular, in the paper[29] it is shown that for YIG films (which is the main material in microwave applications) the exchange coupling is significant for the distances less than 0.01 μm. The liquid-phase epitaxial garnet films which are usually used for microwave purposes have the thicknesses of imperfect surface layers about the 0.005 mm. So it is clear that for the layered structures consisting of such films the interlayer exchange coupling is negligible.

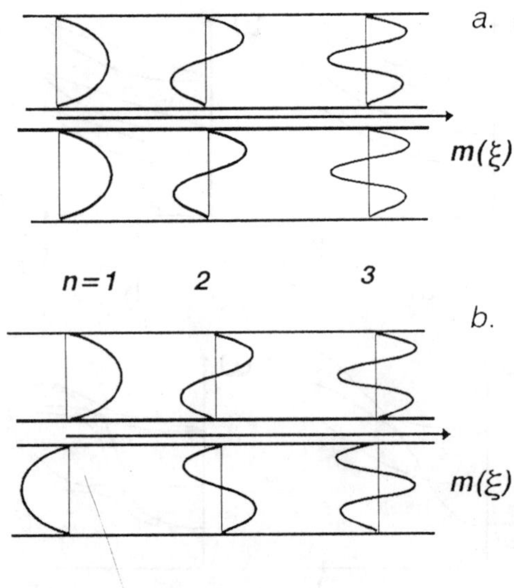

Figure 3: Transverse distribution of the variable magnetization across the structure thickness: a - symmetric, b - antisymmetric

It is useful to consider also the structure consisting of two ferromagnetic films with different parameters. Here also two principal situations are possible. First, the initial dispersion branches of ferromagnetic films formed layered structure do not cross each other. In this case the dipole interaction between ferromagnetic films causes the variation of the dispersion branches shape. Second, the initial spectra of ferromagnetic films formed layered structure cross each other. In this case the dipole interaction in the crosspoints causes the repulsion of spin-wave dispersion branches. This repulsion leads to the appearance of dipole "gaps" in the spin-wave spectrum. The width of these gaps depends on the parameters of the structure and the symmetry of interacting spin- wave modes. Below we shall illustrate this guantitative consideration by rigorous calculations.

The process of the spectrum formation of three-layered ferromagnetic film structure may be considered as a result of interaction of a single film with double-layered structure, etc.

3 Spin-wave spectrum for perpendicularly magnetized double-layered ferrite-dielectric structure

In the previous sections we have introduced our approach to calculate the dipole-exchange spin-wave spectrum of the multilayered ferrite-dielectric structure. We now present the detailed calculations for some important cases. We consider first the double-layered structure ($N = 2$) magnetized perpendicularly to its surface ($\theta_0 = \theta_1 = \theta_2 = \frac{\pi}{2}$). In this case the system of Eqs.(18),(19) becomes

$$(\omega + \Omega_{in'})m^1_{in'} + \frac{1}{2}\sum_{j=1}^{2}\omega_{Mj}\sum_n(m^1_{jn} - m^2_{jn})P^{ij}_{nn'} = 0 \quad , \tag{22}$$

$$(\omega - \Omega_{in'})m^2_{in'} + \frac{1}{2}\sum_{j=1}^{2}\omega_{Mj}\sum_n(m^1_{jn} - m^2_{jn})P^{ij}_{nn'} = 0 \quad . \tag{23}$$

To illustrate the main features of spin-wave spectrum we choose for our calculations the structure with the following parameters: $M_{01} = 1700$ G, $M_{02} = 1600$ G, $L_1 = 10\ \mu$m, $L_2 = 1\ \mu$m, $\alpha_1 = \alpha_2 = 3 \times 10^{-12}$ cm^2, $a = 0.001\ \mu$m. Fig.4 presents the results of numerical calculations obtained with a help of QR-algorithm. The chosen set of parameters provides the situation when the spin wave dispersion branches of the initial ferromagnetic films formed the structure cross each other. In particular, the dispersion branches of thin ($L_2 = 1\ \mu$m) film cross the dispersion branch of the fundamental mode of thick ($L_1 = 10\ \mu$m) film. (Note that this situation has been considered qualitatively in the Section 2.2.) The dipole interaction between crossing spin-wave branches causes their repultion. As a result the spectrum with the dipole gaps appeares. The appearance of these gaps leads to the oscillations on the group velocity versus frequency dependence of spin waves and, as a result, to the notches on the amplitude characteristics of spin wave devices. The width of these gaps depends on the film parameters and on the numbers (types of symmetry) of interacting spin-wave dispersion branches. It can be seen from Fig. 4(a) that in the crosspoints of the dispersion branches of different types of symmetry the dipole interaction is weak (the width of dipole gaps is less than 1 MHz and it cannot be shown in the figure).

Another important aspect is concerned with the influence of exchange interaction on spin-wave spectrum formation. Let us consider the same double-layered

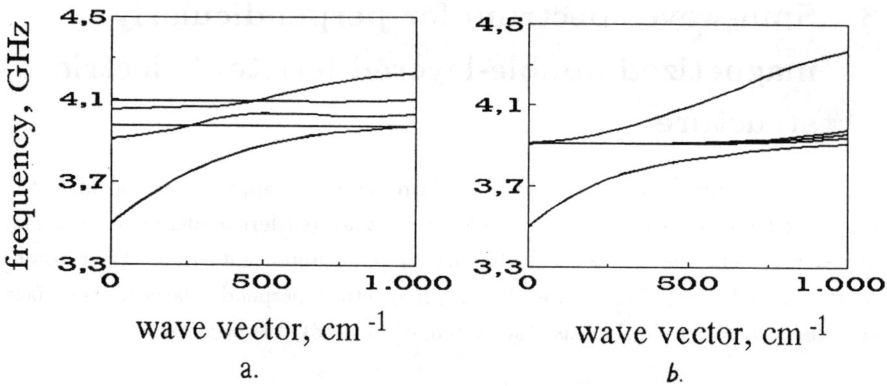

Figure 4: Spin-wave spectra for perpendicularly magnetized double-layered structure with parameters: $L_1 = 10\ \mu m$, $L_2 = 1\ \mu m$, $4\pi M_{01} = 1700$ G, $4\pi M_{02} = 1600$ G; a - $\alpha_1 = \alpha_2 = 3 \times 10^{12}\text{cm}^2$, b - $\alpha_1 = \alpha_2 = 0$.

structure but in non-exchange approximation. Such approximation are widely used now (see e.g.[1,4,14]). Fig.4(b) presents the results of calculations in the non-exchange limit $\alpha_1 = \alpha_2 = 0$. Comparison of Fig.4(a) and Fig.4(b) shows that exchange interaction is very significant for spin-wave spectrum formation process of the layered structure.

The above presented numerical results were obtained with the help of *QR*-algorithm. In other words, for numerical calculations of spin wave spectra the dispersion relations in the form of the vanishing determinant of the infinitive system of Eqs.(22),(23) were used.

In the frames of the suggested theory it is possible to receive the dispersion relations in the form of an infinite series also. Such dispersion relations are:

$$D_{1o}(k_\zeta,\omega)D_{1e}(k_\zeta,\omega)D_{2o}(k_\zeta,\omega)D_{2e}(k_\zeta,\omega) = B_1(k_\zeta,\omega)B_2(k_\zeta,\omega) \quad , \tag{24}$$

where

$$D_{io}(k_\zeta,\omega) = 1 - \frac{k_\zeta}{L_i}\sum_{n=2k}^{\infty}\frac{y_{in}}{d_{in}} \quad , \tag{25}$$

$$D_{ie}(k_\zeta,\omega) = 1 - \frac{k_\zeta}{L_i} \sum_{n=2k+1}^{\infty} \frac{y_{in}}{d_{in}} \quad , \tag{26}$$

$$B_i(k_\zeta,\omega) = \frac{e^{-k_\zeta a}}{L_i}\Big(\big(1 - e^{-k_\zeta L_i}\big) D_{io}(k_\zeta,\omega) \sum_{n=2k}^{\infty} y_{in} d_{in}$$
$$+\big(\big(1 + e^{-k_\zeta L_i}\big) D_{ie}(k_\zeta,\omega) \sum_{n=2k+1}^{\infty} y_{in} d_{in}\Big) \quad , \tag{27}$$

k=0,1,2,3,...

$$d_{in} = \frac{k_{in}^2(\omega + \Omega_{in}) - \frac{k_\zeta^2 \omega_{Mi} \Omega_{in}}{\omega + \Omega_{in}}}{\kappa_{in}\big(1 - (-1)^n e^{-k_\zeta L_i}\big)\omega_{Mi}} \quad , \tag{28}$$

$$y_{in} = \left(1 - \frac{\omega + \Omega_{in}}{\omega - \Omega_{in}}\right)\frac{1}{k_{in}^2} \times \begin{cases} \kappa_{in} & \text{–for pinned surface spins} \\ k_\zeta(1+\delta_{0n})^{-1/2} & \text{–for pinned surface spins} \end{cases} \quad . \tag{29}$$

We now come to the physical role of the terms in Eq.(24). The conditions $D_{io}(k_\zeta,\omega) = 0$ and $D_{ie}(k_\zeta,\omega) = 0$ give the dispersion relations accordingly for odd and even spin waves in single ferromagnetic films[33]. The physical nature of the terms B_i $(i = 1,2)$ is that they describe the dipole interaction between ferromagnetic films.

For the case of pure dipole waves ($\alpha_1 = \alpha_2 = 0$) it is possible to fulfill summation of the infinite series in the system of Eqs.(26),(27). After summation in (25) and (26) we obtain

$$D_{ie} = \frac{1}{2}\big(1 - e^{-2k_\zeta L_i}\big)\big(1 + \beta_i \ \mathrm{ctg}(L_i \beta_i k_\zeta/2)\big) \quad , \tag{30}$$

$$D_{io} = \frac{1}{2}\big(1 + e^{-2k_\zeta L_i}\big)\big(1 - \beta_i \ \mathrm{tg}(L_i \beta_i k_\zeta/2)\big) \quad , \tag{31}$$

where

$$\beta_i = \left(\frac{\omega_{Mi}(\omega_{Hi} + \omega_{Mi}) - \omega^2}{\omega^2 - \omega_{Hi}^2}\right)^{\frac{1}{2}} \quad . \tag{32}$$

It is worth noting that the last expression may be rewritten as

$$\beta_i = \sqrt{-\mu_i} \tag{33}$$

where μ is the diagonal component of the usual magnetic permeability tensor (see e.g.[35]).

We now make some digressions. According to the above given explanations, the equations $D_{io}(k_\zeta,\omega) = 0$ and $D_{ie}(k_\zeta,\omega) = 0$ determine the dispersion relations for symmetrical and antisymmetrical spin waves in single ferromagnetic films. From the Eqs. (30) and (31) follows

$$\mathrm{ctg}(L_i\beta_i k_\zeta/2) = -\frac{1}{\beta_i} \tag{34}$$

and

$$\mathrm{tg}(L_i\beta_i k_\zeta/2) = \frac{1}{\beta_i} \quad . \tag{35}$$

The last two equations coincide exactly with the dispersion relations obtained in[34] by magnetostatic potential method for symmetrical and antisymmetrical modes. By combining of these two dispersion relations for one and the same film we obtain the relation describing the full spin-wave spectrum of the ferromagnetic film (see, e.g. Eq. (4.75a) in[35]):

$$\mathrm{tg}(L_i\beta_i k_\zeta/2) = 2\frac{\beta_i}{\beta_i^2 - 1} \quad . \tag{36}$$

Returning to the double-layered structure we obtain the dispersion relation for non-exchange (purely dipole) spin waves

$$\begin{aligned}\left[1 - \beta_1^2 + 2\beta_1 \mathrm{ctg}(L_1\beta_1 k_\zeta)\right]\left[1 - \beta_2^2 + 2\beta_2 \mathrm{ctg}(L_2\beta_2 k_\zeta)\right] \\ = e^{-2k_\zeta a}(\beta_1^2 + 1)(\beta_2^2 + 1) \quad .\end{aligned} \tag{37}$$

The terms in the left side of Eq.(37) represent the product of dispersion relations for single films. The terms in the right side of Eg.(37) reflect the dipole interaction between ferromagnetic films. We underline that the last equation coincides in the case of $\alpha = 0$ with the dispersion relation obtained by the magnetic potential method[35].

It is worth nothing that one and the same final trigonometric dispersion equation for pure dipole waves (37) is received from both non-exchange dispersion relations originally obtained for different surface spin pinning conditions (see Eq.(29)). It means that exchange boundary conditions which cause the mathematical difference between dispersion equations (30) and (31) even in non-exchange limit $\alpha_1 = \alpha_2 = 0$, do not really influence on the dispersion properties of purely dipole waves (usually called magnetostatic waves[35]).

3.1 Approximate dispersion relation

Derived in the previous section Eqs.(22), (23) as well as non-exchange relation Eq.(37) have an implicit forms of the dispersion low $\omega_n(k_\zeta)$. An approximate, but quite accurate, excplicit dependence of the frequency of the normal wave on its wave number is of interest for practical applications. To obtain such a dependence we apply the perturbation theory method[37] for analysis of the systems of Eqs.(22),(23). For zero-order approximation we choose the situation when the transverse distribution of variable magnetization is described only by one set of SWM

$$\mathbf{m}_i(\xi) = M_{0i}\left[m^1_{in}\mathbf{S}^1_{in}(\xi) + \mathbf{m}^2_{in}\mathbf{S}^2_{in}(\xi)\right] \quad . \tag{38}$$

After calculations we find the fist order approximate dispersion relation in the form

$$(\omega^2_{1n} - \omega^2)(\omega^2_{2m} - \omega^2) - \omega_{M1}\omega_{M2}\Omega_{1n}\Omega_{2m}P^{12}_{nm}P^{21}_{mn} = 0 \quad . \tag{39}$$

In this dispersion relation the subscripts n and m designate the numbers of interacting (due to dipole energy) dispersion branches. These dispersion branches $\omega_{in}(k_\zeta)$ are given by the approximate relation for a single film[27]:

$$\omega^2_{in} = \Omega_{in}(\Omega_{in} + \omega_{Mi}P^{ii}_{nn}) \quad . \tag{40}$$

The equation (39) has an explicit form which enables one to find easily the dependencies $\omega_j(k_\zeta)$.

In obtaining above form of dispersion relation we have assumed that the distribution of variable magnetization across each film is described only by one set of spin wave modes. We now discuss the limits of applicability of such approximation. These limits are determined by the infinitesimal condition for perturbation ($\varepsilon \ll 1$). The expression for the perturbation parameter, being a function of the frequency, the wave vector and the structure parameters, has the form

$$\begin{aligned}\varepsilon_{1,2} \;=\; & \left\{ \frac{(\omega+\Omega_{2m})P^{11}_{nm}}{(\omega+\Omega_{1n'})P^{21}_{mn}} \left(\frac{\omega^2_{2m}-\omega^2}{(\omega+\Omega_{2m})\omega_{M2}-\lambda} \right) \frac{\Omega_{1n'}}{\Omega_{2m}} \right. \\ & \times \left(\frac{\omega^2_{2m'}-\omega^2}{(\omega+\Omega_{2m'})\omega_{M2}-\lambda} \right) - \frac{\Omega_{1m'}P^{12}_{n'm'}}{\omega+\Omega_{1n'}} \left(\frac{\omega^2_{2m'}-\omega^2}{(\omega+\Omega_{2m})\omega_{M2}-\lambda} \right)\end{aligned}$$

$$- \frac{\Omega_{1n'}\Omega_{2m'}P^{12}_{n'm'}P^{21}_{m'n}}{(\omega+\Omega_{1n'})(\omega+\Omega_{2m})P^{12}_{nm}}\left(\frac{\omega^2_{2m}-\omega^2}{(\omega+\Omega_{2m})\omega_{M2}-\lambda}\right)$$
$$+ \left.\frac{\Omega_{1n'}\Omega_{2m'}P^{12}_{n'm'}P^{22}_{mm'}}{(\omega+\Omega_{1n'})(\omega+\Omega_{2m'})}\right\}\frac{\omega_{M2}}{\omega_{M1}}\Big[(\lambda-\lambda_1)(\lambda-\lambda_2)\Big]^{-1} , \qquad (41)$$

where the value λ depends on the numbers n and m and is determined by the relation

$$\lambda_{1,2} = \frac{1}{2}\left\{\frac{\omega^2_{1n}-\omega^2}{(\omega+\Omega_{1n})\omega_{M1}} - \frac{\omega^2_{2m}-\omega^2}{(\omega+\Omega_{2n})\omega_{M2}}\right.$$
$$\left.\pm \left[\left(\frac{\omega^2_{1n}-\omega^2}{(\omega+\Omega_{1n})\omega_{M1}} - \frac{\omega^2_{2m}-\omega^2}{(\omega+\Omega_{2n})\omega_{M2}}\right)^2 + \frac{\omega_{M1}}{\omega_{M2}}\frac{\Omega_{1n}\Omega_{2m}P^{12}_{nm}P^{21}_{mn}}{(\omega+\Omega_{1n})(\omega+\Omega_{2m})}\right]^{\frac{1}{2}}\right\} \qquad (42)$$

Condition $\lambda_{1,2} = 0$ gives the dispersion relation (39) which describes the shapes of two respective dispersion branches. The quantities λ_1 and λ_2 are found as the roots of the polynomial determinant of Eq.(41). They are given by Eq.(42) if the changes of subscripts n to n' and m to m' are made in the latter.

Numerical estimations show that the condition $\varepsilon \ll 1$ is well satisfied for low wave modes (except for the frequency regions situated near the cross points of the dispersion branches). The frequency regions where the condition $\varepsilon \ll 1$ breaks down have values of a few megahertz. As the numbers n and m increase, the frequency regions where the condition $\varepsilon \ll 1$ is not satisfied grow wider.

Now we turn to the comparison between the exact dispersion law computed with the help of QR-algorithm and the approximate dispersion law given by the Eq.(39). For these calculations we choose the structure with the type of spectrum considered in Fig.4 According to the definition given earlier, the accuracy of the approximate dispersion relation Eq.(39) for this case is the worst. The curves $\omega_j(k_\zeta)$ presented in Fig.5 show that even in this case the approximate dispersion relation still has good accuracy.

4 Spin-wave spectrum for double-layered ferrite-dielectric structure magnetized at an arbitrary direction

Consider first a tangentially magnetized double-layered structure. In contrast to a perpendicularly magnetized structure discussed in the Section 3 here we have

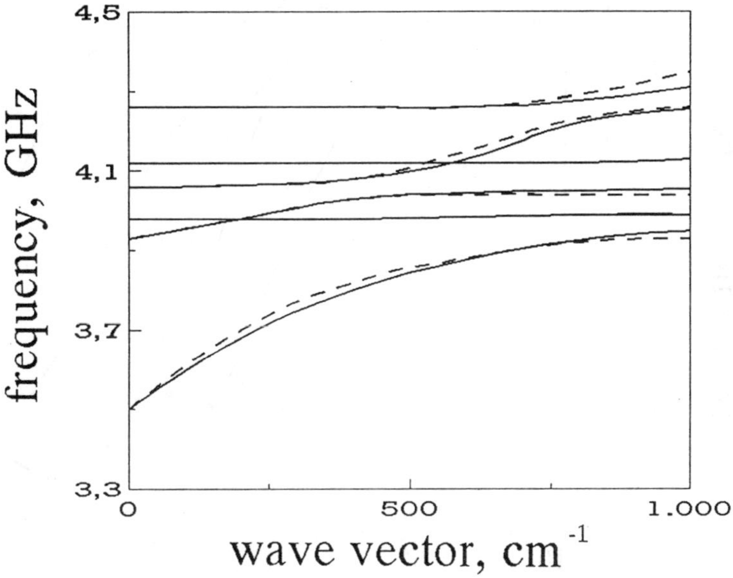

Figure 5: Comparison of exact (dashed lines) and approximate (solid lines) dispersion laws of double-layered structure with parameters: $L_1 = 10\ \mu m$, $L_2 = 1\ \mu m$, $4\pi M_{01} = 1700$ G, $4\pi M_{02} = 1600$ G.

the distinguished direction in the structure plane. It is the direction of bias magnetic field. The propagation characteristics for waves in a tangentially magnetized structure depend on the angle φ between the directions of in-plane wave vector and bias magnetic field.

We start from the case when spin-wave propagation direction coincides with the bias field direction. In this case spin-wave spectra of initial ferromagnetic films formed the structure are described by noncrossing dispersion branches for any set of film parameters. Let us consider the structure consisting of two ferromagnetic films having the initial spectra which cross each other. Because of the dipole coupling between ferromagnetic films the repulsion in spin-wave spectrum takes place (see Fig.6). Such process is similar to that in perpendicularly magnetized structure. In

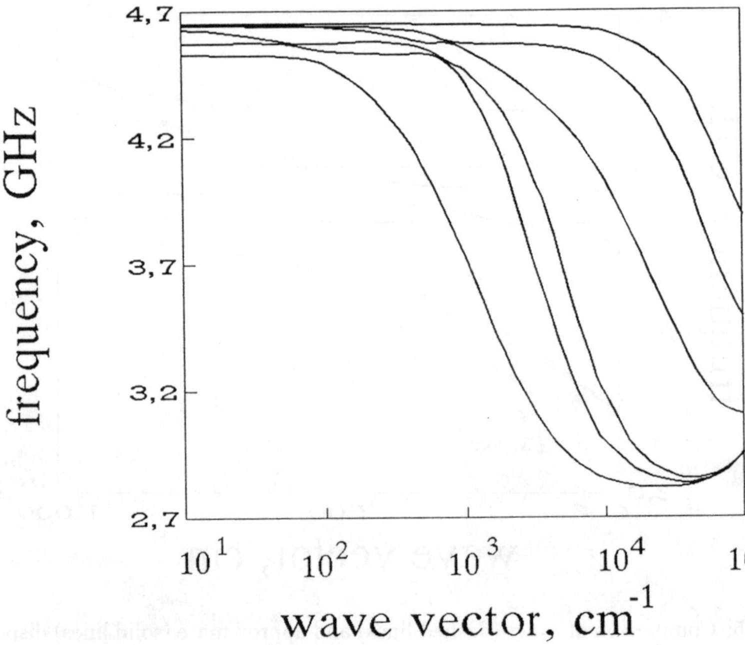

Figure 6: Spin-wave spectrum for tangentially magnetized double-layered ferromagnetic structure with parameters: $L_1 = 20\ \mu m$, $L_2 = 1\ \mu m$, $a = 0.1 \mu m$, $4\pi M_{01} = 1750$ G, $4\pi M_{02} = 1600$ G, $\theta_1 = \theta_2 = 0$, $\varphi = 0$.

propagation experiments such repulsion leads to the appearance of the notches on the amplitude frequency characteristics of the structures which is due to decrease of spin wave group velocity. It should be noted that for the initial films such amplitude frequency characteristics are always smooth.

Now we turn to the case of the transverse wave propagation relative to the direction of bias magnetic field ($\mathbf{k}_\zeta \perp \mathbf{H}_i$). In this case fundamental modes of the initial films have a quasi-surface nature. So it is clear that in this case there will be the dependence of the spin wave dispersion characteristics on the propagation direction ($\varphi = 90^o$ or $\varphi = 270^o$). Such strucrures were first investigated theoretically for pure dipole waves[4,12]. It was found that for the case $a_1 = 0$ and $4\pi M_{01} \neq 4\pi M_{02}$ spin-wave mode with negative dispersion appears. For the case $a \neq 0$ this phenomena depends

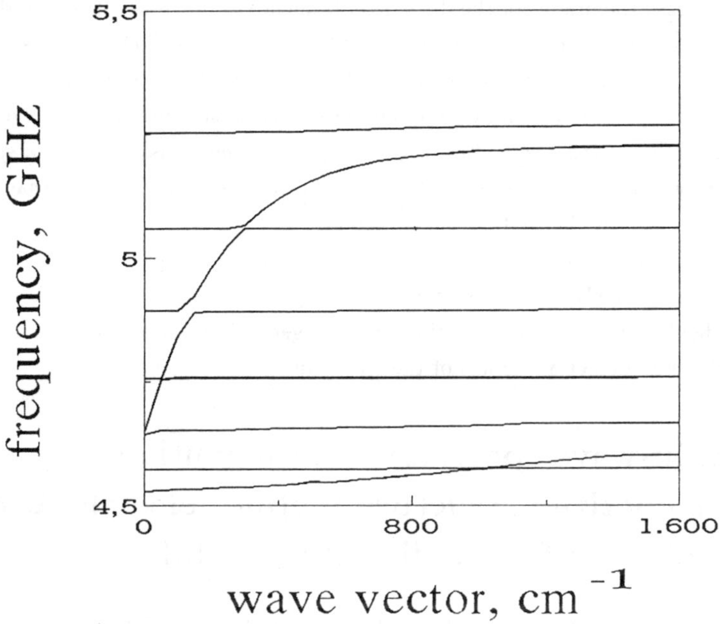

Figure 7: Spin-wave spectrum for tangentially magnetized double-layered ferromagnetic structure with parameters: $L_1 = 20\ \mu m$, $L_2 = 1\ \mu m$, $a = 0.1 \mu m$, $4\pi M_{01} = 1750$ G, $4\pi M_{02} = 1600$ G, $\theta_1 = \theta_2 = 0$, $\varphi = 90^o$.

on the value of saturation magnetisations of ferromagnetic films, film thicknesses and exchange boundary conditions (surface spin pinning conditions). Fig.7 presents spin-wave spectrum for the structure calculated with a help of Eqs.(18),(19). It is seen from the figure that there are no spin-wave modes with negative dispersion.

Let us now discuss briefly the spin-wave spectrum transformation due to the change of external magnetic field direction. It should be stressed that the behavior of the dispersion branches is rather complicated and depends strongly on the particular choice of film parameters. So we shall mention here only some features. When the structure is rotated (when the value of θ_0 is changed) the origins of the dispersion branches are moving along the frequency axis (owing to the variation of the demagnetization factors). Let us consider for certaincy the rotation when $\varphi = 90^o$ (θ_0 - any)

In this case the origins of the dispersion branches of different films moved to each other due to the different demagnetization fields of the films. So there are such values of θ_0 that the origins of the different modes coincide (these values of θ_0 depend on the structure parameters and on the numbers of dispersion branches). The in-plane rotation of the external field ($\theta_0 = 0, \varphi$ - any) causes the transformation of spin-wave spectrum which is shown in Fig.6 into that shown in Fig.7. It is clear that the sign of the dispersion is changed during this rotation.

So as we have considered the rotations in two planes separately. It is possible to imagine qualitatively the reconstruction of the spectrum due to the arbitrary variation of bias field. But these considerations are too general and it is necessary to provide calculations to detect all details of the spectrum.

5 Spin-wave spectrum for tangentially magnetized structure metal-dielectric-ferrite-dielectric-ferrite-dielectric-metal

Previous sections were devoted to the study of free structures: structures without metal screens. The aim of this section is to apply the above described theory to the case of screened layered structures.

Consider a structure consisting of two in-plane magnetized ferromagnetic films ($N = 2$, θ_0 =0). The upper and lower metal screens are separated from ferromagnetic films by dielectric layers of thickness d and b. Applying the tensorial Green function method we obtain the following system for SWM amplitudes

$$\begin{aligned}\frac{\Omega_{in'}+\omega}{\omega_{Mi}}m^1_{in'} \;&+\; \frac{1}{2}(m^1_{in'}+m^2_{in'}) \\ &= \frac{1}{2}\sum_{j=1}^{N}\frac{\omega_{Mj}}{\omega_{Mi}}\sum_n\Big[m^1_{in}(Y^{xx}_{ijn'n}-Y^{xy}_{ijn'n}+Y^{yx}_{ijn'n}-Y^{yy}_{ijn'n}) \\ &+\; m^1_{in}(Y^{xx}_{ijn'n}+Y^{xy}_{ijn'n}+Y^{yx}_{ijn'n}+Y^{yy}_{ijn'n})\Big]\quad, \qquad (43)\end{aligned}$$

$$\frac{\Omega_{in'}-\omega}{\omega_{Mi}}m^2_{in'} \;+\; \frac{1}{2}(m^1_{in'}+m^2_{in'})$$

$$= \frac{1}{2}\sum_{j=1}^{N}\frac{\omega_{Mj}}{\omega_{Mi}}\sum_{n}\Big[m_{in}^{1}\big(Y_{ijn'n}^{xx} - Y_{ijn'n}^{xy} - Y_{ijn'n}^{yx} - Y_{ijn'n}^{yy}\big)$$

$$+ \quad m_{in}^{2}\big(Y_{ijn'n}^{xx} + Y_{ijn'n}^{xy} - Y_{ijn'n}^{yx} - Y_{ijn'n}^{yy}\big)\Big] \quad . \tag{44}$$

Here dipole matrix elements $Y_{ijn'n}^{pq}$ are equal to

$$Y_{ijn'n}^{pq} = \frac{2}{L_i}\int_{-\infty}^{\infty}\int_{-\infty}^{\infty} G_i^{pq}(k_\zeta;\xi,\xi')\Phi_{in}(\xi)\Phi_{in'}(\xi')d\xi d\xi' \quad , \tag{45}$$

where p and q represent coordinate indexes $(p, q = x, y)$. The components of the Green function for the screened structure in coordinate system $\xi\eta\zeta$ are:

$$G_i^{\xi\xi} = -\delta(\xi-\xi') - R\begin{cases} \text{sh } k_\zeta(\xi_{di}-\xi') \text{ sh } k_\zeta(\xi_{bi}+\xi) & ,\text{for}\xi' < \xi \\ \text{sh } k_\zeta(\xi_{bi}+\xi') \text{ sh } k_\zeta(\xi_{di}-\xi) & ,\text{for}\xi' > \xi \end{cases} ,$$

$$G_i^{\xi\zeta} = iR\begin{cases} \text{sh } k_\zeta(\xi_{di}-\xi') \text{ ch } k_\zeta(\xi_{bi}+\xi) & ,\text{for}\xi' < \xi \\ \text{sh } k_\zeta(\xi_{bi}+\xi') \text{ ch } k_\zeta(\xi_{di}-\xi) & ,\text{for}\xi' > \xi \end{cases} ,$$

$$G_i^{\zeta\xi} = iR\begin{cases} \text{ch } k_\zeta(\xi_{di}-\xi') \text{ sh } k_\zeta(\xi_{bi}+\xi) & ,\text{for}\xi' < \xi \\ \text{ch } k_\zeta(\xi_{bi}+\xi') \text{ sh } k_\zeta(\xi_{di}-\xi) & ,\text{for}\xi' > \xi \end{cases} ,$$

$$G_i^{\zeta\zeta} = -R\begin{cases} \text{ch } k_\zeta(\xi_{di}-\xi') \text{ ch } k_\zeta(\xi_{bi}+\xi) & ,\text{for}\xi' < \xi \\ \text{ch } k_\zeta(\xi_{bi}+\xi') \text{ ch } k_\zeta(\xi_{di}-\xi) & ,\text{for}\xi' > \xi \end{cases} ,$$

where

$$R = k_\zeta \text{ sh } k_\zeta(a+b+d+L_1+L_2) \quad ,$$

ξ_{di} and ξ_{bi} are the distances between upper and lower metal screens and the ferromagnetic film centers (taken for each film $i = 1, 2$).

The condition of the vanishing of the determinant of the system (42), (43) yields the exact dispersion relation $\omega(k_\zeta)$ for dipole-exchange spin waves in the structure metal-dielectric-ferrite-dielectric-ferrite-dielectric-metal (MDFDFDM). To discuss spin-wave spectrum we shall use the results of "rigorous" numerical calculations $\omega_n(k_\zeta)$.

We consider first the process of spin-wave spectrum formation for MDFDFDM structure. Fig. 8 presents two independent spectra (shown by solid and dashed

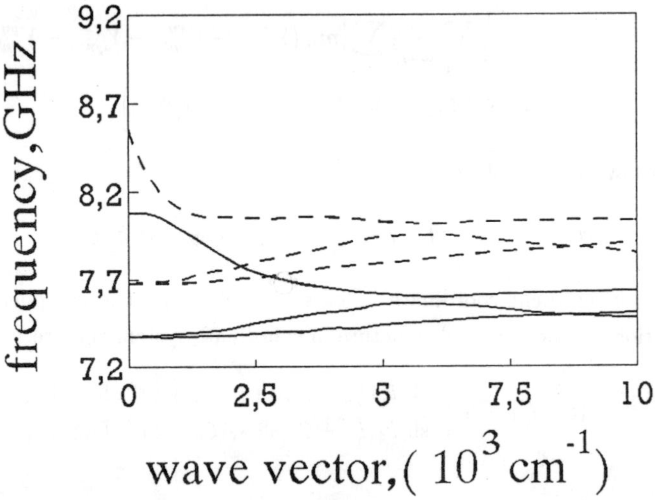

Figure 8: Spin-wave spectra for the structures MDFDM: dashed lines corresponds to the structure with parameters: $a = 6.5$ μm, $b = 5$ μm, $L = 5$ μm, $4\pi M_0 = 1450$ G, $\varphi = 90^o$; solid lines corresponds to the structure with parameters: $a = 1$ μm, $b = 10.5$ μm, $L = 5$ μm, $4\pi M_0 = 1750$ G, $\varphi = 90^o$.

lines), each one corresponding to its own MDFDM structure. In other words, interaction between the ferromagnetic films formed the screened double-film structure under consideration is completely neglected. We note that in Fig.8 only first three spin-wave branches ($n, m = 1, 2, 3$) are presented. Fig.9 shows the spectrum for the MDFDFDM structure which has been obtain "by inserting" two MDFDM structures into each other. It could be seen from Fig.9 that the dipole-dipole interaction between ferromagnetic films in the MDFDFDM structure leads to the reconstruction of the spin-wave spectrum. Especially the behavior of the fundamental mode dispersion branches ($n, m = 1$) is changed in the long wavelength region of spin-wave spectrum. In the short wave-length region $k_\zeta L_i >> 1$ dipole fields of the ferromagnetic films are localized near their surfaces and the dipole-dipole interaction between ferromagnetic films is weak.

It should be stressed that due to the dipole-dipole interaction between ferromagnetic films in MDFDFDM structures the spin-wave resonance frequencies $\omega_n(k_\zeta = 0)$ are shifted (in comparison with the case of unscreened double-film struc-

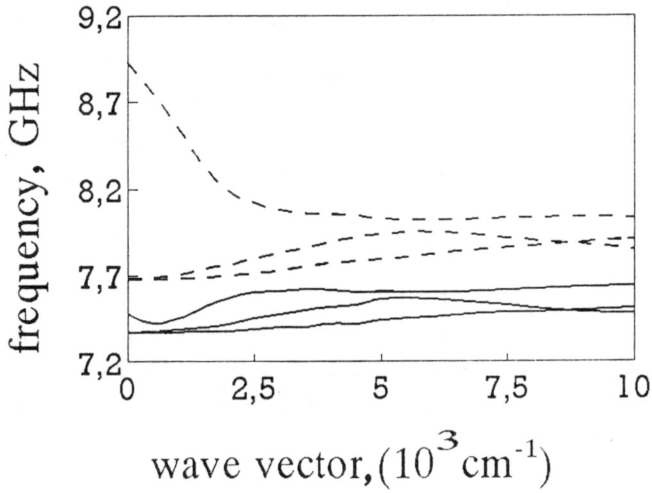

Figure 9: Spin-wave spectrum for the structures MDFDFDM with parameters: $a = 1\ \mu$m, $b = 5\ \mu$m, $L_1 = 5\ \mu$m, $L_2 = 5\ \mu$m, $d = 0.5\mu m$, $4\pi M_{01} = 1450$ G, $4\pi M_{02} = 1750$ G, $\varphi = 90^o$.

ture).

Analysis of spin-wave dispersion properties of the tangential magnetized MDFDFDM structures and their transformation under the influence of the metal screens show that it is possible to obtain a great variety of dispersion laws $\omega_n(k_\zeta)$. In conclusion, we have shown that the delivered method of tensorial Green function can be used not only for free layered structures but also for the structures containing metal screens.

6 Band theory for an infinite film stack

The Green function formalism described in the present Chapter can be applied to the spin-wave band theory for an infinite film stack. Such a theory for a perpendicularly magnetized superlattice consisting of very large number N (effectively N tends to infinity) of identical ferromagnetic films of thickness L, separated by nonmagnetic layers each of thickness a was established in the paper[29]. In this paper the Bloch's theorem[38] was applied to describe the ξ-dependence of the magnetization

$$\mathbf{m}_{stack}(\mathbf{r},t) = \mathbf{m}(\mathbf{r},t)e^{-iK\xi} \quad , \tag{46}$$

where the Bloch wave vector $K = 2\pi n/NL$, $n = -N/2, ..., +N/2$ $(N \to \infty)$, and $\mathbf{m}(\mathbf{r},t)$ is necessarily periodic

$$\mathbf{m}(\xi,\eta,\zeta,t) = \mathbf{m}(\xi + nL,\eta,\zeta,t) \quad . \tag{47}$$

$\mathbf{m}(\mathbf{r},t) = 0$ in all nonmagnetic regions of the stack.

As far as the infinite stack of ferromagnetic films is considered, it is convenient to choose an arbitrary film and set its number i equal to zero. Then after calculations analogous to those in Section 2 the system of algebraic equations for spin-wave mode amplitudes is obtained

$$(\omega + \Omega_{in'})m^1_{in'} + \frac{\omega_M}{2}\sum_{n=1}^{\infty}(m^1_{0n} - m^2_{0n})\left[P^{00}_{n'n} + \sum_{j=-\infty}^{\infty} P^{0j}_{n'n}e^{iK\xi_j}\right] = 0 \quad , \tag{48}$$

$$(\omega - \Omega_{in'})m^2_{in'} + \frac{\omega_M}{2}\sum_{n=1}^{\infty}(m^1_{0n} - m^2_{0n})\left[P^{00}_{n'n} + \sum_{j=-\infty}^{\infty} P^{0j}_{n'n}e^{iK\xi_j}\right] = 0 \quad . \tag{49}$$

This system could be analyzed numerically applying one of the existed algorithms (see, e.g.[32]).

For a discussion of the physical effects, below we present the approximate dispersion equation derived by the perturbation theory method in[29]:

$$\begin{aligned}\omega_n(k_\zeta) &= \Omega_{nk}(\Omega_{nk} + \omega_M P^{ii}_{NN}) - \Omega_{nk} \\ &\times \frac{\omega_M \frac{4}{1+\delta_{0n}}\frac{k_\zeta}{k_n^4}\frac{1-(-1)^n \operatorname{ch}(k_\zeta L)}{k_\zeta L}e^{-k_\zeta a}\cos(Ka) - e^{-k_\zeta a}}{1 - 2\,e^{-k_\zeta a}\cos(Ka) + e^{-2k_\zeta a}} \quad .\end{aligned} \tag{50}$$

This dispersion equation consists of two terms. The first represents the well known approximate dispersion relation for one ferromagnetic film[27]. The second term represents the dipole coupling between spin waves in the different films of the stack. This term contains the band structure, depending on the Bloch wave vector K and decreasing approximately exponentially with the interlayer distance a.

7 Conclusion

The present Chapter emphasizes the application of tensor Green function method for layered ferromagnetic film structures. We may refer to this method is an universal one in the meaning that in the frames of the method all main physical effects could be taken into account. These effects are Zeeman interaction,magnetic dipole-dipole interaction, exchange interaction, magnetic cristallographyc anisotropy, external "driving" magnetic field. In the present Chapter only first three effects were taken into account. It was done to stress the role of dipole coupling in spin-wave theory for ferromagnetic films in multilayered structure. All the other contributions could be incorporated in the theory (see, for example, Chapter 7 where excitation of dipole-exchange spin waves in anisotropic ferromagnetic films is considered). So the theory developed in the present Chapter could be used for further physical study of magnetic properties of layered structures and also for technical application in microwave and magneto optical technology.

References

1. *Circuits, Systems and Signal Processing,* **4** (1985).

2. P.Grünberg and K.Mika, *Phys. Rev.* **B27** (1983) 2955.

3. H.Weizhong and L.Farang, *J. Magn. Magn. Mater.***35** (1983) 153.

4. V.I.Zubkov and V.A.Epanechnikov, *Radiotekhnika i Elektronika* **31** (1986) 656.

5. S.H.Talisa, J.D.Adam, and M.R.Daniel, *Microwave J.* **33** (1990) 105.

6. T.Wolfram, *J. Appl. Phys.* **41** (1970) 4748.

7. H.Pfeiffer, *Phys. Status Solidi (a)* **19** (1974) K75.

8. A.K.Ganguly and C.Vittoria, *J. Appl. Phys.* **45** (1974) 4665.

9. L.R.Adkins and H.L.Glass, *Electronics Letters* **16** (1980) 590.

10. P.Grünberg, *J. Appl. Phys.* **51** (1980) 4338.

11. H.Sasaki and N.Mikoshiba, *J. Appl. Phys.* **52** (1981) 3546.

12. A.V.Vashkovskii and A.A.Stalmakhov, *Radiotekhnika i Elektronika* **29** (1985) 901.

13. P.R.Emtage and M.R.Daniel, *Phys. Rev.* **B29** (1984) 212.

14. V.V.Filippov, *Radiotekhnika i Elektronika* **33** (1988) 1225.

15. R.P.van Stapele, F.J.A.M.Greidanus, and J.W.Smits, *J. Appl. Phys.* **57** (1985) 1282.

16. L.Dobrzynski, B.Djafari-Rouhani, and H.Puszkarski, *Phys. Rev.* **B33** (1986) 3251.

17. E.L.Albuquerque, P.Fulco, E.F.Sarmento, and D.R.Tilley, *Solid State Commun.* **58** (1986) 41.

18. L.L.Hinchey and D.L.Mills, *Phys. Rev.* **B33** (1986) 3329.

19. K.Vayhinger and Kronmüller. *J.Magn. Magn. Mater.* **72** (1988) 307.

20. J.Barnas, *J. Phys. C.* **21** (1988) 4097.

21. P.Grünberg, *J. Appl. Phys.* **57** (1985) 3673).

22. F.Saurenbach, J.Barnas, G.Binasch, M.Vohl, P.Grünberg, and W.Zinn, *Thin Solid Films* **175** (1989) 317.

23. J.Fassbender, F.Nörtemann, R.L.Stamps, R.E.Camley, B.Hillebrands, G.Güntherodt and S.S.P.Parkin, *Phys. Rev.* **B46** (1992) 5810.

24. B.Hillebrands, *Phys. Rev.* **B37** (1988) 9885.

25. M.Vohl, J.Barnas, and P.Grünberg, *Phys. Rev.* **B39** (1989) 12003.

26. O.G.Vendik, B.A.Kalinikos, and D.N.Chartorizhskii, *Sov. Phys. - Solid State* **19** (1977) 222 [*Fiz. Tverd. Tela* **19** (1977) 387].

27. B.A.Kalinikos, *IEE Proc. Pt.H* **127** (1980) 4.

28. B.A.Kalinikos and A.N.Slanin, *J. Phys. C.* **19** (1986) 7013.

29. K.Vayhinger and H.Kronmüller, *J. Magn. Magn. Mater.* **62** (1986) 15.

30. B.A.Kalinikos and P.A.Kolodin, *Sov. J. Radiophysiks and Quantum Electronics* **32** (1989) 957 [*Izv. Vyssh. Uchebn. Zaved., Radiofizika* **32** (1989) 1230].

31. B.A.Kalinikos and P.A.Kolodin, *J. Magn. Magn. Mater.* **83** (1990) 103.

32. J.H.Wilkinson and C.Reinsch, *Handbook for Automation of Linear Algebra* (Springer, Berlin, 1976).

33. O.G.Vendik and D.N.Chartorizyskii, *Sov. Phys. - Solid State* **12** (1970) 1290 [*Fiz. Tverd. Tela* **12** (1970) 387].

34. V.G.Bary'akhtar and M.I.Kaganov, *Ferromagnetic Resonance* ed. S.V.Vonsovskii (Fizmatgiz, Moscow, 1961) 266.

35. D.D.Stancil, *Theory of Magnetostatic Waves,* (Springer-Verlag, New York, 1993).

36. J.P.Parekh and K.W.Chang, *IEEE Trans. on Magn.* **MAG 18** (1982) 1610.

37. L.D.Landau and F.M.Lifshitz, *Quantum Mechanics Non-Relativistic Theory, 2nd edn.* (Pergamon, London, 1965).

38. F.Bloch, *Z.Phys.* **52** (1929) 555.

MAGNETIC RESONANCE IN EXCHANGE COUPLED MULTILAYER SYSTEMS

Zhenyong Zhang and Philip E. Wigen
Department of Physics, The Ohio State University, Columbus, Ohio 43210

I. Introduction

Multilayered structures constructed by alternating layers of ferromagnetic and non-magnetic materials have received much attention in the past several years. The development of sample preparation techniques, including molecular-beam epitaxy (MBE), sputtering and chemical vapor deposition, have made it possible to manufacture a large variety of magnetic layered structures to extraordinarily accurate specifications on a 1-Å scale. Due to a large variety of short- and long-range interactions in these multilayered media, their electrical and magnetic properties can be drastically modified through small changes in the specific chemistry and morphology of the samples and through relatively modest forces such as those arising from magnetic field and uniform and nonuniform stresses. Hence magnetic multilayer systems leads to a large number of useful engineering and technical applications, such as in information storage, integrated microwave circuits, optical communication and neutron interferometer devices.[1]

One of the most noticeable phenomena is the interlayer exchange interaction between magnetic layers mediated by non-magnetic spacers. In the early work on a Fe/Cr/Fe trilayer structure,[2,3] a clear-cut anti-parallel coupling between the Fe layers was observed. Further investigations on Fe/Cr multilayer structures revealed an oscillatory behavior of the coupling strength as a function of the Cr layer thickness.[4] In addition, the anti-parallel coupled samples in this series exhibit a giant magnetoresistance (GMR) effect due to the change in the conduction electron scattering mechanisms as the orientation of the magnetization in the Fe layers is modified by the application of an external magnetic field.[5] Such structures are promising as the new generation of spin valve sensors as well as information readheads.

Another novel behavior was observed in Co/Pt and Co/Pd multilayer structures in which the perpendicular uniaxial anisotropy energy is significantly enhanced with the decrease of the Co layer thickness.[6,7] It was believed that this change is due to a large surface anisotropy energy at the Co/Pt interfaces. Samples with perpendicular magnetic anisotropy were obtained as the Co layer thickness was below 10 Å and could be used as a high density perpendicular magneto-optical recording media.

Out of the many experimental methods that have been employed for the study of the interlayer exchange interaction and magnetic anisotropy energies in these multilayer media, ferromagnetic resonance (FMR) has proved to be one of the most powerful tools to sensitively evaluate both properties at the same time.[8,9] In the

FMR experiment, the position of the uniform (or nearly uniform) resonant mode is sensitive to the magnetic anisotropy energies within each magnetic layer, therefore, its variation with the external applied field angle and frequency provides an accurate determination of the in-plane and out-of-plane anisotropy fields. In addition to the uniform mode, there are higher order exchange coupled modes which correspond to the rf magnetization having either an amplitude or a phase different from one magnetic layer to the next. Since an extra interlayer exchange energy term is involved in the internal energies, the shift of the exchange coupled modes from the uniform mode can be used to determine the interlayer exchange coupling strength for both parallel and anti-parallel coupled multilayer structures.

In section II, the theory of FMR in exchange coupled magnetic trilayer structures is discussed. Starting from the torque equation of motion, the resonance frequencies as a function of the applied field (i.e., dispersion relation) is calculated with the field applied at an arbitrary angle to the film plane. The concepts of both the acoustic mode (uniform mode) and the optic mode (exchange coupled mode) are introduced with emphasis on their roles in the evaluation of the magnetic properties in this trilayer structure.

In order to observe those two modes in the FMR spectra, the microwave pumping field must couple to the dynamic component of the magnetic moment in the system. Therefore, the intensity of each resonance mode is discussed in section III. Unlike the acoustic mode, the optic mode is often difficult to be observed in the FMR spectra. Section III also discusses a few methods that have been used to overcome this difficulty, either through changing the experimental configuration (for example, longitudinal pumping) or through modifying the sample structures (such as asymmetrical trilayer structure).

The resonance theory in magnetic trilayer structures can easily be expanded to magnetic multilayer systems. In section IV, a simple example is given with the external field applied normal to the film plane. The general resonance spectra in multilayer media is discussed with special attention to the importance of the surface layer conditions on the wave-vectors of the exchange coupled resonance modes.

At the end, an overall review of the FMR results in magnetic trilayer and multilayer films are given. Several active research topics in the field of interlayer exchange coupling and magnetic anisotropy energy are also addressed along with a brief introduction of some microscopic theories related to these FMR results.

II. Resonance Theory In Magnetic Trilayer Systems

II.1. Torque Equation

A typical magnetic trilayer structure is shown in Fig. 5.1 which includes two ferromagnetic layers separated by a non-magnetic spacer. For most trilayer structures, the thickness of the ferromagnetic layer is in the order of a few nm, therefore, each layer can be represented by an uniform magnetization vector $\mathbf{M}_1$ or $\mathbf{M}_2$.[8]

Several magnetic energy terms are involved in such a trilayer system. Within the ith ferromagnetic layer, where $i = 1$ or 2, the energies that contribute to the

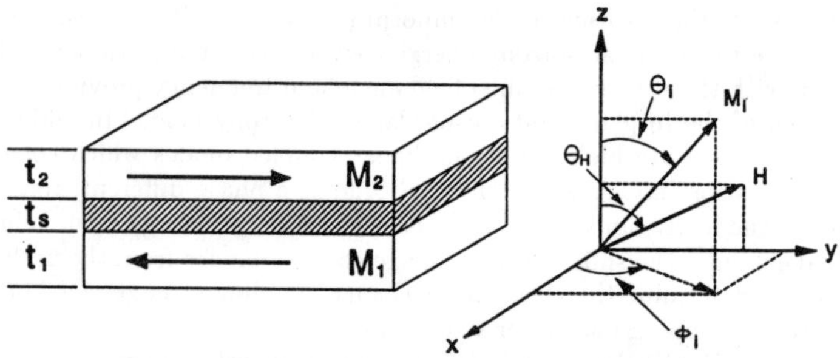

Figure 5.1: A typical trilayer structure and the coordinate system used in the model.

dispersion relation normally include the Zeeman energy,

$$E_{Z,i} = -\mathbf{M}_i \cdot \mathbf{H} = -M_s H[cos\theta_H cos\theta_i + sin\theta_H sin\theta_i cos(\phi_H - \phi_i)] \ , \qquad (5.1)$$

the demagnetization energy,

$$E_{D,i} = 2\pi M_{z,i}^2 = 2\pi M_s^2 cos^2\theta_i \ , \qquad (5.2)$$

and the magnetocrystalline anisotropy energy $E_{K,i}$. The coordinate system used here is shown in Fig. 5.1 with the z-axis perpendicular to the film plane. $\mathbf{M}_i$ is the magnetization of the ith ferromagnetic layer and M_s is its saturation value. $\mathbf{H}$ is the external field applied at the direction described by the polar and azimuth angles θ_H and ϕ_H.

The expression of the magnetocrystalline anisotropy energy depends on the crystal structure as well as the crystal axis perpendicular to the film plane. For an hcp single crystal film with its c-axis perpendicular to the film plane, this energy term is written as

$$E_{K,i} = K_{u2,i} \sin^2\theta_i + K_{u4,i} \sin^4\theta_i + ... \ , \qquad (5.3)$$

where the higher order terms are usually very small and can be neglected. For a cubic polycrystalline film with its crystal axis randomly oriented, the above expression is also used as a good approximation. However if the film has a single cubic crystal structure, the magnetocrystalline energy normally has the form

$$E_{K,i} = K_1(\alpha_{1,i}^2\alpha_{2,i}^2 + \alpha_{2,i}^2\alpha_{3,i}^2 + \alpha_{3,i}^2\alpha_{1,i}) + K_{u2} sin^2\theta_i \ , \qquad (5.4)$$

where $\alpha_{1,i}$, $\alpha_{2,i}$ and $\alpha_{3,i}$ are the direction cosines of the magnetization $\mathbf{M}_i$ with respect to the cubic axes.

When the non-magnetic layer thickness is reduced to a few nm or less, an interlayer exchange interaction arises between the two ferromagnetic layers which can be represented by an exchange energy[10]

$$E_{ex} = A_{12} cos\theta_{12} + B_{12} cos^2\theta_{12} + ... \ . \qquad (5.5)$$

Here A_{12} and B_{12} are bilinear and biquadratic exchange coupling coefficients and θ_{12} is the angle between the magnetization vectors $\mathbf{M}_1$ and $\mathbf{M}_2$. Except for a few specific structures,[11,12] the biquadratic term and other higher order terms are normally very small and neglected. A positive (or negative) sign of A_{12} indicates that $\mathbf{M}_1$ and $\mathbf{M}_2$ prefer to stay anti-parallel (or parallel) to each other in the absence of the applied field.

The torque equation of motion for this trilayer structure includes a set of two equations,

$$\begin{aligned} \frac{1}{\gamma}\frac{d\mathbf{M}_1}{dt} &= \frac{\mathbf{M}_1}{M_s} \times (\frac{1}{t_1}\nabla_1 \mathcal{E}) \\ \frac{1}{\gamma}\frac{d\mathbf{M}_2}{dt} &= \frac{\mathbf{M}_2}{M_s} \times (\frac{1}{t_2}\nabla_2 \mathcal{E}) , \end{aligned} \tag{5.6}$$

where

$$\nabla_i = \mathbf{e}_{\theta_i}\frac{\partial}{\partial\theta_i} + \mathbf{e}_{\phi_i}\frac{1}{sin\theta_i}\frac{\partial}{\partial\phi_i} \qquad (i = 1, 2) , \tag{5.7}$$

$\mathcal{E}$ is the magnetic energy per unit surface area,

$$\mathcal{E} = \sum_{i=1}^{2} t_i(E_{Z,i} + E_{D,i} + E_{K,i}) + E_{ex} , \tag{5.8}$$

and t_i is the thickness of the ith ferromagnetic layer. At equilibrium, both $\mathbf{M}_1$ and $\mathbf{M}_2$ are time independent, therefore, the equilibrium positions of the magnetization vectors can be obtained by taking $\partial\mathcal{E}/\partial\theta_i = 0$ and $\partial\mathcal{E}/\partial\phi_i = 0$, where $i = 1$ and 2.

Expanding Eq.(5.6) around the equilibrium positions of $\mathbf{M}_1$ and $\mathbf{M}_2$ and assuming that the time variation of the resonant moments follows the $e^{-i\omega t}$ form, four linear equations to the first order of m_{θ_i} and m_{ϕ_i} are obtained:

$$\begin{pmatrix} \frac{iw}{\gamma} + \frac{E_{\theta_1\phi_1}}{t_1M_s sin\theta_1} & \frac{E_{\phi_1\phi_1}}{t_1M_s sin^2\theta_1} & \frac{E_{\theta_2\phi_1}}{t_1M_s sin\theta_1} & \frac{E_{\phi_1\phi_2}}{t_1M_s sin\theta_1 sin\theta_2} \\ -\frac{E_{\theta_1\theta_1}}{t_1M_s} & \frac{iw}{\gamma} - \frac{E_{\theta_1\phi_1}}{t_1M_s sin\theta_1} & -\frac{E_{\theta_1\theta_2}}{t_1M_s} & -\frac{E_{\theta_1\phi_2}}{t_1M_s sin\theta_2} \\ \frac{E_{\theta_1\phi_2}}{t_2M_s sin\theta_2} & \frac{E_{\phi_1\phi_2}}{t_2M_s sin\theta_1 \sin\theta_2} & \frac{iw}{\gamma} + \frac{E_{\theta_2\phi_2}}{t_2M_s sin\theta_2} & \frac{E_{\phi_2\phi_2}}{t_2M_s sin^2\theta_2} \\ -\frac{E_{\theta_1\theta_2}}{t_2M_s} & -\frac{E_{\theta_2\phi_1}}{t_2M_s sin\theta_1} & -\frac{E_{\theta_2\theta_2}}{t_2M_s} & \frac{iw}{\gamma} - \frac{E_{\theta_2\phi_2}}{t_2M_s sin\theta_2} \end{pmatrix} \begin{pmatrix} m_{\theta_1} \\ m_{\phi_1} \\ m_{\theta_2} \\ m_{\phi_2} \end{pmatrix} = 0 , \tag{5.9}$$

where m_{θ_i} and m_{ϕ_i} are small deviations of $\mathbf{M}_i$ along $\mathbf{e}_{\theta_i}$ and $\mathbf{e}_{\phi_i}$ orientations of the equilibrium positions. Here $E_{\theta_i\theta_j} = \partial^2\mathcal{E}/\partial\theta_i\theta_j$, $E_{\theta i\phi_j} = \partial^2\mathcal{E}/\partial\theta_i\phi_j$ and $E_{\phi_i\phi_j} = \partial^2\mathcal{E}/\partial\phi_i\phi_j$ are energy derivatives that can be calculated using the expression in Eq.(5.8). The roots of the determinant of the 4×4 matrix on the left hand side of Eq.(5.9) will give the dispersion relation of this trilayer structure.

In general, four solutions of ω can be obtained for a given applied field $\mathbf{H}$. However, only two of them have positive values and represent the two resonance modes (the acoustic mode and the optic mode) generally observed in the trilayer structures. Depending on the orientation of the applied field, the dispersion relation can be quite different from one to the other as discussed in the following several sections.

II.2. The Dispersion Relation With The External Field Applied In The Film Plane

Applying the external field in the film plane is the most frequently used experimental configuration. Except for a few perpendicular orientated multilayer structures, the equilibrium magnetization vector of each ferromagnetic layer is confined to move within the film plane in this case.

In order to emphasis the significant changes of the dispersion relation due to the exchange interaction, the magnetocrystalline anisotropy term is assumed to include only the first term of Eq.(5.3). Since the angular variation of the K_{u2} term is the same as the demagnetization energy term, an effective demagnetization field is introduced

$$4\pi M_{eff,i} = 4\pi M_s - 2K_{u2,i}/M_s \,. \tag{5.10}$$

which has a positive value for each ferromagnetic layer if the equilibrium magnetization vectors stay in the film plane.

If the system is parallel coupled ($A_{12} < 0$) and the two ferromagnetic layers are identical, i.e., $t_1 = t_2 = t_F$, $4\pi M_{eff,1} = 4\pi M_{eff,2} = 4\pi M_{eff}$, the in-plane dispersion relation has a simple form:

$$\begin{aligned} (\frac{\omega}{\gamma})_{ac} &= \sqrt{H(H + 4\pi M_{eff})} \\ (\frac{\omega}{\gamma})_{op} &= \sqrt{(H - 2A_{12}/t_F M_s)(H - 2A_{12}/t_F M_s + 4\pi M_s)} \end{aligned} \tag{5.11}$$

and is shown in Fig. 5.2a. The $(\frac{\omega}{\gamma})_{ac}$ mode (acoustic mode), which corresponds to the rf components of the two magnetization vectors resonating in-phase as illustrated in Fig. 5.2b, has the same dispersion relation as the uniform mode of a single layer system.[13] This is due to the fact that the exchange energy, which is constant during the precession, does not produce any dynamic contribution to the resonance condition. The $(\frac{\omega}{\gamma})_{op}$ mode (optic mode), however, corresponds to the rf component of the magnetization vector in each layer resonating out-of phase (also shown in Fig. 5.2b). The exchange energy, in this case, introduces an extra field to the dispersion relation and shifts the resonance field by a value of $2A_{12}/t_F M_s$ from that of the acoustic mode at the same frequency. Since the two ferromagnetic layers are parallel coupled, the optic mode increases the exchange energy as the magnetization vectors deviate from the parallel orientation and therefore is observed at the higher frequency or lower magnetic field side of the acoustic mode.

Unlike the case in parallel coupled systems, $\mathbf{M}_1$ and $\mathbf{M}_2$ are not always saturated along the external field orientation for an anti-parallel coupled (i.e., $A_{12} > 0$)

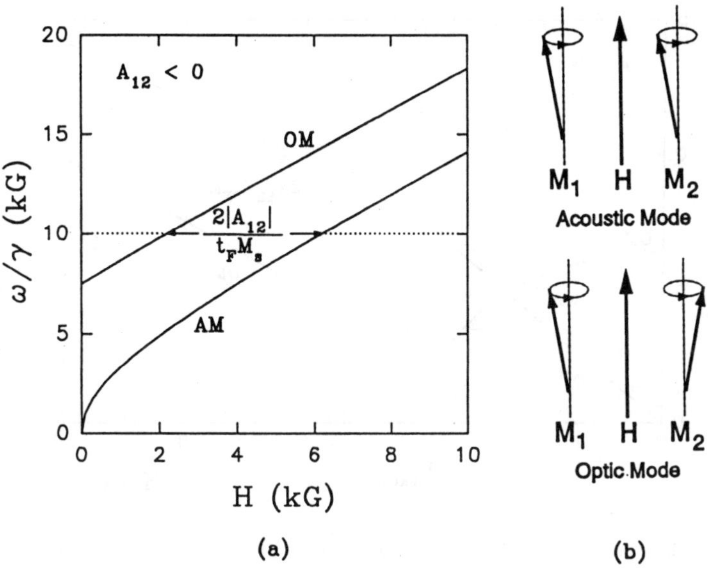

Figure 5.2: (a) The in-plane dispersion relation for a parallel coupled symmetrical trilayer structure. The parameters are: $4\pi M_{eff}$ = 10 kG, $2A_{12}/t_F M_s$ = −4.0 kG. AM and OM refer to the acoustic mode and the optic mode respectively. (b) A schematic diagram of the phase relation between the resonant moments of the two magnetization vectors in the acoustic mode and the optic mode.

symmetrical trilayer structure. As a result, the dispersion curve (see Fig. 5.3) shows two different regions which are separated by the in-plane saturation field

$$H_{sat,\|} = 2A_{12}/t_F M_s \ . \tag{5.12}$$

In the saturated region, the expression of the dispersion relation has the same form of Eq.(5.11). Similar to a parallel coupled system, the acoustic mode in this anti-parallel coupled system is independent of the exchange coupling coefficient A_{12} and the optic mode is shifted by a field value of $2A_{12}/t_F M_s$ from the acoustic mode. However in contrast to the case for parallel coupled systems, the optic mode is observed at the lower frequency or higher magnetic field side of the acoustic mode. This shift is due to the decrease in the exchange energy as the magnetization vectors deviate from the parallel orientation as opposed to the increase in the exchange energy for the parallel coupled system.

When the applied field is reduced below the saturation value $H_{sat,\|}$, the two magnetization vectors deviate from the external field orientation as the exchange

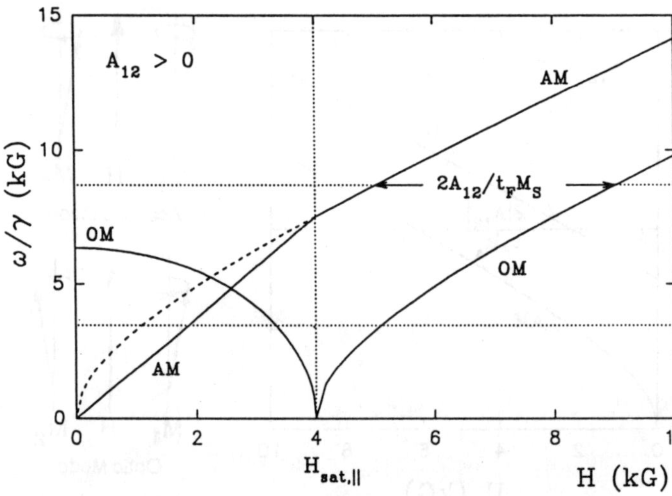

Figure 5.3: The dispersion relation for an anti-parallel coupled symmetrical trilayer structure. The parameters are: $4\pi M_{eff}$ = 10 kG, $2A_{12}/t_F M_s$ = 4 kG. The broken line is the dispersion relation for a non-coupled system (i.e., $A_{12} = 0$). AM and OM refer to the acoustic mode and the optic mode respectively.

energy starts to dominate the Zeeman energy in this system. The canted angle between $\mathbf{M}_1$ (or $\mathbf{M}_2$) and $\mathbf{H}$ has a simple relation with the field value H,

$$\phi_i - \phi_H = \pm cos^{-1}(H/H_{sat,\|}) \,, \tag{5.13}$$

where $\pm$ indicates that $\mathbf{M}_1$ and $\mathbf{M}_2$ are on the opposite side of the $\mathbf{H}$ direction. Due to the decrease of the net magnetic moment along the field orientation, the dynamic demagnetization field along the z-axis (perpendicular to the film plane) is reduced. As a result, the resonance frequency of the acoustic mode is smaller than that of a non-coupled (or single layer) system which is shown by the dashed line in Fig. 5.3.[14] In the FMR experiment where the resonance frequency is constant, the resonance field of the acoustic mode will be significantly higher than the expected value for a saturated systems.

Some typical in-plane FMR spectra for both parallel and anti-parallel coupled symmetrical trilayer structures are shown in Fig. 5.4 at different frequencies.[15] The optic modes have much weaker intensity than the acoustic modes and their position are labeled by the arrows in this figure. For the parallel coupled system, i.e., the Ru(16 Å) sample, the optic mode is on the lower field side of the acoustic mode while for anti-parallel coupled systems, i.e., the Ru(12 Å), Ru(20 Å) and Ru(24 Å)

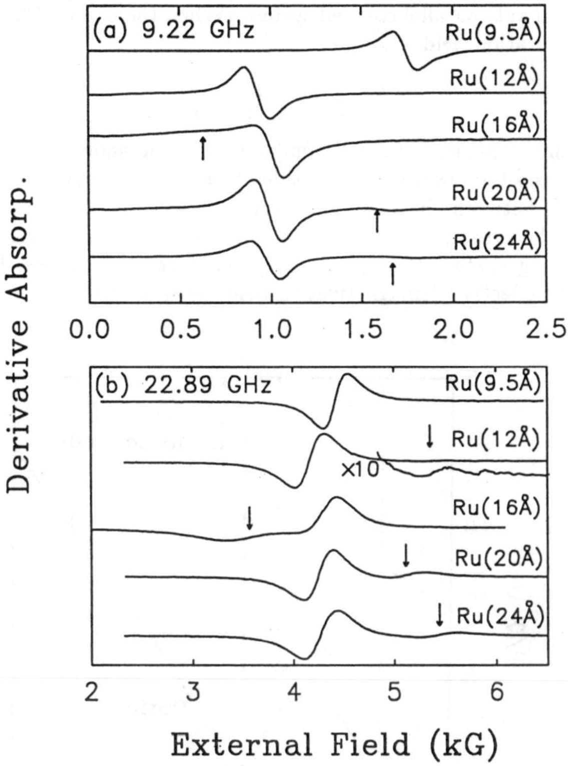

Figure 5.4: In-plane FMR spectra at (a) 9.22 GHz and (b) 22.89 GHz for some Co(32 Å)/Ru(t_{Ru})/Co(32 Å) samples at room temperature. The arrows indicate the positions of the optic mode under high resolution. (from Ref.8)

samples, the optic mode is on the higher field side of the acoustic mode. Except the Ru (9.5 Å) sample, the position of the acoustic mode is nearly constant from sample to sample, in agreement with the prediction that the dispersion relation of the acoustic mode only depends on the magnetic energies within the ferromagnetic layer if the film is saturated at resonance. For the Ru(9.5 Å) sample, a large in-plane saturation field, about 4.5 kG, was observed in the in-plane magnetization measurement, suggesting that the magnetization vectors are in the canted state at the X-band resonance condition. The substantial shift of the acoustic resonance field for this sample in Fig. 5.4a, therefore, provides a direct evidence of the dispersion calculation discussed previously.

If the two ferromagnetic layers are not identical, i.e., $t_1 \neq t_2$ and $4\pi M_{eff,1} \neq 4\pi M_{eff,2}$, the dispersion relation could be more complicated. One example is shown in Fig. 5.5 for an anti-parallel coupled system. When the external field is larger than the in-plane saturation field

$$H_{sat,\parallel} = \frac{A_{12}}{M_s}\left(\frac{1}{t_1} + \frac{1}{t_2}\right), \tag{5.14}$$

the shape of the dispersion curve is similar to the one shown in Fig. 5.3 with the optic mode also on the lower frequency side of the acoustic mode. However, the field separation $\Delta H_\parallel$ between the two modes (at constant frequency) is no longer equal to $H_{sat,\parallel}$ as in the symmetrical trilayer system. Instead, it depends not only on the exchange coupling coefficient A_{12} and magnetic layer thicknesses, but also on the difference of the effective demagnetization field between the two magnetic layers.

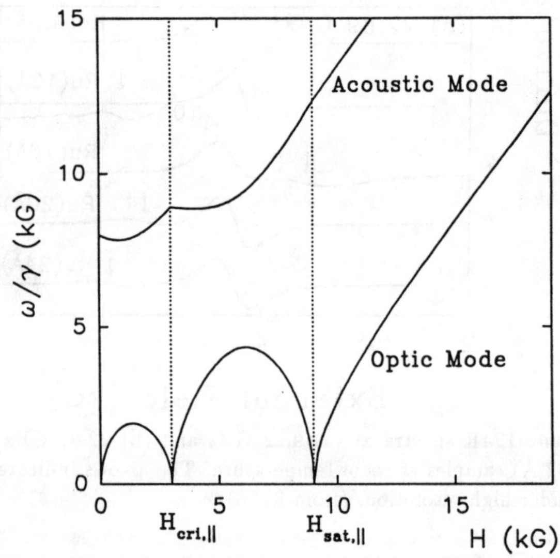

Figure 5.5: The in-plane dispersion relation for an asymmetrical trilayer system. The parameters are: $t_1/t_2 = 2$, $4\pi M_{eff,1} = 9.5$ kG, $4\pi M_{eff,2} = 4.5$ kG and $A_{12}/t_1 M_s = 3$ kG.

Unlike in the saturated state, the dispersion relation below the saturation field is significantly different from that of the symmetrical trilayer system. On one hand, there is a frequency gap between the upper and lower branches of the dispersion curve, suggesting that no acoustic mode can be observed over a certain frequency range. On the other hand, due to the thickness difference between the two ferromagnetic layers, the magnetization vectors will remain anti-parallel to each other until the field is above an critical value

$$H_{cri,\parallel} = \frac{A_{12}}{M_s}\left|\frac{1}{t_1} - \frac{1}{t_2}\right|. \tag{5.15}$$

As a result, an extra region, i.e., $0 \leq H \leq H_{cri,||}$, exits on the dispersion curve which corresponds to this anti-parallel state.

II.3. The Dispersion Relation With The External Field Applied Normal To The Film Plane

In most trilayer structures, the two ferromagnetic layers are usually not exactly identical, i.e., their magnetic properties could differ from each other due to the variation in the layer thickness or in the crystal quality of the layer. In-plane FMR measurement, in this case, could not provide enough data to evaluate the exchange coupling coefficient unless the magnetic anisotropy energies within each layer can be accurately determined from other experiments. One effective way to overcome this problem is to measure the FMR spectra with the external field applied perpendicular to the film plane.

The perpendicular dispersion relation for both parallel and anti-parallel coupled symmetrical trilayer systems is shown in Fig. 5.6. Two regions exist in both systems which correspond to the saturated region and the canted region. In the saturated region, the resonance frequency of the acoustic mode and the optic mode can be solved explicitly as

$$\begin{aligned} (\frac{\omega}{\gamma})_{ac} &= H - 4\pi M_{eff} \\ (\frac{\omega}{\gamma})_{op} &= H - 4\pi M_{eff} - 2A_{12}/t_F M_s \,. \end{aligned} \qquad (5.16)$$

Similar to the in-plane dispersion relation, the acoustic mode is degenerate with that of a non-coupled system while the field of the optic mode is shifted by a value of $|2A_{12}/t_F M_s|$ toward the higher (or lower) field side of the acoustic mode for an anti-parallel (or parallel) coupled system.

The variation of the dispersion relation in the canted region indicates that one mode always stays at zero frequency. This mode corresponds to the free rotation of $\mathbf{M}_1$ and $\mathbf{M}_2$ around the z-axis while keeping their relative position unchanged. The other mode stays at a relatively high frequency, resulting in a frequency gap that exists even for symmetrical trilayer structures. For the anti-parallel coupled system, this frequency gap is extremely important for the understanding of the resonance behavior because unlike a single layer system, a minimum frequency is required to observe the acoustic mode.

For other systems in which the two ferromagnetic layers are not identical, the perpendicular dispersion curve has very similar shape as the one shown in Fig. 5.6. One noticeable difference, however, is the field separation $\Delta H_\perp$ between the acoustic mode and the optic mode in the saturated region which is very sensitive to the difference of the effective demagnetization field between the two magnetic layers. In general, $\Delta H_\perp$ is larger than both $\Delta H_{||}$ and the exchange field $|\frac{A_{12}}{M_s}(\frac{1}{t_1} + \frac{1}{t_2})|$. Detailed description can be found in the literature.[8]

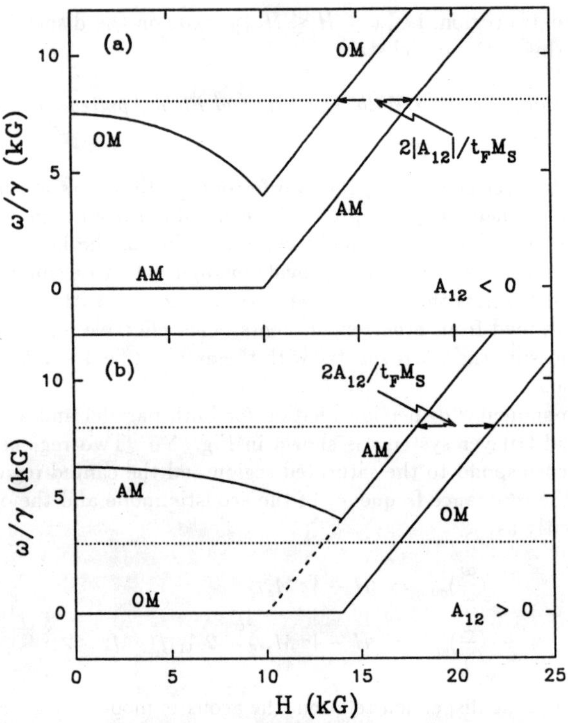

Figure 5.6: The perpendicular dispersion relation for (a) a symmetrical parallel coupled trilayer systems and (b) a symmetrical anti-parallel coupled trilayer systems. The parameters used in the calculations are the same as in Fig. 5.2 and Fig. 5.3. AM and OM refer to the acoustic mode and the optic mode respectively.

II.4. Angular Dependence Of The Dispersion Relation

From the previous discussion, it was found that the dispersion relation of the exchange coupled trilayer structure is significantly modified from that of a single layer (or non-coupled trilayer) system. In particular, for an anti-parallel coupled system, the acoustic mode might not be observed in the vicinity of the perpendicular configuration due to the existence of the frequency gap in the unsaturated region. In order to investigate the development of the frequency gap with the applied field orientation, the angular dependence of the dispersion relation for a symmetrical anti-parallel coupled trilayer structure is shown in Fig. 5.7a.

When the external field is applied in the film plane ($\theta_H = 90^0$), the dispersion curves of the acoustic mode and the optic mode across each other at a particular

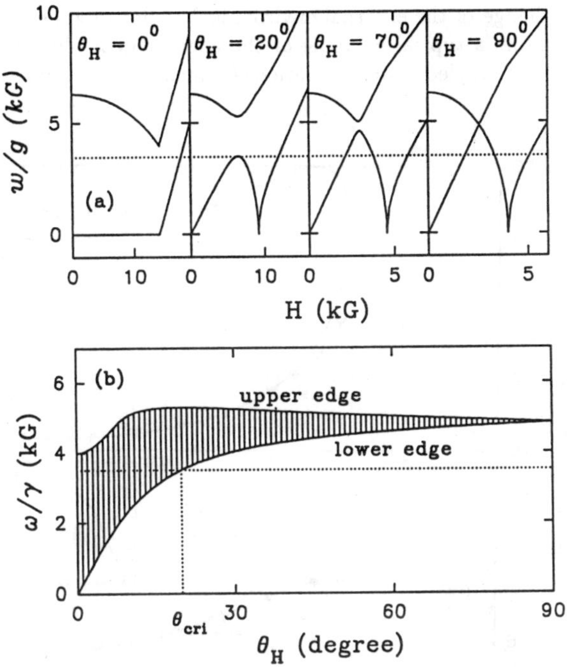

Figure 5.7: (a) The dispersion relation for a symmetrical anti-parallel coupled trilayer system with the external field applied at $\theta_H = 0^0$, 20^0, 70^0, and 90^0 with respect to the film normal orientation. (b) The frequency gap (shadowed area) as a function of θ_H for this system. The parameters are the same as in Fig. 5.3.

field value and no frequency gap is observed. As the external field is rotated away from the film plane ($\theta_H \leq 90^0$), a finite gap appears near the crossover region. With decreasing θ_H, the minimum edge of the gap decreases while the maximum edge remains relatively flat as also shown in Fig. 5.7b. If the resonance frequency is lower than the maximum of the upper gap edge, a critical value of θ_H will be obtained below which the "acoustic" mode can not be observed in the FMR spectra.

With a closer look at the dispersion relation near the perpendicular orientation ($\theta_H \sim 0^0$), it can be found that the slope of the upper branch in the unsaturated region is very small in comparison with that for $\theta_H \sim 90^0$. This suggests that at a constant frequency, the resonance field is sensitive to any small changes in the shape of the dispersion curve. Therefore, by choosing suitable resonance frequency for each individual system, angular dependent FMR experiment can provide a very sensitive way to evaluate any systematic variation of the exchange coupling coefficient or anisotropy energy with the change of the external environment such as temperature.[16]

Fig. 5.8 shows a typical example of the angular dependent FMR data for a strongly anti-parallel coupled system at various frequencies.[8] At the X-band frequency

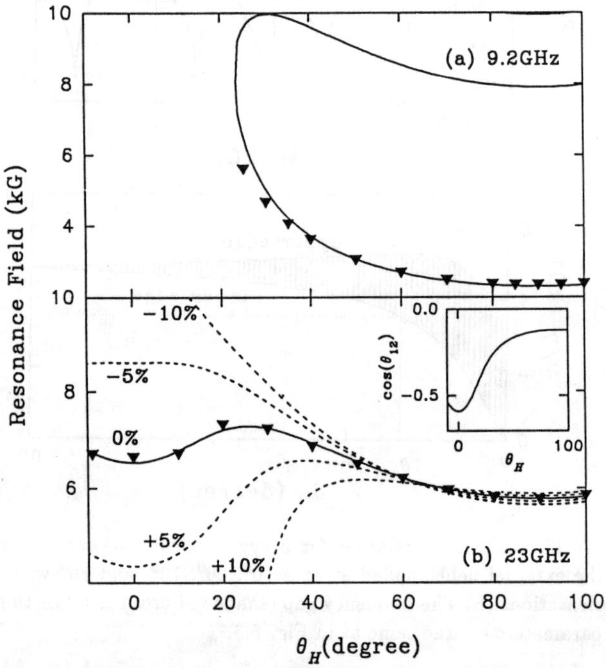

Figure 5.8: Angular dependence of the resonance field for the Co(32 Å)/ Ru(9 Å)/Co(32 Å) sample at X-band (a) and K-band (b) frequencies and room temperature. The solid lines are theoretical fits using the parameters: $4\pi M_{eff} = 5.8$ kG and $2A_{12}/t_{Co}M_s = 8.7$ kG. The dashed lines in (b) are predictions corresponding to 5 and 10% changes in A_{12}. The inset of (b) is the variation of $\cos\theta_{12}$ with θ_H at the K-band resonance condition, where θ_{12} is the angle between $\mathbf{M}_1$ and $\mathbf{M}_2$. A constant A_{12} used in this fit indicates that the biquadratic exchange term B_{12} is negligible in this sample. (from Ref.8)

(9.2 GHz), the acoustic mode was observed only when $\theta_H \geq 25^0$, in agreement with the theoretical prediction of 24^0 at which 9.2 GHz intercepts with the frequency gap of this sample. The K-band frequency is located just above the upper frequency edge of this sample, resulting a sensitive change in the resonance field near the perpendicular orientation with small variations in A_{12} or $4\pi M_{eff}$ as shown by the dashed lines in Fig. 5.8b.

III. Intensity And Longitudinal Pumping

One important criteria for the observation of the resonance mode is that the system must be able to couple to the microwave pumping field which is usually applied parallel to the film plane and perpendicular to the bias field in the traditional FMR spectrometer. The intensity of the microwave absorption depends directly on the net resonant component of the magnetic moment along the pumping field orientation and can be written as[17]

$$I \propto \frac{(t_1 m_{h,1} + t_2 m_{h,2})^2}{(t_1 + t_2)(t_1 m_1^2 + t_2 m_2^2)} \tag{5.17}$$

for the trilayer system, where $\mathbf{m}_1$ and $\mathbf{m}_2$ are the resonant components of $\mathbf{M}_1$ and $\mathbf{M}_2$ respectively and $m_{h,1}$ and $m_{h,2}$ are their projections along the pumping field $\mathbf{h}$ orientation.

For a symmetrical magnetic trilayer system, only the acoustic mode can be observed in the FMR spectra. The optic mode, as illustrated in Fig. 5.2b, does not have a net resonant moment along the microwave field orientation and therefore it can not absorb the microwave power. This, unfortunately, limits the application of the FMR technique in the evaluation of the exchange coupling coefficient in magnetic multilayer structures.

Many efforts have been devoted to enhance the intensity of the optic mode. One of the most commonly used methods is to break the symmetry of the magnetic anisotropy energies between the two magnetic layers. By doing this, the resonant amplitude of the magnetic moment $t_i m_{h,i}$ will be different from one layer to the other and a relatively small net moment can be obtained for the optic mode.

For most magnetic multilayer systems in which the layer thickness is on the order of 1nm, an important contribution to the magnetic anisotropy energies comes from the surface anisotropy term. As a result, the change of the average anisotropy energy within each layer can be achieved by changing the layer thickness[18] or changing the material of the protective (or buffer) layer[19] to which the magnetic layer is directly contacted. Once a difference is built in the anisotropy energy between the two magnetic layers, the intensity ratio between the optic mode and the acoustic mode can be adjusted to some extent by varying the bias field orientation from in the film plane to perpendicular to the film plane[8] where the ratio reaches its maximum value. Since $\mathbf{m}_1$ and $\mathbf{m}_2$ are still in-phase for the acoustic mode and out-of-phase for the optic mode, the ratio is always less than one as shown in Fig. 5.4.

If the two magnetic layers are anti-parallel coupled, longitudinal pumping technique[16] can be used to observe the optic mode even for symmetrical trilayer systems. Consider the case where $\mathbf{M}_1$ and $\mathbf{M}_2$ are not saturated at resonance due to their anti-parallel exchange coupling. In the traditional transverse pumping configuration, only the acoustic mode can be observed as shown in Fig. 5.9a. However it can be seen that the resonant magnetization vectors of $\mathbf{M}_1$ and $\mathbf{M}_2$ also have components along the bias field $\mathbf{H}$ orientation. Therefore, when the microwave pumping field $\mathbf{h}$ is applied along the bias field direction (longitudinal pumping), the resonant movement can also be excited. In contrast to the transverse pumping configuration, only

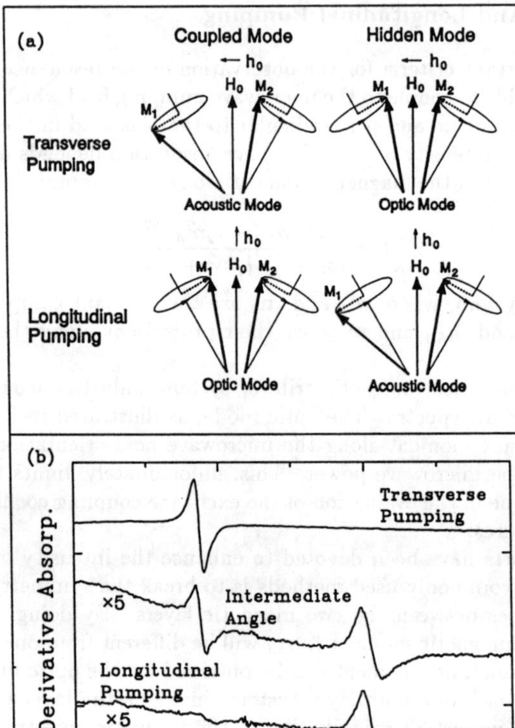

Figure 5.9: (a) A schematic diagram of longitudinal and transverse pumping in the FMR setup. $\mathbf{h_0}$ and $\mathbf{H_0}$ are microwave pumping field and bias field, respectively. M_1 and M_2 are magnetization vectors in the two layers. (b) The in-plane FMR spectra of the Co(32 Å)/Ru(9.5 Å)/Co(32 Å) sample at X-band frequency and room temperature. The rf field is applied at 90^0 (top), 20^0 (middle), and 0^0 (bottom) with respect to the bias field. (from Ref.16)

the optic mode in this case has a net resonant component along the **h** orientation and produces a noticeable absorption at resonance. The acoustic mode, on the other hand, becomes the hidden mode. If **h** is applied at an arbitrary angle with respect to the bias field orientation, both modes are observable in the FMR spectra. An experimental example is shown in Fig. 5.9b with the bias field **H** applied in the film plane.

When the above mentioned system is saturated at resonance, $\mathbf{M}_1$ and $\mathbf{M}_2$ do not have has resonant component along the bias field orientation. As a result, neither the acoustic mode nor the optic mode can be observed in the longitudinal pumping FMR experiment unless a non-linear phenomena occurs.[20] This is the reason why the third mode (in the saturation region) predicted by the in-plane dispersion relation calculation (Fig. 5.3) has not been observed in Fig. 5.9b.

The change of FMR experiment from transverse pumping to longitudinal pumping is relatively simple if the resonant cavity has a rectangular shape. Fig. 5.10 is a schematic diagram of a typical TE_{102} cavity placed in the center of a DC magnet.[21] The circles inside the cavity indicate the directions of the microwave pump-

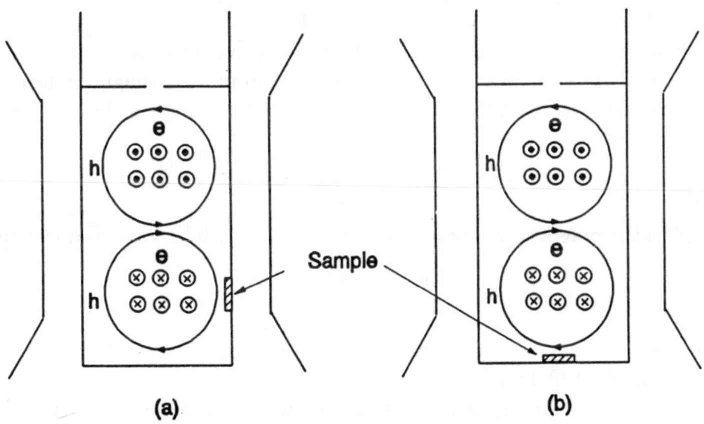

Figure 5.10: The electrical (**e**) and magnetic (**h**) field configuration and the sample position in a rectangular TE_{102} cavity for (a) the traditional FMR experiment and (b) the longitudinal pumping experiment.

ing field **h**. It is assumed that the magnet can rotate along the vertical axis of the cavity. In the traditional transverse pumping experiment, the sample can be mounted on the side wall of the cavity and at the position where the microwave field reaches its maximum value. By rotating the magnet, angular dependence of the FMR experiment is performed while the pumping field is always in the film plane and perpendicular to the bias field. In order to get the spectra shown in Fig. 5.9b, the sample is placed in the center of the bottom plate of the cavity. Through the rotation of the magnet, both **H** and **h** remain in the film plane, however, the relative angle between these two fields continuously changes.

IV. Magnetic Resonance In Multilayer Systems

In reality, many magnetic artificial structures include more than two magnetic layers because in most cases, those systems are more favorable in terms of application purposes. For examples, anti-parallel coupled multilayer structures have larger magnetoresistance than trilayer films. In Co/Pt and Co/Pd systems, multilayer structures also show larger Kerr rotation than the trilayer structures because of a thicker interactive region with the photons. Even from the consideration of magnetic resonance experiment, multilayer structures have some advantages under certain circumstances. From previous discussion, it can be seen that for parallel coupled trilayer systems, a minimum frequency is required for the observation of the optic mode no matter which direction the external field is applied. If the coupling between the magnetic layers is strong, this frequency might exceeds most of the commercially used microwave frequencies. By adding more magnetic layers into the system, more exchange coupled modes will be filled in between the acoustic mode and the optic mode, therefore, the minimum frequency value can be significantly reduced.

The resonance theory in magnetic trilayer systems can easily be expanded to magnetic multilayer structures. Considered a system of N magnetic layers separated by non-magnetic spacers, the energy per unit surface area can be written as

$$\mathcal{E} = \sum_{i=1}^{N} t_i E_i + \sum_{i=1}^{N-1} A_{i,i+1} \frac{\mathbf{M}_i \cdot \mathbf{M}_{i+1}}{M_s^2} \tag{5.18}$$

where E_i is the magnetic energies within the ith layer. The torque equations of motion now becomes

$$\frac{d\mathbf{M}_i}{dt} = \gamma \frac{\mathbf{M}_i}{M_s} \times \left(\frac{1}{t_i} \nabla_i \mathcal{E}\right) \tag{5.19}$$

with i = 1,2, ... , N. The dispersion relation is then obtained from the roots of the determinant of a $(2N) \times (2N)$ matrix similar to that on the left hand side of Eq.(5.9). Half of these solutions give positive values which corresponds to the N exchange modes split from the uniform mode of a single layer system.

The general behavior of the dispersion relation is very complicated for the multilayer system. Therefore in this section, only a simplified system is considered. In this system, the thickness of each ferromagnetic layer is assumed to be constant, $t_i = t_F$, and the separation between successive layers is the same, therefore, the interlayer exchange coupling coefficient is constant (i.e., $A_{i,i+1} = A_{12}$) across the film. Each layer is also assumed to have the same effective demagnetization field, $4\pi M_{eff,i} = 4\pi M_{eff,b}$, except for the first and the Nth magnetic layers which might have a different surface anisotropy energy on one of the surfaces. Furthermore, it is assumed that the external field is applied normal to the film plane ($\mathbf{H} \parallel z$) and large enough so that the magnetization vector in each layer is saturated along the film normal orientation.

Under these conditions, the resonant components of the magnetization vectors are restricted to stay within the film (or x-y) plane and precess in a circular motion around the z-axis, i.e., $|m_{\theta_i}|/|m_{\phi_i}| = 1$. Using the new notation:

$$m_i^{\pm} = m_{\theta_i} \pm i m_{\phi_i} \tag{5.20}$$

the equations of motion can be expressed explicitly:

$$\begin{aligned} &(\frac{\omega}{\gamma} - H_b)m_i^+ - H_{ex}(m_{i-1}^+ + m_{i+1}^+) = 0 \qquad\qquad for\ i = 2, ..., N-1 \\ &(\frac{\omega}{\gamma} - H_1)m_1^+ - H_{ex}m_2^+ = 0 \\ &(\frac{\omega}{\gamma} - H_N)m_N^+ - H_{ex}m_{N-1}^+ = 0\ , \end{aligned} \tag{5.21}$$

where $H_b = H - 4\pi M_{eff,b} - 2H_{ex}$, $H_1 = H - 4\pi M_{eff,1} - H_{ex}$, $H_N = H - 4\pi M_{eff,N} - H_{ex}$ and $H_{ex} = A_{12}/t_F M_s$. The solution of m_i^+ will have the form of

$$m_i^+ \propto cos(i\delta + \delta_0)e^{-iwt}\ , \tag{5.22}$$

and the dispersion relation can be obtained directly from the first relationship of Eq.(5.21),

$$\frac{\omega}{\gamma} = H - 4\pi M_{eff,b} - 4H_{ex}sin^2\frac{\delta}{2}\ , \tag{5.23}$$

where the parameters, δ and δ_0, depend on the surface resonance conditions:

$$\begin{aligned} &(4\pi M_{eff,1} - 4\pi M_{eff,b} - H_{ex})cos(\delta + \delta_0) + H_{ex}cos\delta_0 = 0 \\ &(4\pi M_{eff,N} - 4\pi M_{eff,b} - H_{ex})cos(N\delta + \delta_0) + H_{ex}cos[(N+1)\delta + \delta_0] = 0 \end{aligned} \tag{5.24}$$

If the first and the Nth layer have the same anisotropy field as in the inner layers (i.e., $4\pi M_{eff,1} = 4\pi M_{eff,b} = 4\pi M_{eff,N}$), the parameter, δ, will have N possible values:

$$\delta_n = n\pi/N\ , \tag{5.25}$$

where n = 0,1,2,...N-1. Each δ_n corresponds to an exchange coupled mode across the film thickness which is similar to the spin wave mode in a single layer system with free boundary conditions.[22] However in contrast to a continuous variation of the resonant amplitude across the film thickness as in the single layer system, the resonant amplitude and phase of the magnetization vector are nearly constant within each magnetic layer, but vary significantly from one layer to another in the multilayer system as shown in Fig. 5.11. The $n = 0$ mode is the uniform precession mode, therefore, the resonance field does not depend on the exchange coupling coefficient A_{12}. The $n > 0$ modes are on the higher field side of the uniform mode for anti-parallel coupled systems and on the lower field side of the uniform mode for parallel coupled systems as clearly indicated in Eq.(5.23). These modes, however, can not

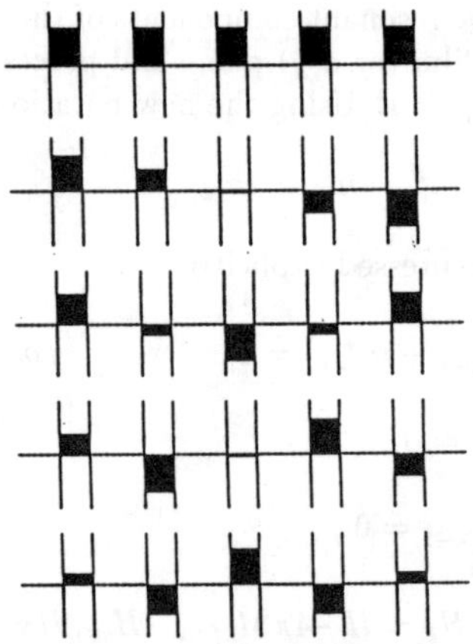

Figure 5.11: The spatial variation of the resonant moment of each exchange coupled mode for a parallel coupled multilayer system having 5 identical magnetic layers equally separated by non-magnetic spacers.

be observed in the FMR experiment because their net resonant moment along the pumping field orientation is zero.

In general, the anisotropy fields of the surface layers are different from those of the inner layers. As a result, the $n = 0$ mode is usually not a uniform mode, but a lowest order bulk mode or a surface mode. The surface mode corresponds to an imaginary solution of δ and the resonant amplitude is large in the surface layer but exponentially decays into the inner layers. For parallel coupled systems, the surface mode exists if $4\pi M_{eff,1} > 4\pi M_{eff,b}$ and/or $4\pi M_{eff,N} > 4\pi M_{eff,b}$. For anti-parallel coupled systems, the surface mode exists if $4\pi M_{eff,1} < 4\pi M_{eff,b}$ and/or $4\pi M_{eff,N} < 4\pi M_{eff,b}$.

Fig. 5.12 shows the dispersion relation of a non-identical multilayer system in which $4\pi M_{eff,1} > 4\pi M_{eff,b} > 4\pi M_{eff,N}$. The surface mode can be observed in both the parallel and anti-parallel coupled systems under this condition. However for $H_{ex} > 0$, the surface mode is localized to the Nth layer while for $H_{ex} < 0$, the surface mode is associated with the first layer. With increasing the exchange coupling strength $|A_{12}|$ the field separation between the surface mode and the first order bulk mode initially decreases, reaches a minimum value, and then increases for strongly coupled systems as shown in Fig. 5.12b. Both the surface mode and the first order bulk mode are easy to observe in the FMR spectra although the intensity ratio between them depends sensitively on the exchange coupling coefficient A_{12}. For weakly coupled systems, the first order bulk mode ($n = 2$ mode) is the dominate mode while for strongly coupled systems the surface mode normally have larger intensity in the FMR spectra. An experimental example is shown in Fig. 5.13.[19]

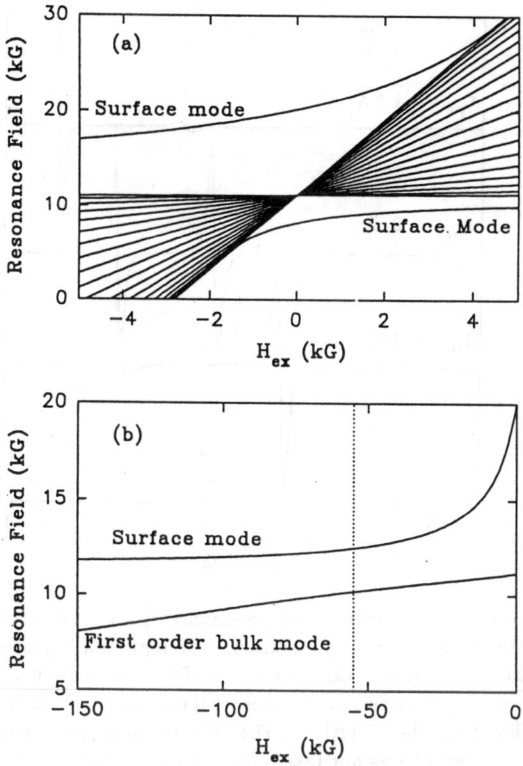

Figure 5.12: The perpendicular resonance fields as a function of H_{ex} ($= A_{12}\ /t_F M_s$) for a magnetic multilayer system at X-band frequency (ω/γ = 3.5 kG). This system include 20 ferromagnetic layers and the effective demagnetization field $4\pi M_{eff}$ is 17 kG for the first layer, 5 kG for the last layer and 8 kG for the other layers. (b) The variation of the surface mode and the first order bulk mode. The horizontal axis is significantly expanded from (a) in order to show the position (dotted line) at which a minimum separation between these two modes is achieved.

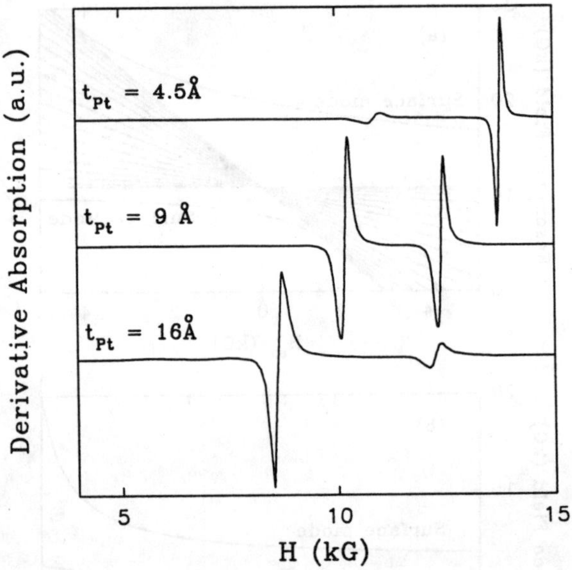

Figure 5.13: The perpendicular ferromagnetic resonance spectra for several [Co25Å)/Pt(t_{Pt})]$_{20}$ samples at 9.8 GHz and room temperature. The high and low field modes are the surface mode associated with the first Co layer and the first order bulk mode. At $t_{Pt} = 9$ Å, the field separation between these two modes reaches its minimum value which corresponds to the dotted line shown in Fig. 5.12(b). (The effective demagnetization field in these samples changes with t_{Pt} which was not taken into account in Fig. 5.12). (from Ref.19)

An analogue to a perfect pinned boundary condition would require that $(4\pi M_{eff,1}, 4\pi M_{eff,N}) \rightarrow -\infty$ for parallel coupled systems and $(4\pi M_{eff,1}, 4\pi M_{eff,N}) \rightarrow \infty$ for anti-parallel coupled systems. Under these situations, the solution of the phase change δ is similar to Eq.(5.25) except that δ of the first and the last modes are π/N and π respectively. From the intensity calculation similar to Eq.(5.17), it is obvious that only the odd numbered modes can couple to the microwave pumping field and their intensity is inversely proportional to n^2.

Such an ideal system does not exist in reality. However some multilayer systems could have a nearly pinned boundary condition. In these cases, a family of exchange coupled resonance modes can to be observed in the FMR spectra. An experimental example is shown in Fig. 5.14.[23]

V. Ferromagnetic Resonance Experiments

Almost all of the FMR experiments on magnetic trilayer and multilayer films

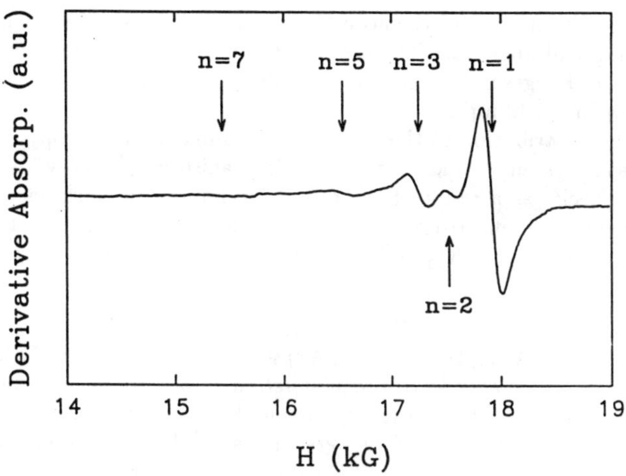

Figure 5.14: The perpendicular FMR spectra for the [Fe(11 Å)/Pt(15 Å)]$_{35}$ sample at 9.2 GHz and room temperature. The arrows indicate the positions of the exchange coupled modes. (from Ref.23)

are centered by two purposes. One is to study the magnetic anisotropy energies within each magnetic layer and the other is to evaluate the interlayer exchange coupling strength between adjacent magnetic layers.

Even though the linewidth of magnetic resonance signals, in general, includes a lot of information about the inhomogeneity[24] and relaxation parameters[25] for normal magnetic materials, it does not provide significant information in the layered structures except serving as a supplementary judgement about the quality of the structure. As a result, the discussion of the FMR linewidth in magnetic multilayer systems is totally neglected.

Due to numerous variations in the composition of magnetic multilayer structures and the nature of the flexibility in FMR experiments, i.e., different resonance frequencies and different magnetic field orientations, the amount of work that has been done over the past several years is enormous and the results are very fruitful and diverse. It is, therefore, quite difficult to give a comprehensive overview of all of the work associated with magnetic resonance in multilayer media. The purpose of this section, instead, will concentrate on some typical experiments that has been used to gain insights into the properties of multilayer structures.

V.1. Magnetic Anisotropy Energy

Two kinds of magnetic anisotropy energies can be evaluated in magnetic multilayer films. One is the perpendicular uniaxial anisotropy energy which can be obtained by measuring the variation of the acoustic resonance field as a function of the out of plane external field angle θ_H. The other is the in-plane anisotropy energy which is reflected by the changes in the acoustic resonance field with the change of the external field angle ϕ_H in the film plane.

A large contribution to the perpendicular uniaxial anisotropy energy in magnetic multilayer systems comes from the surface anisotropy energy[26] which was first discussed by Nèel[27] as a result of the reduced symmetry in surfaces and interfaces. The average uniaxial anisotropy constant K_{u2}, therefore, can be written in terms of a volume anisotropy term K_{u2}^v and a magnetic layer thickness t_F dependent surface term K_s[28]

$$K_{u2} = K_{u2}^v + 2K_s/t_F \, . \tag{5.26}$$

From a series of (275 Å Ag/t_{Fe} Fe/550 Å Ag) structures in which the Fe layer thickness t_{Fe} varied from 24 Å to 80 Å, Hurdequint, *et. al.*[29] observed the linear relationship between $t_F K_{u2}$ and t_F. From the experimental data, they determined the surface anisotropy energy constant K_s of 1.05 erg/cm^2 at each Fe/Ag interface. Similar experiments has been done on a series of Co/Pd(111) structures for which a 9 monolayer (ML) film of Pd was deposited on Co layers that varied from 5 to 12.5 ML.[30] The surface anisotropy coefficient K_s in this system was found to be 0.8 erg/cm^2. Angular dependent of FMR experiment was also performed on a series of $(\mathrm{Fe/Pt})_N$ multilayer films by Wigen, *et. al.*[23] where the repetition number N was 35 or 50. The Fe layer thickness varied from 5 Å to 15 Å while the Pt layer thickness was constant at 15 Å. An average surface anisotropy energy of 0.14 erg/cm^2 was obtained at each Fe/Pt interface.

The existence of a large positive surface anisotropy energy suggests that when the ferromagnetic layer thickness t_F is thinner than a critical value

$$t_{cri} = 2K_s/(4\pi M_s - K_{u2}^v) \, , \tag{5.27}$$

the film perpendicular orientation will become the easy axis, i.e., the magnetization vector will align perpendicular to the film plane at zero external field. This critical thickness is about a few monolayer thick using the above mentioned surface anisotropy energy data. Unfortunately when the magnetic layer thickness is below its critical value t_{cri}, the intensity of the uniform resonance mode is significantly reduced and no signal has been observed in the FMR spectra.

Zhang, *et. al.*[19,31] investigated the development of the induced perpendicular uniaxial anisotropy energy as a function of the Pt thickness in Co/Pt multilayer films. For a series of Co/Pt structures having 20 periods, the Co layers were 25 Å thick and the Pt layers varied from 1 Å to 18 Å. The evolution of the effective out-of-plane uniaxial anisotropy energy $H_{eff} = -4\pi M_{eff}$ was observed to depend sensitively on the Pt thickness t_{Pt} as shown in Fig. 5.15. Under a suitable scale adjustment, it was found (also shown in Fig. 5.15) that the change of H_{eff} with t_{Pt} follows a similar relation as that of the average lattice spacing d determined from the high angle X-ray diffraction measurement. These results suggested that the large surface anisotropy energy is gradually developed at each Co/Pt interfaces and is closely related to the

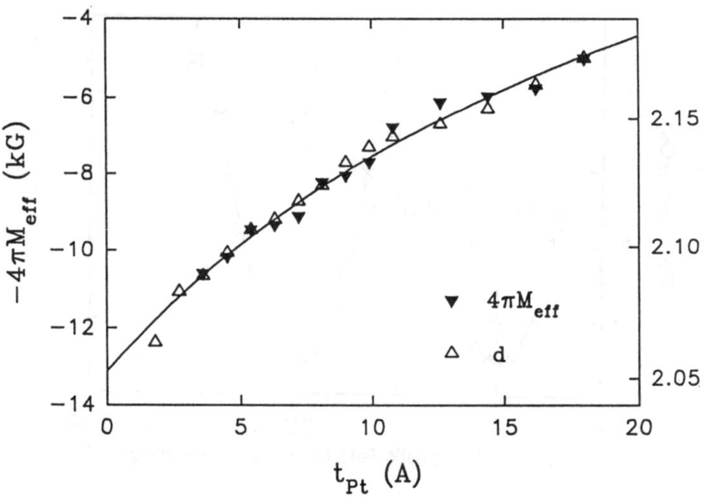

Figure 5.15: The effective demagnetization field $4\pi M_{eff}$ (solid triangles) and the average lattice spacing d (open triangles) as a function of the Pt layer thickness t_{Pt} for a series of $[Co(25\text{ Å})/Pt(t_{Pt})]_{20}$ samples. $4\pi M_{eff}$ and d are evaluated from FMR and X-ray diffraction measurements, respectively. (from Ref.21)

magnetoelastic anisotropy energy associated with the large lattice mismatch between bulk Co and Pt.

With the development of MBE technique, it becomes possible to grow single crystalline multilayer structures. Magnetocrystalline anisotropy energy which is often hidden in polycrystalline films makes a significant contribution to the anisotropy energies in single crystalline systems. Depending on the crystal axis perpendicular to the film plane, the symmetry in the film plane is totally different from one to the other and has been studied extensively by in-plane FMR experiments.

For a film grown along the (001) orientation, the cubic crystalline anisotropy energy shows a four-fold symmetry in the film plane. Heinrich, *et. al.* studied various kinds of single and multilayer thin films using in-plane FMR methods, including Fe(001)/Ag(100),[32] Ni(001)/Fe(001)/Ag(001),[33] and Co(001)/Cu(001)[34] structures. A clear four-fold symmetry was observed in these samples as shown by the variation of the in-plane resonance field with the external field angle ϕ_H in the film plane (Fig. 5.16). In the case of a single crystal Fe layer deposited on Ag substrate, K_1 is close to the bulk Fe value for the samples with the Fe layer thickness larger than 40 Å. Upon decreasing Fe layer thickness, K_1 decreases slowly for $t_{Fe} \geq 20$ Å and more dramatically when the Fe layer thickness is below 10 Å. The deposition of an extra Ni overlayer (> 6 monolayer) on top of the Fe layer enhanced the cubic crystalline

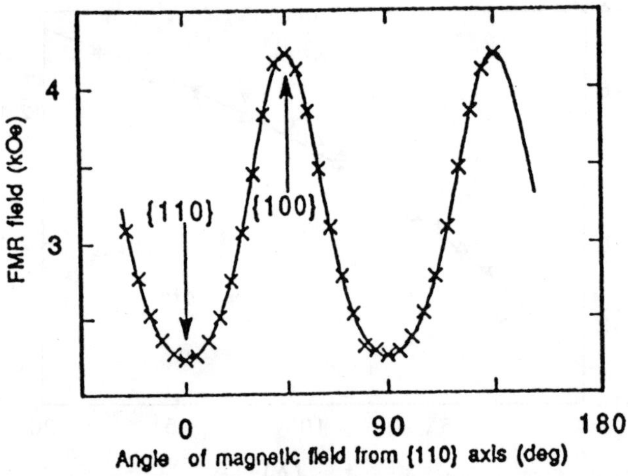

Figure 5.16: The in-plane angular dependence of the FMR field observed for the Co(4.3 ML)/Cu(6 ML)/Co(4 ML) sample. The solid line represents a theoretical fit using the following parameters: $4\pi M_{eff} = -40.8$ kG, $2K_1/M_s = -1.05$ kG, $f = 36.3$ GHz, and $g = 2.16$. (from Ref.34)

anisotropy energy by an order of magnitude due to lattice reconstruction in the Fe/Ni bilayers.

The interest in fcc Co(001)/Cu(001) films comes from the fact that fcc Co is a new metallic phase and only exists in strained ultrathin layered structures. The magnetocrystalline energy K_1 was found to have a negative value in these new systems, i.e., the easy axis is along the [110] orientation in the film plane. This result is confirmed by R. Naik *et. al.*[35] who has also compared the Co(001)/Cu(001) structures with thin fcc Ni(001) and bcc Fe(110) grown on the same fcc Cu(001) substrate. Like the fcc Co films, the fcc Ni layer also has a negative K_1 value, in agreement with the bulk Ni value. The bcc Fe layer, on the other hand, shows a positive value of K_1 which is consistent with the bulk Fe properties.

The first series of single crystal (110) orientated films (in this case, Fe) was grown by Prinz, *et. al.*[36] on GaAs substrates. In-plane FMR measurements show a two-fold symmetry in the film plane which is expected from the expression of the cubic crystalline anisotropy energy. However, in order to explain the variation of the resonance field with the in-plane external field angle and frequency, an extra in-plane uniaxial anisotropy energy is needed and its value varies with the change of the Fe layer thickness t_{Fe}. The easy axis is along $[1\bar{1}0]$ for $t_{Fe} < 50$ Å and [001] for $t_{Fe} > 50$ Å, resulting from the competition between the in-plane uniaxial anisotropy energy and the cubic crystalline anisotropy energy. An induced magnetocrystalline surface anisotropy energy was suggested as the mechanism of this thickness dependent in-plane uniaxial anisotropy term observed in this experiment.

For an (111) orientated single crystal system, the first order cubic crystalline anisotropy energy does not vary within the film plane. As a result, the six-fold variation of the in-plane resonance field with the applied field angle is very small even with the existence of a large K_1 term. Although such a structure is not an ideal system for the evaluation of K_1, it is a perfect system to study small contributions to the anisotropy energies which may also be important for the understanding of the physics in magnetic thin films. Using FMR on several single and multilayer layer Fe(111)/Cu(111) structures, Rezende *et. al.*[37] demonstrated that the in-plane angular dependent FMR spectra can be used to accurately evaluate any in-plane uniaxial anisotropy energy, second order cubic anisotropy energy K_2 and/or misorientation (see Fig. 5.17) that are normally hidden under the large K_1 term in the (001) and (110) orientated multilayer films. Their results indicated that the second order cubic anisotropy term K_2 is about two orders of magnitude smaller than K_1 in the Fe/Cu samples. Very small in-plane uniaxial anisotropy fields, on the order of 10 Oe, and small misorientation of about 1^0 were also observed in those thin films.

V.2. Interlayer Exchange Coupling

FMR is not only a powerful technique to evaluate magnetic anisotropy energies in single and multilayer thin films, but also one of the most effective methods to evaluate interlayer exchange coupling in magnetic multilayer systems. As described in the previous several sections, an important difference between exchange coupled multilayer systems and isolated systems is the existence of the exchange coupled modes in the FMR spectra. Many multilayer systems have been investigated using the FMR technique, such as Fe/NM, where NM = Cr,[9] Cu,[38] Pd,[39] Pt,[23] C,[40] Ag and Au;[41] Co/NM, where NM = Cu,[34] Pt[31] and Ru;[8] and $Ni_{.8}Fe_{.2}$/NM, where NM = Ag,[42] Cu,[43] Cr and Pd.[44] The investigations include both the bilinear and biquadratic exchange coupling coefficients and how they vary with the non-magnetic layer thickness, the magnetic layer thickness and temperature.

(a) variation of A_{12} with non-magnetic layer thickness

Fe/Cr multilayer structure is the first system that generated a lot of interest because of the discovery of anti-parallel exchange coupling[2] and the GMR effect.[5] Using FMR on a series of [40 Å Fe(001)/t_{Cr} Cr(001)/40 Å Fe(001)] samples at different frequencies, Krebs, *et. al.*[9] observed multi-resonance modes in some samples with the external field applied along the $[1\bar{1}0]$ orientation in the film plane. These modes, as shown in Fig. 5.18, correspond to the exchange coupled modes (acoustic mode and optic mode) in the unsaturated region due to the combination of the four-fold cubic anisotropy energy and the strong anti-parallel exchange coupling between the Fe layers. By fitting the resonance data with the exchange coupled FMR theory, they were able to evaluate the bilinear exchange coupling coefficient A_{12} as a function of the Cr spacer thickness. Their results indicated that the two Fe layers are anti-parallel

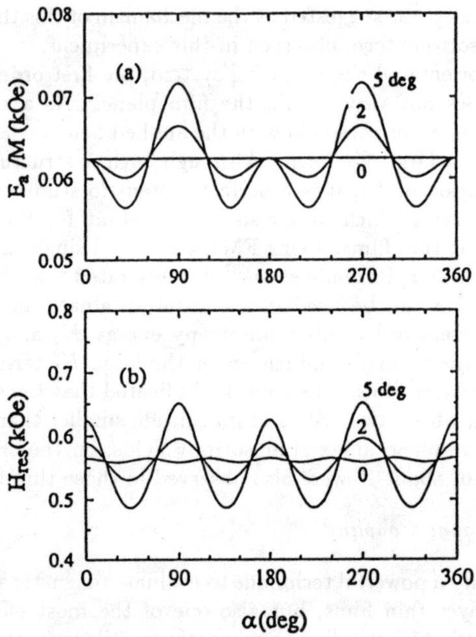

Figure 5.17: Demonstration of the effect of small misorientations of the film plane with respect to the (111) crystal plane in an Fe film with $4\pi M_{eff}$ = 18 kG, $2K_1/M_s$ = 500 Oe. The angle (0, 2^0, or 5^0) is between the film plane and the (111) crystal plane. The vertical axis is the cubic anisotropy energy in (a) and the X-band (9.4 GHz) resonance field in (b). The horizontal axis is the in-plane angle α of the bias field. (from Ref.37)

coupled when the Cr layer thickness is between 12 Å and 25 Å. The coupling strength depends sensitively on the Cr layer thickness in this region and has a maximum value of $A_{12}/t_{Fe}M_s$ = 600 Oe (or A_{12} = 0.4 erg/cm^2) at t_{Cr} = 16 Å. Subsequent studies not only confirmed this result, but also indicated that A_{12} oscillates both in sign[45] and amplitude[4] with the Cr layer thickness and has a long oscillation period of about 18 Å.

Celinski, *et. al.*[41] investigated the interlayer exchange coupling of Fe layers separated by noble metals, i.e., Cu, Ag, Au. By fabricating asymmetrical Fe/NM/Fe trilayer structures in which the two Fe layers have different thicknesses, a difference in the effective magnetic anisotropy energy was produced between the Fe layers which is due to the existence of the surface anisotropy energy at the Fe/MN interface. As a result, both the acoustic mode and the optic mode were observed in the FMR spectra.

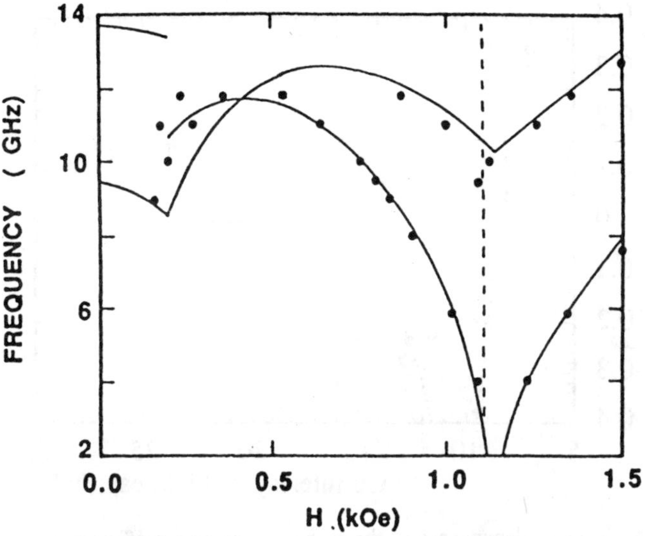

Figure 5.18: Multi-frequency FMR data for the Co(40 Å)/Cr(13 Å)/Co(40 Å) sample with the external field parallel to the hard axis in the film plane [1$\bar{1}$0]. The solid curves are the result of resonance mode calculations using the following parameters: $2K_1/M_s$ = 520 G, $A_{12}/t_{Fe}M_s$ = 300 G, and in-plane uniaxial anisotropy field, $H_{u,i}$ = 10 G. The dashed line marks the saturation field. (from Ref.9)

The magnetic anisotropy energies in each Fe layer was calibrated by growing a single Fe layer structure with the same layer thickness. FMR experiments was performed at 36 GHz, therefore all of the samples were saturated at resonance. From the in-plane field separation between the acoustic mode and the optic mode along with the anisotropy parameters in each Fe layer, A_{12} was evaluated as a function of Cu, Ag, and Au layer thickness. In the bcc Fe(001)/Cu(001)/Fe(001) series, an oscillatory exchange coupling with the Cu layer thickness was observed as shown in Fig. 5.19. The Fe layer are parallel coupled for $t_{Cu} < 8$ ML and anti-parallel coupled for $t_{Cu} > 8$ ML except when t_{Cu} is around 16 ML thick where a weak parallel coupling was observed. A long oscillation period was also observed in this series but the oscillation is added on an overall anti-parallel coupling background. The maximum anti-parallel coupling strength, about 0.2 erg/cm^2 at t_{Cu} = 11 ML is smaller than that observed in the Fe/Cr system.

Similar to the Fe/Cu/Fe system, they also observed the parallel coupling between the two Fe layers in the Fe/Ag/Fe and Fe/Au/Fe systems when the spacer layer thickness is thinner that 8 ML. However, the oscillatory behavior in the exchange coupling was too small to be observed in the latter two series, in particular,

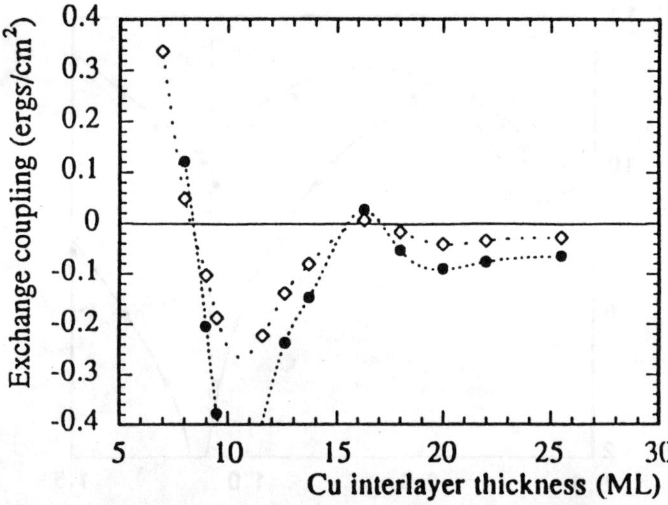

Figure 5.19: The thickness dependence of the exchange coupling coefficient J $(= -A_{12})$ in Fe(5.7 ML)/Cu/Fe(9.7 ML) trilayers. Closed and open symbols correspond to 77 and 295 K, respectively. (from Ref.41)

there was no evidence of anti-parallel coupling in those films. Instead, the coupling strength decreases with increasing spacer layer thickness and becomes too small to be detected when the spacer layer thickness is above 10 ML. Further detailed investigations have shown that the exchange coupling also oscillates with Ag[46] and Au[47] and the oscillation amplitude is indeed very small, in agreement with the FMR results.

Co/Ru is another system that has been extensively investigated because of its large anti-parallel exchange interaction at small Ru layer thickness. Zhang, *et. al.*[8] has studied a symmetrical (32 Å Co/t_{Ru} Ru/32 Å Co) structures using FMR technique. For strongly anti-parallel coupled system in which the magnetization of the Co layer is not saturated at resonance, angular dependence of the acoustic resonance mode as well as longitudinal pumping techniques were used for the evaluation of the interlayer exchange coupling. In some of the samples, the optic mode was observed even with the traditional transverse pumping FMR technique as shown in Fig. 5.4 which is due to a slight variation in the film quality between the two Co layers. Instead of growing separated single layer structures for the evaluation of the anisotropy energies within each layer, those parameters can be automatically justified along with the exchange coupling coefficient by fitting the out-of-plane angular dependent FMR data. A_{12} was found to oscillate both in sign and amplitude with the change of the Ru layer thickness and the oscillation period is about 10 Å. Unlike the coupling behavior through noble metals, the two Co layers are anti-parallel coupled in ultrathin Ru layer samples

($t_{Ru} < 12$ Å). The maximum anti-parallel coupling strength A_{12} was observed to be 1.8 erg/cm^2 at $t_{Ru} = 9$ Å which is the largest value among currently investigated multilayer systems. Similar to the Fe/Cu system, it was found that the oscillatory behavior is built on top of an anti-parallel coupling background in this Co/Ru/Co structures.

Even though the main interest on Co/Pd and Co/Pt multilayer systems is in their perpendicular anisotropy energy and magneto-optical properties, interlayer exchange coupling has also been investigated in these films using FMR techniques. In a series of (25 Å Co/t_{Pt} Pt)$_{20}$ structures reported by Zhang, *et. al.*,[31] the first Co layer was deposited on a Ru buffer layer. This removes one of the Pt interface contributions to the uniaxial anisotropy energy in layer one, which is on the order of 5 KG for $t_{Pt} \geq 10$ Å. As a result, the effective anisotropy energy in layer one is different from those of bulk layers. This produces an unpinned surface condition when the external field is normal to the film plane. Two modes, the surface mode associated with the first Co layer and the first order exchange coupled bulk mode, were observed in the films when $t_{Pt} > 4$ Å. The intensity ratio between these two modes decreases with increasing t_{Pt} as shown in Fig. 5.13, indicating a change in the coupling strength between the Co layers. From the analysis of the positions and intensities of the modes in the FMR spectra, the coupling coefficient, A_{12}, was determined as a function of t_{Pt} which follows an exponential relation

$$A_{12}(t_{Pt}) = A_{12}^0 exp(-t_{Pt}/t_0) \tag{5.28}$$

in the range 4 Å $\leq t_{Pt} \leq$ 18 Å. The maximum coupling strength $|A_{12}^0| = 70$ erg/cm^2 is more than an order of magnitude larger than the maximum anti-parallel coupling value observed in any other systems but is still only 25% of the exchange energy in bulk Co. The decay length t_0 was found to be about 7 Å. A supplemental experiment was performed by Bloemen, *et. al.*[48] on a series of asymmetrical (50 Å Co/t_{Pt} Pt/20 Å Co) trilayer systems where t_{Pt} varied from 4 Å to 100 Å. Both the acoustic mode and the optic mode were observed in the in-plane FMR spectra when the Pt layer thickness is greater than 20 Å. The optic mode, which is on the lower field side of the acoustic mode, is pushed out of the positive external field range for $t_{Pt} < 20$ Å at 34 GHz resonance frequency. Assuming that the effective anisotropy energies in each layer is independent of the Pt layer thickness, A_{12} was evaluated for the films with $t_{Pt} > 20$ Å. The results confirmed the exponential decay of A_{12} with the Pt layer thickness and the decay length is about 8 Å.

Celinski, *et. al.*[41] investigated an asymmetrical Fe(001)/Pd(001)/Fe(001) trilayer structure in which the first and second Fe layers are about 5 and 10 ML thick. In the range of 4 ML $\leq t_{Pd} \leq$ 20 ML, the evaluation results from the FMR experiments show that the Fe layers are always parallel coupled to each other. A_{12} decreases nearly exponentially with the Pd layer thickness even though the data also show a small oscillatory deviation from a pure exponential decay.

Many experimental observations about the bilinear exchange interaction have been qualitatively explained by microscopic theories. The oscillatory behavior of A_{12} with the spacer layer thickness in most of the multilayer structures is similar to the RKKY interaction first introduced in magnetic diluted alloy systems.[49] The long oscillation period and anti-parallel coupling background which are not predicated from the classical RKKY model can be explained as a RKKY-type interaction determined by the detailed topology of the Fermi surface of the spacer material.[50,51] Although there is no clear explanation for the lack of oscillatory exchange coupling through Pt or Pd spacers, it is generally believed that the polarization of the Pt or Pd atoms in the multilayer structures plays a predominate role in the exchange coupling mechanism.

(b) temperature dependence of A_{12}

One advantage of using FMR is its ability to sensitively evaluate any small changes of the interlayer exchange coupling upon changing the external environment, such as temperature. Previous theoretical models[52,53] have suggested that A_{12}, which oscillates in most multilayer systems, does not change its oscillation period at low temperatures. However its amplitude increases with decreasing temperature and follows approximately the relation

$$A_{12} = A_{12}^{0}(\frac{T}{T_0})/\sinh(\frac{T}{T_0}) , \qquad (5.29)$$

where the characteristic temperature

$$T_0 = \hbar v_F / 2\pi k_B t_{NM} \qquad (5.30)$$

is related to the Fermi velocity v_F of the non-magnetic spacer and its thickness t_{NM}.

Zhang, *et. al.*[16] have systematically studies of the temperature dependence of A_{12} on several (32 Å Co/t_{Ru} Ru/32 Å Co) trilayer samples using longitudinal pumping, angular dependent FMR techniques as well as the traditional in-plane FMR spectra at 9 and 23 GHz frequencies. They observed that only the amplitude of A_{12} increases upon decreasing temperature T and its variation, as shown in Fig. 5.20, follows the theoretical prediction of Eq.(5.29) through the whole temperature range. The characteristic temperature T_0 was found to decrease with increasing Ru layer thickness, also in agreement with the previous predictions. When the Ru layer thickness is between 9 Å and 25 Å, T_0 is about 100 to 200 K which corresponds to a typical Fermi velocity of 10^7 cm/s. This value is about an order of magnitude smaller than the estimation for most metals using a free electron gas model.[54]

Low temperature FMR experiments have also been performed on other trilayer structures, such Fe(001)/NM/Fe(001) where NM = Pd and Cu.[41] Although there were not sufficient temperature data in those series to verify the theoretical predictions of Eq.(5.29), all of the results have indicated that the amplitude of A_{12} at 77 K is almost twice as large as the value at room temperature.

Experimental investigation of A_{12} with temperature is still at its early stage. The above mentioned results have shown that this experiment can provide fruitful information about the electronic structures in the non-magnetic materials. So far, FMR is the only experimental technique that has the ability of conducting such measurements. Therefore, it will not be surprised to see more FMR reports in this area in the near future.

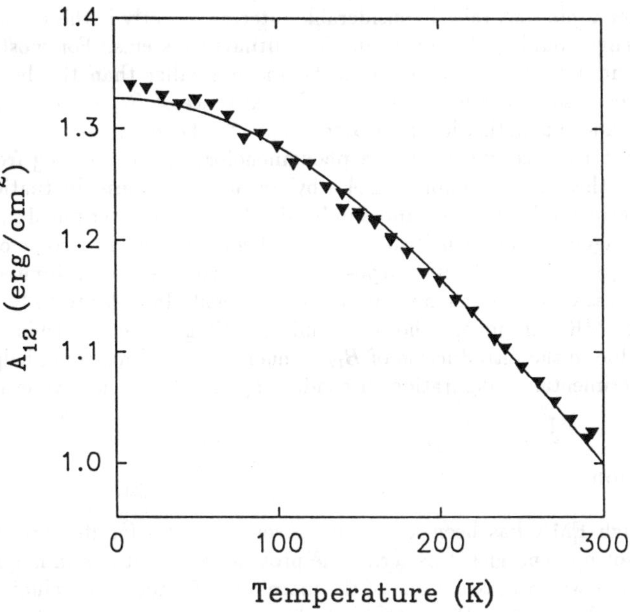

Figure 5.20: The variation of A_{12} with temperature T for the Co(32 Å)/Ru(9.5 Å)/Co(32 Å) sample. The solid line is the best fit using Eq.(5.29) with the parameters: $A_{12}^0 = 1.33$ erg/cm^2 and $T_0 = 225$ K. (from Ref.8)

(c) miscellaneous effects

Currently there is still controversy over the magnetic layer thickness dependence of the interlayer exchange coupling. Theoretically calculation on a symmetrical trilayer system has indicated that A_{12} also oscillates with the magnetic layer thickness as well.[55,56] However, the oscillation amplitude is much smaller than the change with the non-magnetic spacer thickness. Zhou, *et. al.*[57] studied four asymmetrical (32 Å Co/t_{Ru} Ru/t_{Co} Co) trilayer series within which t_{Ru} was constant at 10 Å, 12 Å, 16 Å and 24 Å respectively but the second Co layer thickness t_{Co} varied from 8 Å to 32 Å. From the angular dependent FMR experiments at 9 GHz frequency, they observed the variation of the amplitude of A_{12} with t_{Co} in all four series. However the sign of A_{12} did not alter within each series and the oscillation period was rather large. These experimental results were qualitatively reproduced from microscopic calculation by Stamps[58] using a non-degenerated tight-binding d band model along with some reasonable fitting parameters such as Fermi energy and interfacial overlapping integral.

Another topic that raised considerable interest recently is the biquadratic interlayer exchange coupling B_{12} in magnetic multilayer systems. For most magnetic multilayer structures, $|B_{12}|$ was found to be much smaller than the bilinear term $|A_{12}|$.[8] However in some particular systems, the existence of B_{12} was clearly observed from the in-plane magnetization measurements.[59,60] The mechanism corresponding to this energy term was explained as a phenomenological term arising from spatial fluctuations of the bilinear coupling ΔA_{12} by terraced thickness fluctuations of the spacer.[61] Experimentally, the sign and amplitude of B_{12} were determined mostly from the in-plane magnetization loops. FMR has also been performed on some of the multilayer structures.[12] However, its purpose, at the current stage, is limited to serve only as a verification of the magnetization measurement. In order to fully utilize the advantages of FMR technique, a thorough understanding of the change in the dispersion relation due to the introduction of B_{12} is much needed. This could help to select the right experimental configuration in conducting the most sensitive evaluation of B_{12}.

VI. Conclusion

Although FMR has been known for several decades for its ability to study magnetic anisotropy energies, this technique provides new features in magnetic multilayer structures which makes it one of the more powerful tools to evaluate magnetic properties in such systems. Due to the interlayer exchange coupling, the dispersion relation in magnetic multilayer systems is significantly modified from that of a single layer system. Besides the uniform mode observed (or the acoustic mode) in a single layer system, there also exist a number of exchange coupled bulk modes in multilayer structures each of which corresponds to a nearly uniform precession within each magnetic layer but a significant phase and/or amplitude shift of the resonant moment from one layer to the other. These unique modes in the FMR spectra allows an accurate determination of both the anisotropy energies in each magnetic layer and the interlayer exchange coupling between the magnetic layers.

The applications of the FMR technique in various multilayer structures have been enormous as demonstrated by many reports in the literature. A few of the examples includes the investigation of (1) the perpendicular uniaxial anisotropy energy which helps to understand the surface anisotropy energy, an extremely important phenomena in these multilayer media; (2) the magnetocrystalline anisotropy energy which provides insight to the crystal structure; (3) the interlayer exchange coupling and its variation with the non-magnetic layer thickness, the magnetic layer thickness as well as temperature. These FMR results have provided valuable data for the establishment of microscopic theories, as well as unprecedented experimental support to theoretically predictions.

It is worth to mention that the magnetic properties in ultrathin multilayer systems also depend on the film quality. Different growth conditions can provide different anisotropy energies or different exchange coupling strength even for samples having the same structure. Unfortunately, their are not sufficient experimental data to provide such a comparison which should be important for the foundation of the theoretical modeling.

As more combinations of coupled magnetic/non-magnetic structures are developed and investigated, it is likely that additional new effects and phenomena will be observed. Further application of the FMR technique depends predominately on the development of the resonance theory under new magnetic environments and on the improvement of the experimental configuration in order to optimize the sensitivity of the evaluation.

References

1. L.M. Falicov, S.D. Bader, R. Gronsky, K.B. Hathaway, H.J. Hopster, D.N. Lambeth, S.S.P. Parkin, G. Prinz, M. Salamon, I.K. Schuller, and R.H. Victora, J. Mater. Res., **5**, 1299 (1990).
2. P. Grünberg, R. Schreiber, Y. Pang, M.B. Brodsky, and H. Sowers, Phys. Rev. Lett., **57**, 2442 (1986).
3. F. Saurenbach, U. Walz, L. Hinchey, P. Grünberg, and W. Zinn, J. Appl. Phys., **63**, 3473 (1988).
4. S.S.P. Parkin, N. More, and K.P. Roche, Phys. Rev. Lett., **64**, 2304 (1990).
5. M.N. Baibich, J.M. Broto, A. Fert, F. Nguyen Van Dau, F. Petroff, P. Eitenne, G. Greuzet, A. Friederich, and J. Chazelas, Phys. Rev. Lett., **61**, 2472 (1988).
6. P.F. Carcia, A.D. Meinhaldt, and A. Suna, Appl. Phys. Lett., **47**, 178 (1985).
7. W.B. Zeper, F.J.A.M. Greidanus, P.F. Carcia, and C. R. Fincher, J. Appl. Phys., **65**, 4971 (1989).
8. Z. Zhang, L. Zhou, P.E. Wigen, and K. Ounadjela, Phys. Rev. B, **50**, 6094 (1994).
9. J.J. Krebs, P. Lubitz, A. Chaiken, and G.A. Prinz, Phys. Rev. Lett., **63**, 1645 (1989); J. Appl. Phys., **67**, 5920 (1990).
10. R.P. Erickson, K.B. Hathaway, and J.R. Cullen, Phys. Rev. B, **47**, 2626 (1993).
11. M.E. Filipkowski, C.J. Gutierrez, J.J. Krebs, and G.A. Prinz, J. Appl. Phys., **73**, 5963 (1993).
12. M.E. Filipkowski, C.J. Gutierrez, J.J. Krebs, and G.A. Prinz, presented at the 6th Joint MMM-Intermag conference, Albuquerque, NM, June 20-23, 1994.
13. see, for example, Charles Kittle, *Introduction to Solid State Physics*, p481, John Wiley & Sons, Inc., 1986.
14. P.E. Wigen, Z. Zhang, L. Zhou, M. Ye, and J.A. Cowen, J. Appl. Phys., **73**, 6338 (1993).
15. Z. Zhang, P.E. Wigen, and K. Ounadjela, IEEE Trans. Magn., **29**, 2717 (1993).
16. Z. Zhang, L. Zhou, P.E. Wigen, and K. Ounadjela, Phys. Rev. Lett., **73**, 336 (1994).
17. A. Layadi, and J.O. Artman, J. Magn. Magn. Mater., **92**, 143 (1990).
18. J.F. Cochran, B. Heinrich, and A.S. Arrott, Phys. Rev. B, **34**, 7788 (1988).

19. Z. Zhang, P.E. Wigen, and S.S.P. Parkin, J. Appl. Phys., **69**, 5649 (1991).
20. For some systems, when FMR is operated at very high microwave power, the resonant component **m** of the magnetization vector **M** is on the same order of **M** itself. As a result, the DC component, $\mathbf{M}_H$, of **M** along the bias field orientation is significantly smaller than $|\mathbf{M}|$ (because $\mathbf{M} = \mathbf{M}_H + \mathbf{m}$). If at the same time an oscillatory field is applied along the bias field orientation (so called parallel pumping), it modifies M_H, therefore modifies the precessional amplitude of the system.
21. Z. Zhang, Ph.D. thesis, *The Ohio State University*, 1994.
22. H. Puszkarski, Acta Phys. Polonica, **A38**, 271 (1970).
23. P.E. Wigen, Z. Zhang, S. Iwata, and T. Suzuki, J. Magn. Soc. Jpn., **15**, Supp. **S1**, 33 (1991).
24. J.F. Cochran, J.M. Rudd, M. From, B. Heinrich, W. Bennett, W. Schwarzacher, and W.F. Egelhoff, Jr., Phys. Rev. B, **45**, 4676 (1992).
25. M. Sparks, *Ferromagnetic-Relaxation Theory*, Mcgraw-Hill Inc., New York, 1964.
26. U. Gradmann, J. Magn. Magn. Mater., **54-57**, 733 (1986).
27. L. Nèel, Compt. Rend., **237**, 1468 (1953).
28. H.J.G. Draaisma, F.J.A. den Broeder, and W.J.M. de Jonge, J. Magn. Magn. Mater., **66**, 351 (1987).
29. H. Hurdequint, J. Magn. Magn. Mater., **93**, 336 (1991).
30. S.T. Purcell, H.W. van Kesteren, E.C. Cosman, W.B. Zeper, and W. Hoving, J. Appl. Phys., **69**, 5640 (1991).
31. Z. Zhang, P.E. Wigen, and T. Suzuki, J. Magn. Soc. Jpn., **17**, Suppl. **S1**, 119, 1993.
32. B. Heinrich, K.B. Urquhart, A.S. arrott, J.F. Cochran, K. Myrtle, and S.T. Purcell, Phys. Rev. Lett., **59**, 1756 (1987).
33. B. Heinrich, S.T. Purcell, J.R. Dutcher, K.B. Urquhart, J.F. Cochran, and A.S. Arrott, Phys. Rev. B, **38**, 12879 (1988).
34. B. Heinrich, J.F. Cochran, M. Kowalewshi, J. Krischner, Z. Celinski, A.S. arrott, and K. Myrtle, Phys. Rev. B, **44**, 9348 (1991).
35. R. Naik, C. Kota, J.S. Payson, and G.L. Dunifer, Phys. Rev. B, **48**, 1008 (1993).
36. G.A. Prinz, G.T. Rado, and J.J. Krebs, J. Appl. Phys., **53**, 2087 (1982).
37. S.M. Rezende, J.A. S. Moura, F.M. de Aguiar, and W.H. Schreiner, Phys. Rev. B, **49**, 15105 (1994).
38. B. Heinrich, Z. Celinski, J.F. Cochran, W.B. Muir, J. Rudd, Q.M. Zhong, A.S. Arrott, K. Myrtle, and J. Krischner, Phys. Rev. Lett., **64**, 673, (1990); J.F. Cochran, J. Rudd, W.B. Muir, B. Heinrich, and Z. Celinski, Phys. Rev. B, **42**, 508 (1990).
39. Z. Celinski, B. Heinrich, J.F. Cochran, W.B. Muir, A.S. Arrott, and J. Kirschner, Phys. Rev. Lett., **65**, 1156 (1990).

40. M. Pomerantz, J.C. Slonczewski, and E. Spiller, J. Appl. Phys., **61**, 3747 (1987).
41. Z. Celinski and B. Heinrich, J. Magn. Magn. Mater., **99**, L25 (1991).
42. A. Layadi and J.O. Artman, J. Appl. Phys., **67**, 4451 (1990).
43. J.S.S. Whiting, M.L. Watson, A. Chamber, I.B. Puchalska, H. Niedoba, H.O. Gupta, L.J. Heyderman, J.C.S. Levy, and D. Mercier, IEEE Trans Magn., **26**, 2350 (1990).
44. H. Bosse and H. Gartner, J. Magn. Magn. Mater., **80**, 339 (1989).
45. P. Grünberg, S. Demodritov, A. Fuss, and J.A. Wolf, J. Appl. Phys., **69**, 4789 (1991).
46. J. Unguris, R.J. Celotta, and D.T. Pierce, J. Magn. Magn. Mater., **127**, 205 (1993).
47. S.S.P. Parkin, R.F.C. Farrow, R.F. Marks, A. Cebollada, G.R. Harp, and R.J. Savoy, Phys. Rev. Lett., **72**, 3718 (1994).
48. P.J.H. Bloemen, E.A.M. van Alphen, W.J.M. de Jonge, and E.J.A. den Broeder, proceeding of 13th ICMFS, Glasgow, UK (1991).
49. K. Yosida, Phys. Rev., **106**, 893 (1957).
50. Y. Wang, P.M. Levy, and J.L. Fry, Phys. Rev. Lett., **65**, 2732 (1990).
51. P. Bruno and C. Chappert, Phys. Rev. Lett., **67**, 1602 (1991).
52. D.M. Edwards, J. Mathon, R.B. Muniz, and M.S. Phan, Phys. Rev. Lett., **67**, 493 (1991).
53. P. Bruno and C. Chappert, Phys. Rev. B, **46**, 261 (1992).
54. for example, see N.W. Ashcroft and N.D. Mermin, *Solid State Physics*, p38, Saunders College HRW, 1976.
55. J. Barnas, J. Magn. Magn. Mater., **111**, L215 (1992); J. Magn. Magn. Mater., **128**, 171 (1994).
56. P. Bruno, Europhys. Lett., **23**, 615 (1993).
57. L. Zhou, Z. Zhang, P.E. Wigen, and K. Ounadjela, presented at the 6th Joint MMM-Intermag Conference, Albuquerque, New Mexico, June 20-23, 1994.
58. R.L. Stamps, *et. al.*, unpublished.
59. M. Ruhrig, R. Schafer, A. Hubert, R. Mosler, J.A. Wolf, S. Demokritov, and P. Grünberg, Phys. Status Solidi (a), **125**, 635 (1991).
60. Z. Celinski, B. Heinrich, and J.F. Cochran, J. Appl. Phys., **73**, 5966 (1993).
61. J.C. Slonczewski, Phys. Rev. Lett., **67**, 3172 (1991).

PART II

NONSTATIONARY AND NONLINEAR MICROWAVE PROCESSES

PROPAGATION AND INTERACTION OF MAGNETOSTATIC SPIN WAVES IN PLANAR MAGNETIC STRUCTURES WITH NON-STATIONARY PARAMETERS

Y.K.FETISOV

Department of Physics, Moscow Institute of Radioengineering, Electronics and Automation,

Vernadskogo 78, Moscow 117454, Russia

1. Introduction

Wave processes in non-stationary dispersive media attract attention for a long time because of various possibilities of the waves transformation and interaction under conditions of a slow variation of the medium parameters. List of phenomena that could be observed in the medium with non-stationary parameters includes different kinds of amplitude, phase and polarization modulations, frequency spectrum transformation, space-time focusing of the wave packets, different kinds of parametric interactions of the waves, etc..

The main part of investigations performed up to this time was devoted to the study of propagation of pure electromagnetic waves in non-stationary media. The effects of frequency conversion and reflection of waves in a homogeneous dielectrics at a sudden change of the dielectric permeability were considered theoretically,[1] various kinds of frequency modulation and compression of electromagnetic pulses in a magnetized plasma with slowly time-varying permeability were repeatedly discussed,[2-4] the possibility of electromagnetic wave frequency conversion in a ferroelectrics placed in a time-varying electric field was mentioned.[5]

Propagation of magnetoelastic waves in crystals with controlled acoustic parameters under non-stationary conditions was also investigated. In particular, a specific mutual conversion of acoustic waves and spin waves near magnetoelastic resonance frequencies in ferrites[6] and transformation of magnetoelastic wave frequency spectrum in antiferromagnets[7] placed in a time-varying magnetic field have been observed.

The first investigations of magnetostatic spin wave (MSW) propagation in ferrites under non-stationary conditions were carried out at the end of the seventies. Longitudinally magnetized yttrium iron garnet (YIG) rods were used in experiments. A non-stationarity was created by means of the external bias field modulation. The effects of frequency spectrum transfer of MSW pulses at a sudden change of the magnetic field,[8] serrodyne shift of the MSW frequency at a slow linear field modulation,[9] and compression of short MSW pulses in a non-stationary magnetic field[10] have been observed.

Qualitatively new possibilities for investigation of wave processes in non-stationary media were opened by the studies on MSW propagation in planar magnetic structures based on thin epitaxial YIG films. MSWs in planar structures have substantial advantages for the non-stationary experiments: wide variety of dispersion characteristics, strong dispersion, relatively low excitation and propagation losses, possibility of realization of single-mode and multi-mode propagation regimes, and high degree of homogeneity of the structure parameters in the direction of the wave propagation.

Characteristics of MSW in planar magnetic structures are determined by a set of parameters incoming in the dispersion equation. These parameters depend on the magnetizing field, properties of the ferrite film, and boundary conditions on the film surfaces. A non-stationarity condition in the structure may be created by time-variation of any of the above mentioned parameters. For example, in a planar ferrite-dielectric-conductor structure non-stationarity may be realized by the following methods:

- by changing of the magnitude or direction of the external magnetizing field;
- due to modulation of the dielectric gap between the ferrite and the

conductor;[11]

- by changing the conductivity of the conductor under the action of a light beam;[12]
- by means of modulation of the internal magnetic field using the effect of magnetostriction;[13]
- by changing the saturation magnetization of the ferrite as a result of the pulsed optical heating of the film.

The modulation of the external magnetizing field magnitude (or orientation) is the most natural method to create non-stationarities in such structures. Modulation of the field allows to control effectively the MSW characteristics. The wave frequency can be tuned over tens of GHz, the MSW wavelength, phase and group velocities can be changed by one or two orders in magnitude when the field variation is no more then hundreds of Oe. Thus, our attention below will be mainly focused on this method of non-stationarity creation in planar magnetic structures.

In the first part of this chapter we summarize the theory of propagation of MSW packets in a ferrite film placed in a homogeneous slowly varying external magnetic field. Then, we discuss the most important phenomena accompanying MSW propagation in planar magnetic structures with non-stationary parameters experimentally observed and investigated up to this time. They include different kinds of wave frequency spectrum transformations, compression and spreading of MSW packets, distortion of MSW rays trajectories, and spectral separation of MSW modes. The results of experimental investigations of different parametric interactions of MSWs in non-stationary magnetized ferrite structures are also presented. In conclusion, we discuss perspectives of further investigations and applications of MSW propagation in planar magnetic structures with non-stationary parameters.

2. Theory of MSW packet propagation in a non-stationary medium

Let us consider the characteristic features of MSW propagation in a structure with non-stationary parameters on the example of an MSW packet propagation in a free ferrite film placed in a slowly varying bias magnetic field.[14,15] The choice of the bias magnetic field as a time-varying parameter is not of a principle importance here, it is determined only by the convenience of consideration and interpretation of experimental results.

We consider a ferrite film of the thickness d placed in the **x-y** plane, in such a way that its surfaces coincide with the planes $z = -d/2$ and $z = d/2$. An external homogeneous saturating magnetic field $\mathbf{H}_e$ has an arbitrary direction with respect to the plane of the film. The MSWs are excited by an external microwave magnetic field $\mathbf{h}^{ex}(\mathbf{r},t)$ with a carrier frequency ω.

In the magnetostatic approximation the microwave magnetic field $\mathbf{h}(\mathbf{r},t)$ and magnetization $\mathbf{m}(\mathbf{r},t)$ of MSW are given by the solutions of the magnetostatic equations and the equation of motion for the magnetization when corresponding electromagnetic and exchange boundary conditions on the film surfaces are taken into account:

$$\text{rot } \mathbf{h}(\mathbf{r},t) = 0, \qquad \text{div } [\mathbf{h}(\mathbf{r},t) + 4\Pi\mathbf{M}(\mathbf{r},t)] = 0, \tag{1}$$

$$\frac{\partial \mathbf{M}}{\partial t} = \gamma\, [\mathbf{H}_{eff}, \mathbf{M}\,], \tag{2}$$

where γ is the gyromagnetic ratio, $\mathbf{M} = \mathbf{M}_0 + \mathbf{m}(\mathbf{r},\mathbf{t})$, and $\mathbf{M}_0$ is the equilibrium magnetization of the ferrite.
The effective magnetic field may be written in the form:

$$\mathbf{H}_{eff} = \mathbf{H} + \mathbf{h}(\mathbf{r},t) + \alpha_{ex} \nabla^2 \mathbf{m}(\mathbf{r},t) + \mathbf{h}^{ex}(\mathbf{r},t), \tag{3}$$

where $\mathbf{H}$ is the internal magnetic field in the ferrite, α_{ex} is the constant of non-uniform exchange.

In discussion that follows we are not taking into consideration the field of non-uniform exchange ($\alpha_{ex} = 0$) that is usually acceptable for the film thicknesses $d \geq 5\ \mu m$ and MSW wave numbers $k \leq 10^5\ cm^{-1}$.

For the case of a stationary magnetic field and monochromatic external signals the problem is solved by the method of separation of variables and MSW fields may be represented in the form of the Fourier integral expansion in wave vectors $\mathbf{k}$ in the film plane:

$$\mathbf{h}(\mathbf{k}, t) = \int \mathbf{h}_k(z, t) e^{-i\mathbf{kr}} d\mathbf{k}, \tag{4a}$$

$$\mathbf{m}(\mathbf{r}, t) = \int \mathbf{m}_k(z, t) e^{-i\mathbf{kr}} d\mathbf{k}. \tag{4b}$$

The Fourier components of the field $\mathbf{h}_k(z, t)$ and magnetization $\mathbf{m}_k(z, t)$ are given as a set of eigenmodes of the ferrite film:

$$\mathbf{h}_k(z, t) = \sum_n \mathbf{h}^0_{kn}(z) A_{kn} e^{i\omega_{\mathbf{kn}} t}, \tag{5a}$$

$$\mathbf{m}_k(z, t) = \sum_n \mathbf{m}^0_{kn}(z) A_{kn} e^{i\omega_{\mathbf{kn}} t}, \tag{5b}$$

where $\mathbf{m}^0_{\mathbf{kn}}(z) = \hat{\chi}(\omega_{\mathbf{kn}})\, \mathbf{h}^0_{\mathbf{kn}}(z)$, $\hat{\chi}(\omega)$ is the high-frequency magnetic susceptibility tensor, which specific form depends on the orientation of the magnetic field with respect to the film plane, $\omega_{\mathbf{kn}} = \omega_{\mathbf{n}}(\mathbf{k})$ is the dispersion law for the n-th mode (dependence of the mode frequency on the wave vector $\mathbf{k}$), and $A_{\mathbf{kn}}$ is the mode amplitude. The distributions of the field $\mathbf{h}^0_{\mathbf{kn}}(z)$ and the magnetization $\mathbf{m}^0_{\mathbf{kn}}(z)$ along the film thickness may be easily found using the known distributions of the magnetostatic potential.[16]

When the external bias field is varied in time or when the microwave field which excites MSW is not monochromatic, the method of separation of variables, strictly speaking, can not be used and the expansion (4) in the set of spin wave eigenmodes is incorrect.

However, if the bias magnetic field is varied slowly in comparison with the wave frequency $(\omega\tau)^{-1} \ll 1$ (where τ is the typical time of the field variation) and the microwave excitation field is quasi-monochromatic $\Delta\omega/\omega_0 \ll 1$ (where $\Delta\omega$ is the width of the spectrum of the external signal), the expressions (4), (5) are valid approximately. In such a case MSW fields can be written, taking into consideration the homogeneity of the medium, as before, in the form of a superposition of waves with fixed wave vectors (4) and time-dependent amplitudes and phases

$$\mathbf{h}_k(z,t) = \sum_n \mathbf{h}_{kn}(z,\varepsilon t)\, A_{kn}(\varepsilon t)\, e^{i\frac{\Psi_{kn}(\varepsilon t)}{\varepsilon}}, \tag{6a}$$

$$\mathbf{m}_k(z,t) = \sum_n \mathbf{m}_{kn}(z,\varepsilon t)\, A_{kn}(\varepsilon t)\, e^{i\frac{\Psi_{kn}(\varepsilon t)}{\varepsilon}}, \tag{6b}$$

where $A_{\mathrm{kn}}(\varepsilon t)$ and $\Psi_{\mathrm{kn}}(\varepsilon t)$ are the slowly varying in time amplitudes and nonlinearly depending on time phases of the waves, respectively, and $\varepsilon \sim \Delta\omega/\omega_o \sim (\tau\omega_o)^{-1} \ll 1$ is the dimensionless small parameter of the problem. Using this parameter we are simultaneously taking into account slowness of the bias magnetic field variation and quasi-monochromatic character of the external microwave signal. We consider below the single-mode regime of the MSW propagation and omit the mode number n in all formulae. Peculiarities of the multimode MSW propagation under non-stationary conditions will be discussed in Section 5.

Let us substitute the integral expansions (4) and expressions (6) for $\mathbf{h}(\mathbf{k},t)$ and $\mathbf{m}(\mathbf{k},t)$ in the linearized equation of motion for the magnetization (2). Then, we can solve magnetostatic equations (1) using the Green's formula and taking into consideration electrodynamic boundary conditions on the film surfaces. As a result, we obtain the equation of motion for the high-frequency magnetization in the following integral form:

$$\mathbf{m}_k(z,\varepsilon t) + 4\Pi\hat{\chi}_k\nabla_k \int_{-d/2}^{d/2} \mathbf{m}_k(z',\varepsilon t)\nabla'^{*} G(z,z')\mathrm{d}z' = \mathbf{J}_k, \tag{7}$$

where $$\mathbf{J}_k = \hat{\chi}_k \mathbf{h}_k^{ex} e^{i\omega_0 t - \frac{i}{\varepsilon}\Psi_k} + \varepsilon \hat{T}_k \frac{\partial \mathbf{m}_k}{\partial(\varepsilon t)},$$

$\hat{\chi}_k = \hat{\chi}(\omega_k, \varepsilon t)$ is the high-frequency magnetic susceptibility tensor, the components of which are slowly varying functions of time,

$$\nabla_k = \mathbf{e}_z \frac{\partial}{\partial z} - i\mathbf{k}, \quad G(z,z') = -\frac{1}{2k} e^{-|z-z'|}, \quad \hat{T} = [i\omega \hat{I} - \hat{\kappa}\, \gamma H(\varepsilon t)]^{-1},$$

$G(z,z')$ is the Green's function, $\hat{I}$ and $\hat{\kappa}$ are the symmetrical and anti-symmetrical unit tensors respectively, $\omega_k(\varepsilon t)=\partial\Psi_k(\varepsilon t)/\partial(\varepsilon t)$ is the time-dependent MSW dispersion law, $H(\varepsilon t)$ is the internal time-dependent magnetic bias field in the ferrite, and $\mathbf{h}_k^{ex}(z,t)$ is a Fourier component of the external microwave pulsed signal:

$$\mathbf{h}_k^{ex}(\mathbf{r},t) = A(\varepsilon t) e^{i\omega_0} \int_{-d/2}^{d/2} \mathbf{h}_k^{ex}(z,\varepsilon t) e^{-i\mathbf{kr}} d\mathbf{k}\ , \tag{8}$$

which has the carrier frequency ω_0 and time envelope $A(\varepsilon t)$.

The Fourier component $\mathbf{h}_k(z,\varepsilon t)$ of the MSW field may be calculated as follows:

$$\mathbf{h}_k(z,t) = 4\Pi\nabla_k \int_{-d/2}^{d/2} \mathbf{m}_k(z',z)\nabla'^{*} G_k(z,z')dz' \tag{9}$$

The formulation of the problem presented above is convenient for use in asymptotic perturbation theory. Equation (7) automatically takes into account boundary conditions when the successive terms in the expansion of the Fourier component in the small parameter ε are calculated.

Let us try to find a solution of Eq.(7) in the form of the following expansion in powers of the small parameter ε:

$$\mathbf{m}_k(z,\varepsilon t) = A_k(\varepsilon t)\ \mathbf{m}_k^0(z,\varepsilon t) + \sum_{l=0} \varepsilon^l \mathbf{m}_k^l\ (z,\varepsilon t). \qquad (10)$$

Setting formally $\mathbf{h}^{ex} \sim \varepsilon$, we find that the zero-th order (in ε) approximation corresponds to the case when Eq. (7) is homogeneous ($J_k \to 0$ for $\varepsilon \to 0$). Solutions of this equation give distributions of the magnetization $\mathbf{m}_k^0(z,\varepsilon t)$ that coincide with the distributions obtained in the stationary case. The frequency $\omega_k(\varepsilon t)$ satisfies the usual dispersion equation where the time-dependent bias magnetic field $H(\varepsilon t)$ is introduced.

In the first order approximation in ε a non-trivial solution of the non-homogeneous Eq. (7) is found from the condition of removal of singularity for the first order correction in perturbation theory. This condition, as it was shown in,[15] gives the following equation for the slowly varying amplitudes $A_k(\varepsilon t)$ of the MSW:

$$\varepsilon \frac{\partial A_k}{\partial(\varepsilon t)} + \varepsilon\lambda_k(\varepsilon t)\ A_k + A(\varepsilon t)\ C_k(\varepsilon t)\ e^{i\omega_0 t - \frac{i\Psi_k}{\varepsilon}} A_k = 0, \qquad (11)$$

where $\lambda_k(\varepsilon t) = \frac{1}{b_k} \int_{-d/2}^{d/2} \mathbf{h}_k^{0*}\ \hat{T}_k \frac{\partial \mathbf{m}_k^0}{\partial(\varepsilon t)}\, dz$, $\hat{C}_k(\varepsilon t) = \frac{1}{b_k} \int_{-d/2}^{d/2} \mathbf{h}_k^{0*}\ \hat{\chi}_k\ \mathbf{h}_k^{ex}\, dz$,

and

$$b_k(\varepsilon t) = \int_{-d/2}^{d/2} \mathbf{h}_k^{0*}\ \hat{T}_k\ \mathbf{m}_k^0\, dz\ .$$

The coefficient λ_k takes into account time dependence of the susceptibility and also possible non-stationary nature of the MSW field distribution along the film thickness. The coefficient C_k determines the efficiency of excitation of MSW with wave number k at the time t.

The damping of MSW can be easily introduced in Eq. (11) by adding the item αA_k in which the decay constant α (in dB/μs) is connected with the uniform ferromagnetic resonance linewidth ΔH (in Oe) by the relation $\alpha = 17.5\ \Delta H$.

Integrating Eq. (9) with the phenomenologically introduced dissipation and using equations (3), (5) and (8) we finally obtain the following expression for the high-frequency field of the MSW packet propagating in a non-stationary magnetized ferrite film in an arbitrary direction:

$$\mathbf{h}(\mathbf{r},t) = -\int_{-\infty}^{\infty}\int_{-\infty}^{t} \mathbf{h}_k^0(z,\varepsilon t')A(\varepsilon t')C(t')e^{-\alpha(t-t')}\,e^{i\Psi(\mathbf{k},t')}\,d\mathbf{k}dt', \tag{12}$$

where
$$\Psi(\mathbf{k},t') = \omega_0 t' - \mathbf{kr} + \int_{t'}^{t} \omega(\mathbf{k},t'')dt'' . \tag{13}$$

For the particular case when MSW packet is excited in the plane $y = 0$ of the ferrite film and propagates strictly along the y axis, the expression (12) takes the following form (here and below ε will be omitted in the formulae):

$$\mathbf{h}(y,z,t) = \int_{-\infty}^{\infty}\int_{-\infty}^{t} \mathbf{h}_k^0(z,t')A(t')C_k(t')e^{-\alpha(t-t')}\,e^{i\Psi(k,t')}\,d\mathbf{k}dt' , \tag{14}$$

where the phase shift of the wave after travelling a distance y is:

$$\Psi(k,t') = \omega_0 t' - ky + \int_{t'}^{t} \omega(k,t'')dt'' . \tag{15}$$

The integral (14) can be calculated for sufficiently long initial microwave pulses using the method of stationary phase and neglecting in the expansion of the phase ΨΦ in k and t near the stationary point $(\tilde{k}_j , \tilde{t}_j)$ all terms higher than quadratic.

The stationary points are found from the system of equations:

$$\int_{\tilde{t}'}^{t} V_{gk}(\tilde{k}_j\,,\ t'')dt'' = y, \qquad \omega(\tilde{k}_j, \tilde{t}'_j) = \omega_0\,, \tag{16}$$

where $V_g(k,t) = -\frac{\partial\omega}{\partial k}(k,t)$ is the group velocity of the wave, $\tilde{k}_j$ is the MSW wave number at the stationary point, and j is the number of the stationary point. It is assumed that the dispersion law has no extrema at $k = \tilde{k}_j$. After integrating we find that the envelope of the MSW packet, after travelling a distance y, repeats the form of the initial microwave pulse. However, there is a transformation of the amplitude and time scales which is determined by the relation $\tilde{t}'(t)$:

$$\mathbf{h}(y,z,t) = \sum_j \mathbf{h}_{\tilde{k}_j}(z,t)\ C_{\tilde{k}_j}(\tilde{t}'_j)B_j(y,t)e^{i\Psi(\tilde{k}_j,\tilde{t}'_j)}\,, \tag{17}$$

$$B_j(y,t) = \left(\frac{\partial\tilde{t}'_j}{\partial t}\right)^{1/2} \frac{2\Pi\ A_j(\tilde{t}'_j)\ e^{-\alpha(t-\tilde{t}'_j)}}{\sqrt{V_g(\tilde{k}_j,t)V_g(\tilde{k}_j,\tilde{t}'_j)}}\,, \tag{18}$$

$$\frac{\partial\tilde{t}'_j}{\partial t} = \frac{V_g(\tilde{k}_j,t)\ V_g(\tilde{k}_j,\tilde{t}'_j)}{V_g^2(\tilde{k}_j,\tilde{t}'_j) + \frac{\partial\omega}{\partial t}(\tilde{k}_j,\tilde{t}'_j)\displaystyle\int_{\tilde{t}'_j}^{t}\frac{\partial\omega^2}{\partial k^2}(\tilde{k}_j,t'')dt''}\,.$$

The above expressions describe the main features of the process of MSW packet propagation in a non-stationary medium.

3. Transformation of MSW frequency spectrum in a non-stationary medium

Transformation of a frequency spectrum is one of the most striking effect accompanying the MSW propagation in media with non-stationary parameters. Using the results of the previous section we can now consider different kinds of MSW frequency spectrum transformations in a ferrite film magnetized by a non-stationary bias magnetic field.

As it follows from Eqs. (15) and (17), the instantaneous MSW frequency at the point y at the instant t is given by the expression:

$$\omega(y,t) = \frac{\partial\Psi(\tilde{k},\tilde{t}')}{\partial t} = \omega(\tilde{k},t) \ . \tag{19}$$

Equations (16) and (19) allow us to give the following interpretation to the processes of the MSW frequency conversion under non-stationary magnetic field conditions:

- different fragments of an MSW packet acquire the same frequency $\omega(\tilde{k},\tilde{t}') = \omega_0$, but different wave numbers $\tilde{k}$ at the instant of excitation $\tilde{t}'$;
- during the processes of propagation, because of the spatial homogeneity of the medium, the wave number $\tilde{k}'$ of the fragment is being conserved, while the frequency is being changed;
- the fragment of the packet with a wave number $\tilde{k}$ excited at the instant $\tilde{t}'$, covers the distance y during the time $\tau_d = t-\tilde{t}'$ and its frequency changes from $\omega(\tilde{k},\tilde{t}')$ to $\omega(\tilde{k},t)$.

The character of the MSW frequency transformation depends on the relation between the duration of the wave packet τ and the time of the packet propagation τ_d.

The most simple situation is realized if the packet length τ is smaller than τ_d and all changes of the field start and finish at the instants when the packet is being entirely in the ferrite film. In this case all the fragments of the packet acquire the same wave number and, consequently, the same frequency shift. The magnitude of the shift Δf does not depend on the field modulation law, and is determined only by the magnitude δH of the field variation $\Delta f=\gamma\delta H$.

If the packet length τ is considerably larger than the time of propagation τ_d, then the different fragments of the packet acquire different frequency shifts. In the case of small field changes during the time of packet propagation $((\partial H/\partial t)\tau \ll H)$ the frequency shift is given by the expression:

$$\Delta\omega = \omega(k,t) - \omega(k,t') \simeq \frac{\partial\omega}{\partial t}(t-t') = \left.\frac{\partial k}{\partial H}\right|_{H_0,\,\omega_0} \frac{\partial H}{\partial t}\, y \, . \tag{20}$$

Equations (19) and (20) allow us to calculate the MSW frequency spectrum transformation for all cases of practical interest.

Experimental investigations of the MSW frequency spectrum transformation were carried out using a transmission line configuration. A typical transmission line and the experimental set-up are shown in Fig. 1. The line contained a YIG film of the thickness d = 5-20 μm, saturation magnetization $4\pi M$=380-1750 G and ferromagnetic resonance (FMR) linewidth ΔH≤0.6 Oe, grown by a liquid-phase-epitaxy method on a gallium-gadolinium garnet (GGG) substrate of the dimensions 4x20 mm^2, thickness 0.6 mm, and orientation [111]. Two microstrip transducers of the width of 50 μm deposited on an alumina substrate were used to excite and receive MSWs. The YIG film was placed in a close contact with the transducers. The structure was placed in a homogeneous external DC bias magnetic field H_e, directed normally or tangentially to the film plane

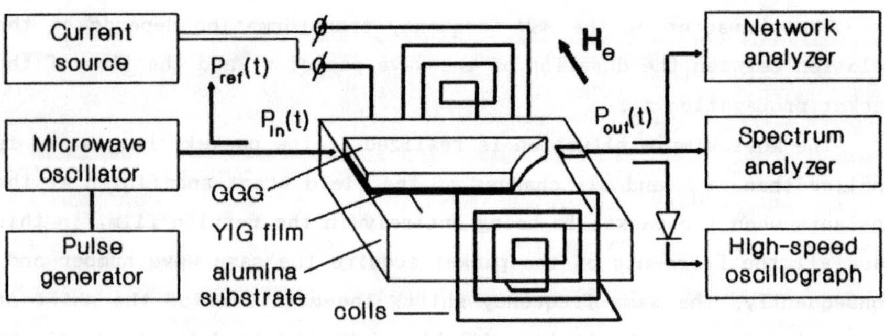

Fig. 1 MSW transmission line and the experimental set-up.

to excite the magnetostatic forward volume wave (MSFVW) or magnetostatic backward volume wave (MSBVW) and magnetostatic surface wave (MSSW), respectively. We used an electro-magnetic system in the form of low inductance coils or a microstrip deposited on the free side of the GGG substrate for the bias field modulation. The AC current in a magnetic system has a magnitude of up to 150 A . This modulation system provided a pulsed variation of the bias field of about 100 Oe at a time-rate of s=0-500 Oe/μs or harmonical modulation of the field with a frequency of up to F=100 MHz and a magnitude of up to h_m=10 Oe. In the frequency range f_0=0.7-10 GHz at the input microwave power levels P_{in}≤10 mW we registered frequency spectra of continuous or pulsed (τ=0.1-50 μs) microwave signals transmitted through or reflected from the MSW transmission line.

We investigated experimentally several characteristic cases of the MSW frequency spectrum transformation in a non-stationary magnetized YIG films corresponding to different values of the ratio between the packet length τ and the time of propagation τ_d and various laws of the field modulation.

3.1 *Sudden change of the field*

Fig.2 explains the principle of the frequency transformation of the MSW packet having duration τ which is smaller than the time of

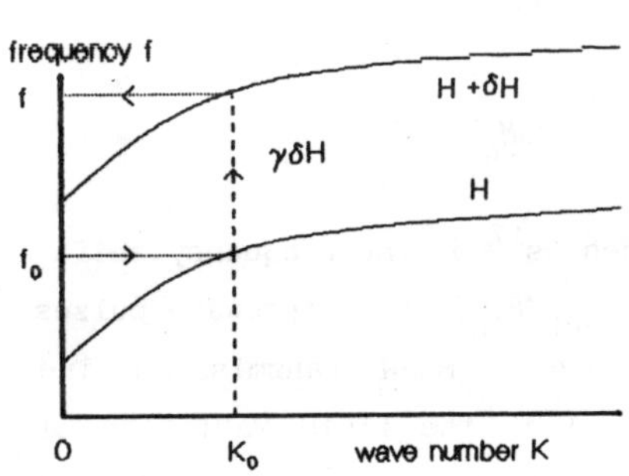

Fig. 2 Conversion of the MSW packet frequency in a pulsed magnetic field.

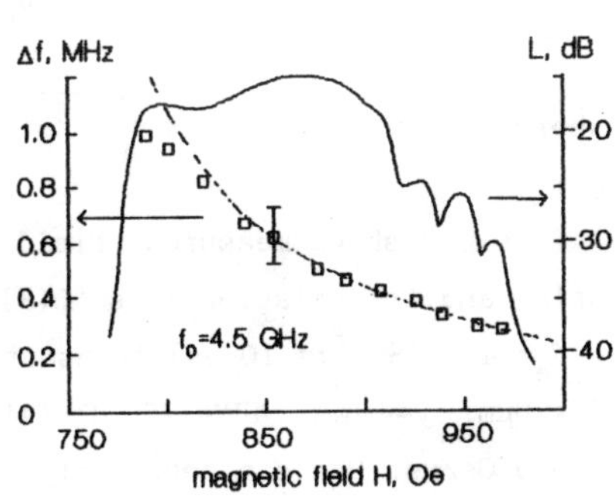

Fig. 3 Dependences of the frequency shift Δf and transmission loss L for MSSW propagating in a YIG film at a bias magnetic feld H. Solid lines and dots are the experimental data, dashed line is the theory.

packet propagation τ_d in the case of a sudden change of the bias magnetic field strength. Two dispersion curves correspond to the initial H and final $H+\delta H$ values of the field. All the frequency spectrum of the packet is shifted by the amount of $\Delta f=\gamma\delta H$. In the experiment the field change of $\delta H \cong 25$ Oe during the time interval $\Delta t \cong 25$ ns has been achieved and corresponding upward and downward frequency shifts of up to $\Delta f=70$ MHz for MSSW packets of the duration $\tau \cong 100$ ns were observed. This value of the frequency shift is not the highest possible. A shift of MSBVW pulses frequency of up to 186 MHz has been previously reported.[8]

3.2 *Linear modulation of the field*

At the linear (or sawtooth-like) modulation of the bias field strength $H(t) = H_0 + st$ the MSW frequency shift is determined by the wave dispersion, the value of H_0, and the rate of the field variation s. For MSSW propagating in a free ferrite film one can obtain (using Eq.(20) and the dispersion equation[16]) the following explicit expression for the frequency shift:

$$2\Pi\Delta f = \frac{y}{d}\,\frac{H_0+2\Pi M_0}{(H_0+2\Pi M_0)^2-(f/\gamma)^2}\,s\ , \tag{21}$$

where $(f/\gamma)-2\Pi M_0 \le H_0 \le \sqrt{(f/\gamma)^2+(2\Pi M_0)^2} - 2\Pi M_0$.

Fig.3 shows measured field dependences[17] of the frequency shift $\Delta f(H)$ and transmission loss $L(H)=10\log(P_{out}/P_{in})$ for microwave pulses ($f_0 = 4.5$ GHz, $\tau = 10$ μs) transmitted through the MSSW transmission line ($d=20\mu$m, $y=4$ mm, $4\Pi M=1750$ G) at the rate of the field variation of $s=3.5$ Oe/μs. One can see, that experimental data are well described by the theoretical curve calculated using Eq.(21). The frequency shift of MSSW excited by a continuous microwave signal was observed in the frequency range f_0=3-10 GHz at a sawtooth-like modulation of the field. Its value increased from Δf=20 Hz to Δf=10 kHz with the increase in the field variation rate. The analogous effect has been reported in,[10] where the shift of MSSW frequency of $\Delta f \cong 40$ MHz was observed at a pulsed modulation of the field.

3.3 *Harmonic modulation of the field*

As it follows from Eq. (20), the harmonic modulation of the bias field results in the harmonical modulation of the MSW frequency. The shape of the wave frequency spectrum in this case depends on the values of the bias field H_0, modulation frequency F, magnitude of modulating field h_m, and is well described by the theory of angular modulation. Fig. 4 shows typical spectra of continuous microwave signals with central frequency f_0=3 GHz at the output of the MSSW transmission line for different frequencies and different magnitudes of modulating field. We observed[15] the appearance of harmonics of frequencies $f_0 \pm lF$ (l=1,2,...) in the frequency spectrum of the output signal and transition from a discrete spectrum to a continuous one with the increase of h_m or decrease of H_0 and F. Experimentally, MSSW frequency modulation with a maximum modulation index of up to m=50 and maximum width of the spectrum of up to $2\Delta f$=80 MHz has been observed.

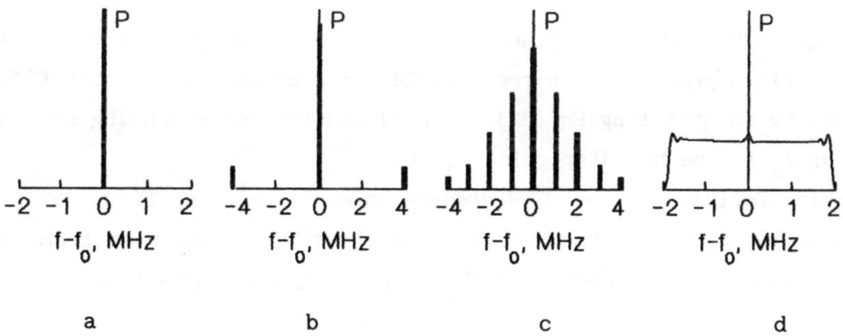

Fig. 4 Frequency spectra of MSSW in a YIG film under the harmonical modulation of the bias magnetic field: (a) input signal, (b) F=4 MHz, (c) F=1 MHz, (d) F=0.14 MHz.

3.4 *Wide-band modulation of the field*

If the magnitude of the magnetic field variation is comparable to the width of the entire field region of MSW existence the MSW packet undergoes a wide-band frequency modulation. This wide-band modulation takes place due to the nonlinearity of the MSW dispersion and the nonlinear character of the bias field modulation. The dependence $\Delta f(t)$ for any law of the field variation can be calculated using Eq.(20) and corresponding dispersion law.

An inverse problem is also of a considerable interest. For any defined smooth law of the MSW frequency modulation it is possible to find a corresponding law of the field variation. This can be demonstrated taking as an example the formation of a MSSW packet with linear frequency modulation (LFM) $f(t)=f_0-s't$, where s' is the rate of the frequency variation.[19]

According to Eq.(20), a slow variation in the bias field results in a change in the instantaneous MSW frequency $2\Pi f(t)=2\Pi f_0-y(\partial k/\partial t)$. Integration of this relation gives the dependence of MSW wave number $k(t)$ on the frequency $f(t)$:

$$k(t) = k_0 - \frac{2\Pi}{y}\int_0^t [f(t')-f_0]dt' , \qquad (22)$$

where k_0 is the wave number at t=0. The time dependence of the bias magnetic field providing the required law of frequency modulation $f(t)$, is found by substituting Eq.(22) for $k(t)$ and the corresponding expression for f_0 in the MSW dispersion equation.

Using Eq.(22) and the MSSW dispersion equation[16] we find that for the carrier frequency f_0 the law of the bias field decrease from the upper value H_U to the lower value H_L is given by the expression:

$$H(t) = \sqrt{(f_0/\gamma)^2 + (2\Pi M_0)^2 e^{-(d/y)\pi s' t^2}} - 2\Pi M_0 . \qquad (23)$$

The calculated dependences shown in Fig.5 demonstrate some

peculiarities of the linear frequency modulation (s'=2.5 MHz/μs) of microwave pulses (f_0=2.5 GHz, τ=4 μs) transmitted through the MSSW transmission line with the following parameters: d = 20μm, $4\Pi M$ = 950 G, ΔH = 0.6 Oe, y=4 mm, H_0=275 Oe. One can see, that LFM requires a significantly nonlinear change in the field $H(t)$. The field dependence of transmission losses $L(H)$ has been calculated following Ref.[20] The field dependence of the transmission losses results in an undesired amplitude modulation of the output pulse. The shape of the output pulse depends, therefore, on the shape of input pulse $P_{in}(t)$, shape of the transmission-loss characteristic $L(H)$, and the function $H(t)$: $P_{out}(t)=P_{in}(t)\ L[H(t)]$. The value of LFM pulse base $B=\Delta f\tau$ is also determined by the $L(H)$ characteristic and the accepted level of insertion loss L_0.

An MSSW transmission line with the parameters close to those used

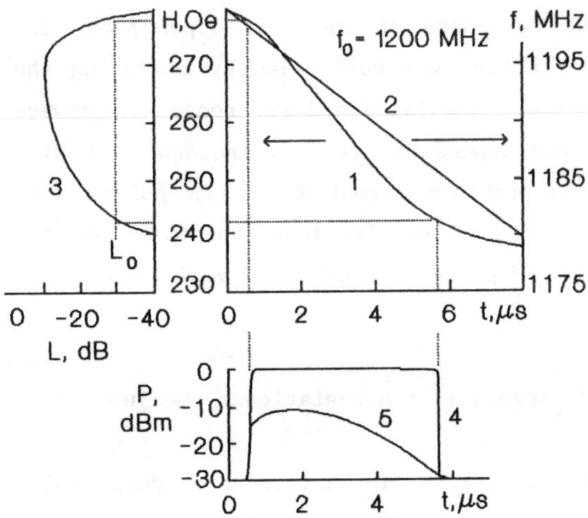

Fig.5 Calculated dependences for MSSW frequency modulation:

1 - magnetic field H vs. time t;

2 -instantaneous frequency f vs. time t;

3 - transmission line loss L vs. field H;

4 - envelope of the input pulse;

5 -envelope of the output pulse.

in the above calculations has been used in experiment where formation of LFM microwave pulses was observed. The bias magnetic field of the transmission line was changed according to the calculated law using a special current source which allowed us to change the current in accordance with any smooth dependence approximated by a broken line consisting of eight straight-line segments.

The frequency spectrum and corresponding time envelope of the LFM pulse at the output of the experimental MSSW line are shown in Fig.6.

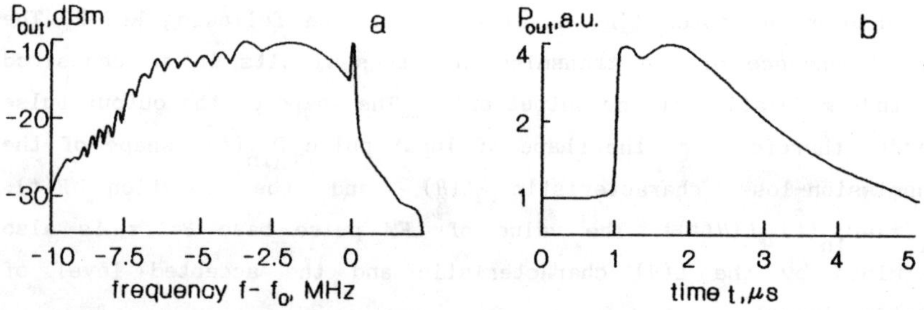

Fig.6 Frequency spectrum (a) and envelope (b) of the LFM MSSW pulse.

In experiments, to provide linear variation of the pulse frequency we adjusted the value of H_0 and the shape of the $H(t)$ dependence. The linearity of the frequency variation was controlled by measuring the period of the interference between modulated and reference microwave signals. Experimentally selected dependence $H(t)$ coincided well with theoretical curve in Fig.5 and provided formation of LFM pulses with divergence of the frequency from the linear law less than 10%. The base and frequency deviation of the LFM pulses formed were equal to $B\simeq40$ and $\Delta f\simeq10$ MHz, respectively.

4. Space-time focusing of MSW packets in a non-stationary medium

A space-time focusing of MSW packets (in particular, a compression or spreading of the packet envelope) in a highly dispersive homogeneous non-stationary medium occurs as a result of the difference in group velocities for the different parts of the packet which enter the medium at the different instants of time.

Let us consider the process of compression of an MSW packet propagating in a dissipative ferrite film placed in a slowly time-varying external magnetic field.[21,22] As it was shown above, the high-frequency field of the packet may be represented in the integral form given by Eq.(14). For sufficiently long input microwave pulses (that are usually used in experiments) the phase factor plays the main role in this integral. After performing integration in Eq.(14) by the method of stationary phase we obtain the envelope of the packet which repeats the envelope of the input pulse with transformation of the amplitude and time scales given by Eq.(17).

For the case of MSSW and the linear time variation of the field $H(t)=H_0+st$ one can use the system (16) and the known dispersion relation[16] to obtain the analytical expression which relates the instant $\tilde{t}'$ of the excitation of the fragment of the packet to the instant t of its arrival at the point y for all the frequencies existing in the wave packet:

$$f(t) = f(t')\mathrm{ch}\left(\frac{(y/d)s}{f^2(t')-f_0^2}\right) + f_0\mathrm{sh}\left(\frac{(y/d)s}{f^2(t')-f_0^2}\right), \qquad (24)$$

where $f(t)=\gamma H(t)+2\Pi M_0$, and $f(t')=\gamma H(t')+2\Pi M_0$.

A plot of this relation for $s>0$ (curve 1) and $s<0$ (curve 2) is shown in Fig.7.

For MSFVW and MSBVW in the case of linear time variation of the bias field, the dependence $t(\tilde{t}')$ is qualitatively similar to the corresponding curve for MSSW.

Maximum compression of the MSW packet, as it follows from equations (15) and (17) and is clear from Fig.7, is obtained by choosing $s>0$ and by making the center of the input microwave pulse to coincide with the time $\tilde{t}'_f$. If the center of the input pulse corresponds to a time $t>\tilde{t}'_f$, than only a partial compression of the packet is possible, while in the region $t>\tilde{t}'_f$ there can be a time inversion of the packet envelope, accompanied by its strong distortion.

The pulse compression is possible only when $|\partial\tilde{t}'/\partial t| > 1$. Thus, according to Eq.(18), compression of MSSW packets and MSFVW packets having $|\partial^2\omega/\partial k^2| < 0$ is possible if the bias magnetic field is increasing with time ($\partial H/\partial t>0$), while compression of MSBVW packets having $|\partial^2\omega/\partial k^2|>0$ is possible if the bias magnetic field is decreasing with time ($\partial H/\partial t<0$).

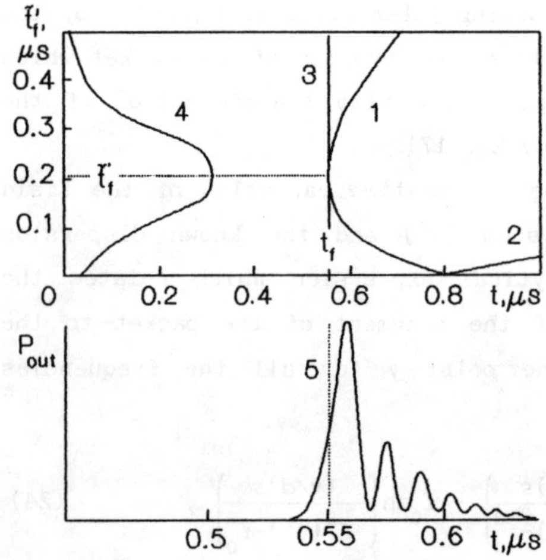

Fig.7 Calculated dependences of MSSW pulse compression in YIG film in the case of linear variation of the bias field: position of stationary point $\tilde{t}'$ vs. time t for increasing field (1); the same curve for decreasing field (2), and for "matched" law of the field variation (3); envelope of the input pulse (4), and envelope of the compressed pulse (5).

As it follows from Fig.7, the stationary points $\tilde{t}'_j$ merge as t approaches $\tilde{t}'_j$ resulting in $\partial\tilde{t}'/\partial t$ becoming infinite. As a consequence, it is necessary to include the cubic terms as well as the quadratic terms in the expansion of the phase when calculating the integral in Eq.(14). The appropriate calculation gives the following expression for the high-frequency magnetic field of the MSW packet near the focusing point $t\cong\tilde{t}'_f$:

$$\mathbf{h}(y,z,t) = \mathbf{h}_{k_f}(z,t)C_{k_f}(\tilde{t}')B_f(y,t)e^{i\Psi(k_f,\tilde{t}'_f)} , \qquad (25)$$

where the explicit form for the factor B_f is given in.[22]

Thus, the form of the MSW packet excited by a long microwave pulse can be found by matching asymptotic expansions (17) and (25) for times near $t \cong \tilde{t}'_f$.

Let us consider in more detail compression of the MSSW packet excited by a microwave pulse of a Gaussian form:

$$A(t) = A_0 e^{-(t-t_0)^2/2\tau_0^2} \quad . \tag{26}$$

Fig. 8 shows the results of calculation of the evolution of a packet with the central frequency $f_0 = 8.2$ GHz and the duration $\tau = 2\Pi\sqrt{\ln 2}\,\tau_0 = 0.1$ μs (at a half maximum) as it propagates in a ferrite film (d=20 μm, $4\Pi M$=1750 G, ΔH=0.5 Oe) when the rate of the bias field variation is $\partial H/\partial t$=100 Oe/μs. The efficiency of the MSSW excitation was assumed to be independent of the wave number k in these calculations.

It can be seen that as a result of the competition between the compression and damping processes, the decrease in the packet amplitude is

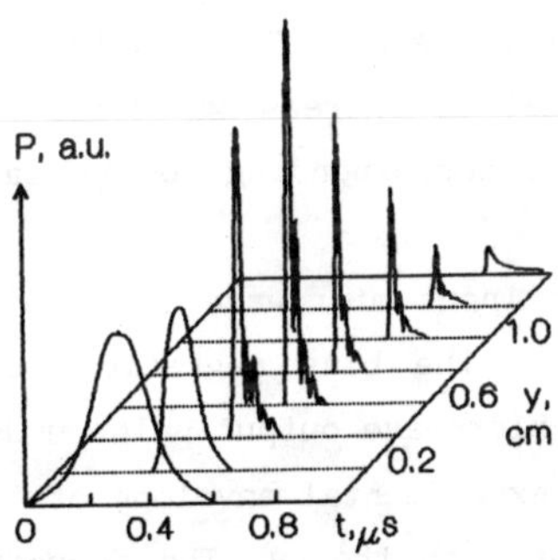

Fig. 8 Evolution of MSW pulse envelope as it propagates in a ferrite film.

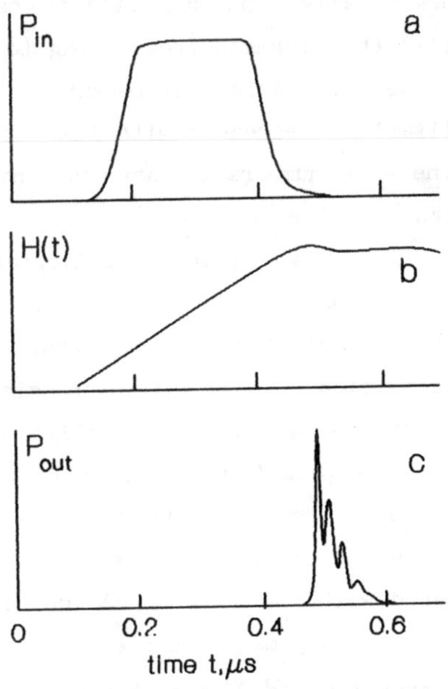

Fig. 9 Oscillograms of input microwave pulse (a), pulse of magnetic field (b), and compressed microwave pulse (c).

initially insignificant. Then, at a certain distance from the point of excitation, a packet of maximum amplitude and minimum duration is formed. After passing the point of maximum compression y_f, the packet amplitude decreases smoothly and its duration once again begins to grow.

A more detailed view of the packet envelope at the point of maximum compression (y_f= 0.6 cm) is shown in Fig. 9. It can be seen that the packet has a characteristic structure. First, the leading edge, independently of the form and input time of the initial microwave pulse, is located in the neighborhood of $t=t_f$. The leading edge has a steep slope whose magnitude does not depend on the pulse duration, but is determined by the dispersion of the medium, distance travelled, and the rate of the magnetic field change. Secondly, there are oscillations on the back edge of the packet. These oscillations are due to interference of different parts of MSW packet having different wave numbers that come simultaneously to the output microstrip transducer. Mathematically, the oscillations can be described if we include in Eq. (17) the terms corresponding to the different stationary point and consequently having different wave numbers $\tilde{k}_j$. The period of oscillations decreases with time as a result of increasing difference in the wave numbers of two stationary point corresponding to the same instant of time.

Fig. 9 shows typical oscillogram of an input microwave pulse with center frequency f_0 =8.2 GHz, the pulse of the bias magnetic field which grows at a rate s=100 Oe/μs, and the microwave output pulse under the conditions of maximum compression. The experimental envelope of the MSW packet shows good qualitative agreement with theory. The decrease in the MSW packet duration during compression is accompanied by the increase of its amplitude by approximately the same factor. In the case of Fig. 9 the total loss of the microwave signal was decreased to -14dB, compared with the loss of -23 dB for the transmitted microwave pulse in a stationary magnetic field.

When the MSW packet is excited after the optimal time, only partial compression and an increase in the irregularity of the transmitted pulse envelope is observed, in accordance with the theoretical prediction. The inversion of the shape of the output pulse envelope when the MSW packet was excited before the optimal compression time was not observed experimentally. This was evidently due to the strong damping of the waves for large delay times.

Fig. 10 shows calculated and experimental dependences of the MSSW packet duration τ_f (at a half maximum) under maximum compression conditions vs. the input microwave pulse duration τ (at a half maximum) for the rate of the bias field variation s = 100 Oe/μs. The dependence

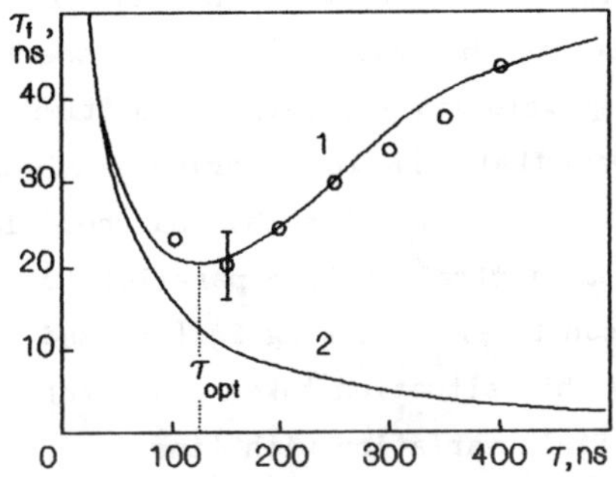

Fig. 10 Theoretical (1) and experimental (circles) dependences of duration τ_f of compressed MSW pulse vs. duration τ of input pulse, (2) theoretical dependence calculated using the quadratic approximation.

$\tau_f(\tau)$ calculated using Eqs. (17) and (25) in the quadratic approximation of the dispersion law is also shown for comparison in the same figure.

It can be seen that for short input pulses the duration of compressed packets is well described by the quadratic approximation. For long pulses difference between the real dispersion and the quadratic approximation leads to the increase in the duration of the compressed packet. As a result, the compressed packet has a minimum duration for a certain duration of the input pulse $\tau=\tau_{opt}$. The compression coefficient $P = \tau_{opt}/\tau_f \simeq 15$ has been obtained experimentally for MSSW packets. The minimum duration of the compressed packet increases with the decrease in the rate of the field variation and is saturated for sufficiently long input pulses. Experiments confirmed existence of a minimum in the dependence $\tau_f(\tau)$. However, a time inversion of the MSW packet envelope was not observed in experiment. This was mainly due to the strong damping of MSWs.

Similar calculations of the space-time evolution of the packet and corresponding experiments have been carried out for MSBVW propagating along the direction of the bias magnetic field. For MSBVW pulses with a center frequency f_0=3.62 GHz and durations τ=0.1-0.5 μs in a case of linear decreasing of the bias field at the rate s=120 Oe/μs the minimum duration of the compressed packet was equal to $\tau_f \simeq 18$ ns and maximum compression coefficient reached was P=11.

The approach presented here may be used for description of the MSW packet evolution in a non-stationary magnetic field whose magnitude changes according to an arbitrary law. The case when the law of the bias field variation is "matched" with the MSW dispersion is of special interest. In this case the MSW group velocity changes at the time of the packet excitation in such a way that all the fragments of the packet cover the distance y in the same time τ_d. For the "matched" law the dependence $\tilde{t}'(t)$ is transformed to a straight line parallel to the $\tilde{t}'$ axis (see curve 3 in Fig.7). It can be shown (using Eq.(16) and the dispersion equation[16]) that for MSSW this situation takes place for the following nonlinear law of the bias field variation with time:

$$H(t) = \frac{1}{\gamma}\left(\sqrt{\frac{(yd/f_2)}{(t-t_f)} + f_0^2} - \gamma 2\Pi M_0 \right) \qquad (27)$$

The compression coefficient for the MSW packet in this case can reach values of up to $P \simeq 10^2$-10^3, and it is limited only by the damping.

5. Anisotropic propagation of MSSW in non-stationary magnetic field

Propagation of MSSW and MSBVW in a tangentially magnetized ferrite film, as it was shown earlier,[23,24] is substantially anisotropic. The wave energy flows along the direction of the group velocity $\mathbf{V}_g$ which, in general, is not collinear with the wave vector $\mathbf{k}$. The direction of $\mathbf{V}_g$ is determined by the frequency f_0 of the microwave excitation signal and by the magnitude and orientation of the internal magnetic field $\mathbf{H}$. Thus, if the spatially homogeneous bias magnetic field is non-stationary, then the wave energy is transmitted along a curved trajectory, tangent to which at every point is parallel to $\mathbf{V}_g$.

In the approximation of geometric optics (as it follows from equations (12) and (13)), trajectories of MSW rays may be found from the system of equations:

$$\mathbf{r}(t) = \mathbf{r}_0 + \int_{t_0}^{t} \mathbf{V}_g(\mathbf{k}_0, t'')dt'', \qquad f(\mathbf{k}_0, t) = f_0 , \tag{28}$$

where f, $\mathbf{k}_0$, $\mathbf{V}_g(\mathbf{k})=-2\Pi\nabla f(\mathbf{k},\mathbf{H})$ are the frequency, wave vector, and the group velocity of the wave fragment excited at the point $\mathbf{r}_0$ at the instant t_0 respectively while $\mathbf{r}(t)$ is the position of the fragment at the instant t. The initial values of $\mathbf{k}_0$ and $\mathbf{V}_g$ are defined by the frequency f_0 of the microwave excitation signal, value of $H(t_0)$, and the orientation of the microstrip transducer.

As the wave continues to propagate, $\mathbf{k}_0$ is conserved because of homogeneity of the medium, while the magnitude and direction of $\mathbf{V}_g$ are changing because of the change in the instantaneous wave frequency $f(t)$ in accordance with the wave dispersion equation. This effect is responsible for the distortion of the MSSW ray trajectories in non-stationary magnetized films.

Let us consider two kinds of field modulation.[25] In the first case magnetic field **H** lies in the film plane **x-y**, is directed along the **x** axis, and its magnitude is linearly changing with the rate s : $H(t)=H_0+st$. The MSSW is exited by a long microstrip transducer, in such a way that its wave vector **k** is directed at the angle φ to the **x** axis. In the second case the field also lies in the **x-y** plane, but being constant in magnitude it is oscillating near the direction of the **x** axis with the angular amplitude $\delta\varphi$, frequency F, and the initial phase ψ : $H(t) = H_0(\cos\varphi, \sin\varphi)$, where $\varphi = \delta\varphi \cos(2\Pi Ft+\psi)$. The MSSW wave vector **k** is directed exactly along the **y** axis.

To explain qualitatively the formation of MSSW ray trajectories in a film magnetized by a non-stationary bias field it is convenient to use constant frequency curves drawn in the wave vector space. The curves corresponding to the fixed wave frequency f_0=3.25 GHz and different field magnitudes (1 - H=500 Oe, 2 - 530 Oe, 3 - 560 Oe) are shown in Fig. 11. The MSSW group velocities, directed in every point perpendicularly to the curves, are marked with the pointers.

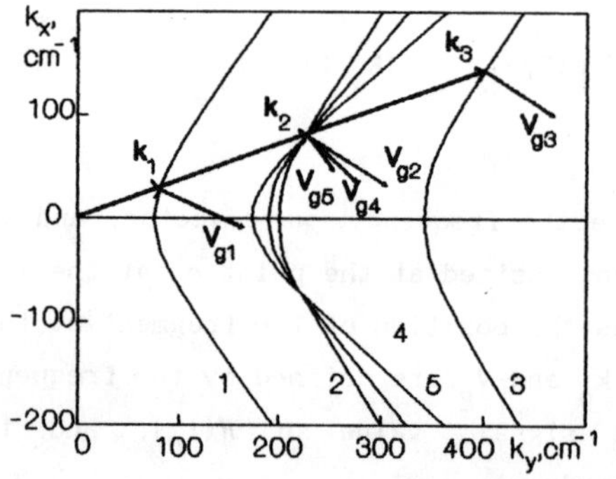

Fig. 11 Constant frequency curves for MSSW in a tangentially magnetized YIG film.

It can be seen, that MSSW fragments exited at instants corresponding to different values of H, have different wave vectors $\mathbf{k}_1, \mathbf{k}_2, \mathbf{k}_3$ and different group velocities $\mathbf{V}_{g1}$, $\mathbf{V}_{g2}$, $\mathbf{V}_{g3}$. Thus, if the direction normal to the transducer and the direction of the wave vector are not collinear, a monochromatic signal excites in the film an MSSW beam with a finite angular divergence.

During propagation the wave vector of the fragment is conserved, while its instantaneous frequency and group velocity are changed because of the field variation. Three constant frequency curves for MSSW of fixed wave vector $\mathbf{k}_2$ at different values of H are shown in Fig. 11. Group velocities of the waves have different directions. Therefore, the magnetic field increase must result in a deflection of the MSSW ray trajectories out from the y axis of the film. The formation of the MSSW ray trajectories in oscillating magnetizing field may be explained analogously.

The trajectories of MSSW rays in a non-stationary magnetic field were calculated by numerical integration of the system of equations (28) with the help of the results obtained in.[24] All calculations were carried out for YIG film of the thickness 10 μm with the saturation magnetization 1750 G ignoring magnetic dissipation.

The trajectories of MSSW rays propagating from the origin of the coordinate system for the case of linear increase in the field $H(t) = H_0+st$ (H_0=600 Oe, s=1000 Oe/μs, t in μs) are shown in Fig. 12. The lower part of Fig. 12 corresponds to MSSW rays of the fixed wave vector orientation $\varphi=80^0$ and different initial frequencies: 1 - f_0= 3.25 GHz, 2- 3.0 GHz, 3- 3.75 GHz, 4- 3.8 GHz. The upper part of Fig. 12 describes MSSW rays of the fixed frequency f_0=3.25 GHz with different wave vector orientations: 5-φ=91^0, 6 -100^0, 7 -110^0, 8 -120^0. It is clear, that variation in the magnetic field magnitude results in a weak distortion of the MSSW ray trajectories. The rays are deflected from the y axis of the film. The larger is the wave frequency and the smaller is the angle φ, the greater is the ray distortion. The instantaneous MSSW frequency f increases while moving along every ray in accordance with the change of the magnetic field.

The trajectories of MSSW rays for the case of modulation of magnetic field direction with the frequency F=2 MHz are shown in Fig. 13. The upper part of Fig. 13 corresponds to fixed amplitude of the

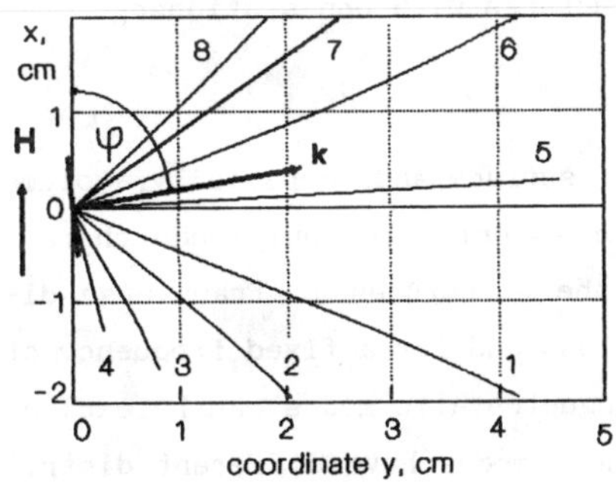

Fig. 12 Trajectories of MSSW rays in amplitude modulated field.

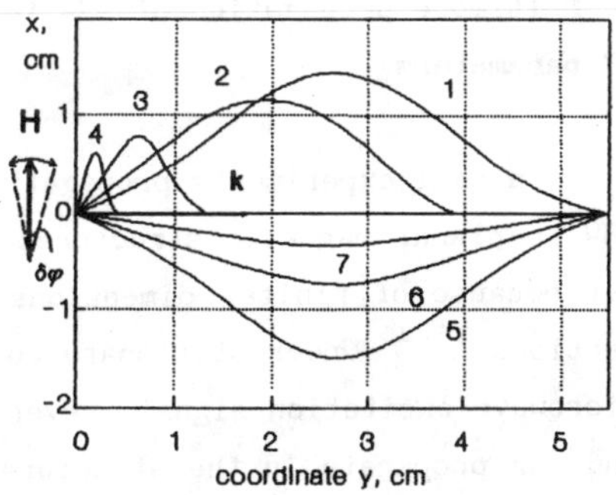

Fig. 13 Trajectories of MSSW rays in direction modulated field.

field modulation ($\delta\varphi=20^0$, $\psi=270^0$) and various frequencies of the microwave excitation signal: 1 - f_0 = 3 GHz, 2 - 3.2, 3 - 3.4, 4 - 3.8. The lower part of Fig. 13 demonstrates the effect of amplitude of the field modulation ($\psi = 90^0$) on the trajectories of MSSW rays exited by a signal of the fixed frequency f_0=3.0 GHz: 5 - $\delta\varphi$ =20^0, 6 - 10^0, 7 - 5^0.

The MSSW ray trajectories, as was expected, have a form of periodic curves. The distance that the wave covers during one period of the magnetic field variation (t=T) decreases sharply with the increase in the signal frequency f_0, while the maximum ray deviation from the y axis of the film decreases more slowly. The MSSW instantaneous frequency $f(t)$ is also changing periodically during propagation with the period which is two times smaller than the period of the field modulation. The frequency $f(t)$ becomes equal to f_0 when the bias magnetic field $\mathbf{H}_e$ passes through its position at the instant of the wave excitation and the ray crosses the film axis x=0.

Analogous calculations have been carried out for MSBVW propagation in a non-stationary magnetized YIG film and they gave similar results.

Note, that in the case of MSFVW propagation in a normally magnetized YIG film dispersion characteristics are isotropic in the film plane and no distortions of the rays trajectories take place under the non-stationary conditions.

6. Multimode propagation of MSW in structures with non-stationary parameters

In real experiments propagation of surface and, especially, volume MSW in planar magnetic structures is of essentially multimode character because of finite dimensions of the structures in transverse directions.[16,26] Under stationary conditions and for a fixed frequency of microwave excitation signal several magnetostatic modes can be excited and can propagate in the structure. These modes have different distributions of their high-frequency magnetization along the thickness or the width of the ferrite film and they also have different wave numbers and group velocities.

For theoretical description of multimode MSW propagation in the structures with non-stationary parameters one can use the approach described in Section 2. After making necessary conversions, we obtain equations for slowly varying amplitudes A_{kn} of different modes that are analogous to Eq. (11) :

$$\varepsilon \frac{\partial A_{kn}}{\partial(\varepsilon t)} + \varepsilon\lambda_{kn}(\varepsilon t)A_{kn} + A(\varepsilon t)C_{kn}(\varepsilon t)e^{i\omega_0 t - i\frac{\Psi_{kn}}{\varepsilon}} A_{kn} +$$

$$+ \sum_{n\neq n'} \varepsilon\lambda_{knn'}A_{kn'}e^{i(\Psi_{kn}-\Psi_{kn'})} = 0 , \tag{29}$$

where. $\lambda_{knn'}(\varepsilon t) = \frac{1}{b_{kn}} \int_{-d/2}^{d/2} \mathbf{h}^{0*}_{kn} \hat{T}_{kn} \frac{\partial \mathbf{m}^0_{kn'}}{\partial(\varepsilon t)} dz, \qquad \lambda_{knn} = \lambda_{kn}.$

Eq. (29) was derived using the condition of the mode orthogonality:

$$\int_{-d/2}^{d/2} \mathbf{h}^{0*}_{kn} \hat{T}^{*}_{kn}\mathbf{m}^{0}_{kn'} dz = \delta_{nn'} b_{kn}(\varepsilon t) , \tag{30}$$

where $\delta_{nn'}$ is the Kroneker's symbol.

The fact that the matrix $\lambda_{knn'}$ is non-diagonal reflects intermode interactions arising from the non-stationarity of the structure parameters. An explicit form of the matrix may be determined in every particular case for a specific type of interaction and a specific law of the field modulation. In the lowest approximation in ε all the results obtained for the frequency spectrum transformation and compression of packets can be applied to each individual mode.

In particular, in a linearly varying magnetic field each mode has its own frequency shift, the value of which is determined by the dispersion of the mode at the frequency ω :

$$\Delta\omega = \left.\frac{\partial k_n}{\partial H}\right|_{H_0,\omega_0} \frac{\partial H}{\partial t} y . \tag{31}$$

The frequency spectrum of MSW propagating in a ferrite structure in a linearly modulated field splits into several frequency components even in the case when the waves are excited by monochromatic pumping.

Therefore, a frequency separation of MSW modes is possible under these conditions. Intensities of the corresponding frequency components characterize the efficiency of excitation and propagation losses for each MSW mode in a specific structure.

In a case of a wideband magnetic field modulation an independent compression of packets corresponding to different MSW modes will take place. Due to the difference in dispersion characteristics, a space-time focusing of the packets will be happening at various instants t_{fn} and at various distances y_{fn} from the input transducer. Therefore, a space-time separation of MSW modes is also possible.

A transmission line containing a YIG film of thickness 10μm and dimensions 2.5 x 25 mm^2, and saturation magnetization 1750 G was used for experiments on multimode MSW propagation under non- stationary conditions. We registered frequency spectra of MSFVW and MSBVW excited by microwave pulses of the carrier frequency f_0=3-10 GHz and duration τ=10 μs at simultaneous linear variation of the field with the rate of up to s=1 Oe/μs and magnitude of up to $h_m \simeq$ 30 Oe.

In accordance with theoretical predictions a dynamic splitting of the frequency spectrum of transmitted signal has been observed. A considerable difference in dispersion characteristics of different MSW modes allowed us to separate them by the individual frequency shifts. Fig. 14 and Fig. 15 show measured field dependences of the frequency shift $\Delta f_n(H)$ and insertion loss $L_n(H)$ for MSFVW and MSBVW modes respectively. Dashed curves in Fig. 14 and Fig. 15 demonstrate the field dependences of $\Delta f_n(H)$ calculated using Eq. (31) and the known dispersion equations for the corresponding wave modes.

We observed propagation of the lowest (curve 1) and two higher (curves 2 and 3) MSFVW modes. The maximum frequency shift for the third mode was equal to $\Delta f_3 \simeq 2$ MHz at $\partial H/\delta t \simeq 1$ Oe/μs. In the case of MSBVW we were able to observe only propagation of the lowest (curve 1) and one higher (curve 2) volume modes. The frequency shift of MSBVW modes depended weakly on the field magnitude and exhibited a maximum value $\Delta f_2 \simeq$ 1.1 MHz at $\partial H/\partial t$;0.8 Oe/μs for the second mode. It follows from Fig. 14 and Fig. 15 that amplitudes of the higher MSVW modes are on average 10dB smaller than the amplitudes of the lowest modes. The result agrees with the calculated efficiency of MSVW modes excitation by microstrips.[27]

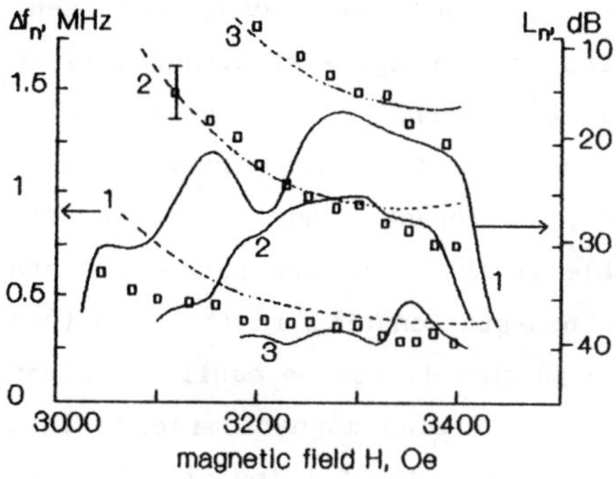

Fig. 14 Dependences of the frequency shift Δf_n and transmission loss L_n for MSFVW modes in YIG film vs. bias magnetic field H (n =1,2,3 is the mode number).

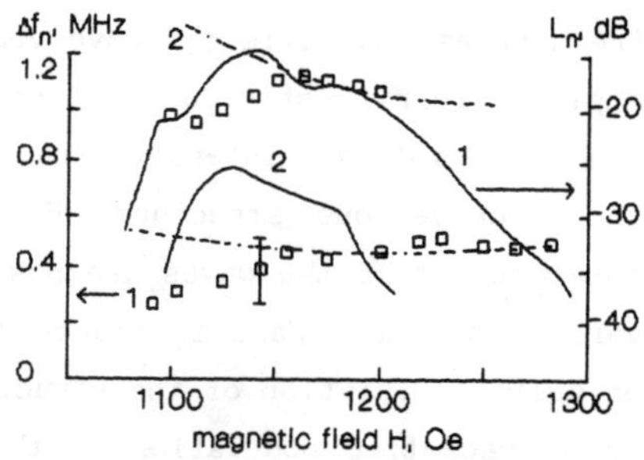

Fig. 15 Dependences of the frequency shift Δf_n and transmission loss L_n for MSBVW modes in YIG film vs. bias magnetic field H (n = 1,2 is the mode number).

The compression of MSVW packets in a ferrite film at multimode propagation has also been investigated. However, in that case we did not observe a space-time separation of the packets corresponding to different volume modes. As our calculations showed, for a fixed distance between transducers a simultaneous optimal focusing of all modes is impossible. Besides that, the difference between instants of focusing t_{fn} of packets for different modes is smaller than the duration of the packet compressed. Therefore, a multimode character of MSW propagation results only in additional distortion in the shape of compressed pulse due to the interference of different modes at the output of the transmission line.

Note, that the effect of frequency separation of MSW modes in linearly varying magnetic field may be used for investigation of the mode structure of MSWs. It allows us to separate a direct electromagnetic leakage, the frequency of which is not shifted, and to measure the ratio of the modes intensities in real experimental conditions.

7. Parametric interaction of MSWs in a ferrite film magnetized by a homogeneous non-stationary magnetic field.

As it has been mentioned above, several MSWs having different frequencies and different wave vectors can propagate simultaneously in a planar ferrite structure placed in an external bias magnetic field. These waves do not interact with each other under stationary conditions in a homogeneous structure. However, different kinds of parametric interactions of the waves are possible if the structure parameters are varied in time or/and in space. The non-stationarity in the structure or spatial variation of the structure parameters can be easily realized by a space-time modulation of the external bias magnetic field. This idea opens interesting possibilities for experimental investigation of parametric interactions of MSWs.

Let us consider the simplest case of parametric interaction of two MSWs of different frequencies propagating in a homogeneous ferrite structure magnetized by a non-stationary bias field. As it follows from a general theory for waves interaction in media with space-time periodic properties,[28] an effective interaction of MSWs takes place when the following phase-matching conditions are satisfied :

$$\mathbf{k}_1 = \mathbf{k}_2, \qquad f_1 + F = f_2 , \tag{32}$$

where ($\mathbf{k}_1$, f_1) and ($\mathbf{k}_2$, f_2) are the wave vectors and the frequencies of the interacting waves respectively, and F is the frequency of the magnetic field modulation.

A variety of MSW dispersion characteristics and multi-mode character of MSW propagation in planar magnetic structures opens a lot of ways to realize the conditions (32). As two waves having different frequencies and the same wave vectors one can choose, for example:

- different modes of MSFVW or MSBVW propagating in a normally or tangentially magnetized ferrite film;
- MSSW and main mode of MSBVW propagating at the angle to the bias field in a tangentially magnetized ferrite film;
- two MSSW or two MSVW propagating in a layered structure consisting

of two or more ferrite films;

- main modes of MSFVW and MSBVW propagating in a ferrite film magnetized at the angle to the normal to the film.

Theoretical analysis showed that the case of interaction of two MSVWs in a ferrite film magnetized at the angle to the film normal (see Fig. 16) provides the most effective interaction and is the preferable from the experimental point of view. Therefore, this kind of parametric interaction of MSWs will be considered below in more detail.

There are two kinds of MSVWs that can propagate along the projection of the bias magnetic field in a ferrite film magnetized at an arbitrary angle to the film normal : forward volume wave (MSFVW) and backward volume wave (MSBVW). Dispersion curves for both these types of waves are well described by the simple equation:[29,30]

$$kd = \frac{|\mu\sin\beta+\cos\beta|}{\sqrt{-\mu}}\left(\frac{\Pi}{2}N + \mathrm{arctg}\left(\frac{\mu+1}{\sqrt{-\mu}}\,\mathrm{sgn}(\mu\sin\beta+\cos\beta)\right)\right), \quad (33)$$

where: $\mu = 1+(f_H f_M)/(f_H^2-f_M^2)$, $f_H=\gamma H$, $f_M=\gamma 4\Pi M_0$, γ=2.8 MHz/Oe, N=0, 1, 2... is the volume mode number, and H and β are the magnitude and the angle of internal magnetic field in the film. Note, that H and β for the internal field differ from the similar quantities for the external field (H_e and β_e) due to the demagnetization effects and can be easily calculated.[30] MSFVW and MSBVW having parallel wave vectors have oppositely directed group velocities.

In experiment described in Ref.[31] we used a YIG-GGG structure with the parameters : thickness d=20 μm, dimensions 3x25 mm^2, $4\Pi M_0$ = 1750 G, y =4 mm. The structure was placed in the external DC homogeneous magnetic field with the magnitude H_e=1.95 kOe applied at an angle β_e=17^0 to the normal to the film in the plane containing the normal to the film and the wave propagation direction. Field modulation with the frequency F=20-300 MHz and magnitude of up to h_m=10 Oe was applied using an electromagnetic coil. The MSVW was excited by an AC microwave signal of frequency f_0=3-4 GHz and power P_{in} =1 mW. The spectra of reflected $P_{ref}(f)$ and transmitted $P_{out}(f)$ signals were registered using a directional coupler and a narrow-band electronically tuned filter.

Under stationary conditions, the measured amplitude response $L(f)$ and dispersion $k(f)$ for MSVWs in the transmission line are presented in Fig. 16. The shape of the amplitude response is determined by the MSVW excitation efficiency and MSVW propagation losses at each frequency.

The theoretical dispersion curve (dashed line) was calculated using Eq. (33) for the main mode and the parameter values H=0.8 kOe and $\beta=45^0$. Propagation of the waves with wave numbers of up to $k \cong 300\ \mathrm{cm}^{-1}$ was observed. Minimum insertion loss was -15 dB for both MSFVW and MSBVW of the wave numbers $k \simeq 125\ \mathrm{cm}^{-11}$.

When the bias field was modulated we observed two kinds of transformation of spectra of reflected and transmitted microwave signals.

In the first case, when the frequency f was placed in the middle of the MSFVW or MSBVW frequency band and the frequency of modulation was equal to $F \leq 50$ MHz, spectral satellites having frequencies $f \pm lF$ (l is an integer) appeared in the spectrum of both reflected and transmitted signals. The intensities of these satellites and their number depended on the frequency F and magnitude of modulation. Such kind of MSW spectrum transformation appears due to the phase modulation of the wave that was described in Section 3.3.

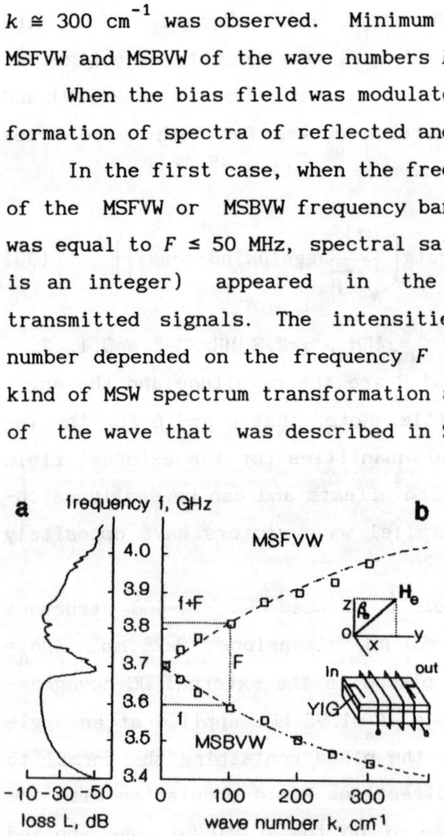

Fig. 16 Amplitude response (a), dispersion characteristic (b) of MSVW transmission line, and geometry of parametric MSVW interation.

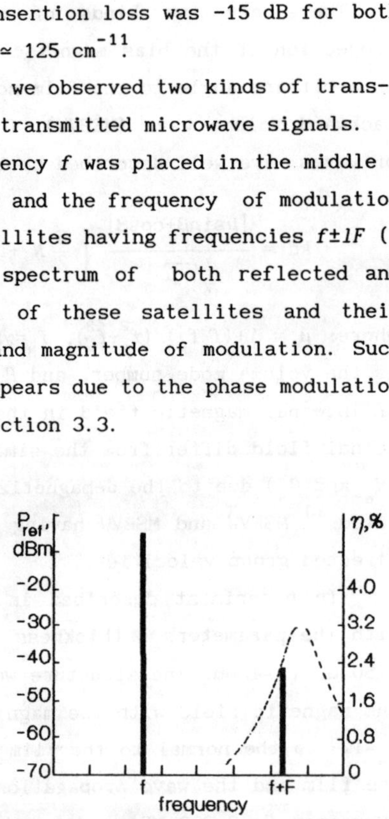

Fig. 17 Frequency spectrum of reflected microwave signal for F = 220 MHz.

In the second case, when the frequency f belonged to the MSBVW (or MSFVW) branch, the frequency $f \pm F$ was in the other branch, and the frequency of modulation was equal to $F \simeq$ 100-300 MHz, one spectral satellite of the frequency $f+F$ (or $f-F$) appeared only in the spectrum of the reflected signal. The spectrum of the reflected microwave signal is shown in Fig. 17. The power of the satellite increased nonlinearly with increase in the magnitude of modulation field. The appearance of this spectral satellite is a direct evidence of the generation of a new MSVW due to parametric interaction between the waves.

The asymmetry of reflected and transmitted signal spectra may be easily understood taking into account the fact that the group velocities of initial and generated waves have opposite directions. The input signal of the frequency f_1 excites in the film a wave of the wave vector $\mathbf{k}_1$. The power flow of this wave is directed to the output transducer. In the region between the transducers a wave of the frequency $f_2=f_1+F$ (or f_1-F) and wave vector $\mathbf{k}_2= \mathbf{k}_1$ is generated parametrically. The power flow of this wave is directed in the opposite direction. It is this newly generated wave that is registered by the input transducer and appears in the spectrum of the reflected signal.

The MSVW interaction efficiency can be described by a coefficient:

$$\eta(f, \pm F) = \frac{P_{MSW}(f \pm F)}{P_{MSW}(f)} *100\% , \tag{34}$$

where $P_{MSW}(f)$ and $P_{MSW}(f \pm F)$ are the powers of the initial and generated waves near the input transducer respectively. The efficiency η can be approximately expressed in terms of the measured amplitude response of the transmission line and the powers of microwave signals $P_{in}(f)$ and $P_{ref}(f \pm F)$:

$$\eta(f, \pm F) = \frac{P_{ref}(f \pm F)}{P_{in}(f)\sqrt{L(f)L(f \pm F)}} *100\% , \tag{35}$$

where $L(f)$ and $L(f \pm F)$ should be expressed in relative units.

The dashed line in Fig. 17. shows the dependence of the interaction efficiency η on the field modulation frequency F for the wave number $k \simeq 125\ \text{cm}^{-1}$, calculated using Eq. (35) and experimentally obtained data. Maximum value of efficiency $\eta \simeq 3.2\%$ was obtained under the conditions of exact synchronization of interacting waves ($f_2 = f_1 + F$). The interaction bandwidth at -3 dB level was equal to $\delta F \simeq 100$ MHz. Under the conditions of exact synchronization, the efficiency η increased and the bandwidth δF decreased with the increase of k, i.e. with the increase of the ratio of interaction length y to the MSW wavelength λ. Maximum value of interaction efficiency η was reached for $y/\lambda \simeq 8$ and then for $k > 125\ \text{cm}^{-1}$ η decreased again.[1] This decrease was due to the violation of the phase-matching conditions (32) because of non-uniformity in the internal bias field in the film and also due to the decrease in modulation field magnitude at higher frequencies.

We believe, that the efficiency of parametric MSVW interactions may be increased to 20-30% by increasing the modulating field magnitude and by using more homogeneous ferrite structures.

8. Scattering of MSVW in a spatially periodic non-stationary magnetic field

If the parameters of a planar magnetic structure are varied not only in time, but also in space, then the MSW scattering in the structure may be observed. The scattering is accompanied by changes in both frequency and the wave vector of the propagating wave.[32] The most appropriate way to produce a non-stationarity of this sort is to modulate a DC bias magnetic field. It may be done, for example, by passing an AC current through a meander-line conductor placed on the ferrite film surface (see insert in Fig. 19). The flowing current modulates the normal and longitudinal components of the internal magnetic field in the ferrite and forms a "magnetic lattice", which causes the most effective scattering of MSFVW and MSBVW.[33]

Let us expand the internal magnetic field in the film $H(y,t) = H_0 + h_m(y,t)$ in the set of spatially-periodic harmonics. The first-order (and every higher-order) harmonic may be represented in the form of two "magnetic waves" propagating in opposite directions:

$$h_m(y,t)=h_m\cos(\Omega t)\cos(py)=(h_m/2)\cos(\Omega t-py)+(h_m/2)\cos(\Omega t+py) , \quad (36)$$

where $\Omega=2\Pi F$, $p=2\Pi/\Lambda$, F is the frequency of AC current, Λ is the period of the meander, and h_m is the amplitude of the field modulation.

Scattering of MSVW may take place on each of these "magnetic waves" when the phase-matching conditions are satisfied:

$$\mathbf{k}_2 = \mathbf{k}_1 + \mathbf{p}, \qquad f_2 = f_1 \pm lF , \quad (37)$$

where $(\mathbf{k}_1,f_1)$, $(\mathbf{k}_2,f_2)$ and $(\mathbf{p},F)$ are the wave vectors and frequencies of the incident, scattered, and "magnetic wave", respectively, l=1,2...

Fig. 18 shows a qualitative diagram for a collinear scattering of the main MSBVW mode on the "magnetic wave" in a ferrite film. If the frequency of the incident MSBVW f_1 is small, so its wave number $k_1 > p$, then only a forward scattering takes place and frequencies of the generated waves lie on both sides of the incident frequency. If f_1 is close to the upper limit of the MSBVW frequency region and $k_1 < p$, then both forward and backward propagating waves appear and frequencies of both waves may be smaller then the frequency of the incident wave. The frequencies of the backward scattered waves lie on the upper part of the synchronization curve.

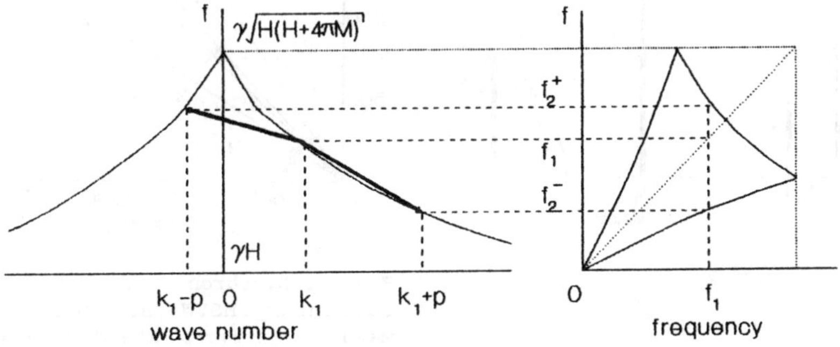

Fig. 18 Diagram of collinear MSBVW scattering in a spatially-periodic non-stationary magnetic field.

In experiments we used a structure containing a YIG film and an alumina substrate placed in close contact to each other. The YIG film had the parameters: d=20 μm, dimensions 4x20 mm^2, $4\Pi M$=1750 G. A meander-line of 50 μm conductor width, 5 mm aperture, 6 mm length, and 200, 500, 1000 μm period was deposited between microstrip transducers on the alumina substrate. The structure was uniformly magnetized by an external DC magnetic field H_0= 600 Oe. For the field modulation with a magnitude up to $h_m \simeq$ 20 Oe we applied to the meander-line an AC current with the frequency F=20-150 MHz and magnitude of up to I=0.2 A.

Under modulation, two satellite harmonics appeared in the reflected and transmitted signal spectra (see Fig. 19). The power of the harmonics depended on the carrier frequency position inside the MSBVW frequency band and the shape of the amplitude response $L(f)$ of the transmission line. The power of harmonics increased linearly with the increase in the modulating current I and with the decrease of the meander-line period Λ. The dashed line in Fig. 19 shows typical dependence of the harmonics power on the modulation frequency F. The maxima of the harmonics power were positioned at different distances from the carrier frequency.

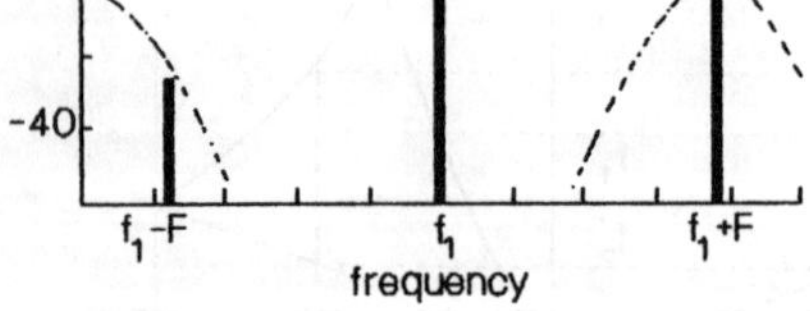

Fig. 19 Frequency spectrum of transmitted microwave signal for F = 80 MHz.

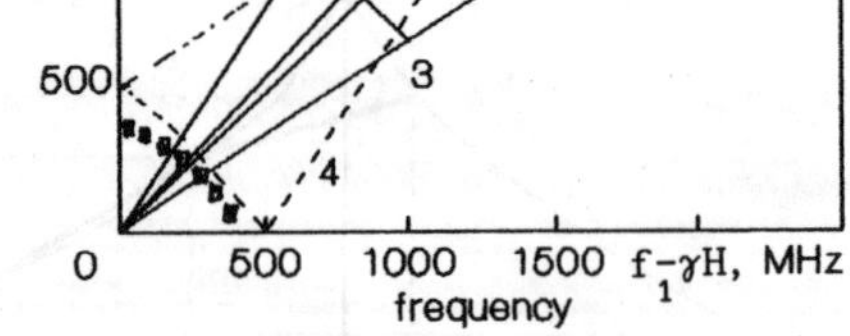

Fig. 20 Synchronization curves for collinear MSVW scattering on magnetic lattices. MSBVW: 1 - H = 600 Oe, Λ = 1000 μm, 2 - H = 600 Oe, Λ = 100 μm, 3 - H = 200 Oe, Λ = 100 μm. MSFVW: 4 - H = 600 Oe, Λ = 200 μm.

Fig.20 shows dependences of frequencies $f_2^{\pm}$ of the scattered waves on the incident wave frequency f_1 (synchronization curves) for a collinear MSBVW and MSFVW scattering on magnetic lattices with different periods. Theoretical curves were calculated for the main volume mode using the phase-matching conditions (37) and the known dispersion equations. It can be seen, that the decrease in Λ (compare curves 1 and 2) results in the increase in the distance between the incident and scattered waves frequencies, and the broadening of the backward scattering band. With the decrease of the field strength H (compare curves 2 and 3) the frequency band of interaction becomes narrower, but synchronization curves corresponding to different magnetic fields partly coincide with each other.

In experiment we observed both forward and backward scattering of MSBVW and only backward scattering of MSFVW. The frequencies of the scattered waves were measured at maximum powers of the lateral harmonics. Some discrepancy between measured and calculated frequencies for MSFVW in Fig.20 is, possibly, due to the influence of the transmission line amplitude response. The bandwidth of interaction depends on the magnetic lattice length y and is approximately given by $\Delta f \simeq 2V_g/y \simeq$ 50 MHz. This result agrees with the experimental data in Fig.19. The efficiency of the scattering estimated using the definition (34) for the data of Fig.19 was equal to $\eta(+F) \simeq 6$ % and $\eta(-F) \simeq 11$ %, respectively. We believe, that after optimization of parameters, the efficiency of the MSVW scattering in a space-periodic non-stationary magnetic field can be increased at least up to $\eta \simeq$ 20-30 %.

9. Phase synchronization of MSWs in a non-stationary magnetic field

Let us consider now the situation when several MSWs, having different frequencies, wave vectors and group velocities propagate in a planar ferrite structure simultaneously. If the frequencies of the waves are equidistant (or nearly equidistant), then a multi-wave interaction can be realized by means of modulation of the structure parameters with a frequency close to the difference between the frequencies of the adjacent waves. Note, that the analogous phenomenon is well known for electromagnetic waves in laser physics and is used for generation of short light pulses.[34]

We have realized simultaneous excitation and propagation of MSWs with nearly equidistant frequencies by including a MSBVW transmission line in a loop of a feedback oscillator.[35,36] The oscillator contained a GaAs-monolithic transistor amplifier (frequency range f=3-4 GHz, voltage controlled gain coefficient G=18-25 dB and output power saturation level P_s=3-5 mW), planar directional coupler (β=-4 dB) and wide-band MSBVW transmission line (d=20 μm, $4\Pi M$=1750 G, H_0=715 Oe, y=4 mm) in the feedback loop. The MSBVW transmission line had an amplitude response bandwidth of Δf=380 MHz at -30 dB level with a minimal insertion loss of -13 dB and the variation in the phase of the transmitted signal of up to 32Π rad. For harmonic modulation of the field with the frequency F=2-30 MHz and magnitude up to h_m=20 Oe we used an electromagnetic coil wound around the ferrite structure.

For the gain coefficient G < 20 dB, a single-mode oscillation of the frequency f=3535 MHz existed, and only one MSVW propagated in the ferrite structure. For G = 20-23 dB, a stable multi-frequency oscillation at up to 15 frequencies (see Fig.21a) was observed, and up to 15 MSVWs propagated in the ferrite structure. The distance between wave frequencies increased from 17 MHz to 22 MHz.

The frequencies of the waves may be found from the phase-balance condition:

$$\Phi_1(f_N) + \Phi_2(f_N) + \Phi_3(f_N) = 2\Pi N \, , \tag{38}$$

where Φ_1=$k(f)$y, Φ_2, and Φ_3 are the phase shifts in the transmission line, in the connection lines, and in the amplifier respectively, $k(f)$ is the MSBVW wave number, N = 1,2... . Phase shifts $\Phi_2(f)$ and $\Phi_3(f)$ have been measured and approximated within the 3.3-3.8 GHz frequency band by analytical expressions.

Figure 21 demonstrates good agreement between the measured and calculated spectra of MSBVWs. Note, that nonlinearity of the spectrum is mainly caused by the nonlinearity of the MSBVW transmission line phase response. The real ratio of the waves intensities arises from the wave competition process and is determined by the real amplitude and phase responses of the oscillator components.

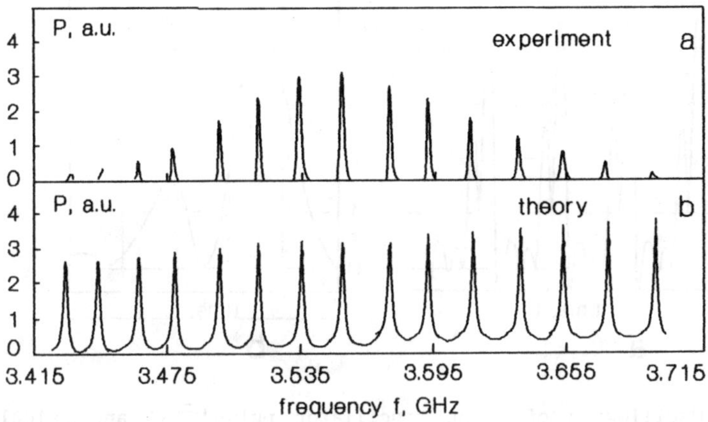

Fig. 21 Measured (a) and calculated (b) MSBVW frequency spectra.

The interaction of the waves was initiated by the harmonic modulation of the field with the frequency F=19 MHz. Under modulation, the spectrum spread up to 330 MHz, intensities of the waves decreased by 1-2 dB and the distance between the frequencies became equal to the modulation frequency F. Under these conditions the oscillator generated shot pulses of asymmetric shape, shown in Fig. 22a. The pulse repetition period was equal to T = 53 ns, the pulses duration was τ = 7-10 ns.

We believe, that the mechanism of wave interaction in this case is based on the specific dependence of MSW properties on the magnetic field strength. As shown above, the harmonic field modulation $H(t)=H_0+h_m\cos(2\Pi Ft)$ resulted in the MSW phase modulation and gave rise to the appearance of harmonics of frequencies $f_N=lF$ (where l=1,2... is the number of a harmonic) at the output of the transmission line.

In the multi-frequency oscillation regime the harmonics arise near each frequency f_N. Provided lF is equal to the inter-frequency interval Δf_N, the harmonics act as an external force and provide the synchronization of the waves. The number of the waves synchronized is

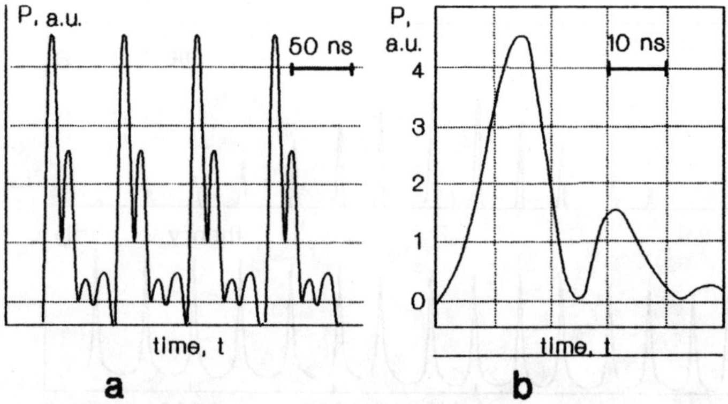

Fig. 22 Oscillogram of output oscillator pulses (a) and calculated pulse envelope (b).

determined by the magnitude h_m of the modulating field and is limited by the nonlinearity of the MSBVW dispersion.

The envelope of the MSBVW pulses is given by the expression:

$$P(t) = \left(\sum_{N_1}^{N_2} A'_N \, Re \left(e^{i(2\Pi FN + \Phi'_N)} \right) \right)^2 , \tag{39}$$

where $N_1 < N < N_2$, A'_N and Φ'_N are the amplitudes and the phases of the synchronized waves having numbers from N_1 to N_2. It should be also taken into account that the external pumping defines the non-zero initial phases Φ'_N for the waves under consideration. These phases may be calculated using the equations for amplitude $L(f)$ and phase $\Phi(f)$ responses of the oscillator loop:

$$L(f) = \left(1 + e^{2\alpha'(f)y} - 2e^{\alpha'(f)y}\cos[k(f)y]\right)^{-1/2}, \qquad (40)$$

$$\Phi(f) = \mathrm{arctg}\left(\frac{e^{\alpha'(f)y}\sin[k(f)y]}{1-e^{\alpha'(f)y}\cos[k(f)y]}\right), \qquad (41)$$

where $V_g=2\Pi(\partial f/\partial k)$ and $\alpha'(f)=2\Pi\gamma\Delta H/V_g$ are the group velocity and attenuation coefficient of MSVW respectively.

Fig.22 shows the theoretical MSBVW pulse envelope for measured amplitudes A'_n and calculated phases Φ'_n for five central frequencies from the spectrum of Fig.21a. It can be seen, that nonlinearity of the MSBVW transmission line phase response explains well the asymmetry of the pulse envelope.

Note, that in principle a passive synchronization of MSWs with different frequencies may be realized by including a MSW signal-to-noise enhancer in the oscillator loop. In both cases, to increase the number of interacting waves and to obtain the narrowing of the MSW pulses, one should use a non-dispersive MSW transmission line and increase the magnitude of the field modulation.

10. Conclusion

The results presented above demonstrate that MSWs in planar magnetic structures are a unique object for experimental investigation of wave processes in dispersive media with non-stationary parameters. We were able to observe and study the following phenomena: different kinds of MSW frequency spectrum transformation, space-time focusing of MSW packets, spectral separation of MSW modes, parametric interaction of MSVW in homogeneous media and scattering of MSVW in a spatially-periodic non-stationary magnetic field, and, finally, synchronization of phases of MSW with different frequencies..

Besides the above mentioned phenomena, there are some new effects that has been observed recently and require further investigations.

These are the shift of the frequency of MSW, propagating in a moving ferrite film,[37] and the increase of the parametric instability threshold of MSW under the action of non-stationary magnetic field.[38]

We believe that the three-wave thresholdless processes of combined MSW scattering where one of the waves play the role of non-stationarity are of some special interest. Such processes were considered for the first time in Ref.[39] and then the theory of combined MSW scattering which takes into account real dispersion characteristics of the waves in ferrite films has been developed in Ref.[40] The calculations in Ref.[41] showed a high efficiency of the three-wave MSW coupling processes. This theoretical result has been recently confirmed experimentally.[42]

The effects accompanying MSW propagation under non-stationary conditions may form a basis for new experimental methods. For example, the effect of the wave frequency shift in a linear-modulated magnetic field has been used recently for measuring individual characteristics of MSW modes,[14] as well as for measuring MSW saturation power level.[43] These results help to explain the behavior of the amplitude and phase responses of the MSW transmission line at high power levels.

The above described effects have been used in the development of a new class of "dynamic" microwave signal processing devices[44] such as: serrodyne, microwave mixers, compressor of monochromatic microwave pulses, coherent microwave pulses oscillator etc..

11. Acknowledgments

The author would like to thank Professor V. Preobrazhenskii and Professor A. Slavin for many helpful discussions and for technical assistance. The research described in this publication was made possible in part by grant N M6Q000 from the International Science Foundation.

12. References

1. F.R.Morgenthaler, *IRE Trans.* **MTT-6** (1958) 167.
2. N.S.Stepanov, *Izv. VUZov - Radiofiz.* **11** (1968) 700.
3. Y.A.Kravtsov, L.A.Ostrovskii and N.S.Stepanov, *Proc. of IEEE* **62** (1974).
4. W.M.Manheimer and B.H.Ripin, *Phys. Fluids.* **29** (1986) 2283.
5. *Ferroelectrics in Microwave Technology*, ed O.G.Vendik (Sov. Radio, Moscow, 1979).
6. S.M.Rezende and F.R.Morgenthaler, *Appl. Phys. Lett.* **10** (1967) 184.
7. V.V.Berezhnov, N.N.Evtikhiev, V.L.Preobrazhenskii and N.A.Ekonomov, *Radiotekh. and Electr.* **28** (1983) 376.
8. B.A.Auld, J.H.Collins and H.R.Zapp, *Electr. Lett.* **3** (1967) 35.
9. D.C.Webb and R.A.Moore, *Proc. of IEEE.* **54** (1965) 685.
10. S.M.Rezende nd F.R.Morgenthaler, *J. Appl. Phys.* **40** (1969) 537.
11. V.I.Zubkov, V.N.Kildishev and B.A.Murmuzhev, *Radiotech. and Electr.* **18** (1973) 811.
12. A.V.Vashkovskii, V.I.Zubkov, R.I.Kuznetsov, A.D.Morenov and T.A.Tsarukian, *Sov. Tech. Phys. Lett.* **4** (1978) [*Pis'ma Zh. Tekh.* Fiz. **4** (1978) 1231].
13. A.M.Mednikov and A.F.Popkov, *Sov. Tech. Phys. Lett.* **9** (1983) [*Pis'ma Zh. Tech.* Fiz. **9** (1983) 485].
14. N.N.Evtikhiev, V.V.Medvedev, V.L.Preobrazhenskii and Y.K.Fetisov, *Sov. Phys. Solid State* **27** (1985) 52 [*Fiz.Tverd.Tela.* **27** (1985) 90].
15. V.L.Preobrazhenskii and Y.K.Fetisov, *Sov. Phys. J.* **31** (1988) 898 [*Izv. VUZov-Fizika.* **31** (1988) 54].
16. R.W.Damon and J.H.Eshbach, *J. Phys. Chem. Solids.* **19** (1961) 308.
17. V.V.Medvedev and Y.K.Fetisov, in *Problems of Cybernetics. Devices and Systems* (MIREA, Moscow, 1973).
18. Y.F.Ogrin and V.V.Tarasenko, *Sov. Tech. Phys. Lett.* 10 (1984) [*Pis'ma Zh. Tekh.* Fiz. **10** (1984) 940].
19. S.N.Dunaev and Y.K.Fetisov, *Electr. Lett.* **29** (1993) 1216.
20. B.A.Kalinikos, *Sov. Phys. J.* **24** (1981) 719 [*Izv. VUZov-Fizika*, **24** (1981) 42].

21. V.L.Preobrazhenskii, V.P.Rubakov and Y.K.Fetisov, *JETP Lett.* **46** (1987) 114 [*Pis'ma Zh. Eksp. Teor. Fiz.* **46** (1987) 94].

22. S.N.Dunaev, V.L.Preobrazhenskii, V.P.Rubakov and Y.K.Fetisov, *Sov. Phys. Tech. Phys.* **36** (1991) 1198 [*Zh. Th. Fiz.* **61** (1991) 6].

23. J.H.Collins and F.A.Pizzarello, *Int. J. Electr.* **34** (1973) 319.

24. Y.K.Fetisov and V.L.Preobrazhenskii, *Sov. Phys. Tech. Phys.* **32** (1987) [*Zh. Tekh. Fiz.* **57** (1987) 564].

25. Y.K.Fetisov, *Digests of INTERMAG'93*, *BR-10*, April 13-16, 1993, Stockholm, Sweden.

26. T.W.O'Keeffe and R.W.Patterson, *J.Appl. Phys.* **49** (1978) 4886.

27. J.P.Parekh and H.S.Tuan, *IEEE Trans.* **MAG-16** (1980) 1165.

28. C.Elachi, *Proc. of IEEE.* **64** (1976) 1666.

29. Y.K.Fetisov, *Sov. Phys. Tech. Phys.* **32** (1987) 1451 [*Zh. Tekh. Fiz.* **57** (1987) 2393].

30. A.N.Slavin and Y.K.Fetisov, *Sov. Phys. Tech. Phys.* **33** (1988) 1343 [*Zh. Tekh. Fiz.* **58** (1988) 2210].

31. S.N.Dunaev and Y.K.Fetisov, *Electr. Lett.* **28** (1992) 1998.

32. A.N.Myasoedov and Y.K.Fetisov, *Sov. Phys. Tech. Phys.* **34** (1989) 666 [*Zh. Tekh. Fiz.* **59** (1989) 133].

33. A.V.Voronenko, S.V.Gerus and V.P. Haritonov, *Sov. Phys. J.* **31** (1988) [*Izv. VUZov-Fizika.* **31** (1988) 76].

34. A.Yariv, *Introduction to Optical Electronics* (Holt, Rinahart & Winston, New York, 1976).

35. S.N.Dunaev and Y.K.Fetisov, *Electr. Lett.* **28** (1992) 789.

36. S.N.Dunaev and Y.K.Fetisov, *Proc. 22 EuMC*, 588, 24-27 August 1992, Espoo, Finland.

37. N.L.Koros, N.Belaniuk and D.D.Stancil, J. *Appl. Phys.* **67**(1990)511.

38. V.G.Gaivoron and Y.F.Ogrin, *Sov. Phys. J.* **30** (1987) 75 [*Izv.VUZov-Fizika*, **30** (1987) 113].

39. R.E.Camley and A.A.Maradudin, *Phys. Rew. Lett.* **49** (1982) 168.

40. V.L.Preobrazhenskii, V.P.Rubakov and Y.K.Fetisov, *Sov. J. Electr. Commun.* **33** (1988) 110 [*Radiotekh. and Elektr.* **33** (1988) 1218].

41. V.L.Preobrazhenskii, V.P.Rubakov and Y.K.Fetisov, *Sov. Phys. Solid State* **29** (1987) 1456 [*Fiz. Tverd. Tela* **29** (1987) 2525].

42. N.G.Kovshikov, P.A.Kolodin and M.P.Kostylev, *Proc. of 22 EuMC*, 257, 24-27 August 1992, Espoo, Finland.

43. A.N.Myasoedov, A.V.Maryakhin, B.P.Nam and Y.K.Fetisov, *Sov. Phys. Tech. Phys.* **36** (1991) 71 [*Zh. Tech. Fiz.* **61** (1991) 118].

44. Y.K.Fetisov, *Proceedings of the 20th European Magnetic Conference*, 1142, 10-13 Sept. 1990, Budapest, Hungary.

SIZE EFFECTS IN PARAMETRIC EXCITATION OF SPIN WAVES IN FERRITES

V.B. CHEREPANOV *

Department of Physics of Complex Systems, Weizmann Institute of Science, Rehovot, 76100, Israel

A.N. SLAVIN

Department of Physics, Oakland University, Rochester, Michigan, 48309, USA

1. Introduction

There are two ways to excite spin waves in the microwave frequency range in ferrites: direct generation and parametric excitation. As a rule, spin waves directly generated by the microwave field have the wavelength λ of order of the sample size D. In contrast, parametrically excited spin waves have the wavelength λ several orders of magnitude smaller. Usually, the wavelength of parametrically excited spin waves is of the order of 10^{-4} – 10^{-5} *cm* while the sample size is of order of 10^{-1} *cm*. The purpose of this review is to discuss the physical mechanisms of influence of sample boundaries on excitation and interaction of short spin waves.

The influence of the sample boundaries on the spin wave dynamics can be expressed in terms of boundary conditions for ac magnetization. The boundaries make the spin waves modes non-planar and modify the spectrum of these modes in comparison with that in an unlimited medium.[1,2] Let us estimate the role of the boundary conditions in dynamics of spin waves with the wavelength λ. The boundary conditions result in a discontinuous spin wave spectrum, therefore the energy owing to the boundary condition can be estimated as follows:

$$\Delta E_s = \pi \hbar v_k / D, \tag{1}$$

where v_k is the group velocity and D is the size of a sample. The energy of the spin wave is

$$E_k = \hbar \omega_k, \tag{2}$$

where ω_k is the spin wave frequency. The ratio

$$\frac{\Delta E_s}{E_k} = \frac{v_k k}{2\omega_k}\left(\frac{\lambda}{D}\right) \sim \frac{\lambda}{D}, \tag{3}$$

*Present address: School of Physics and Astronomy, Tel Aviv University, Tel Aviv, 69978, Israel, on leave from the Institute of Automation & Electrometry, 630090, Novosibirsk, Russia where k is the wave number, gives us an estimate of the boundary condition contribution to the spin wave dynamics. If the wavelength is small in comparison with the sample size the size effect on dynamics of individual spin waves is negligible. However, the boundaries can play a very important role in collective behavior of short spin waves. We consider these effects in this review.

We need to discuss the collective processes in spin waves in some detail prior to further analysis of the size effects. Usually, the frequency of nonlinear spin waves is in the GHz range, i. e., it is a fraction of Kelvin degree, and the experiments are performed at significantly higher temperatures: 4 — 400 K. In this case, there are a lot of thermal spin waves (magnons) with relatively high energy. The frequency of thermal magnons is very high, their relaxation time is very short in comparison with those of spin waves in the microwave frequency range. Thus the thermal magnons are a source of high frequency noise for the microwave frequency spin waves. According to the fluctuation-dissipation theorem this noise causes spin waves damping with the relaxation frequency $\gamma_{\mathbf{k}}$. Thus the characteristic energy of the spin wave interaction with thermal magnons can be estimated as follows:

$$E_t \approx \hbar\gamma_k. \tag{4}$$

More precisely, E_t is the mean spin wave energy fluctuation due to its interaction with thermal magnons.

As far as the spin waves in the microwave range are concerned, they can have large amplitude and they interact with each other. Since the frequencies of such spin waves are comparable, these spin waves cannot be treated as a source of noise in the microwave frequency range. The energy of spin wave nonlinear interaction can be estimated as follows:

$$E_{nl} \approx \hbar k v_k \frac{m_z}{M_0}. \tag{5}$$

Here m_z is the longitudinal ac magnetization of the spin waves and M_0 is the saturation magnetization. Nonlinear interactions between spin waves result in various spin wave instabilities, harmonic generation, soliton formation, collapses, etc.

In general, one has to consider both processes of nonlinear interaction of spin waves in the microwave range and the relaxation owing to the interaction of spin waves with thermal magnons. Thus, there are three characteristic scales of energy in a finite system: E_s, E_t and E_{nl}. One can introduce characteristic length scales instead of the energies. The length for the size effects is the size of a sample D, the mean free path $l = v_k/\gamma_k$ is the scale of the spin wave envelope decay, and the length $\Lambda = \hbar v_k/E_{nl} = k^{-1}M_0/m_z$ gives a scale for nonlinear spin wave interaction. Λ can be either the wave length of the most unstable mode in the case of the modulation instability, or the size of a soliton, or the spatial scale of the phase interference caused by nonlinear processes. A qualitative consideration of various limiting cases gives an idea of what kind of size effects can be expected.

1. In the case when $E_s \gg E_t$ and $E_s \gg E_{nl}$ ($D \ll l$ and $D \ll \Lambda$) there are inhomogeneous spin wave modes with large frequency gaps between the modes. The

relaxation and spin wave interaction can be considered as small perturbations of the inhomogeneous harmonic modes.

2. If $E_t \gg E_s$ and $E_t \gg E_{nl}$ ($l \ll D$ and $l \ll \Lambda$) the spin wave damping is very strong. The nonlinear effects can be neglected in this case. The spin wave dynamics can be described in terms of a linear equation for spin wave envelope $A_{\mathbf{k}}(\mathbf{r}, t)$
$$\frac{\partial A_{\mathbf{k}}}{\partial t} + \mathbf{v}_{\mathbf{k}} \nabla A_{\mathbf{k}} + \gamma_{\mathbf{k}} A_{\mathbf{k}} = 0. \tag{6}$$
This equation describes exponential decay of the envelope with the distance from the source.

3. The case when $E_{nl} > E_s$ and $E_{nl} > E_t$ ($\Lambda < D$ and $\Lambda < l$) is the richest one. In this case, the size effects interfere with strong nonlinear effects. Since many nonlinear effects are owing to a balance of nonlinear forces and dispersion (e. g., soliton formation) or nonlinear forces and dissipation (instabilities) the size effects can play a critical role in such nonlinear phenomena intervening in the balance.

The nonlinear spin wave interactions were found to result in three kinds of instabilities: parametric decay instability,[3–5] collective oscillation instability,[6] and kinetic instability.[7] The parametric decay instability is responsible for the parametric excitation of short spin waves by homogeneous or long wavelength external pumping. The intensive packet of parametrically excited spin waves arising from the decay instability can, in its turn, demonstrate auto-oscillation or kinetic instability when the external pumping amplitude increases. The above mentioned instabilities differ in frequency range and phase correlations of arising oscillations.

In the case of parametric decay instability there is a strong phase coupling in pairs of parametrically excited spin waves. The kinetic instability results in growth of wide ($\Delta\omega_k \ll \gamma_k$) spin wave packets in the microwave range with random phases of individual spin waves. The collective oscillation instability leads to the auto-oscillations of magnetization with frequencies of the order of $E_{nl}/\hbar$. Typically, it is in the MHz frequency range.

In this chapter we discuss the influence of the sample-size effects on two kinds of nonlinear spin wave processes: (i) parametric excitation of primary spin waves by the homogeneous parallel pumping and (ii) the collective oscillation instability of parametrically excited spin waves above the threshold of parametric excitation. As far as the kinetic instability is concerned, the sample-size effect on it is usually negligible, as this instability results in excitation of a wide packet of spin waves in **k** space, with the packet width Δk of the order of $10^2 — 10^3\ D^{-1}$, where D is the sample size.

In a homogeneous unlimited medium, the threshold of the parametric decay instability is reached when two conditions are fulfilled: (i) the frequencies of spin waves obey the parametric resonance condition and (ii) the phase shift between the pumping field and the phases of spin waves in pairs allow for the maximum energy influx in the spin wave pairs. In a finite sample, the role of the boundary conditions is twofold: they describe an additional channel of the energy dissipation on the sample boundaries and they give constraints on phases of spin waves. The phases of the spin waves cannot correspond to the maximum energy influx to the spin waves in each point of the sample simultaneously because the complex spin wave amplitudes are inhomogeneous in the sample. Therefore in the case of the parametric decay instability the threshold amplitude of the microwave pumping field h_{th} in a finite magnetic sample is higher than the threshold h_0 in an unlimited medium[8,9]:

$$\left(\frac{h_{th}}{h_0}\right)^2 - 1 \propto \left(\frac{\Lambda}{D}\right)^2. \tag{7}$$

This formula was confirmed experimentally in detailed studies[10,11] in a wide temperature range on high quality ferrite samples.

The second of these two processes, the collective oscillation instability, results from nonlinear interaction of parametrically excited primary spin waves when their amplitudes are sufficiently large. This instability takes place above the threshold of parametric excitation and manifests itself as auto-oscillations of magnetization. These auto-oscillations can be both periodic and chaotic and their characteristic frequency is several orders of magnitude lower than the microwave frequency of parametric spin waves. The influence of finite size of a sample on the threshold of this instability and the frequency of auto-oscillations is the main subject of this chapter.

Theoretical and experimental investigations of the properties of auto-oscillations of magnetization above the spin wave instability threshold have been carried out for more than thirty years and considerable experimental experience is gained in this field.[6,12–28] The physical mechanism of spin wave auto-oscillations in the case of parallel pump parametric instability[6] was explained by L'vov, Musher and Starobinets.[13] The authors of[13] demonstrated theoretically that the auto-oscillations of magnetization observed in spin wave experiments are nothing else but the unstable collective oscillations in the system of parametrically excited spin waves analogous to the second sound in liquid helium. The idea of unstable collective oscillations was further developed in a series of subsequent theoretical papers.[9,29,30] The frequency spectra of both spatially-homogeneous[13] and spatially inhomogeneous[9] spin wave collective oscillations were calculated in the framework of this approach, and the conditions for stability of these oscillations were determined.

The spectrum of spatially-homogeneous collective oscillations against the background of spatially-homogeneous parametrically excited magnons has the form[13,31]

$$\Omega = i\gamma \pm \sqrt{\Delta^2 - \gamma^2}, \quad \Delta^2 = 4S(2T + S)N^2, \tag{8}$$

where γ is the equilibrium parameter of magnon dissipation (or the magnon relaxation frequency), S and T are the amplitudes of nonlinear interaction between magnons and N is the number of parametrically excited magnons.

It is clear from the Eq. (8) that if the product $S(2T+S)$ is positive the collective oscillations decay with time and the stationary state of the system of parametric magnons is stable. In the opposite case when $S(2T+S) < 0$ the stationary state is unstable for arbitrary N and the amplitude of collective oscillations increases exponentially with time.

It is worth noting, that the amplitudes S, T of nonlinear interaction between magnons demonstrate a very strong dependence on the crystallographic orientation of the sample even in "almost isotropic" cubic magnetic crystals like yttrium-iron garnet (YIG), where the magnon spectrum is practically not influenced by the crystallographic anisotropy.[14] This property of the coefficients S and T explains the giant crystallographic anisotropy of spin wave auto-oscillations observed.[12,14,32] The expression (8) also explains such characteristic features of spin wave auto-oscillations as the resonant properties of radio-frequency susceptibility and the dependence of the auto-oscillation frequency on the input power.[12,14–16]

However, the theory of auto-oscillations[13] based on the expression (8) does not explain the following important qualitative properties of auto-oscillations observed in experiments[12,14,33,34]:

(i) The difference between the thresholds of parametric excitation of spin waves h_{th} and the threshold of spin wave auto-oscillations h_{osc}. In experiment this difference is about 0.1- 2.0 dB and depends on the wave number of the parametrically excited spin waves, while in the theory[14] these two thresholds coincide;

(ii) The finite frequency of auto-oscillations at the threshold. In the experiment this frequency is about 10 - 1000 kHz and it depends on the spin wave wave number, while in the theory[13] this frequency is exactly zero at the threshold of auto-oscillations;

(iii) The dependence of the auto-oscillation threshold and the auto-oscillation frequency at the threshold on the size of the experimental sample. Experiments[33,34] have demonstrated a sample size effect on both the auto-oscillation threshold and frequency in yttrium-iron garnet (YIG) spheres of the orientation [111] while the theory[13] does not take into account the finite size of a sample.

An attempt to develop a general theory of spin wave auto-oscillations (for the cases of both parallel and perpendicular pumping) was undertaken by Zhang and Suhl in their paper.[30] They considered a rough (compared to S-theory[13,14]) model for the spin wave nonlinear dynamics and obtained not only the auto-oscillation spectrum, but also the stability condition for the periodic auto-oscillations of finite amplitude. In the case of perpendicular pumping the theory[30] gives the different threshold values for the spin wave excitation and spin wave auto-oscillations ($h_{osc}/h_{th} = 10$ dB) as well as the finite frequency of auto-oscillations at the threshold ($\Omega = 2\gamma$, where γ is the spin wave relaxation parameter). In the case of parallel pumping the theory[30] gives an estimate for the auto-oscillation frequency $|\Omega| \sim \gamma$ when $SN \sim \gamma$. This result agrees with the formula (8). Once again, as in the model[13,14] the influence of the sample size was not taken into account in the theory.[30]

The pronounced size dependence of the auto-oscillation frequency observed in experiments[33,34] suggests, however, that the above mentioned differences between the experimental results for auto-oscillations and the results of the theories[14,30] can be explained by the influence of the sample boundaries.

The main aim of this chapter is to present a theory of spin wave auto-oscillations that takes into account the effect of sample boundaries and explains all the above mentioned peculiarities of spin wave auto-oscillations observed in the experiments.[35,36]

The outline of the chapter is as follows. In section 2 we formulate a system of equations for the envelopes of spin wave packets. These equations describe parametric excitation and interaction of spin waves in the spatially inhomogeneous case and can be used in finite-size magnetic samples. In this section we also introduce phenomenologically the boundary conditions for the spin wave envelopes at the sample boundaries.

We consider the theory of the sample size effect in parametric excitation of spin waves in section 3. In the same section we obtain the expression for the threshold of parametric excitation of spin waves in a finite-size sample, and discuss related experimental results.

In section 4 we derive the system of equations for the amplitudes of small spatially inhomogeneous perturbations on the background of a stationary distribution of parametrically excited spin waves (or magnons) and obtain the boundary conditions for these perturbations. It is shown that when the characteristic size of the sample D is much greater than the magnon mean free path l the stationary distribution of parametric magnons is spatially homogeneous everywhere in the sample with the exception of a relatively small region (of the size l) near the sample boundaries. As a result, in this limiting case ($D >> l$) we arrive at a closed set of equations for the amplitudes of collective oscillations in the system of parametrically excited magnons. We obtain the frequency spectrum of collective oscillations in a sample of a finite size. We consider the case when the distribution of parametrically excited magnons is quasi-one-dimensional and the ferromagnetic sample has a finite size D in the direction of propagation of parametric magnons. We study in detail the particular case when the product $S(2T + S)$ is negative. This condition can be fulfilled, for example, in a cubic ferromagnet magnetized along the "easy" axis [111]. In an infinite ferromagnetic medium the condition $S(2T + S) < 0$ leads to the instability of spin wave collective oscillations for any arbitrary small number N of parametric magnons.[14] We show that contrary to that in a sample of a finite size D the spin wave collective oscillations become unstable only when the number N of parametric magnons is large enough to obey the condition

$$4|S(2T + S)|N^2 > 2\pi^2 v^2/D^2 , \tag{9}$$

where v is the group velocity of parametric magnons. In the same section we find the frequency of spin wave collective oscillations at the threshold of their instability, which turned out to be finite and dependent on the size of the sample. In the samples of a relatively large size ($D >> l$) this frequency is inversely proportional to the square of the sample size

$$\Omega \sim \frac{v^2}{\gamma D^2}, \tag{10}$$

while in a relatively small samples ($D << l$) the frequency of collective oscillations is inversely proportional to the sample size itself.

$$\Omega \sim \frac{v}{D}. \tag{11}$$

In section 5 we compare the above developed theory of unstable spin wave collective oscillations in finite-size samples with experimental results obtained in yttrium-iron garnet (YIG) spheres.

In section 6 we develop a microscopic theory of spin wave scattering on surface roughnesses of the sample. This theory provides a basis for the dissipative boundary conditions of the spin wave envelops used in the phenomenological approach presented in section 2.

Conclusions are given in section 7.

2. Formulation of the Problem and Basic Equations

In a ferromagnetic sample of a finite size D the distribution of parametric magnons above the threshold of their excitation is spatially inhomogeneous. Two alternative ways to describe the spin wave propagation in finite samples can be suggested: (i) one can introduce new eigenfunctions corresponding to the precise boundary conditions, (e. g. free spins or fixed spins, etc.) or (ii) one can develop an approximate theory considering the spin wave amplitudes smooth functions of the space coordinates, i. e. introducing wave envelopes. Since the packets of parametrically excited waves are narrow in the **k** space both representations must yield the same results, of course. However, the envelope representation is preferable because it allows one to escape many difficult problems irrelevant to collective properties of spin waves such as the boundary conditions for spins, details of statistical properties of surface roughnesses, etc.

The theoretical tool for description of spatially inhomogeneous effects in the system parametrically excited spin waves was developed by V. L'vov and A. Rubenchik.[9] They combined the nonlinear theory for the homogeneous state of parametrically excited spin waves[14] and the envelope approximation for narrow in the **k**-space (smooth in the coordinate space) wave packets. The S-theory equations for the homogeneous distribution of parametrically excited spin waves take the form[14]:

$$\begin{aligned} &\left(\frac{1}{2}\frac{\partial}{\partial t} + \gamma_{\mathbf{k}}\right) n_{\mathbf{k}} + Im\,(P^*_{\mathbf{k}}\sigma_{\mathbf{k}}) = 0, \\ &\left[\frac{1}{2}\frac{\partial}{\partial t} + \gamma_{\mathbf{k}} + i\,(\tilde{\omega}_{\mathbf{k}} - \omega_p/2)\right]\sigma_{\mathbf{k}} + iP_{\mathbf{k}}n_{\mathbf{k}} = 0, \end{aligned} \tag{12}$$

where $\gamma_{\mathbf{k}}$ is the spin wave relaxation frequency, $n_{\mathbf{k}}$ and $\sigma_{\mathbf{k}}$ are normal and anomalous spin wave correlation functions:

$$n_{\mathbf{k}} = \langle a_{\mathbf{k}} a^*_{\mathbf{k}} \rangle, \quad \sigma_{\mathbf{k}} = \langle a_{\mathbf{k}} a_{-\mathbf{k}} \rangle \exp{(i\omega_p t)}, \tag{13}$$

and the effective spin wave spectrum $\tilde{\omega}_{\mathbf{k}}$ and the pumping amplitude $P_{\mathbf{k}}$ depend on the spin wave occupation numbers $n_{\mathbf{k}}$ and phase correlations in pairs of spin waves with opposite momenta $\sigma_{\mathbf{k}}$:

$$\begin{aligned} \tilde{\omega}_{\mathbf{k}} &= \omega_{\mathbf{k}} + 2\sum_{\mathbf{k}'} T_{\mathbf{k},\mathbf{k}'} n_{\mathbf{k}'} \\ P_{\mathbf{k}} &= h\, V_{\mathbf{k}} + \sum_{\mathbf{k}'} S_{\mathbf{k},\mathbf{k}'} \sigma_{\mathbf{k}'}. \end{aligned} \tag{14}$$

Here $T_{\mathbf{k},\mathbf{k}'}$ and $S_{\mathbf{k},\mathbf{k}'}$ are the amplitudes of four magnon interaction, they can be found from expansion of the energy of the magnet in a series in spin wave amplitudes $a_{\mathbf{k}}$ and $a^*_{\mathbf{k}}$, and $\sum_{\mathbf{k}'}$ is a shorthand notation for integration over $\mathbf{k}'$:

$$\sum_{\mathbf{k}'} \longrightarrow \frac{\mathcal{V}}{(2\pi)^3} \int d\mathbf{k}', \tag{15}$$

where $\mathcal{V}$ is the sample volume. Thus it is assumed that an infinite number of spin waves is excited.

This system of equations is, in a certain sense, analogous to the BCS equations in the theory of superconductivity. In particular, in a spatially homogeneous case the threshold of parametric excitation of spin waves is reached when $|hV_{\mathbf{k}}| = \gamma_{\mathbf{k}}$, which is equivalent to the condition of the superconducting transition $T = T_c$. In the spatially homogeneous case the distribution of parametric magnons above the threshold of parametric excitation has a form such that the renormalized pumping $P_{\mathbf{k}}$ (14) is fixed at its threshold level $|P_{\mathbf{k}}| = \gamma_{\mathbf{k}}$ and the renormalized spin wave frequency $\tilde{\omega}_{\mathbf{k}}$ is equal to one half of the frequency of the pumping field (i. e., it stays in the exact parametric resonance). The homogeneous stationary solution of these equations has been studied in detail[14,17] and it has the form

$$\begin{aligned} |a^0_{\mathbf{k}}| &= |a^0_{-\mathbf{k}}|, \\ N = \sum_{\mathbf{k}} n_{\mathbf{k}} &= \frac{|V_k|}{|S|}\sqrt{h^2 - h^2_{th}}, \\ \arg{(hV\sigma_{\mathbf{k}})} &= \arcsin{(\gamma_k/|hV_k)}. \end{aligned} \tag{16}$$

Here S is the amplitude of four magnon interaction, $S_{\mathbf{k},\mathbf{k}}$, averaged over the spin wave distribution $n_{\mathbf{k}}$, N is the number of parametrically excited magnons, h_{th} is the threshold pumping field. The packet of parametrically excited spin waves is very narrow. Its width in the $\mathbf{k}$-space is equal to zero in the S-theory approximation. To describe the spin wave dynamics in the spatially inhomogeneous case one needs to to make two further steps: (i) to introduce the effective spin wave amplitudes $A_{\mathbf{k}}$ such that the spin wave distribution functions $n_{\mathbf{k}}$ and $\sigma_{\mathbf{k}}$ can be factorized:

$$n_{\mathbf{k}} = A_{\mathbf{k}}A^*_{\mathbf{k}}, \quad \sigma_{\mathbf{k}} = A_{\mathbf{k}}A_{-\mathbf{k}}, \tag{17}$$

and (ii) to apply the envelope approximation to the amplitudes $A_{\mathbf{k}}$.[9]

We are not presenting a rigorous derivation of the spin wave envelope equations here. This derivation is given in refs.[14,18] Instead we formulate the main ideas and approximations that form the basis for these envelope equations. If both the external pumping and the nonlinear interaction of spin waves are ignored the evolution of a narrow wave packet can be described in terms of smooth envelope dynamics (6). In the particular case when the energy of spin wave interaction has the form $E_{nl} = T|a(\mathbf{r})|^4$ the nonlinear frequency shift due to the nonlinear spin wave interaction is local and the shifted frequency can be written in the form: $\tilde{\omega}_{\mathbf{k}} = \omega_{\mathbf{k}} + 2\sum_{\mathbf{k}'} T|A_{\mathbf{k}'}(\mathbf{r})|^2$. In general, the amplitude of four magnon interaction depends on all four wave vectors of the magnons participating in the interaction process. Nevertheless, if the packet is narrow all the wave vectors are almost equal and one can neglect the $\mathbf{k}$ dependence of the four magnon amplitude. Thus, the nonlinear frequency shift for narrow spin wave packets can always be considered approximately local. The coupling of spin waves with opposite wave vectors through the homogeneous pumping (that leads to the renormalization of the pumping term $P_{\mathbf{k}}$) can also be considered local in the case of a narrow spin wave packet. Therefore the forms of the frequency shift and the renormalized pumping in the spatially inhomogeneous case for a narrow spin wave packet have the same form as in the spatially homogeneous case, and the equations for the envelope $A_{\mathbf{k}}$ have the form[9]:

$$\left[\frac{\partial}{\partial t} + \mathbf{v}_{\mathbf{k}}\frac{\partial}{\partial \mathbf{r}} + \gamma_{\mathbf{k}} + i\left(\tilde{\omega}_{\mathbf{k}} - \frac{\omega_p}{2}\right)\right] A_{\mathbf{k}}(\mathbf{r}) + iP_{\mathbf{k}}A^*_{-\mathbf{k}}(\mathbf{r}) = 0, \tag{18}$$

$$P_{\mathbf{k}} = hV_{\mathbf{k}} + \sum_{\mathbf{k}'} S_{\mathbf{k},\mathbf{k}'}A_{\mathbf{k}'}(\mathbf{r})A_{-\mathbf{k}'}(\mathbf{r}), \tag{19}$$

$$\tilde{\omega}_{\mathbf{k}} = \omega_{\mathbf{k}} + 2\sum_{\mathbf{k}'} T_{\mathbf{k},\mathbf{k}'}|A_{\mathbf{k}'}(\mathbf{r})|^2, \tag{20}$$

where the only explicit term describing spatial inhomogeneity is $\mathbf{v}_{\mathbf{k}}\frac{\partial}{\partial \mathbf{r}}A_{\mathbf{k}'}(\mathbf{r})$. Here $\gamma_{\mathbf{k}}$ and $\mathbf{v}_{\mathbf{k}}$ are the dissipation parameter and the group velocity of parametric spin waves, $\omega_{\mathbf{k}}$ is the spectrum of spin waves in the sample when the pumping field is absent, $h(t) = h\exp(\omega_p t)$ is the magnetic field of parametric pumping which is parallel to the bias magnetic field $\mathbf{H_0}$ ($\mathbf{h} \parallel \mathbf{H_0}$), $V_{\mathbf{k}}$ is the coupling coefficient of parametric magnons with the pumping field, and $S_{\mathbf{k},\mathbf{k}'}$, $T_{\mathbf{k},\mathbf{k}'}$ are the amplitudes of four magnon interactions calculated in.[14]

We consider in our present paper the simplest (but important for experiments) case when the magnetic sample is a thin mono-crystalline ferrite film of the width D. The constant bias magnetic field $\mathbf{H}_0$ and the ac pumping magnetic field $\mathbf{h}$ lie in the film plane, are parallel to each other, and are both oriented along the crystallographic axis [111] of the film. In this case the parametrically excited spin waves form a narrow packet with wave vectors oriented perpendicularly to the axis [111] (along which the pumping field **h** is aligned). This experimental configuration was used in the recent measurements.[32]

The simple quasi-one-dimensional model described above should also give a good qualitative description of the distribution of parametrically excited spin waves in spherical magnetic samples as in these samples above the threshold of parametric excitation this distribution is also highly anisotropic and consists of several narrow packets of parametric spin waves (see[37,38]).

In order to study the problem we need to know the boundary conditions for the spin wave envelopes in addition to the dynamic equations (18) – (20). The boundary conditions for $A_{\mathbf{k}}(\mathbf{r})$, $A_{-\mathbf{k}}(\mathbf{r})$ are, strictly speaking, dependent on the form and properties of the sample surface and can be derived rigorously by means of the diagrammatic technique for nonequilibrium processes.[39] However, in the case when the sample has a rough surface (so that the roughnesses are of the order of the wavelength of a parametric spin wave) and a spectrally narrow packet of spin waves is excited in the sample the spin waves scattered from the rough surface do not, as a rule, return back to the packet and simply dissipate in the volume of the sample[40] (see also section 6.) In such a case the boundary condition for the envelope of the spin wave packet is that the envelope amplitude of the spin wave packet reflected from the sample surface vanishes. We note, that the boundary conditions for the individual spin waves forming a narrow wave packet have no direct relation with the above formulated boundary condition for the packet envelope $A_{\mathbf{k}}(\mathbf{r})$.

Our aim is to study the behavior of the spin wave packet envelope $A_{\mathbf{k}}(\mathbf{r})$ taking into account the boundary conditions of absence of reflections on the sample boundaries:

$$A_{\mathbf{k}}|_{x=-D/2} = 0, \quad A_{-\mathbf{k}}|_{x=D/2} = 0. \tag{21}$$

We give a detailed consideration of the situation when the sample size D is much larger than the mean free path of a spin wave in the sample $l = v_k/\gamma_k$ ($D >> l$) and the material of the sample along with the magnetic field in the sample are homogeneous. The inhomogeneity arises from the sample boundaries only. This condition corresponds to a typical experimental situation.[10,18,34,33]

3. Threshold of Parametric Excitation of Spin Waves in Finite-size Samples

In this section we discuss the threshold of parametric instability in finite size homogeneous samples placed in a homogeneous microwave pumping field $h(t) = h\ \exp(-i\omega_p t)$. Since in this section we are interested only in calculation of the threshold of parametric excitation of spin waves we assume that the amplitudes of spin waves are small at the threshold and, therefore, we neglect the nonlinear terms in the equations (12)–(14). In that case the equation for the spin wave envelope $A_{\mathbf{k}}$

will have the form:

$$\left[\frac{\partial}{\partial t} + \mathbf{v_k}\frac{\partial}{\partial \mathbf{r}} + \gamma_\mathbf{k} + i\left(\omega_\mathbf{k} - \frac{\omega_p}{2}\right)\right] A_\mathbf{k} + ih\, V_\mathbf{k} A^*_{-\mathbf{k}} = 0. \tag{22}$$

Here $h(t) = h\exp(\omega_p t)$ is the magnetic field of parametric pumping which is parallel to the bias magnetic field $\mathbf{H_0}$ ($\mathbf{h} \parallel \mathbf{H_0}$).

In a homogeneous unlimited ferromagnet amplitudes of spin wave envelopes $A_{\mathbf{k_0}}(\mathbf{r}), A^*_{\mathbf{k_0}}(\mathbf{r})$ are[3,5,9]:

$$A_{\mathbf{k_0}}(\mathbf{r}) \sim \exp(\nu_\mathbf{p} t)\exp(i\mathbf{pr}),\ A^*_{\mathbf{k_0}}(\mathbf{r}) \sim \exp(\nu_\mathbf{p} t)\exp(-i\mathbf{pr}), \tag{23}$$

where the instability increment $\nu_\mathbf{p}$ has the form:

$$\nu_\mathbf{p} = -\gamma_{\mathbf{k_0}} + [|hV_{\mathbf{k_0}}|^2 - (\mathbf{v_{k_0}p})^2]^{1/2},\ \ \omega_{\mathbf{k_0}} = \omega_p/2. \tag{24}$$

Here $\mathbf{k_0}$ corresponds to the center of the packet and $\mathbf{k} = \mathbf{k_0} + \mathbf{p}$ is the current wave vector. In an unlimited medium we ignore the boundary conditions, therefore the solution (23) is a pair of spin waves with opposite wave vectors.

The instability takes place if the increment $\nu_\mathbf{p}$ is positive. The threshold value of microwave field h_{th} is reached when the increment (24) is positive for at least one pair of wave vectors $(\mathbf{k}, -\mathbf{k})$. This happens when the wave vector $\mathbf{k}$ is located on the resonant surface in $\mathbf{k}$-space: $\omega_{\mathbf{k_0}} = \omega_p/2,\ \ \mathbf{p} = \mathbf{0}$. Thus, in a homogeneous unlimited medium the threshold value h_{th_0} is:

$$h_{th_0} = \min_{\omega_{\mathbf{k_0}} = \omega_p/2}\left(\frac{\gamma_{\mathbf{k_0}}}{|V_{\mathbf{k_0}}|}\right). \tag{25}$$

The threshold, i. e. the minimum value in (25) is reached for a certain direction of the wave vector $\mathbf{k_0}$. If the bias magnetic field is directed parallel to the crystallographic axis [111] in the cubic ferromagnet the minimum of (25) is reached when $\mathbf{k_0}$ is perpendicular to the bias magnetic field. If $h > h_{th}$ the absorption of energy of microwave pumping increases sharply.[5,12,14]

In a ferromagnetic sample of a finite size D the distribution of parametric magnons above the threshold of their excitation is spatially inhomogeneous. The boundary conditions in this case still have the form (21). The problem of parametric instability of waves in a layer of finite thickness was studied in a general formulation in.[8] For the case of parametric excitation of spin waves by a homogeneous pumping field this problem was solved in.[9] We present below a short account of this solution.

Since we are considering the one-dimensional case we may assume that $\mathbf{r} = (x, 0, 0)$, where $\mathbf{x}$ is the direction of the wave vector $\mathbf{k_0}$ corresponding to the minimum in Eq. (25). In that case the solution of the Eq. (22) with the boundary conditions (21) can be found in the form

$$A_{\mathbf{k_0}}(\mathbf{r}) = A\sin(p(x + D/2)),\ A^*_{-\mathbf{k_0}}(\mathbf{r}) = A^*\sin(p(x - D/2)), \tag{26}$$

where $p = \psi/D$ is the characteristic wave number of the solution and ψ is a phase determined from the boundary conditions. Substituting Eqs. (26) for $A_{\mathbf{k_0}}$ and $A^*_{-\mathbf{k_0}}$ in Eq. (22) and its complex conjugate, assuming that at the threshold $\omega_{\mathbf{k_0}} = \omega_p/2$, and using the boundary conditions Eq. (21) we get the following equations for the phase $\psi = pD$ and the threshold value of the pumping field amplitude h_{th}

$$h_{th}\, V_{\mathbf{k}} \sin\psi = pv_{\mathbf{k}} = \gamma_{\mathbf{k}} \frac{l}{D}\psi,$$
$$h_{th}\, V_{\mathbf{k}} \cos\psi = -\gamma_{\mathbf{k}}. \tag{27}$$

Solving Eqs. (27) for ψ and h_{th} we get

$$\cos^2\psi = \left[1 + \left(\frac{\psi v_{\mathbf{k_0}}}{D\gamma_{\mathbf{k_0}}}\right)^2\right]^{-1}, \tag{28}$$

$$h_{th} = \min_{\omega_{\mathbf{k_0}}=\omega_p/2} \left(\frac{\gamma_{\mathbf{k_0}}}{|V_{\mathbf{k_0}}|} \frac{1}{|\cos\psi|}\right) = \min_{\omega_{\mathbf{k_0}}=\omega_p/2} \left(\frac{\gamma_{\mathbf{k_0}}{}^2 + (v_{\mathbf{k_0}} p)^2}{|V_{\mathbf{k_0}}|}\right) \tag{29}$$

Analysis of Eq. (27) for ψ shows that the value of ψ depends on the ratio of the mean free path of a parametric spin wave $l = v_{\mathbf{k}}/\gamma_{\mathbf{k}}$ and the sample size D. In the case of a relatively large sample ($D \gg l$) $\psi = \pi$, and $\psi = \pi/2$ in the opposite limiting case of small samples ($D \ll l$).[9]

Comparing Eq. (25) for the infinite medium with Eq. (29) we see that the threshold in the case of a finite sample is higher because of the phase shift ψ determined by the boundary conditions. Introducing the mean free path of a parametric spin wave $l = v_{\mathbf{k}}/\gamma_{\mathbf{k}}$ and a dimensionless parameter $\zeta = \psi/\pi$ (which is a smooth function of the ratio l/D and changes from 1/2 at $l/D \to 0$ to 1 at $l/D \to \infty$ in Eq. (28)) we can rewrite Eq. (29) in the form

$$h_{th} = h_{th_0} \left[1 + \left(\zeta\pi \frac{l}{D}\right)^2\right]^{1/2} \tag{30}$$

In the case when $l \ll D$ (large sample) the threshold h_{th} in a finite sample is close to the threshold in an infinite medium h_{th_0}. In the opposite limiting case $l \gg D$ (which takes place either for very small samples or at low temperatures when $\gamma_{\mathbf{k}}$ is small and the mean free path of spin waves is large) Eq. (30) shows that the threshold is independent of dissipation in the material and it is inversely proportional to the sample size D

$$h_{th} = \min\left(\frac{\pi v_{\mathbf{k_0}}}{2D|V_{\mathbf{k_0}}|}\right). \tag{31}$$

This theoretical conclusion was confirmed experimentally.[10,11] The threshold of parametric excitation of spin waves under parallel pumping was measured as a function

Table 1. Comparison of experimental and theoretical values of $|h_{th}\, V_{\mathbf{k}}|$ for YIG spheres of different diameter D at the temperature $T = 1.8K$

D, mm	$\lvert h_{th}\, V_{\mathbf{k}}\rvert$, $MHz \pm 10\%$, experiment	$\lvert h_{th}\, V_{\mathbf{k}}\rvert$, theory, from Eq. (31)
1.8	0.182	0.175
1.5	0.21	0.21
1.2	0.25	0.26

of temperature in the temperature interval $1.8K < T < 5K$ in a YIG sphere which was polished down from the diameter $D_{\max} = 1.8\ mm$ through $D_{mid} = 1.5\ mm$ to the diameter $D_{\min} = 1.2\ mm$. The YIG sphere was magnetized along the [100] axis and the pumping frequency was $\omega_p/2\pi = 9370$ MHz. The threshold was measured for spin waves with wave number $k \sim 10^5 cm^{-1}$ propagating perpendicularly to the direction of the bias magnetic field. The coupling coefficient $V_{\mathbf{k}}$ in this case has the magnitude $V_{\mathbf{k}} = g\omega_M/2\omega_p$, where g is the gyromagnetic ratio for the electron spin ($g = 2.8$ MHz$/Oe$), $\omega_M = g4\pi M_0$, and the M_0 is the saturation magnetization ($\omega_M/2\pi = 4.9$ GHz in YIG).

The results of these experiments are presented in Fig. 1 for the diameters $D_{\max} = 1.8\ mm$ and $D_{\min} = 1.2\ mm$. It is clear from Fig. 1 that with the decrease of temperature (and the increase of mean free path for spin waves) the thresholds for spheres of different diameters approach different limiting values. These values are in a good numerical agreement with the results of calculations using Eq. (31) (see Table 1). It is important to note not only a good numerical agreement between theory and experiment, but also inversely proportional dependence of the threshold on the sample diameter which allows us to conclude that the experiments[10,11] clearly demonstrate sample-size effect in parametric excitation of spin waves.

4. Auto-oscillations in Finite-size Samples above the Threshold of Parametric Excitation

In the previous section we considered the process of parametric excitation of spin waves in a finite size magnetic sample, i. e. the process of decay instability of the homogeneous parallel pumping field into pairs of parametric spin waves. It turns out that the system of parametrically excited spin waves above the threshold of their excitation often demonstrates the auto-oscillation instability, i. e. the instability with respect to collective oscillations of these parametric spin waves. Below we consider the collective oscillation instability in the system of parametrically excited spin waves in a finite size ferromagnetic sample. We assume that the collective oscillations in a finite-size sample could be spatially inhomogeneous and that they are responsible for auto-oscillations of magnetization observed experimentally above the threshold of parametric excitation in YIG spheres (see e. g.[34,33]). To describe collective oscillations of spin waves in this case we must take into account interactions between parametric spin waves ($S_{\mathbf{kk'}}$, $T_{\mathbf{kk'}} \neq 0$) and use the full system of equations and boundary conditions (18) – (21) for the envelopes of spin wave packets.

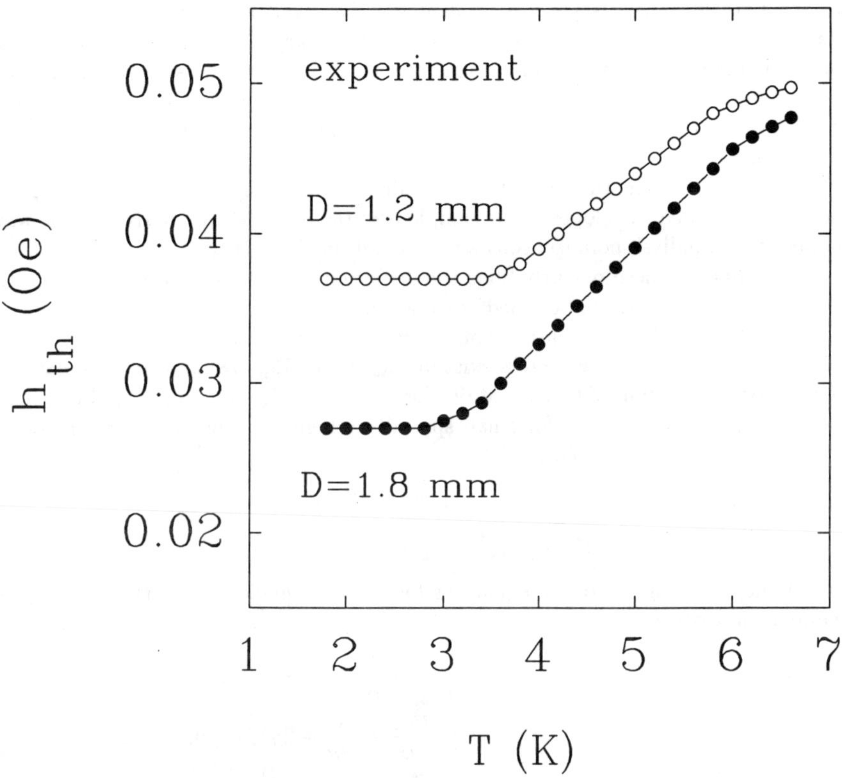

Fig. 1. Dependence of the threshold h_{th} of parametric excitation of spin waves under parallel pumping on the temperature T for YIG spheres of different diameters: $1 - D = 1.8$ *mm*, $2 - D = 1.2$ *mm*.[10]

We give a detailed consideration of the situation when the sample size D is much larger than the mean free path of a spin wave in the sample $l = v_k/\gamma_k$ ($D >> l$). This condition corresponds to a typical experimental situation.[18,33]

To solve this problem we need to know the stationary state of the the system of parametric spin waves in the sample beyond the threshold of parametric excitation. This stationary state was studied by L'vov and Rubenchik.[9] It was shown that when the condition $D >> l$ holds the stationary state in the volume of the sample is approximately spatially homogeneous and almost the same as in the case of unbounded ferromagnetic medium. The deviations from the spatial homogeneity of the stationary state appear only in a layer of the thickness l near the boundaries of the sample.

At the same time the characteristic scale of spatial inhomogeneity of the perturbation of spin wave envelopes in the sample is of the order of sample size D (the wave number of spatially-inhomogeneous auto-oscillations is usually of the order of $1/D$). It means that for these perturbations the deviations of the stationary state from the spatially homogeneous "unbounded" case are negligible.

In the case when two narrow spin wave packets with opposite directions ($\mathbf{k}$ and $-\mathbf{k}$) of the spin wave wave vectors exist in the sample Eq. (18) and its conjugate can be reduced to a system of four equations for the values $A_{\mathbf{k}}, A_{-\mathbf{k}}, A^*_{\mathbf{k}}, A^*_{-\mathbf{k}}$. Linearizing the dynamic equations (18) for small spatially inhomogeneous perturbations of the homogeneous stationary solution

$$\begin{aligned} A_{\pm k} &= A^0_{\pm k} + \alpha_{\pm k} \exp(-i\Omega t + ipx), \\ A^*_{\pm k} &= {A^0_{\pm k}}^* + \alpha^\dagger_{\pm k} \exp(i\Omega t + ipx) \end{aligned} \qquad (32)$$

we get the following system of equations for the renormalized perturbations $\{\phi_n\}$ of the magnon distribution

$$\begin{aligned} \frac{\partial \phi_1}{\partial t} + v\frac{\partial \phi_2}{\partial x} + 2iSN\phi_3 &= 0, \\ v\frac{\partial \phi_1}{\partial x} + (\frac{\partial}{\partial t} + 2\gamma)\phi_2 &= 0, \\ 2i(T^+ + T^- + S)N\phi_1 + (\frac{\partial}{\partial t} + 2\gamma)\phi_3 + v\frac{\partial \phi_4}{\partial x} &= 0, \\ 2i(T^+ - T^-)N\phi_2 + v\frac{\partial \phi_3}{\partial x} + \frac{\partial \phi_4}{\partial t} &= 0, \end{aligned} \qquad (33)$$

where

$$\begin{aligned} \phi_1 &= A^{0*}_{\mathbf{k}}\alpha_{\mathbf{k}} + A^0_{\mathbf{k}}\alpha^\dagger_{\mathbf{k}} + A^{0*}_{-\mathbf{k}}\alpha_{-\mathbf{k}} + A^0_{-\mathbf{k}}\alpha^\dagger_{-\mathbf{k}}, \\ \phi_2 &= A^{0*}_{\mathbf{k}}\alpha_{\mathbf{k}} + A^0_{\mathbf{k}}\alpha^\dagger_{\mathbf{k}} - A^{0*}_{-\mathbf{k}}\alpha_{-\mathbf{k}} - A^0_{-\mathbf{k}}\alpha^\dagger_{-\mathbf{k}}, \\ \phi_3 &= A^{0*}_{\mathbf{k}}\alpha_{\mathbf{k}} - A^0_{\mathbf{k}}\alpha^\dagger_{\mathbf{k}} + A^{0*}_{-\mathbf{k}}\alpha_{-\mathbf{k}} - A^0_{-\mathbf{k}}\alpha^\dagger_{-\mathbf{k}}, \\ \phi_4 &= A^{0*}_{\mathbf{k}}\alpha_{\mathbf{k}} - A^0_{\mathbf{k}}\alpha^\dagger_{\mathbf{k}} - a^{0*}_{-\mathbf{k}}\alpha_{-\mathbf{k}} + A^0_{-\mathbf{k}}\alpha^\dagger_{-\mathbf{k}}. \end{aligned} \qquad (34)$$

Here we introduced the notations

$$T^{+}=T_{\mathbf{k},\mathbf{k}},\;\; T^{-}=T_{\mathbf{k},-\mathbf{k}},\;\; S=S_{\mathbf{k},\mathbf{k}}=S_{\mathbf{k},-\mathbf{k}}\;\; v=v_k\;\; \gamma=\gamma_k. \tag{35}$$

We note that in the chosen geometry of the problem $T^{-}=S$. The coefficients T^{+}, T^{-} and S in Eq. (33) can be expressed in terms of Fourier harmonics S_0, T_0, $S_{\pm 2}$, $T_{\pm 2}$ of the coefficients of the four-magnon Hamiltonian presented in review[14] (see Eqs.(4.5)-(4.9)[14])

$$\begin{aligned} T=\frac{1}{2}\left(T^{+}+T^{-}\right)&=T_0+T_2+T_{-2},\\ S&=S_0+S_2+S_{-2}. \end{aligned} \tag{36}$$

Calculation of the determinant of the above derived system of linear equations (33) leads to the following dispersion equation for the spatially inhomogeneous collective oscillations of the envelope of the spin wave packet

$$\Omega\left(\Omega+2i\gamma\right)=\left[\frac{\Omega_0^2}{2}+(vp)^2\right]\pm\sqrt{\left[\frac{\Omega_0^2}{2}+(vp)^2\right]^2-(vp)^2\left[\delta^2-(vp)^2\right]}, \tag{37}$$

where

$$\Omega_0^2=4S(2T+S)N^2,\quad \delta^2=4S(T-S)N^2. \tag{38}$$

In accordance with the boundary conditions for the spin wave envelopes $A_{\pm\mathbf{k}}$, $A^{*}_{\pm\mathbf{k}}$ (21) the boundary conditions for the small perturbations of the envelope $\alpha_{\pm\mathbf{k}}$, $\alpha^{\dagger}_{\pm\mathbf{k}}$ have the following form

$$\begin{aligned} \alpha_k|_{x=-D/2}=0,\qquad \alpha_{-\mathbf{k}}|_{x=D/2}=0,\\ \alpha^{\dagger}_{\mathbf{k}}|_{x=-D/2}=0,\qquad \alpha^{\dagger}_{-\mathbf{k}}|_{x=D/2}=0. \end{aligned} \tag{39}$$

In terms of the renormalized perturbations ϕ_1, ϕ_2, ϕ_3, ϕ_4 (34) the boundary conditions (39) can be reduced to

$$\begin{aligned} (\phi_1+\phi_2)|_{x=-D/2}=0,\qquad (\phi_1-\phi_2)|_{x=D/2}=0,\\ (\phi_3+\phi_4)|_{x=-D/2}=0,\qquad (\phi_3-\phi_4)|_{x=D/2}=0. \end{aligned} \tag{40}$$

The linear equations (33) in combination with the boundary conditions (40) form a closed system of equations from which the spectrum of the collective spin wave oscillations in a finite size magnetic sample can be determined.

The general solution of the linear system of equations (33) for the amplitudes of spatially inhomogeneous perturbations in magnon distribution can be written as a sum of exponentials

$$\phi_i(x)=\Phi_i^{+}\,e^{ip_1x}+\Phi_i^{-}\,e^{ip_1x}+\Psi_i^{+}\,e^{ip_2x}+\Psi_i^{-}\,e^{ip_2x}, \tag{41}$$

where $i = 1, 2, 3, 4$ and p_1 and p_2 are the roots of the Eq. (37) corresponding to the same frequency Ω

$$(vp_{1,2})^2 = \Omega(\Omega + 2i\gamma) + \frac{1}{2}\delta^2 \pm \left[\frac{1}{4}\delta^4 + \Omega(\Omega + 2i\gamma)\left(\Omega_0^2 + \delta^2\right)\right]^{1/2}. \tag{42}$$

Substituting Eqs. (41) for the perturbation amplitudes ϕ_i in the boundary conditions (40) we obtain the expression for the spectrum of the spin wave collective oscillations in a finite size magnetic sample as a condition of existence of a nontrivial solution of the linear system (40). The (4×4) matrix of boundary conditions (40) splits into two (2×2) matrices in the variables $(\phi_1 + \phi_2,\ \phi_3 + \phi_4)$ and $(\phi_1 - \phi_2,\ \phi_3 - \phi_4)$. Introducing the dimensionless variables

$$\begin{gathered} q_{1,2} = \frac{1}{2}p_{1,2}D, \quad \tilde{\Omega} = \frac{\Omega}{2\gamma}, \\ \Lambda^2 = i\tilde{\Omega}(1 - i\tilde{\Omega}), \quad \Delta^2 = \delta^2/(2\gamma)^2, \\ \rho^2 = \frac{\Omega_0^2}{\delta^2} \equiv \frac{2T + S}{T - S}, \quad \xi = \frac{v}{\gamma D} \ll 1 \end{gathered} \tag{43}$$

we can rewrite the determinant of the first of these (2×2) matrices in the form

$$\begin{aligned} \det M_{n,m} &= 0 \quad (n,\ m = 1,\ 2), \\ M_{1,m} &= (1 - i\tilde{\Omega})\cos q_m - \xi q_m \sin q_m, \\ M_{2,m} &= (\Lambda^2 - \xi^2 q_m^2)\tilde{\Omega}\cos q_m - i(\Lambda^2 - \xi^2 q_m^2 + \Delta^2)\xi q_m \sin q_m. \end{aligned} \tag{44}$$

The second matrix differs from the first one only by substitution

$$\cos q_{1,2} \to \sin q_{1,2}, \quad \sin q_{1,2} \to -\cos q_{1,2}. \tag{45}$$

The problem of calculation of the threshold of spin wave auto-oscillations in a finite size magnetic sample is thus reduced to finding the minimum value of $|\Delta|$ (which is proportional to the number N of parametrically excited magnons) for which the imaginary part of the frequency Ω of spin wave collective oscillations is positive

$$Im\,\Omega \geq 0. \tag{46}$$

The fulfillment of the inequality (46) corresponds to the appearance of the non-decaying spatially inhomogeneous perturbations of the stationary distribution of parametric magnons in the sample, i. e., to the appearance of the auto-oscillations.

The general analytical solution of the transcendental equation (44) is difficult to find, but in the case of relatively small value of the coefficient $|\rho|^2 < 8$ the solution of Eq. (44) can be obtained in a closed form and the maximum of $|\Delta|^2$ is reached when

$$p_1 \simeq \frac{\pi}{D}, \quad p_2 \simeq i\frac{\sqrt{2}\pi}{\rho^2 D} \tag{47}$$

We note, that in the experimental situation which is most interesting for us (YIG sample magnetized along the [111] axis and pumped at the frequency $\omega_p/2\pi = 9.4\ GHz$) the value of $\rho^2 = 0.78$.

The threshold value of the number of parametric magnons corresponding to the appearance of unstable (undamped) perturbations is

$$4\,|S\,(2T+S)|\,N_{osc}^2 = 2\pi^2\frac{v^2}{D^2}. \tag{48}$$

The condition (48) means that the spin wave collective oscillations become unstable only when the characteristic energy of interaction between magnons δE exceeds the energy splitting in the magnon spectrum caused by the influence of the sample boundaries $\delta E > \hbar\pi v/D$. The condition (48) also demonstrates that the threshold of parametric excitation of spin waves h_{th} and the threshold of spin wave auto-oscillations h_{osc} do not coincide in a finite sample and the difference between these two thresholds depends both on the sample size D and on the wave number of parametric magnons (through the magnon group velocity v).

Using Eqs. (29) we can get from the Eq. (48) the expression for the threshold amplitude of the pumping field h_{osc} corresponding to the appearance of unstable spin wave collective oscillations i.e. the threshold pumping amplitude for the spin wave auto-oscillations

$$\frac{h_{osc}^2}{h_{th}^2} - 1 = \frac{1}{2}\left|\frac{S}{2T+S}\right|\left(\frac{\pi v}{\gamma D}\right)^2. \tag{49}$$

The frequency of the collective oscillations at the threshold of their instability is equal to

$$\frac{\Omega}{\gamma} = \frac{1}{2}\sqrt{\frac{1+2\rho^2}{\rho^2}}\left(\frac{\pi v}{\gamma D}\right)^2. \tag{50}$$

The calculation of the determinant of the second (2×2) matrix (analogous to Eq. (44)) yields a higher value for the magnon number N_{osc} corresponding to the threshold of auto-oscillations. Thus, the most unstable are the collective oscillations having the eigenvectors

$$(\phi_1+\phi_2),\quad (\phi_3+\phi_4) \tag{51}$$

and the instability threshold determined by the Eq. (49).

The influence of the sample boundaries is the strongest in the case when the size of a sample D is small compared to the magnon mean free path l, so that the dimensionless parameter ξ is large ($\xi = l/D \gg 1$). Unfortunately, in this case the stationary distribution of parametric magnons is not known. However, it is clear that the characteristic spatial scale of this stationary distribution of magnons is of the order of D and the dissipation of magnons is caused mainly by their scattering on the sample boundaries. This means that only one parameter having the dimension of frequency exists in the problem:

$$\Omega_D = \frac{v}{D}. \tag{52}$$

Thus, in the case $\xi \gg 1$ the threshold of instability of spin wave collective oscillations and their frequency at the threshold can be estimated as follows:

$$4\left|S\left(2T+S\right)\right|N_{osc}^2 \approx \Omega_D^2, \tag{53}$$

$$\Omega \approx \Omega_D. \tag{54}$$

It is easy to prove that in the intermediate case, when the size of the sample D is comparable to the mean free path of parametric magnons l ($D \sim l$), the results of the exact solution for the threshold and the frequency of spin wave collective oscillations Eqs. (48) and (50) and the results of estimations Eqs. (53) and (54) are smoothly transformed one into another.

5. Comparison of Theory and Experiments for Spin Wave Auto-oscillations in YIG Spheres

To compare predictions of the above developed theory of spin wave auto-oscillations in finite-size samples with experiment in detail it is necessary to carry out systematic measurements where the influence of the size of the sample on the spin wave auto-oscillations is studied in a wide range of sample sizes and relaxation parameters of parametric spin waves.

However, the experimental data available in the literature (see e.g.[12,14]) and, especially, the results of the recent measurements[33,34,41] enable us to make some estimations. The most detailed study of the influence of the sample size on the spin wave auto-oscillations was done by Rezende et al.[33,34] In these experiments the threshold of parametric excitation of spin waves under parallel pumping, as well as the threshold and the frequency (at the threshold) of spin wave auto-oscillations, were measured at the room temperature as functions of bias magnetic field in YIG spheres of different size D magnetized along the axis [111]. Calculations show that for the conditions of the experiments[33,34] the ratio of the spin wave mean free path l to the sample size D was about $l/D \approx 0.3 - 0.5$, while our expressions for the characteristics of spin wave auto-oscillations (Eqs. (49), (50)) were obtained in the limit $l/D \ll 1$. Thus *a priori* we can expect only qualitative agreement of our theory with the results of the experiments.[33,34]

To make an accurate comparison between theory and experiment we must calculate spin wave interaction coefficients T and S for the conditions of the experiments.[33,34] We also must find the spin wave group velocity v and the relaxation parameter γ as functions of the bias magnetic field H_0 under the same conditions.

Coefficients of four-wave interaction between parametrically excited spin waves were calculated approximately[14,18] for the case of a ferromagnetic ellipsoid magnetized along the crystallographic axes [100], [111] and [110]. In the case of a spherical sample of cubic ferromagnet these expressions have the form

$$\frac{S}{2\pi g^2} = \left(\frac{\omega_M}{\omega_p}\right)^2 \left[\left(1+\frac{\omega_M^2}{\omega_p^2}\right)^{1/2} + \beta\frac{\omega_a}{\omega_M} + N_z - 1\right], \tag{55}$$

$$\frac{T}{2\pi g^2} = S + \beta\frac{\omega_a}{\omega_M} + N_z - \frac{3}{2}, \tag{56}$$

where N_z is the demagnetization factor of the magnetic sample in the direction of the bias magnetic field ($N_z = 1/3$ for a sphere), $\omega_a = gH_a$, $H_a = 2K_1/M_0$ is the effective anisotropy field ($H_a = -84Oe$ for YIG at a room temperature), K_1 is the constant of cubic crystallographic anisotropy, and β is a coefficient dependent on the orientation of the saturation magnetization $\mathbf{M}_0$ relative to the crystallographic axes of the sample

$$\beta = \begin{cases} 9 \text{ when } \mathbf{M_0}||[100], \\ -4 \text{ when } \mathbf{M_0}||[111], \\ -9/2 \text{ when } \mathbf{M_0}||[110]. \end{cases} \tag{57}$$

The values of coefficients S and T calculated using Eqs. (55), (56) for the conditions of the experiments[33,34] ($4\pi M_0 = 1760\ Gs$, $\omega_p/2\pi = 9.5\ GHz$, $\beta = -4$) are $S/2\pi g^2 = 0.354$ and $T/2\pi g^2 = -0.503$, that gives the ratio of these coefficients $R = -1.42$. It is clear that for these values of S and T the necessary condition for the instability of spin wave collective oscillations $S(2T + S) < 0$ (or $(2R + 1) < 0$) is fulfilled.

We note, that both values and signs of the four-wave interaction coefficients S and T depend strongly on the crystallographic orientation of the sample even in the case of cubic magnetic crystals like YIG where the spin wave spectrum is only weakly anisotropic. This property of the coefficients S and T explains the experimentally observed "giant" crystallographic anisotropy of spin wave auto-oscillations under parallel pumping (see e. g.[6,13,14,32]). We also would like to note that the expressions for the four magnon interaction amplitudes (55) and (56) strictly speaking are correct only in the low temperature limit, when one can neglect the spin wave thermal fluctuations that renormalize the spin wave interaction. This renormalization cannot be expressed through the temperature dependence of the magnetization M_0. As to the dynamic equations for the spin wave amplitudes we started with, they are correct in a wide temperature range when the spin wave relaxation rate is small compared to their frequency. The spin wave interaction amplitudes can be considered as phenomenological constants.

To calculate the dependence of the spin wave group velocity v on the bias magnetic field H_0, $v = \partial\omega_k/\partial k = f(H_0)$ we used the dispersion equation for the case of spin waves propagating perpendicularly to the bias magnetic field ($\theta_k = \pi/2$) in a sphere of cubic ferromagnet magnetized along the axis [111] (see e. g.[2,14])

$$\omega_k^2 = \left(\omega_0 + \alpha\omega_M k^2\right)\left(\omega_0 + \alpha\omega_M k^2 + \omega_M\right), \tag{58}$$

where $\omega_0 = \omega_H - N_z\omega_M + \epsilon\omega_a$, k is the wave number, $\omega_H = gH_0$, H_0 is the external bias magnetic field, α is the exchange constant ($\alpha = 3 \cdot 10^{-12} cm^2$ for YIG), $N_z = 1/3$ for a sphere, and $\epsilon = 2/3$ for the orientation [111].

Spin wave relaxation parameter $\gamma(H_0)$ was calculated from the experimental dependences of the threshold of parametric excitation $h_{\text{th}}(H_0)$("butterfly curves") presented by Rezende et al.[33,34] These calculations were done using the threshold formula Eq. (29) in the case $\theta_k = \pi/2$ and $l/D \ll 1$. Substituting the results of these calculations for v, γ, S and T in Eqs. (49), (50) we can compare our theoretical results with the results of the experiments[33,34] for the threshold and the frequency (at the threshold) of spin wave auto-oscillations. The results of this comparison are presented in Fig. 2 for the YIG sphere of diameter $D = 1$ mm. It is clear from Fig. 2 that our theory is in reasonable quantitative agreement with the experimental data on the auto-oscillation frequency $f_0 = \Omega/2\pi$, but only qualitatively agrees with the experimental data on auto-oscillation threshold h_{osc}.

We attribute this quantitative discrepancy in the theoretical and experimental values of $h_{osc}(H_0)$ to the fact that our simple estimations of the spin wave interaction coefficients S and T (using Eqs. (55), (56)) are only accurate at low temperature while the experiments were performed at the room temperature, which is about 55% of the Curie temperature. Fortunately, it turned out that using our theory we can estimate the ratio of these coefficients at the room temperature directly from the experimental data.[33,34]

We would like to note one important property of Eq.(49), (50). In the axially symmetric case (sphere) the ratio of these two expressions is independent of both the sample size D and the bias magnetic field H_0

$$G = \frac{(h_{osc}^2/h_{th}^2 - 1)}{\Omega/\gamma} = |(2R+1)(5R+1)|^{-1/2}, \tag{59}$$

For the previously calculated value of $R = -1.42$ Eq. (59) gives $G = 0.3$. Using the experimental data[33,34] we calculated "experimental" values for the same ratio G for two YIG spheres of diameters $D_1 = 1mm$ and $D_2 = 0.52$ mm. The results of comparison of the theoretical (solid line) and the experimental values (points) of G are presented in Fig. 3.

It is clear from Fig. 3 that although the numerical values of theoretical and experimental ratio G are different, the behavior of G with variation of D and H_0 is similar: in both theory and experiment G is independent of both the sample size D and the bias magnetic field H_0. Using Fig. 3 we determined the value of the parameter R_{exp} for which Eq. (59) gives the values for the ratio G that is equal to the average of experimental values $G_{\rm exp} = 0.8$. This value turned out to be equal to $R_{exp} = -0.765$. We note, that the necessary criterion of instability of spin wave collective oscillations $(2R+1) < 0$ is still fulfilled for this "experimental" value of the ratio R. Below we used $R_{exp} = -0.765$ in Eqs. (49), (50) to calculate the theoretical curves $h_{osc}(H_0)$ and $\Omega/2\pi = f_0(H_0)$. The results of comparison of these calculations with the experimental data from[34] are presented in Fig. 4.

It can be seen from Fig. 4 that our theoretical model for $R = R_{exp} = -0.765$ demonstrates good quantitative agreement with the results of the experiment[33] for both diameters of YIG spheres ($D_1 = 1$ mm and $D_2 = 0.52$ mm) used in the measurements.[34] Thus this model gives correct description of the sample-size effect in spin wave auto-oscillations under parallel pumping in YIG spheres.

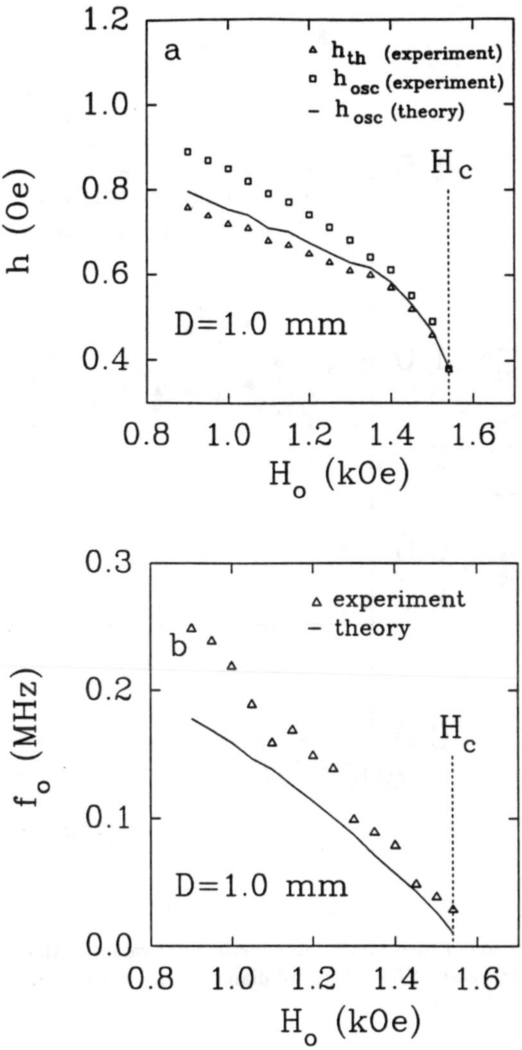

Fig. 2. Comparison of the results of experimental measurements of auto-oscillation threshold h_{osc} (a) and frequency f_0 (b) [33] at room temperature on a YIG sphere of the diameter $D = 1\ mm$ with the results of theoretical calculations of these characteristics done using Eqs. (49), (50) for $R_{theor} = -1.42$.

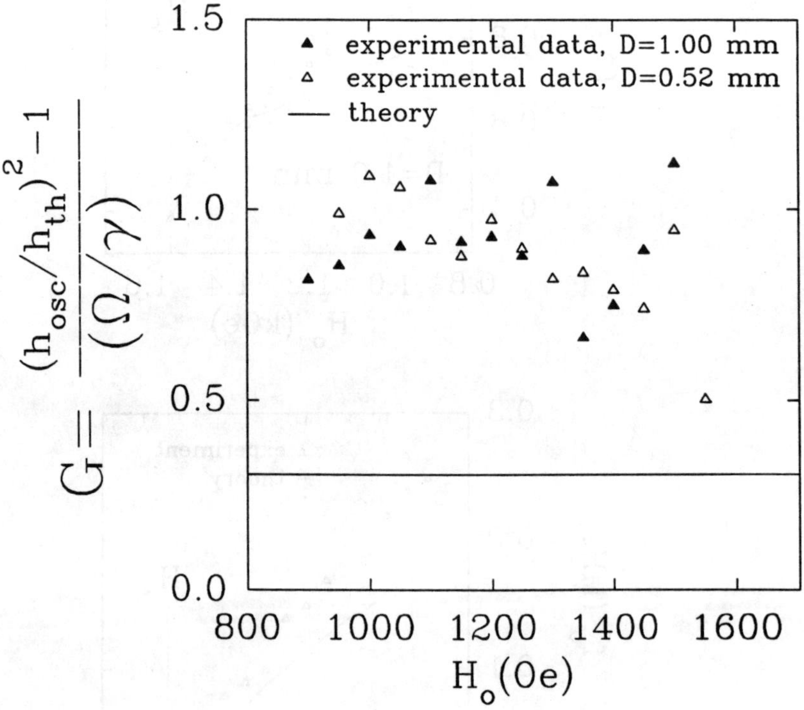

Fig. 3. Comparison of theoretical and experimental values of the ratio G (Eq. (59)) in the axially symmetric case (YIG spheres of diameters $D_1 = 1\ mm$ and $D_2 = 0.52\ mm$.[33]

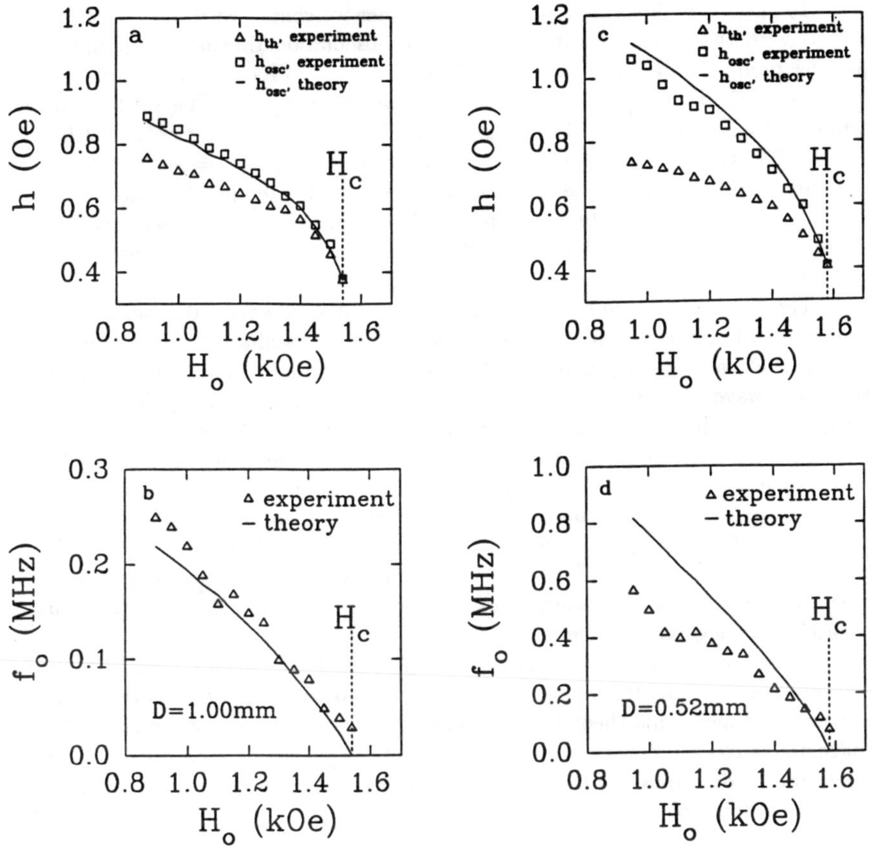

Fig. 4. Comparison of the results of experimental measurements of auto-oscillation threshold h_{osc} (a, c) and frequency f_0 (b, d) [33] at room temperature on two YIG spheres ($D_1 = 1\ mm$ and $D_2 = 0.52\ mm$) with the results of theoretical calculations of these characteristics done using Eqs. (49) and (50) for $R_{exp} = -0.765$ ($G_{exp} = 0.8$).

To check our theoretical model even further we compared results of our calculations with the experimental results for auto-oscillation threshold $h_{osc}(H_0)$ under parallel pumping obtained by Yamazaki et al.[41] in a YIG sphere (diameter $D = 0.84\ mm$, orientation [111]) at the temperature $\Theta = 4.2\ K$. The ratio R at this temperature was calculated from Eqs. (55), (56) for the values $4\pi M_0 = 2400\ Gs$, $H_a = -280\ Oe$ (see e.g.[42]) and turned out to be $R = R_{theor} = -2.84$. Results of comparison of experimental and theoretical data at this temperature are presented in Fig. 5 and demonstrate excellent agreement without any adjustment as the expressions (55) and (56) for the coefficients S and T give good accuracy at this low temperature.

The small difference between the threshold of spin wave auto-oscillations and the threshold of parametric excitation of spin waves as well as the small (but finite) auto-oscillation frequency experimentally observed in the region of relatively small spin wave wave vectors $k < 10^5\ cm^{-1}$ are caused, in our opinion, by the effects that were neglected in the S-theory of spin wave interaction.[44] Among these effects we can name the scattering of parametric magnons on one another[43] and scattering on magnetic impurities.[45] The finite auto-oscillation frequency caused by these effects is proportional to a negative power of the product (kv), and due to this dependence these small effects can play a dominant role in the region of relatively low spin wave wave vectors. It was shown[29,43] that the above mentioned small effects of magnon scattering may result in appearance of new modes of spin wave collective oscillations with very low eigenfrequencies.

The above presented comparison of theoretical and experimental results for the threshold and the frequency of spin wave auto-oscillations demonstrate that our simple one dimensional theoretical model of spin wave auto-oscillations in finite-size magnetic samples[35,36] gives an adequate qualitative (and in some cases even good quantitative) description of the sample-size effect on spin wave auto-oscillations under parallel pumping observed in YIG spheres at both room and liquid helium temperatures.

It was pointed out[14,33,34] that one of the principal questions in the theory of auto-oscillations in a system of parametric spin waves was the question of a physical nature of a finite frequency of spin wave auto-oscillations at the threshold of their appearance. This frequency was measured experimentally to be 1 to 2 orders of magnitude lower that the magnon relaxation frequency γ, the only characteristic frequency in the infinite system. The auto-oscillation frequency also demonstrated a pronounced dependence on the size of the experimental sample.

In the present paper we took into account the finite size of a sample in the framework of the approximations of the S-theory of magnon interaction.[14,18] As a result we were able to eliminate the main contradictions between the existing theories of spin wave auto-oscillations[14,30] and the experimental data on auto-oscillations obtained in finite size magnetic samples:

(i) The threshold of spin wave auto-oscillations in a finite system h_{osc} turned out to be

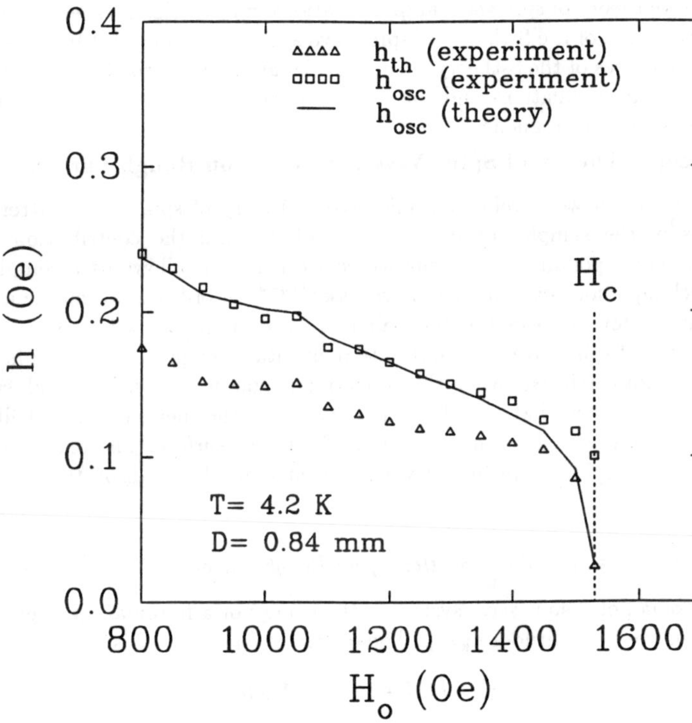

Fig. 5. Comparison of the results of experimental measurements of auto-oscillation threshold h_{osc} [41] at the temperature $\Theta = 4.2\ K$ on the YIG sphere ($D = 0.84\ mm$) with the results of theoretical calculations of these characteristics done using Eqs. (49) and (50) for $R_{theor} = -2.84$.

higher than the threshold of excitation of parametric spin waves h_{th}. The difference between these two thresholds depends on the ratio of the size of the experimental sample D to the mean free path of parametric magnons l (see Eq. (49));
(ii) The frequency of spin wave auto-oscillations at the threshold of their appearance is finite and it is determined by the size of the experimental sample (see Eq. (50));
(iii) The above theory of spin wave auto-oscillations in finite-size magnetic samples is a natural extension of the S-theory of spin wave auto-oscillations.[13,14] It preserves all the positive features of the S-theory such as explanation of the giant crystallographic anisotropy of auto-oscillations and the correct description of the power dependence of the auto-oscillation frequency.

6. Microscopic Theory of Spin Wave Scattering on Rough Surface

In this section we develop the microscopic theory of spin wave scattering on roughnesses of the sample surface. First of all, we find the contribution to the Hamiltonian corresponding to the surface scattering. This allows us to employ the diagram technique for nonequilibrium processes[39,45,46] in order to formulate a closed set of equations for the response and pair correlation functions taking into account strong coupling of spin waves with opposite momenta through the pumping field. We evaluate the parametric excitation threshold in a magnetic film with a rough surface. In particular, we show that in a thick (compared to the mean free path) film the threshold coincides with that in a film with absorptive surface. The reason for that coincidence was explained in Sec. 2 when we introduced the absorptive boundary conditions.

6.1. Hamiltonian of Spin Wave Scattering on Rough Surface

The Hamiltonian of a spin wave system is the energy of a ferromagnet expressed in terms of spin wave amplitudes $a_{\mathbf{k}}, a^*_{\mathbf{k}}$. The energy of ferromagnet is

$$E = -\int (\mathbf{H} + 1/2\alpha\Delta\mathbf{M})\mathbf{M}dV, \tag{60}$$

where $\mathbf{H}$ is the magnetic field, $\mathbf{M}$ is the magnetization, α is the exchange constant, and V is the sample volume. Here the magnetic anisotropy and the magnetic dipole interaction have been neglected. Let the face of the film $z = D$ be smooth and the opposite face $z = \eta(x, y)$ be rough. This means that $\eta(x, y)$ is a random function with zero mean value: $\langle\eta\rangle = 0$. The influence of roughness is twofold: (i) the form of a sample is changed compared to the case of a plane face $z = 0$, and (ii) the roughness distorts the magnetic field and the magnetization:

$$\mathbf{H} = \mathbf{H_0} + \mathbf{H}', \quad \mathbf{M} = \mathbf{M_0} + \mathbf{M}', \tag{61}$$

$\mathbf{H_0}$ and $\mathbf{M_0}$ being the undisturbed field and magnetization. One can rewrite the part of energy (60) caused by surface roughness in the following form:

$$E_s = -\int dxdy \int_0^{\eta(x,y)} dz(\mathbf{H_0} + 1/2\alpha\Delta\mathbf{M_0})\mathbf{M_0}$$

$$-\int dxdy \int_0^D dz(\mathbf{H_0 M'} + \mathbf{H' M_0} + 1/2\alpha \mathbf{M_0} \Delta \mathbf{M'}) = E_{s1} + E_{s2}. \quad (62)$$

After the Holstein - Primakoff and Fourier transformations one can obtain:

$$H_{s1} = D^{-1} A^{-1/2} \sum_{\mathbf{k},\mathbf{k}'} (\omega_H + \alpha \mathbf{k}\mathbf{k}') \eta_{\mathbf{q}-\mathbf{q}'} a_{\mathbf{k}} a^*_{\mathbf{k}'}. \quad (63)$$

Here $\mathbf{q} = (k_x, k_y)$, $\omega_H = gH$, g is the gyromagnetic ratio, and A is the area of the film surface. The second term in (62) is small as compared to H_{s1} if the angular size of the spin wave packet is small. It is definitely the case in a thick film ($D \ll l$). For a wide wave packet these two terms are of the same order. Consequently, one can estimate:

$$H_s \sim D A^{-1/2} \sum_{\mathbf{k},\mathbf{k}} \omega_{\mathbf{k}} \eta_{\mathbf{q}-\mathbf{q}'} a_{\mathbf{k}} a^*_{\mathbf{k}'}. \quad (64)$$

6.2. Diagram Technique

A consistent way to evaluate the threshold of parametric instability in a system with fluctuating parameters is to develop a diagram technique.[39,46] The problem of the description of parametric instability in a film with a rough surface differs from that for a randomly inhomogeneous ferromagnet in two ways. First, the Green's functions are non-diagonal in k-space in contrast with the case of a volume ferromagnet, where $G_{\mathbf{k}\mathbf{k}'} \sim \delta(\mathbf{k} - \mathbf{k}')$. Second, the normal of wave vector is quantized. The Hamiltonian of spin waves in a film is:

$$H = H_0 + H_p + H_s, \quad (65)$$

where H_0 is the Hamiltonian of free spin waves:

$$H_0 = \sum_{\mathbf{k}} \omega_{\mathbf{k}} a^*_{\mathbf{k}} a_{\mathbf{k}}, \quad (66)$$

H_p describes interaction of spin waves with parametric pumping:

$$H_p = 1/2 \sum_{\mathbf{k}} [h \exp(-i\omega_p t)\, V_{\mathbf{k}} a^*_{\mathbf{k}} a^*_{-\mathbf{k}} + c.\,c.], \quad (67)$$

and H_s is estimated in previous subsection (64).

The anomalous Green's function should be taken into account because the pumping field forces anomalous correlations of phases of spin waves with opposite wave vectors. If the roughness of the surface is statistically homogeneous the normal $G_{\mathbf{k}\mathbf{k}'}$ and anomalous $L_{\mathbf{k}\mathbf{k}'}$ Green's functions are diagonal in projections of wave vectors $\mathbf{q}$ and $\mathbf{q}'$:

$$G^{ij}_{\mathbf{k}\mathbf{k}'} = G^{ij}_{\mathbf{q}}(p, p')\delta(\mathbf{q} - \mathbf{q}'), \quad L^{ij}_{\mathbf{k}\mathbf{k}'} = L^{ij}_{\mathbf{q}}(p, p')\delta(\mathbf{q} + \mathbf{q}'). \quad (68)$$

Here p and p' are the normal components of the wave vectors $\mathbf{k}$ and $\mathbf{k}'$, and i, j are Keldysh indices. For retarded Green's functions $G_{\mathbf{q}}(p,p')$, $L_{\mathbf{q}}(p,p')$ the Dyson's equations are:

$$G_{\mathbf{q}}(p,p') = G^0_{\mathbf{q}}(p)\{\Delta(p-p') + \sum_r [M_{\mathbf{q}}(p,r)G_{\mathbf{q}}(r,p') + \Pi_{\mathbf{q}}(p,r)L^*_{\mathbf{q}}(r,p')]\},$$

$$L^*_{\mathbf{q}}(p,p') = G^0_{\mathbf{q}}(p)^* \sum_r [\Pi^*_{\mathbf{q}}(p,r)G_{\mathbf{q}}(r,p') + \Sigma^*_{\mathbf{q}}(p,r)L^*_{\mathbf{q}}(r,p')],$$

$$G^0_{\mathbf{q}}(p) = (\omega_p/2 - \omega_{\mathbf{q}\mathbf{p}} + i\gamma_{\mathbf{q}\mathbf{p}})^{-1}, \ \Delta(0) = 1, \ \Delta(p \neq 0) = 0. \tag{69}$$

The pair correlation functions, sometimes called the distributive Green's functions, are proportional to the pair correlation functions

$$n_{\mathbf{k}\mathbf{k}'} = \langle a^*_{\mathbf{k}} a_{\mathbf{k}'} \rangle, \ \ \sigma_{\mathbf{k}\mathbf{k}'} = \langle a_{\mathbf{k}} a_{\mathbf{k}'} \rangle exp(i\omega_p t). \tag{70}$$

One can obtain the set of equations for the correlation functions from Dyson's equations:

$$\hat{N}_{\mathbf{q}}(p,p') = \sum_{r,r'} \hat{G}_{\mathbf{q}}(p,r)\hat{\Phi}_{\mathbf{q}}(r,r')\hat{G}^*_{\mathbf{q}}(p',r'),$$

$$\hat{N} = \begin{pmatrix} n & \sigma \\ \sigma^* & n \end{pmatrix}, \hat{G} = \begin{pmatrix} G & L \\ L^* & G^* \end{pmatrix}, \hat{\Phi} = \begin{pmatrix} \Phi & \Psi \\ \Psi^* & \Phi^* \end{pmatrix}. \tag{71}$$

The normal mass operators M and Φ can be obtained in the second order in amplitude of scattering

$$M_{\mathbf{q}}(p,p') = \sum_{\sigma,r,r'} \int d\mathbf{q}' g^{\sigma}_{\mathbf{q}p,\mathbf{q}'r} g^{\sigma*}_{\mathbf{q}p',\mathbf{q}'r'} \eta^{\sigma(2)}_{\mathbf{q}-\mathbf{q}'} G_{\mathbf{q}'}(r,r'),$$

$$\Phi_{\mathbf{q}}(p,p') = \sum_{\sigma,r,r'} \int d\mathbf{q}' g^{\sigma}_{\mathbf{q}p,\mathbf{q}'r} g^{\sigma*}_{\mathbf{q}p',\mathbf{q}'r'} \eta^{\sigma(2)}_{\mathbf{q}-\mathbf{q}'} n_{\mathbf{q}'}(r,r') \tag{72}$$

Here the index $\sigma = 1, 2$ numbers the film surfaces.

There is no correlation between roughnesses of the opposite faces, therefore there are no terms involving waves arriving at opposite faces. Consequently, diagrams with anomalous Green's functions are equal to zero, i. e.

$$\Psi_{\mathbf{q}}(p,p') = 0, \ \ \Pi_{\mathbf{q}}(p,p') = hV_{\mathbf{q}}(p)\delta(p+p') \tag{73}$$

The characteristic scale in $\mathbf{k}$-space for the scattering amplitude $g_{\mathbf{k}\mathbf{k}'}$ to change is of order of k. Thus the mass operators $M_{\mathbf{q}}(p,p')$ and $\Phi_{\mathbf{q}}(p,p')$ are slowly changing functions of p and p' near the resonant surface $\omega_{\mathbf{k}} = \omega_p/2$. Therefore the equations (69) and (71) are integral equations with a degenerate kern. After some transformations one can obtain:

$$\sum_{p,p'} n_{\mathbf{q}}(p,p') = n_{\mathbf{q}} = (D_{\mathbf{q}}\epsilon_{\mathbf{q}})^{-2}\{(c^2_{\mathbf{q}} + b^2_{\mathbf{q}})(u^{-2}_{\mathbf{q}} - u^{+2}_{\mathbf{q}})/2b^2_{\mathbf{q}} + m_{\mathbf{q}}u^-_{\mathbf{q}}u^+_{\mathbf{q}}[(u^-_{\mathbf{q}} - u^+_{\mathbf{q}})c_{\mathbf{q}}/b_{\mathbf{q}} + m_{\mathbf{q}}u^-_{\mathbf{q}}u^+_{\mathbf{q}}\} \int |\, g_{\mathbf{q},\mathbf{q}'} \,|^2 \, \eta^2_{\mathbf{q}-\mathbf{q}'} n_{\mathbf{q}'} d\mathbf{q}'. \tag{74}$$

Here

$$\epsilon_{\mathbf{q}} = 2v_{\mathbf{q}}/d, c_{\mathbf{q}} = \gamma_{\mathbf{q}}/\epsilon_{\mathbf{q}}, b_{\mathbf{q}}^2 = (\mid hV_{\mathbf{q}} \mid^2 - \gamma_{\mathbf{q}}^2)/\epsilon_{\mathbf{q}}^2,$$
$$a_{\mathbf{q}} = \{(\omega_{\mathbf{qp}} - \omega_p/2)/\epsilon_{\mathbf{q}}\}, u_{\mathbf{q}}^{\pm} = ctg(a_{\mathbf{q}} - b_{\mathbf{q}}),$$
$$m_{\mathbf{q}} = -ImM_{\mathbf{q}}/\epsilon_{\mathbf{q}}, D_{\mathbf{q}} = 1 + (u_{\mathbf{q}}^- - u_{\mathbf{q}}^+)c_{\mathbf{q}}m_{\mathbf{q}}/b\mathbf{q} + m_{\mathbf{q}}^2 u_{\mathbf{q}}^- u_{\mathbf{q}}^+, \tag{75}$$

$v_{\mathbf{q}}$ is the group velocity, $\{x\}$ is the fractional part of the number x. The threshold value h_{th} is the minimum magnitude h provided that the equation (74) has a nontrivial solution $n_{\mathbf{q}}$.

6.3. Threshold of Parametric Instability

In a thick film $(D \gg l = v/\gamma)$ the frequency of elastic scattering of spin waves on surface roughnesses is small compared to the damping. According to the equation (74) the correlation function $n_{\mathbf{q}}$ is nonzero in a region of $\mathbf{q}$ near the point where the denominator $D_{\mathbf{q}}$ is minimal. If $d >> l$ than $c_{\mathbf{q}} >> m_{\mathbf{q}}$. Consequently,

$$D_{\mathbf{q}} = 1 + (u_{\mathbf{q}}^- - u_{\mathbf{q}}^+)c_{\mathbf{q}}m_{\mathbf{q}}/b_{\mathbf{q}}. \tag{76}$$

$D_{\mathbf{q}}$ reaches its minimum value at $a_{\mathbf{q}} = \pi/2$. In the case of a thick film $\Phi_{\mathbf{q}} \to 0$ and $m_{\mathbf{q}} \to 0$ and, therefore, $b_{\mathbf{q}} = \pi/2$. This results in the threshold value:

$$\mid h_{th}V_{\mathbf{q}} \mid^2 = \gamma_{\mathbf{q}}^2 + (\pi\epsilon_{\mathbf{q}}/2)^2 = \gamma_{\mathbf{q}}^2 + (\pi v_{\mathbf{q}}/D)^2. \tag{77}$$

The result coincides with that for a layer with absorptive faces.[9] The reason for the coincidence is that a wave that arrives at a face with an optimal direction, which provides the minimum in (25), leaves the surface with an arbitrary direction. The scattered wave disappeared in the volume of the film and does not reach the opposite surface. It is equivalent to the absorption of the arriving wave.

In the thin film $(D \ll l)$ the elastic scattering of spin waves on surface roughnesses is the strongest one, i. e. $m_{\mathbf{q}} \gg c_{\mathbf{q}}$. The roughnesses are assumed to be high $\langle(\nabla\eta)^2\rangle > 1$, consequently $m_{\mathbf{q}} \gg 1$. This results in

$$D_{\mathbf{q}} = 1 + m_{\mathbf{q}}^2 \cot^2 a_{\mathbf{q}} \tag{78}$$

Thus one can obtain from (75):

$$h_{th}^2 = (2\gamma/\epsilon)/\langle\mid V \mid^2 /\epsilon^3\rangle,$$
$$\langle\mid V \mid^2 /\epsilon^3\rangle = (4\pi)^{-1} \int \mid V_\Omega \mid^2 /\epsilon^3 \Omega d\Omega, \tag{79}$$

where Ω is the space angle. So, one can estimate:

$$\mid h_{th}V \mid^2 \approx \gamma\epsilon^2/\beta \approx \gamma v/\bar{\eta},$$
$$\beta^2 = \int \mid g_{\mathbf{kk'}} \mid \eta_{\mathbf{q}-\mathbf{q'}}^2 d\mathbf{q'} \approx (\omega_p^2/16\pi)(\bar{\eta}/d). \tag{80}$$

The frequency of elastic scattering γ_s is of order ϵ^2/β therefore

$$| h_{th} V |^2 \sim \gamma\gamma_s. \tag{81}$$

This result coincides with the threshold for a volume sample with elastically scattered defects.[45] The reason for the coincidence is that distribution of spin waves in both a film and a volume are isotropic, so the distribution is insensitive to details of spatial inhomogeneities of scale smaller than the mean free path. It should be noted that the existence of the parameter $\langle(\nabla\eta)^2\rangle$ in the denominator in (80) is not absurd, because it is implied that $\langle(\nabla\eta)^2\rangle > 1$. It should also be noted that if the elastic scattering is strong, it is necessary to take into account higher diagrams to obtain the exact value of the threshold. Nevertheless, the estimate (80) is correct.

It should be noted that the problem of spin wave parametric excitation in a thin film differs from the problem of parametric excitation of spin waves in three dimensional small ($D \ll l$) samples significantly. In a thin film, the region where the waves are located is infinite, the spin waves are distributed homogeneously in the film plane. Therefore waves reflected from roughnesses return in the instability region again. In a three dimensional sample, e. g., in a sphere, the wave distribution is spatially inhomogeneous, the waves are located in the equatorial region of the sample where the size D in the direction of wave propagation is maximum. If a spin wave arriving at the surface scatters on a roughness, it reflects at an arbitrary direction and, most probably, leaves the equatorial region. The waves decay out of the equatorial region and they do not return in the instability region again. It can be shown that the problem of the spin wave scattering in three dimensional samples can be described by the boundary condition of dissipation of arriving wave envelope up to very small sample sizes: $(D/l) < (\lambda/D)^{3/2}$. Thus, in three dimensional samples, the absorbing boundary conditions for the spin wave envelopes (21) are valid for $D < (\lambda^3 l^2)^{1/5}$, i. e. practically always.

7. Conclusion

The main results of the above presented theory of sample-size effects on the properties of parametrically excited spin waves can be formulated as follows:

1. Sample-size effects play an important role in the nonlinear dynamics of parametrically excited spin waves even in the case when the wavelengths of these waves are much smaller than the size of a magnetic sample. Sample-size effects influence the collective properties of short parametric spin waves.

2. The sample size effect on parametrically excited spin waves can be described in terms of dynamic equations for spin waves envelopes and boundary conditions for the envelopes. The nonlinear dynamic equations in the mean field (S-theory) approximation and the boundary conditions for spin wave envelopes describing dissipation of the spin wave packets arriving at a sample boundary are introduced. The boundary conditions lead not only to spin wave dissipation but also to the phase interference of parametrically excited spin waves.

3. The threshold of parametric excitation of spin waves h_{th} calculated by the above described method in a finite-size magnetic sample is higher than the threshold h_{th_0} in an infinite magnetic medium. The threshold h_{th} depends on the phase interference between the parametrically excited spin waves and at low temperatures it is inversely proportional to the sample size. Experiments performed in YIG spheres at low temperature quantitatively confirm these theoretical conclusions.

4. Influence of the finite size of a sample explains the experimentally observed difference between the thresholds of parametric excitation of spin waves h_{th} and the threshold of spin wave auto-oscillations h_{osc}, as well as the finite value of the frequency Ω of auto-oscillations at the threshold of their appearance. However, theory gives the quantitative description of these effects only at low temperatures, when we can accurately calculate the coefficients of the four-wave nonlinear interaction responsible for the spin wave auto-oscillations. At the room temperature our theory gives good qualitative account of the observed effects, but the values of the four-wave interaction coefficients must be determined from experiment to achieve the quantitative agreement between experimental and theoretical results for the parameters of spin wave auto-oscillations.

5. The rigorous calculation done for the finite-size magnetic samples with rough surface by means of a diagrammatic technique gives a justification for the absorptive boundary conditions for the envelopes of spin wave packets introduced earlier phenomenologically and used in our calculations of sample-size effects.

Ferromagnetic films and multilayers (as well as other finite-size magnetic systems) demonstrate such a wide variety of interesting and important nonlinear phenomena that, in our opinion, they will attract close attention of both experimentalists and theorists for many years to come.

8. Acknowledgements

The authors are pleased to thank Prof. Victor S. L'vov, Prof. Michael G. Cottam, Prof. Philip E. Wigen, and Prof. Gopalan Srinivasan for many helpful discussions at various stages of this project. We are grateful to Prof. S.M. Rezende for making available the tables of his experimental data.

The authors are grateful to the University of Western Ontario, the University of Manitoba and Oakland University for creating the possibility of our collaboration. The authors are also grateful to the NSERC of Canada, the Winnipeg Institute for Theoretical Physics and the Weizmann Institute of Science for support of this project. This work was also supported by the grant C-3401 from the Research Corporation and the grant from the Oakland University Foundation.

References

1. C. Kittel, *Phys.Rev.*, **110**, (1958) 1295.
2. *Magnetism* (ed. G. T. Rado and H. Suhl), Vol. 1, (1963) 299.
3. H. Suhl, *J. Phys. Chem. Solids* **1**, (1957) 209.
4. F. R. Morgenthaler, *J. Appl. Phys.* **31**, (1960) 955.
5. E. Schlömann, J. Green, and V. Milano, *J. Appl. Phys.* **31**, (1960) 3865.
6. T. S. Hartwick, E. R. Peressini, and M. T. Weiss, *J. Appl. Phys.* **32** (161) 223.
7. A. V. Lavrinenko, V. S. L'vov, G. A. Melkov, and V. B. Cherepanov, *Sov.Phys.-JETP,* **54** (1981) 542.
8. L. M. Gorbunov, *Sov.Phys. - JETP* **40** (1975) 689.
9. V. S. L'vov and A. M. Rubenchik, *Zh. Exp. Teor. Fiz.* **72** (1977) 127 [*Sov. Phys. JETP* **45**, (1977) 67].
10. V. S. Zhitnyuk, G. A. Melkov and V. B. Cherepanov, *Fiz. Tverd. Tela* **22** (1980) 215 [*Sov. Phys. - Solid State* **22** (1980) 124].
11. V. S. Zhitnuk, G. A. Melkov, and V. B. Cherepanov, *Kinetics of parametric instability of spin waves in ferrites* (Preprint #138, Institute of Automation and Electrometry, Siberian Branch of the USSR Academy of Sciences, 1980) (in Russian).
12. Ya. A. Monosov, *Nonlinear Ferromagnetic Resonance* (Nauka, Moscow, 1971) (in Russian).
13. V. S. L'vov, S. L. Musher, and S. S. Starobinets, *Zh. Exp. Teor. Fiz.* **64** (1973) 1084 [*Sov. Phys. JETP* **37** (1973) 546].
14. V. E. Zakharov, V. S. L'vov, and S. S. Starobinets, *Usp. Fiz. Nauk* **114** (1974) 609 [*Sov. Phys. - Usp.* **17** 1975) 896].
15. V. V. Zautkin, V. S. L'vov, B. I. Orel , and S. S. Starobinets, *Zh. Eksp. Teor. Fiz.* **72** (1977) 272 [*Sov. Phys. JETP* **45** (1977) 143].
16. V. V. Zautkin and B. I. Orel, *Zh. Eksp. Teor. Fiz.* **79** (1980) 281 [*Sov. Phys. JETP* **52** (1980) 142].
17. V. S. L'vov, Ch. 5 in *Solitons*, ed. S. E. Trullinger, V. E. Zakharov, and V. L. Pokrovsky (Elsevier, Amsterdam, 1986).
18. V. S. L'vov, *Nonlinear Spin Waves,* (Nauka, Moscow, 1987) (in Russian); V. S. L'vov, *Wave Turbulence under Parametric Excitation* (Springer Verlag, Berlin, Heidelberg, 1992).
19. G. Gibson and C. Jeffries, *Phys. Rev.* **A19** (1984) 811.
20. S. M. Rezende, O. F. de Alcantra Bonfim, and F. M. de Aguiar, *Phys. Rev.* **B33** (1986) 5153.
21. F. M. de Aguiar and S. M. Rezende, *Phys. Rev. Lett.* **56** (1986) 1070.
22. A. I. Smirnov, *Zh. Eksp. Teor. Fiz.* **90** (1986) 385. [*Sov. Phys. JETP* **63**, (1986) 222].
23. T. L. Carroll, L. M. Pecora, and F. J. Rachford, *Phys. Rev. Lett.* **59** (1987) 2891.
24. C. D. Jeffries, P. H. Bryant, and K. Nakamura, *J. Appl. Phys.* **64** (1988) 5382.
25. H. Yamazaki, *J. Appl. Phys.* **64** (1988) 5391.
26. P. E. Wigen, H. Doetsch, Y. Ming, L. Baselgia, and F. Waldner, *J. Appl. Phys.* **63** (1988) 4157.

27. G. Srinivasan, M. Chen, and C. E. Patton, *J. Appl. Phys.* **64** (1988) 5480.
28. R. D. McMichael and P. E. Wigen, *Phys. Rev. Lett.* **64** (1990) 64.
29. V. S. L'vov and V. B. Cherepanov, *Zh. Exp. Teor. Fiz.* **76** (1979) 2266 [*Sov. Phys. JETP* **49** (1979) 1145].
30. X. Zhang and H. Suhl, *Phys. Rev.* **B38** (1988) 4893.
31. V. G. Morozov and A. I. Mukhai, *Zh. Teor. Math. Fiz.* **51** (1982) 234.
32. G. Srinivasan, *Solid State Comm.* **80** (1991) 5.
33. S. M. Rezende, F. M. de Aguiar, and A. Azevedo, *J. Appl. Phys.* **67** (1990) 5624.
34. S. M. Rezende and A. Azevedo, *Phys. Rev.* **B45**, (1992) 10387.
35. V. B. Cherepanov and A. N. Slavin, Phys. Rev. B **47** (1993) 5874.
36. V. B. Cherepanov and A. N. Slavin, J. Appl. Phys. **73** (1993) 6811.
37. A. S. Bakai, I. V. Krutsenko, G. A. Melkov and G. G. Sergeeva, *Zh. Exp. Teor. Fiz.* **76** (1979) 231. [*Sov. Phys. JETP* **49** (1979) 118].
38. I. V. Krutsenko and G. A. Melkov, *Fizika Nizkikh Temperatur* **7** (1981) 770.
39. L. V. Keldysh, *Zh. Exp. Teor. Fiz.* **47** (1964) 1513. [*Sov. Phys. JETP* **20** (1965) 1018].
40. V. B. Cherepanov, *J. Appl. Phys.* **69** (1991) 5733.
41. H. Yamazaki, Y. Yunoki, M. Mino, S. Mitsudo and V. L. Safonov, *J. Magn. Magn. Materials* **104-107** (1992) 1059.
42. *Handbook of microwave materials*, Ed. W. H. von Aulock (Academic Press, New-York - London, 1965).
43. V. S. L'vov, *Zh. Exp. Teor. Fiz.* **69** (1975) 2079. [*Sov. Phys. JETP* **42** (1975) 1057].
44. I. V. Krutsenko, V. S. L'vov, and G. A. Melkov, *Zh. Exp. Teor. Fiz.* **75** (1978) 1114 [*Sov. Phys. JETP* **48** (1978) 561].
45. V. E. Zakharov and V. S. L'vov, *Sov. Phys.- Solid State* **14** (1973) 2513.
46. H. Wyld, *Ann. Phys.* **14** (1961) 143.

MULTISTABILITY OF PARAMETRICALLY EXCITED SPIN WAVES

G. WIESE and H. BENNER
Institut für Festkörperphysik, Technische Hochschule Darmstadt, Hochschulstr. 6 D-64289 Darmstadt, Federal Republic of Germany

Parametric excitation of spin waves is observed in high power ferromagnetic resonance (FMR) experiments when the amplitude of the pumping microwave field exceeds a certain instability threshold well-known as the first- or second-order Suhl instability[1] or as the related parallel pumping instability[2]. Parametrically excited spin waves show a variety of nonlinear phenomena, such as low-frequency autooscillations, period doublings, quasiperiodicity, intermittency, irregular spiking and chaos which have been explained to result from nonlinear couplings between them and have been widely studied since a decade[3-10]. A first and very important step for understanding the observed nonlinear phenomena is to identify those critical spin waves which are the first to become unstable directly at the threshold. Suhl's theory[1], which is based on a plane-wave approximation, yields an appropriate description only for the field and frequency range where spin waves with large wave numbers k (in comparison with the inverse sample size d^{-1}) are expected to become critical. However, the plane-wave approximation is definitely inadequate for the regime where long-wave modes $\boldsymbol{k} \rightarrow 0$ are predicted to become critical. Nevertheless, this second case is of special interest since more interesting nonlinear phenomena have been observed within this regime.

The aim of this chapter is twofold: First, to give a general theoretical guideline how to describe spin-wave instabilities in terms of the proper eigenmodes of the sample instead of using the conventional plane-wave approximation which neglects boundary conditions. These ideas were developed in a series of recent papers aiming at the detailed understanding of the first-order spin-wave instability threshold in films[11,12] and spheres[13] including e.g. the finestructure observed at subsidiary absorption and within the coincidence regime[14,15]. The presented guideline has been extended to arbitrary sample shapes and various types of interactions and allows a realistic prediction of thresholds, critical modes, and the behaviour above threshold. The second aim is to review recent experimental observations in YIG spheres [10,16–18] and films [19] which exhibit a very complex multistability. Our general formalism provides the basis for a theoretical interpretation of this specific nonlinear phenomenon, attributing it to the excitation of certain magnetostatic modes through resonant three-magnon confluence processes of parametrically excited spin waves.

1. General Formalism

The classical interpretation of spin-wave instabilities was given by Suhl [1] in terms of three- or four-magnon decay processes of the externally excited uniform mode into spin waves. Although it was clear from the beginning that the underlying plane-wave approximation, since it neglects the boundary conditions of the sample, would be inadequate for a detailed description of the experimental data, successful attempts to describe spin-wave instabilities in terms of more appropriate eigenmodes were made only recently: Progress was made to relate the finestructure of the threshold in films [20] and spheres [13] to specific modes of the discrete spectrum. It is also possible to understand differences in the thresholds of films magnetized parallel and perpendicular to the film plane [12] and to explain the frequencies of autooscillations in circular YIG films by nonlinear couplings between specific modes [21].

1.1. Linearized Equation of Motion, Eigenmodes and Orthogonality Relations

The magnetization field of an insulating ferromagnet is described by the following equation of motion:

$$\frac{\partial}{\partial t}\vec{\mathfrak{M}}(\vec{r},t) = \gamma\,\vec{H}_{eff}[\vec{\mathfrak{M}}(\vec{r},t)] \times \vec{\mathfrak{M}}(\vec{r},t) + \textit{dissipation}\,. \tag{1.1}$$

Here, $\vec{\mathfrak{M}}(\vec{r},t)$ is the magnetization field normalized to the saturation magnetization M_S, i.e. $|\vec{\mathfrak{M}}(\vec{r},t)| = 1$. γ is the gyromagnetic ratio, and $\vec{H}_{eff}$ denotes an effective field describing the torque on the magnetization vector $\vec{\mathfrak{M}}(\vec{r},t)$, which includes contributions from the external magnetic field $\vec{H}$, the driving microwave field $\vec{h}(t)$, the demagnetization field, and from dipolar, exchange, and anisotropy fields. Since the latter themselves depend on the magnetization field, the effective field can be written as a functional of $\vec{\mathfrak{M}}(\vec{r},t)$, i.e. $\vec{H}_{\mathit{eff}}[..]$ can be considered to be an operator. Dissipation, which results in a relaxation towards the thermal equilibrium state $\vec{\mathfrak{M}}_0(\vec{r})$, is taken into account by including phenomenological damping terms of Bloch or Landau-Lifshitz type. The effective field $\vec{H}_{\mathit{eff}}$ corresponding to the equilibrium state $\vec{\mathfrak{M}}_0(\vec{r})$ will be called the *internal equilibrium field* $\vec{H}_i(\vec{r}) = \vec{H}_{\mathit{eff}}[\vec{\mathfrak{M}}_0(\vec{r})]$. For ellipsoidal samples and an external field parallel to one of the principal axes of the sample, $\vec{H}_i$ is independent of $\vec{r}$ and parallel to the external field. However, the formalism presented here is quite general and allows to describe more complicated situations where $\vec{H}_i$ is not parallel to the external field (e. g. in films magnetized in oblique orientation) or becomes inhomogeneous. Nevertheless, $\vec{H}_i(\vec{r})$ and $\vec{\mathfrak{M}}_0(\vec{r})$ are parallel to each other as follows from Eq. (1.1) and the equilibrium condition $d\vec{\mathfrak{M}}/dt = 0$.

For the further discussion it is useful to introduce the transverse elongation $\vec{m}(\vec{r},t)$ of the magnetization field from the equilibrium state $\vec{\mathfrak{M}}_0(\vec{r})$. Then the magnetization $\vec{\mathfrak{M}}(\vec{r},t)$ can be expressed in lowest order of $\vec{m}$ by

$$\vec{\mathfrak{M}} \simeq \hat{e}_i + \vec{m} \ , \tag{1.2}$$

where $\hat{e}_i$ is defined by the direction of $\vec{H}_i(\vec{r}) \equiv \hat{e}_i H_i$. Note that in general both $\hat{e}_i$ and H_i are depending on $\vec{r}$. Close to the equilibrium state the effective field $\vec{H}_{\mathit{eff}}[\vec{\mathfrak{M}}]$ can be linearized according to

$$\vec{H}_{\mathit{eff}}[\vec{\mathfrak{M}}] \simeq \vec{H}_i + \vec{H}_{dip}[\vec{m}] + \vec{H}_{ex}[\vec{m}] + \vec{H}_{an}[\vec{m}] \equiv \vec{H}_i + \vec{H}_{dip+ex+an}[\vec{m}] \ . \tag{1.3}$$

Here, $\vec{H}_{dip}[..]$, $\vec{H}_{ex}[..]$, and $\vec{H}_{an}[..]$ are linear operators describing the dipolar, exchange and anisotropy fields, respectively, of a given magnetization. From Eqs. (1.1) – (1.3) the following linearized equation of motion for $\vec{m}(\vec{r},t)$ is derived:

$$\mathrm{i}\frac{\partial}{\partial t}\vec{m} = \mathrm{i}\gamma\, \hat{e}_i \times (H_i\vec{m} - \vec{H}_{dip+ex+an}[\vec{m}]) \equiv \mathcal{H}[\vec{m}] \ , \tag{1.4}$$

where $\mathcal{H}[\vec{m}]$ is, by definition, a linear operator. Its eigenvectors $\vec{m}_\kappa$ and eigenvalues ω_κ

$$\mathcal{H}[\vec{m}_\kappa] = \omega_\kappa \vec{m}_\kappa \, . \tag{1.5}$$

determine the eigenmodes of Eq. (1.4) which is simply shown by means of the following complex ansatz for the eigenmodes (labeled by κ)

$$\vec{m}_\kappa(\vec{r},t) \;=\; \vec{m}_\kappa(\vec{r})\, \mathrm{e}^{-\mathrm{i}\,\omega_\kappa t} \; . \tag{1.6}$$

This eigenvalue problem can only analytically be solved in the case of additional approximations. Assuming an infinitely extended bulk sample, translation symmetry is retained, and we obtain eigensolutions in the form of plane waves (*'spin waves'*). For the Hamiltonian $\mathcal{H}[\vec{m}]$ under consideration, this problem was solved by Holstein and Primakov [22]. If the wavelength of the spin waves becomes comparable with the sample dimension one has to solve the problem of magnetostatic modes, which means to neglect the exchange interaction but to consider the sample shape. The magnetostatic mode problem was solved for ellipsoidal and spherical samples [23,24] and for films magnetized in parallel [25] or perpendicular [26] orientation with respect to the film plane.

We shall discuss the spin-wave solutions of Eq. (1.6.) more in detail. For spheres and films, neglecting the anisotropy, the internal equilibrium field reads $\vec{H}_i = \hat{z}(H - N_z 4\pi M_S)$, where the external field $\vec{H}$ defines the z-direction and the demagnetization factors N_z are given by 0, $\frac{1}{3}$ and 1 for in-plane magnetized films, spheres, and out-of-plane magnetized films, respectively. In the case of isotropic exchange, the exchange operator is given by $\vec{H}_{ex}[\vec{m}] = D\nabla^2 \vec{m}$, where D is the stiffness constant. The dipolar field of a plane wave is described by the operator [1] $\vec{H}_{dip}[\vec{m}_k] = -4\pi M_S \vec{k}\,(\vec{m}_k \cdot \vec{k})/k^2$, replacing the general mode index κ by the wavevector $\vec{k}$. By means of these relations and of the ansatz $\vec{m}_k(\vec{r}) = \hat{m}_k \exp(\mathrm{i}\,\vec{k}\vec{r})$ we arrive at the well-known solution

$$\omega_k^2 \;=\; \gamma^2\,(H_i + Dk^2)\,(H_i + Dk^2 + 4\pi M_S \sin^2\theta_k)\,, \tag{1.7}$$

$$\vec{m}_k = \frac{1}{\sqrt{\epsilon_k}}\,(\hat{\mathrm{e}}_{k\parallel} + \mathrm{i}\,\epsilon_k \hat{\mathrm{e}}_{k\perp})\,\mathrm{e}^{\mathrm{i}\,\vec{k}\vec{r}}\,, \quad \epsilon_k = \frac{\omega_k}{\gamma(H_i + Dk^2)}\,, \tag{1.8}$$

where θ_k is the angle between $\vec{H}$ and $\vec{k}$, $\hat{e}_{k\parallel}$ and $\hat{e}_{k\perp}$ are two unit vectors within the xy-plane, the first one parallel to the projection of $\vec{k}$ onto this plane and the second one perpendicular to it: $\hat{e}_{k\perp} = \hat{z} \times \hat{e}_{k\parallel}$. ϵ_k finally denotes the ellipticity factor of the spin wave. Propagation effects have been neglected: At a given time t_0 the dipolar field $\vec{H}_{dip}(\vec{r}, t_0)$ depends only on the magnetization field $\vec{m}(\vec{r}, t_0)$ at the same time. This approximation holds the better, the smaller the wavelength of the spin wave, but can still be used for wavelengths up to 1 mm and frequencies of 10 GHz. On the other hand, for long wavelengths we have to take the boundary conditions of the sample into account, while the exchange interaction can be neglected. This results in the so-called *magnetostatic approximation*, which allows to describe the mode m_κ by a scalar potential ψ_κ via the relation [27] $\vec{H}_{dip}[\vec{m}_\kappa] = \nabla\psi_\kappa$. A partial differential equation for ψ_κ is obtained, which for given boundary conditions yields a discrete set of eigenvalues ω_κ. The most important magnetostatic mode is the uniform mode, which reads for different geometries [28]:

$$\vec{m}_0 = (\hat{x} + \mathrm{i}\,\hat{y})\ , \quad \omega_0 = \gamma\,(H - \tfrac{4\pi}{3} M_S)\ , \qquad \text{sphere} \tag{1.9}$$

$$\vec{m}_0 = \frac{1}{\sqrt{\epsilon_0}}\,(\hat{x} + \mathrm{i}\,\epsilon_0\,\hat{y})\ , \quad \epsilon_0 = \frac{\omega_0}{\gamma H}\ , \quad \omega_0^2 = \gamma^2 H(H + 4\pi M_S)\ , \text{film}\ (\vec{H} \parallel \text{plane}) \tag{1.10}$$

$$\vec{m}_0 = (\hat{x} + \mathrm{i}\,\hat{y})\ , \quad \omega_0 = \gamma\,(H - 4\pi M_S)\ , \qquad \text{film}\ (\vec{H} \perp \text{plane}) \tag{1.11}$$

In (1.10) the x-direction is perpendicular and the y-direction parallel to the film plane.

The eigenmodes of (1.6) satisfy the following orthogonality relation [23,29]:

$$< \vec{m}_\kappa \mid \vec{m}_{\kappa'} > \equiv \int\limits_{\text{sample}} d^3r\,(\mathrm{i}\,\hat{e}_i \times \vec{m}_\kappa(\vec{r}))^* \cdot \vec{m}_{\kappa'}(\vec{r}) \;=\; 2\,V\,\delta_{\kappa,\kappa'}\ , \tag{1.12}$$

This equation also defines a normalization for the modes (V is the volume of the sample). In order to prove (1.12) one has to show that $\mathcal{H}$ is hermitean with respect to the given scalar product: $<\mathcal{H}[\vec{m}_\kappa] \mid \vec{m}_{\kappa'}> \,=\, <\vec{m}_\kappa \mid \mathcal{H}[\vec{m}_{\kappa'}]>$. This problem can be easily reduced to the proof that the operators $\vec{H}_{dip}[..]$, $\vec{H}_{ex}[..]$, and $\vec{H}_{an}[..]$ have to be hermitean with respect to the usual scalar product $\int d^3r\ \vec{m}^*_\kappa \cdot \vec{m}_{\kappa'}$. For the dipolar operator using the integral form $\vec{H}_{dip}[\vec{m}] = 4\pi M_S \int d^3r' \{3[\vec{m}(\vec{r}')\cdot(\vec{r}-\vec{r}')]\,(\vec{r}-\vec{r}') - \vec{m}(\vec{r}')\,|\vec{r}-\vec{r}'|^2\}/|\vec{r}-\vec{r}'|^5$ this can be shown in a few lines. The hermiticity of the exchange operator can be proven by means of Green's second integral identity including the forces at the sample surface:

$$\vec{H}_{ex}[\vec{m}] = D\nabla^2\vec{m} - D\frac{\partial\vec{m}}{\partial\hat{n}}\,\delta_{surface}\,. \tag{1.13}$$

$\partial\vec{m}/\partial\hat{n}$ is the derivative with respect to the direction perpendicular to the surface and $\delta_{surface}$ denotes the delta function at the surface. The eigenvalue equation (1.6) implies that (1.13) leads to the well known exchange boundary condition[30] $\partial\vec{m}/\partial\hat{n} = 0$. The hermiticity of the anisotropy operator follows from the fact that $\vec{H}_{an}[..]$ is given by a symmetric matrix[31]. Surface anisotropy, which is described by the operator $\vec{H}_{surf\text{-}an}[\vec{m}] = -K_s\,\vec{m}\,\delta_{surface}$, leads to surface pinning $\vec{m} = 0$ and also satisfies hermiticity. From $\vec{H}_{ex}[\vec{m}]$, $\vec{H}_{surf\text{-}an}[\vec{m}]$ and the eigenvalue equation (1.6) follows that the magnetic moment at the surface has to satisfy the well-known mixed boundary condition[30] $D\,\partial\vec{m}/\partial\hat{n} + K_s\vec{m} = 0$.

1.2. *Hamilton Function and Coupling Coefficients*

The magnetization field $\vec{m}(\vec{r},t)$ can be expanded with respect to the eigenmodes

$$\vec{m}(\vec{r},t) = \mathrm{Re}\left\{\sum_{\kappa} c_\kappa(t)\;\vec{m}_\kappa(\vec{r})\right\}, \tag{1.14}$$

where the c_κ are complex amplitudes. The most convenient way to derive equations of motion for the amplitudes c_κ starts from the Hamilton function $\mathcal{H}(c_{\kappa_1}, c_{\kappa_2}\dots, c^*_{\kappa_1}, c^*_{\kappa_2}\dots)$ which is connected with the equations of motion by the following relations:

$$\dot{c}_\kappa = -\mathrm{i}\,\frac{\partial\mathcal{H}}{\partial c^*_\kappa} - \eta_\kappa c_\kappa\,, \quad \dot{c}^*_\kappa = \mathrm{i}\,\frac{\partial\mathcal{H}}{\partial c_\kappa} - \eta_\kappa c^*_\kappa\,. \tag{1.15}$$

Note that the usual canonical equations have been supplemented by phenomenological damping terms $-\eta_\kappa c_\kappa$ and $-\eta_\kappa c^*_\kappa$, respectively, to account for dissipation. The following Hamiltonian includes all terms essential for describing spin-wave instabilities and the behaviour above threshold:

$$
\begin{aligned}
\mathcal{H} = & \sum_{\kappa} \omega_\kappa \, c_\kappa^* c_\kappa \\
& + \sum_{\kappa_1 \kappa_2 \kappa_3} (\rho_{1,23} \, c_{\kappa_1} c_{\kappa_2}^* c_{\kappa_3}^* + \text{c.c.}) + \sum_{\kappa_1 \kappa_2 \kappa_3} (\hat{\rho}_{123} \, c_{\kappa_1}^* c_{\kappa_2}^* c_{\kappa_3}^* + \text{c.c.}) \\
& + \sum_{\kappa_1 \kappa_2 \kappa_3 \kappa_4} W_{12,34} \, c_{\kappa_1} c_{\kappa_2} c_{\kappa_3}^* c_{\kappa_4}^* \\
& + \gamma (h_{eff}^{+} \, c_0^* \, \mathrm{e}^{-\mathrm{i}\omega_p t} + h_{eff}^{-} \, c_0^* \, \mathrm{e}^{\mathrm{i}\omega_p t}) + \text{c.c.} \\
& + \sum_{\kappa_1 \kappa_2} (V_{12} \, c_{\kappa_1}^* c_{\kappa_2}^* \, \mathrm{e}^{-\mathrm{i}\omega_p t} + \text{c.c.}) \; . \qquad (1.16)
\end{aligned}
$$

The two-magnon term in the first line represents the energy of the eigenmodes resulting in linear expressions of the equations of motion: $\dot{c}_\kappa = -i\,\omega_\kappa\, c_\kappa$. The first term in the second line includes three-magnon splitting and confluence processes. These processes are responsible e.g. for the first-order Suhl instability where the uniform mode decays into a pair of spin waves at half the pumping frequency. The second three-magnon process formally describes the simultaneous creation or annihilation of three modes[1]. The four-magnon processes in the third line are the interactions between spin waves and are also responsible for the second-order Suhl instability, where the uniform mode decays into a pair of degenerate spin waves at exactly the pumping frequency. The fourth line of Eq. (1.16) describes the coupling between the mode c_0 and the transverse microwave field oscillating with the pumping frequency ω_p. We have confined ourselves to the excitation of a single mode, which is the case for a homogeneous microwave field. However, the generalization to an inhomogeneous microwave field exciting several non-uniform modes is straighforward. Note that for a linearly polarized driving field both driving terms (corresponding to positive and negative pumping frequency) are present whereas for a circularly polarized driving field only the first term is obtained. The last line of Eq. (1.16) finally describes the parallel pumping instability where the microwave component parallel to the external field couples directly to spin-wave pairs at half the pumping frequency.

The coupling coefficients in Eq. (1.6) are determined from a comparison of Eqs. (1.15) with the equations of motion for c_κ resulting directly from Eq. (1.1). To

[1] Note that the *annihilation* of a counter-rotating mode with frequency $-\omega_\mathrm{p}$ is formally equivalent to the *creation* of a mode with the same positive frequency, and vice versa.

this end the expansion

$$\vec{m}(\vec{r},t) = \tfrac{1}{2}\Big(\sum_{\kappa} c_\kappa(t)\, \vec{m}_\kappa(\vec{r}) + c^*_\kappa(t)\, \vec{m}^*_\kappa(\vec{r})\Big) \tag{1.17}$$

is inserted in Eqs. (1.2) and (1.1). Then, the effective field $\vec{H}_{eff}[\vec{\mathfrak{M}}]$ can be expressed by

$$\vec{H}_{eff}[\vec{\mathfrak{M}}] \;\simeq\; \vec{H}_i + \vec{H}_{dip}[\delta\vec{\mathfrak{M}}] + \vec{H}_{ex}[\delta\vec{\mathfrak{M}}] + \vec{H}_{an}[\delta\vec{\mathfrak{M}}] \;, \tag{1.18}$$

where $\delta\vec{\mathfrak{M}}$ is defined by

$$\delta\vec{\mathfrak{M}} \;=\; \vec{\mathfrak{M}} - \vec{\mathfrak{M}}_0 \;=\; \vec{m} - \tfrac{1}{2}\,\vec{m}^2\,\hat{e}_i + \ldots \quad . \tag{1.19}$$

The derivation of the following results is essentially based on the assumption that $\vec{H}_{dip}[..]$, $\vec{H}_{ex}[..]$, and $\vec{H}_{an}[..]$ in Eq. (1.18) are still considered as being linear in $\delta\vec{\mathfrak{M}}$. This assumption is exact for the dipolar operator $\vec{H}_{dip}[..]$ and still holds for the exchange operator as long as the second order expansion of $\vec{\mathfrak{M}}$, Eq. (1.19), is sufficient. For cubic anisotropy it was shown[31] that the anisotropy field can be written in the linearized form $\vec{H}_{an}[\delta\vec{\mathfrak{M}}]$, if $\vec{H}_{an}[\vec{\mathfrak{M}}]$ is represented by a symmetric 3×3 matrix. In other cases the calculation of the coupling coefficients is more complicated and requires a detailed treatment of the specific nonlinear form of $\vec{H}_{an}[\vec{\mathfrak{M}}]$. The technical procedure for obtaining the equations of motions is the following: Inserting Eqs. (1.2), (1.17) and (1.18) into (1.1) the l.h.s. of the resulting equation reads $\frac{1}{2}\sum_\kappa(\dot{c}_\kappa\vec{m}_\kappa + \dot{c}^*_\kappa\vec{m}^*_\kappa) - \frac{1}{2}\,\hat{e}_i(\vec{m}^2)^{\cdot}$. Multiplying then both sides with $<\vec{m}_{\kappa'}|$ and making use of the orthogonality relations (1.12) the l.h.s. simplifies to[2] $\dot{c}_{\kappa'}\cdot V$. By means of this projection technique a set of equations of motions for all c_κ is derived in a very simple way. The comparison of these equations of motion with the canonical equations (1.15) results in the following coupling coefficients[3]:

[2] Note that the conjugate complex eigenmodes $\vec{m}^*_\kappa$, too, are eigenvectors of Eq. (1.6) with the eigenvalue $-\omega_\kappa$ and, thus, orthogonal to all $\vec{m}_\kappa$. The quadratic term, finally, vanishes since $\hat{e}_i$ is perpendicular to all transverse components $\vec{m}_\kappa$ and $\vec{m}^*_\kappa$.

[3] In order to verify this calculation one has to apply the vector relation $(\vec{a}\times\vec{b})\cdot(\vec{c}\times\vec{d}) = (\vec{a}\,\vec{c})\cdot(\vec{b}\,\vec{d}) - (\vec{a}\,\vec{d})\cdot(\vec{b}\,\vec{c})$, the orthogonality of $\hat{e}_i$ and $\vec{m}_\kappa$, and to consider the hermiticity of $\vec{H}_{dip}[..]$, $\vec{H}_{ex}[..]$, and $\vec{H}_{an}[..]$ with respect to the usual scalar product $\int d^3r\, \vec{m}^*_\kappa \cdot \vec{m}_{\kappa'}$.

$$\rho_{1,23} = \frac{\gamma}{8V} \int_{\text{sample}} d^3r (\vec{m}_1 \vec{m}_2^*) (\hat{e}_i \vec{H}_{dip+an}[\vec{m}_3^*]) + (\vec{m}_1 \vec{m}_3^*) (\hat{e}_i \vec{H}_{dip+an}[\vec{m}_2^*]) + (\vec{m}_2^* \, \vec{m}_3^*) (\hat{e}_i \vec{H}_{dip+an}[\vec{m}_1]) \tag{1.20}$$

$$\hat{\rho}_{123} = \frac{\gamma}{8V} \int_{\text{sample}} d^3r (\vec{m}_1^* \vec{m}_2^*) (\hat{e}_i \vec{H}_{dip+an}[\vec{m}_3^*]) + (\vec{m}_1^* \vec{m}_3^*) (\hat{e}_i \vec{H}_{dip+an}[\vec{m}_2^*]) + (\vec{m}_2^* \, \vec{m}_3^*) (\hat{e}_i \vec{H}_{dip+an}[\vec{m}_1^*]) \tag{1.21}$$

$$W_{12,34} = \frac{\gamma}{8V} \int_{\text{sample}} d^3r (\vec{m}_1 \vec{m}_2) (\vec{m}_3^* \vec{H}_{dip+ex+an}[\vec{m}_4^*]) + (\vec{m}_1 \vec{m}_3^*) (\vec{m}_2 \vec{H}_{dip+ex+an}[\vec{m}_4^*]) - (\vec{m}_1 \vec{m}_2) (\hat{e}_i \vec{H}_{dip+ex+an}[\hat{e}_i (\vec{m}_3^* \vec{m}_4^*)]) - (\vec{m}_1 \vec{m}_3^*) (\hat{e}_i \vec{H}_{dip+ex+an}[\hat{e}_i (\vec{m}_2 \vec{m}_4^*)]) \tag{1.22}$$

$$h_{eff}^{+} = -\frac{1}{2V} \int_{\text{sample}} d^3r \, \vec{m}_0^* \, \vec{h}(\vec{r}) \ , \quad h_{eff}^{-} = -\frac{1}{2V} \int_{\text{sample}} d^3r \, \vec{m}_0^* \, \vec{h}^*(\vec{r}) \tag{1.23}$$

$$V_{12} = \frac{\gamma}{8V} \int_{\text{sample}} d^3r \, (\vec{m}_1^* \, \vec{m}_2^*) \, (\hat{e}_i \vec{h}(\vec{r})) \ . \tag{1.24}$$

Note that the exchange interaction does not contribute to the three-magnon coupling coefficients (1.20) and (1.21), which corresponds to the fact that, according to its definition (1.13), the exchange field resulting from a transverse magnetization has only transverse components. Thus any term $\hat{e}_i \cdot \vec{H}_{ex}[\vec{m}_\kappa^*]$ is vanishing. For calculating the coupling coefficients h_{eff}^+, h_{eff}^- and V_{12} we have chosen a driving microwave field of the form $\mathrm{Re}\{\vec{h}(\vec{r})\exp(-\mathrm{i}\omega_p t)\} = \frac{1}{2}\{\vec{h}\exp(-\mathrm{i}\omega_p t) + \vec{h}^*\exp(\mathrm{i}\omega_p t)\}$, where $\vec{h}(\vec{r})$ is a complex field. In the case of a circular driving field $\vec{h}(\vec{r}) = h(\hat{x} + \mathrm{i}\,\hat{y})$ and a spherical sample we obtain $h_{eff}^+ = h$ and $h_{eff}^- = 0$ whereas for a linear driving field we have $h_{eff}^+ = h_{eff}^- = h/2$. The expression for V_{12}, Eq. (1.24), shows that a direct excitation of spin-wave pairs can only arise from microwave components parallel to $\mathfrak{M}_0$ ('*parallel pumping instability*').

1.3. Threshold of the First-Order Spin-Wave Instability

The spin-wave instabilities of lowest order are (i) the first-order Suhl process where the uniform mode is driven by a transverse microwave field and decays into spin-wave pairs at half the pumping frequency and (ii) the parallel pumping instability where spin-wave pairs at half the pumping frequency are directly excited by a microwave field parallel to the external field. It is also possible to combine both instabilities by applying the microwave field at an arbitrary angle with respect to the external field (*'oblique pumping'*)[32].

In order to determine the threshold for the first-order Suhl instability one has to investigate the stability of the linearized equations of motion for the driven uniform mode c_0 and for some spin wave with $\omega_\kappa \simeq \omega_p/2$. Since the uniform mode is generally driven by counter-rotating components ($+\omega_p$ and $-\omega_p$, see Eq. (1.16)) we start from the following ansatz for the amplitudes of the uniform mode and of the spin wave

$$c_0 = b_0^+ \, \mathrm{e}^{-\mathrm{i}\omega_p t} + b_0^- \, \mathrm{e}^{\mathrm{i}\omega_p t} \quad , \qquad c_\kappa = b_\kappa \, \mathrm{e}^{-\mathrm{i}(\omega_p/2)\,t} \, . \tag{1.25}$$

The amplitudes b_0^+, b_0^- and b_κ are expected to be slowly varying functions of time. If we further assume c_0 and c_κ to be driven on resonance, we may neglect all rapidly oscillating terms in the equations of motion for b_0^+, b_0^-, and b_κ, (*'resonance approximation'*) and obtain

$$\dot{b}_0^+ = -\,[\eta_0 + \mathrm{i}\,(\omega_0 - \omega_p)]\; b_0^+ \; - \; \sum_{\kappa\kappa'} \mathrm{i}\,\rho^*_{0,\kappa\kappa'}\, b_\kappa\, b_{\kappa'} \; - \; \mathrm{i}\,\gamma\, h^+_{\mathit{eff}}$$

$$\dot{b}_0^- = -\,[\eta_0 + \mathrm{i}\,(\omega_0 + \omega_p)]\; b_0^- \; - \; \sum_{\kappa\kappa'} \mathrm{i}\,\hat{\rho}_{0\kappa\kappa'}\, b^*_\kappa\, b^*_{\kappa'} \; - \; \mathrm{i}\,\gamma\, h^-_{\mathit{eff}}$$

$$\dot{b}_\kappa = -\,[\eta_\kappa + \mathrm{i}\,(\omega_\kappa - \omega_p/2)]\; b_\kappa \; - \; \sum_{\kappa'} 2\mathrm{i}\,(\rho_{0,\kappa\kappa'}\, b_0^+ + \hat{\rho}_{0\kappa\kappa'}\, b_0^{-*} + V_{\kappa\kappa'})\; b^*_{\kappa'} \; . \tag{1.26}$$

Below the threshold, where all spin-wave amplitudes b_κ are vanishing, the uniform mode is described by

$$b_0^+ = \frac{-\,\mathrm{i}\,\gamma\, h^+_{\mathit{eff}}}{\eta_0 + \mathrm{i}\,(\omega_0 - \omega_p)} \quad , \qquad b_0^- = \frac{-\,\mathrm{i}\,\gamma\, h^-_{\mathit{eff}}}{\eta_0 + \mathrm{i}\,(\omega_0 + \omega_p)} \; . \tag{1.27}$$

These equations show that b_0^- becomes negligible in comparison with b_0^+ when the uniform mode is driven on resonance ($\omega_p = \omega_0$). According to Eqs. (1.24) and (1.27), below the threshold b_0^+, b_0^-, and V_{12} are proportional to the microwave amplitude h. By introducing the abbreviation

$$h \cdot C_{\kappa\kappa'} \equiv \; -2\mathrm{i}\,(\rho_{0,\kappa\kappa'}\, b_0^+ + \hat{\rho}_{0\kappa\kappa'}\, b_0^{-\,*} + V_{\kappa\kappa'}) \;, \tag{1.28}$$

the equations of motion for the spin waves simplify to

$$\dot{b}_\kappa \;=\; -\;[\eta_\kappa + \mathrm{i}\,(\omega_\kappa - \omega_p/2)]\; b_\kappa \;+\; \sum_{\kappa'} h\, C_{\kappa\kappa'}\, b^*_{\kappa'} \;. \tag{1.29}$$

In order to determine the critical microwave field h_{crit} for parametric excitation one has to study the stability of the trivial solution $b_\kappa = 0$ of this equation. Since it only contains linear terms of b_K it can be expressed in the form $\dot{\vec{b}} = \mathbb{M} \cdot \vec{b}$, where vector $\vec{b}$ represents the real and imaginary parts of all modes and $\mathbb{M}$ is a real matrix depending on h. The critical field h_{crit} is defined by the value of h where the conjugate pair of eigenvalues of $\mathbb{M}$ with the largest real part crosses the imaginary axis. The corresponding eigenvector yields the corresponding critical modes excited at threshold.

Let us first discuss the threshold for conventional plane spin waves neglecting boundary conditions of the sample. It is convenient to consider standing spin waves $(\vec{m}_k + \vec{m}_{-k})/\sqrt{2}$ rather than travelling waves, since for standing waves all coupling coefficients $\rho_{0,\,kk'}$, $\hat{\rho}_{0kk'}$, $V_{kk'}$ and $C_{kk'}$ are diagonal and the sum in Eq. (1.29) is reduced to only one term. Thus, the critical field for the excitation of a specific standing wave b_k is given by

$$h_k = \frac{|\,\eta_k + \mathrm{i}\,(\omega_k - \omega_p/2)\,|}{|\,C_{kk}|} \;. \tag{1.30}$$

The next step is to determine the spin wave with the lowest critical field, which means the threshold for this process. Due to the resonance term in (1.30) the critical spin wave satisfies exactly the resonance condition $\omega_k = \omega_p/2$. The threshold field then reads

$$h_{crit} = \mathrm{Min}\Big\{ \frac{\eta_k}{|C_{kk}|} \Big\}_{\omega_k = \omega_p/2} \;, \tag{1.31}$$

where minimization refers to all spin waves at half the pumping frequency.

The coupling coefficient C_{kk} for a standing spin wave is determined by straight-forward calculation. We consider the case of oblique pumping in a spherical sample. The microwave field reads $\mathrm{Re}\{\, h\,(\cos\psi\,\hat{z} + \sin\psi\,\hat{x})\exp\,(-\mathrm{i}\,\omega_p\, t)\}$, where ψ is the angle between the static field $\vec{H}$ and the microwave field $\vec{h}$. Expressing the unit vector $\hat{e}_{k\parallel}$ in Eq. (1.8) in terms of the corresponding azimuthal spin-wave angle: $\hat{e}_{k\parallel} = \cos\phi_k \hat{x} + \sin\phi_k \hat{y}$, combining Eqs. (1.8) and (1.9) with the expressions for the coupling coefficients (1.20) – (1.24), and making use of the relations $\vec{H}_{dip}[\vec{m}_k] = -4\pi M_S \vec{k}(\vec{m}_k \vec{k})/k^2$ and $\vec{H}_{dip}[\vec{m}_0] = -4\pi M_S \vec{m}_0/3$ we arrive at the following expressions:

$$\rho_{0,\,kk} = -\omega_M \sin\theta_k \cos\theta_k \frac{1+\epsilon_k}{4\epsilon_k} e^{i\phi_k} \tag{1.32}$$

$$\hat{\rho}_{0kk} = -\omega_M \sin\theta_k \cos\theta_k \frac{1-\epsilon_k}{4\epsilon_k} e^{-i\phi_k} \tag{1.33}$$

$$h^{+}_{eff} = h^{-}_{eff} = -\frac{h}{2}\sin\psi \tag{1.34}$$

$$V_{kk} = -\gamma h \cos\psi \frac{1-\epsilon_k^2}{8\epsilon_k} = -\gamma h \cos\psi \frac{\omega_M \sin^2\theta_k}{8\omega_k}\,, \tag{1.35}$$

where ω_M is defined by $\omega_M \equiv \gamma 4\pi M_S$. Inserting these expressions into definition (1.28) we obtain the general solution for C_{kk}, which includes the behaviour on resonance as well as off-resonance excitation ($|\omega_p - \omega_0| \gg \eta_0$). For the latter case we finally have

$$C_{kk} = \frac{i\gamma\omega_M}{4}\left\{\cos\psi \frac{\sin^2\theta_k}{\omega_k} - \sin\psi \sin\theta_k \cos\theta_k \left(\frac{\epsilon_k+1}{\epsilon_k}\frac{e^{i\phi_k}}{\omega_p-\omega_0} + \frac{\epsilon_k-1}{\epsilon_k}\frac{e^{-i\phi_k}}{\omega_p+\omega_0}\right)\right\} \tag{1.36}$$

A discussion of the threshold field considering the proper eigenmodes of the sample instead of plane waves has been worked out so far for spherical modes in spheres[13] and for discrete standing modes in films[11]. In order to evaluate the coupling coefficients $\rho_{0,\kappa\kappa'}$ and $\hat{\rho}_{0\kappa\kappa'}$ for these modes, it is convenient to start from the potential ψ_κ mentioned above. The advantage is that then the dipolar field of the modes, necessary to determine $\rho_{0,\kappa\kappa'}$ and $\hat{\rho}_{0\kappa\kappa'}$, can simply be obtained by the relation $\vec{H}_{dip}[\vec{m}_\kappa] = \nabla\psi_\kappa$.

Transverse pumping in spheres – To this end, instead of spin waves, we introduce eigenmodes of appropriate symmetry, the so-called '*spherical modes*'. These are magnetostatic modes of a sphere which include the exchange interaction in lowest order. They are characterized by the following three components: $\kappa \equiv (n, m, \theta_r)$. i.e. the angular-momentum quantum number m, the index n which determines the effective wave number of the mode $k_{eff} = n \cdot \pi/d$ (d diameter of the sphere), and the angle θ_r which corresponds to the direction of propagation of a plane wave[13]. Spherical modes follow the same dispersion relation as plane waves if k is replaced by the effective wave number k_{eff} and θ_k by θ_r. For the case of transverse pumping it was shown[13] that the coupling coefficients $\rho_{0,\kappa\kappa'}$ follow the selection rules $n = n'$, $m + m' = 1$ and $\theta_r \simeq \theta'_r$. This means that the uniform mode decays into pairs of spherical modes, where the selection rule $m + m' = 1$ describes the conservation of angular momentum (the uniform mode corresponds to $m = 1$). A stability analysis of Eq. (1.29) results in the following critical field for the parametric excitation of a pair of modes (κ, κ')

$$h_{\kappa\kappa'} = \frac{|\eta_\kappa + \mathrm{i}\left(\frac{\omega_\kappa + \omega_{\kappa'}}{2} - \frac{\omega_p}{2}\right)|}{|C_{\kappa\kappa'}|}, \tag{1.37}$$

which strongly resembles the case of plane spin waves, Eq. (1.30). Again one has to determine the coupling coefficients $C_{\kappa\kappa'}$ and to minimize this expression in order to find the pair with the lowest threshold. This has been worked out in detail in a previous paper[13] and was successfully applied for the re-explanation of the finestructure of subsidiary absorption observed earlier by Jantz and Schneider[14]. From the spacings of the finestructure lines we were able[13,15] to identify the corresponding critical pairs of spherical modes, i.e. to determine all their parameters n, m and θ_r.

Parallel pumping in spheres – The coupling coefficient $V_{\kappa\kappa'}$ for parallel pumping follows different selection rules[13]: $n = n'$, $m + m' = 0$ and $\theta_r = \theta'_r$. The critical modes are no longer pairs of two different modes ($m' \neq m$) but of identical modes ($m' = m = 0$) satisfying exactly the condition $\theta_r = \theta'_r$. Again this model could explain[13] previous finestructure data from parallel pumping experiments in spheres[33]. It is also successful in explaining oblique pumping data on high quality YIG spheres, which showed that the parallel and perpendicular pumping processes work independently from each other[32]. Accordingly, the threshold is given by $\mathrm{Min}\{h^{\parallel}_{crit}/\cos\psi, h^{\perp}_{crit}/\sin\psi\}$, where $h^{\parallel}_{crit}$ and $h^{\perp}_{crit}$ denote the thresholds for parallel pumping and for the corresponding first-order Suhl instability, respectively. The independence of both parametric processes results from the fact that the coupling coefficients $\rho_{0,\kappa\kappa'}$ and $V_{\kappa\kappa'}$ follow different selection rules and the corresponding resonance conditions $2\omega_\kappa = \omega_p$ and $\omega_\kappa + \omega_{\kappa'} = \omega_p$, respectively, cannot simultaneously be fulfilled if the spherical modes in the latter case show some small frequency detuning[19] $\omega_\kappa \neq \omega_{\kappa'}$.

Films – The discrete standing modes of a film are described by a tangential wavevector $\vec{k}_{\|}$ and a normal component $k_\perp$, which is quantized in integer multiples of π/d: $k_\perp = n \cdot \pi/d$, where d is the film thickness. For a film magnetized in normal direction, $\rho_{0,kk'}$ follow the selection rules[11] $k'_\perp = k_\perp \pm \pi/d$ and $\vec{k}'_{\|} = \vec{k}_{\|}$ giving rise to a one-dimensional set of coupled modes with increasing index n. The determination of the critical microwave field for this special structure of the coupling matrix $C_{kk'}$ requires a numerical stability analysis of equation (1.29) which has been presented in Ref. 12. For tangentially magnetized films we have a different situation: Here, the critical modes are propagating in the film plane and the usual bulk approximation applies again. Our model was successfully applied to understand the different behaviours of the related 'butterfly curves' $h_{crit}(H)$ of normally and tangentially magnetized films[12]. In another work the coupling coefficient $\rho_{0,kk'}$ was determined for a finite size rectangular shaped film. In this case the two in-plane wavevector components follow the selection rules $k'_z = k_z \pm \pi/L_z$ and $k'_y = k_y \pm \pi/L_y$ (instead of $\vec{k}'_{\|} = \vec{k}_{\|}$ in the case of an infinitely extended film). L_z and L_y are the length and width of the film[34]. This additional quantization leads in k-space to a set of two-dimensional lattices of coupled modes and requires again a numerical stability analysis of Eq. (1.29). We applied this model to understand the size dependence of the experimental butterfly curves of a tangentially magnetized rectangular film[34].

Second-order Suhl instability – This higher-order instability can only be observed in absence of the first-order instability which is the case if $\omega_p/2$ falls below the spin-wave manifold. At threshold the uniform mode decays into degenerate spin-wave pairs according to the the four-magnon process $W_{00,k\text{-}k}\, c_0 c_0 c_k^* c_{-k}^*$. For spherical samples and the simple plane-wave approximation the coupling coefficients read

$$|W_{00,k\text{-}k}| = \frac{1}{4\omega_k}\left[\omega_k + \gamma H - \frac{\omega_M}{3} + \gamma D k^2 + \frac{\omega_M}{2}\sin^2\theta_k\right]\left[\omega_M \cos^2\theta_k + \gamma D k^2 + \frac{\omega_M}{3}\right]. \tag{1.38}$$

If the uniform mode is driven on resonance ($\omega_p = \omega_0$) the amplitude b_0^- of the counter-rotating component can be neglected, and the threshold can simply be described in terms of the critical amplitude

$$| b_0^+ |_{crit} = \mathrm{Min}\Big\{ \sqrt{\frac{\eta_k}{|2\, W_{00,\, k\text{-}k}|}} \Big\}_{\omega_k=\omega_p} = \sqrt{\frac{\eta_k}{\omega_M}} \,. \tag{1.39}$$

According to Eq. (1.38) the critical spin waves with the lowest threshold are those with $\theta_k = 0$, which means[1] propagating parallel to $\vec{H}$. In the case of nonresonant excitation, the counter-rotating component b_0^-, too, becomes important, and one has to include the processes $\hat{W}_{00\,k\text{-}k}\, c_0^* c_0^* c_k^* c_{-k}^*$ and $\tilde{W}_{0,0\,k\text{-}k}\, c_0\, c_0^* c_k^* c_{-k}^*$ in the calculation of the critical microwave field.

1.4. *Behaviour above the Threshold*

The equations of motion describing the behaviour above the threshold of the first-order spin-wave instability have to include both amplitude components of the uniform mode b_0^+ and b_0^- and a set of inhomogeneous modes b_κ at half the pumping frequency. But these are not the only modes excited above threshold. Two parametrically excited modes c_1 and c_2, for example, can excite a third mode c_α by means of a three-magnon confluence process $\rho^*_{\alpha,\,12}\, c_1\, c_2\, c_\alpha^*$. In general, the spin-wave dispersion relation does not allow the simultaneous conservation of energy and quasi-momentum: $\omega_\alpha \neq \omega_1 + \omega_2$ for $\vec{k}_\alpha = \vec{k}_1 + \vec{k}_2$. This means that under normal conditions such three-magnon confluence processes can only occur in a nonresonant way. In the following section we shall see that due to the specific spectrum of magnetostatic modes it is possible to excite certain magnetostatic modes c_α on resonance. In our understanding this mechanism is the reason for the complex multistability observed in experiment[10]. Under normal conditions, however, such processes occur off resonance, and the weakly excited modes c_α can adiabatically be eliminated from the equations of motion. There remains the interesting question how these modes influence the dynamics of the other spin-wave modes at $\omega_p/2$. It turns out that their main effect is a renormalization of four-magnon interactions between different parametrically excited modes.

This can be shown by the following example: Let us consider the modes c_1 and c_2 at $\omega_p/2$ exciting the third mode c_α through a three-magnon confluence process $\rho^*_{\alpha,12}\, c_1\, c_2\, c_\alpha^*$. This mode can decay again through a three-magnon splitting process $\rho_{\alpha,34}\, c_\alpha\, c_3^*\, c_4^*$ into two other modes c_3 and c_4 at $\omega_p/2$. Virtually these two three-magnon processes are acting on the parametrically excited modes like a four-magnon process $W_{12,34}\, c_1\, c_2\, c_3^*\, c_4^*$. Assuming that due to the process $\rho^*_{\alpha,12}\, c_1\, c_2\, c_\alpha^*$ the mode c_α oscillates with the sum of the eigenfrequencies of c_1 and c_2, i.e. $\dot{c}_\alpha \simeq -\mathrm{i}(\omega_1+\omega_2)\,c_\alpha$, the mode c_α can be adiabatically eliminated from the equations of motion, which leads to the following renormalization of $W_{12,34}$:

$$T_{12,34} \simeq W_{12,34} + \frac{\rho^*_{\alpha,12}\,\rho_{\alpha,34}}{\omega_1+\omega_2-\omega_\alpha} \quad . \tag{1.40}$$

In order to obtain the complete renormalized four-magnon interaction, apart from the upper process $(1,2 \to \alpha \;;\; \alpha \to 3,4)$ one has to include further three-magnon processes

$$\begin{array}{l} 2 \to \alpha,4 \;;\; 1,\alpha \to 3 \\ 1 \to \alpha,4 \;;\; 2,\alpha \to 3 \\ 1 \to \alpha,3 \;;\; 2,\alpha \to 4 \\ 2 \to \alpha,3 \;;\; 1,\alpha \to 4 \\ \to \alpha,3,4 \;;\; 1,2,\alpha \to \quad , \end{array} \tag{1.41}$$

which lead to the following result

$$\begin{aligned} T_{12,34} = \; & W_{12,34} + \sum_\alpha \Big\{ \frac{\rho^*_{\alpha,12}\,\rho_{\alpha,34}}{\omega_1+\omega_2-\omega_\alpha} + \frac{\rho_{2,\alpha 4}\,\rho^*_{3,1\alpha}}{\omega_2-\omega_4-\omega_\alpha} + \frac{\rho_{1,\alpha 4}\,\rho^*_{3,2\alpha}}{\omega_1-\omega_4-\omega_\alpha} \\ & + \frac{\rho_{1,\alpha 3}\,\rho^*_{4,2\alpha}}{\omega_1-\omega_3-\omega_\alpha} + \frac{\rho_{2,\alpha 3}\,\rho^*_{4,1\alpha}}{\omega_2-\omega_3-\omega_\alpha} + \frac{\hat{\rho}_{\alpha 34}\,\hat{\rho}^*_{12\alpha}}{-\omega_3-\omega_4-\omega_\alpha} \Big\} \; . \end{aligned} \tag{1.42}$$

For yttrium iron garnet (YIG) and a typical pumping frequency of 10 GHz the renormalization terms are of the same order of magnitude[35] as $W_{12,34}$. In the case of plane spin waves the sum over α is reduced to a single term, since α is determined by the selection rule $\vec{k}_\alpha = \vec{k}_1 + \vec{k}_2$. Including this renormalization the equations of motion for the uniform and for the parametrically excited modes read

$$\begin{aligned} \dot{b}_0^+ &= -\,[\eta_0 + \mathrm{i}\,(\omega_0-\omega_p)]\; b_0^+ \;-\; \sum_{\kappa\kappa'} \mathrm{i}\,\rho^*_{0,\kappa\kappa'}\; b_\kappa\, b_{\kappa'} \;-\; \mathrm{i}\,\gamma\, h^+_{eff} \\ \dot{b}_0^- &= -\,[\eta_0 + \mathrm{i}\,(\omega_0+\omega_p)]\; b_0^- \;-\; \sum_{\kappa\kappa'} \mathrm{i}\,\hat{\rho}_{0\kappa\kappa'}\; b^*_\kappa\, b^*_{\kappa'} \;-\; \mathrm{i}\,\gamma\, h^-_{eff} \\ \dot{b}_\kappa &= -\,[\eta_\kappa + \mathrm{i}\,(\omega_\kappa-\omega_p/2)]\; b_\kappa \;-\; \sum_{\kappa'} 2\mathrm{i}\,(\rho_{0,\kappa\kappa'}\, b_0^+ + \hat{\rho}_{0\kappa\kappa'}\, b_0^{-*} + V_{\kappa\kappa'})\; b^*_{\kappa'} \\ & \quad + \sum_{\kappa'\kappa_1\kappa_2} 2\mathrm{i}\; T_{\kappa_1\kappa_2,\kappa\kappa'}\; b_{\kappa_1} b_{\kappa_2} b^*_{\kappa'} \quad . \end{aligned} \tag{1.43}$$

This equation should be an appropriate starting point for the discussion of autooscillations.

The most important problem in the discussion of low-frequency collective oscillations arising from Eq. (1.43) is that there is no clear idea which of the modes satisfying $\omega_\kappa \simeq \omega_p/2$ really participate in the dynamics. Several authors therefore considered simplified models with only two or three coupled spin-wave modes [36–39]. Another problem is the large number of four-magnon scattering contributions which have to be summed up in Eq. (1.43). In order to reduce their number Zakharov et al. [35] applied a random phase approximation which reduces all four-magnon processes to the relevant terms $\sum_{k,k'} T_{k,k'} c_k^* c_k c_{k'}^* c_{k'} + \frac{1}{2}\sum_{k,k'} S_{k,k'} c_k^* c_{-k}^* c_{k'} c_{-k'}$. The random phase approximation, however, is only justified for plane spin waves in bulk samples when boundary conditions can be neglected.

Other attempts to calculate the frequency of autooscillations in infinitely large bulk samples including the complete spin-wave manifold [40,41] were not successful to explain the experimental data [42]. One reason could be the finite size of the sample in every experiment. Mc Michael *et al.* [21] were able to explain the frequencies of autooscillations in very thin circularly shaped YIG films by four-magnon interactions between certain discrete modes. Recent investigations of Cherepanov and Slavin have verified the importance of including the boundary conditions of the sample in any realistic model (see preceding chapter of this book). All these results underline that a better understanding of spin-wave instabilities above threshold in spin-wave instabilities requires models which are based on the proper eigenmodes of the sample.

2. Multistability by Resonant Three-Magnon Confluence Processes between Parametrically Excited Spin Waves and Magnetostatic Modes

In recent publications we have presented a magnetic system – a YIG sphere pumped on resonance at the first-order Suhl instability – showing a unique and very complex multistability [16–18]. This multistability only occurs within a parameter range where the resonant excitation of a magnetostatic mode, which is degenerate with the uniform mode, through parametrically excited spin waves is possible. A similar multistability was also observed in films very recently [19]. We suggest a mechanism for these phenomena which is based on the resonant excitation of degenerate magnetostatic modes by parametrically excited spin waves in three-magnon confluence processes.

2.1. Multistability in YIG Spheres within the Coincidence Regime of the First-Order Suhl Instability

Ferromagnetic resonance experiments were performed on highly polished spheres of yttrium iron garnet (YIG) within the coincidence regime of the first-order Suhl instability [16-18]. At the first-order Suhl instability spin waves are parametrically excited by the uniform mode which is driven by a homogeneous microwave field $h \cos(\omega_p t)\,\hat{x}$ applied perpendicularly to the static magnetic field $H\hat{z}$. As soon as the amplitude of b_0 exceeds a critical value, the uniform mode decays through three-magnon splitting processes into pairs of spherical modes of half the pumping frequency. Coincidence means that both the spherical modes ($\omega_\kappa = \omega_p/2$) and the uniform mode ($\omega_p = \omega_0$) are simultaneously driven on resonance, which is limited to a certain range of field and frequency ($H \approx 690$-$1280\,\mathrm{Oe}$ for YIG at room temperature and $\vec{H} \parallel (100)$ [10]). For the excitation and detection of the uniform mode we used two microcoils with perpendicular orientation as shown in Fig. 1. The input power P_{in} in the driving coil is propor-

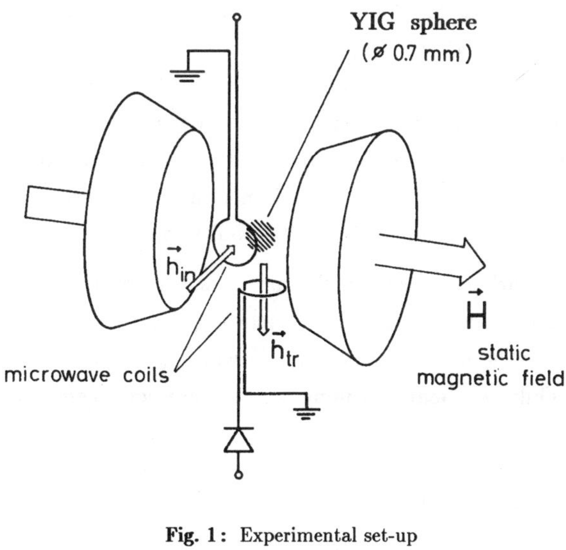

Fig. 1: Experimental set-up

tional to h^2 and the transmitted power in the pick-up coil P_{tr} is proportional to $|b_0|^2$. By means of a digital oscilloscope and an integrating voltmeter we recorded both the time dependence of $P_{tr}(t)$ and its average value $\overline{P_{tr}}$ with respect to P_{in}.

An example of the complex multistable behaviour observed at room temperature in the field range 760-1025 Oe (for $\vec{H} \parallel (100)$) is shown in Fig. 2. At small input powers P_{in} below the threshold of parametric excitation, where only the uniform mode is excited, $\overline{P_{tr}}$ increases linearly with P_{in}, that means b_0 is proportional to h. Above the threshold the system shows a variety of coexisting levels $\overline{P_{tr}}(P_{in})$, which all are well reproducible. If a certain level becomes unstable on variation of P_{in}, the system jumps to another one, as indicated by the dashed lines and arrows. In addition, we found autooscillations within a wide frequency range ($\omega_{\rm osc}/2\pi \simeq$ 10-350 kHz) with amplitudes smaller than 10% of the time average $\overline{P_{tr}}$. The autooscillations show a variety of different types (simple periodic, quasiperiodic, mode-locked, intermittent and chaotic) changing abruptly, when $\overline{P_{tr}}$ jumps from one level to the other. For further details see Ref. 10.

The behaviour at low temperature differs significantly from that at room temperature. Figure 3 shows the typical behaviour of the system at $T = 1.6$ K within the whole field range of the coincidence regime ($H \simeq 1070 - 1900$ Oe, for $\vec{H} \parallel (100)$ and $T = 1.6$ K). Neither multistability nor autooscillations were observed[16]. Above the threshold $\overline{P_{tr}}$ remains constant with increasing P_{in} (dashed line in Fig. 2; the slight increase of $\overline{P_{tr}}$ with P_{in} above threshold represents an experimental artifact resulting from imperfect orientation of the microcoils). This behaviour is unexpected since the (linear) dissipation terms are decreasing with temperature, while the (nonlinear) dipolar couplings essentially remain constant. Therefore, nonlinear phenomena such as the observed multistability were expected to be even more pronounced at low temperature. This inconsistent experimental result plays an important role in the discussion on multistability in section 2.3.

Our interpretation is based on the specific spectrum of magnetostatic modes in a sphere[43]. There are certain modes showing a permanent degeneration for arbitrary magnet fields, e.g. the (4,3,0) mode which is nearly degenerate with the uniform mode. Only a slight detuning occurs between them which is due to propagation effects[44,45]:

$$\omega_0 - \omega_{430} \sim M_S(\omega_0 d)^2 \ . \tag{2.1}$$

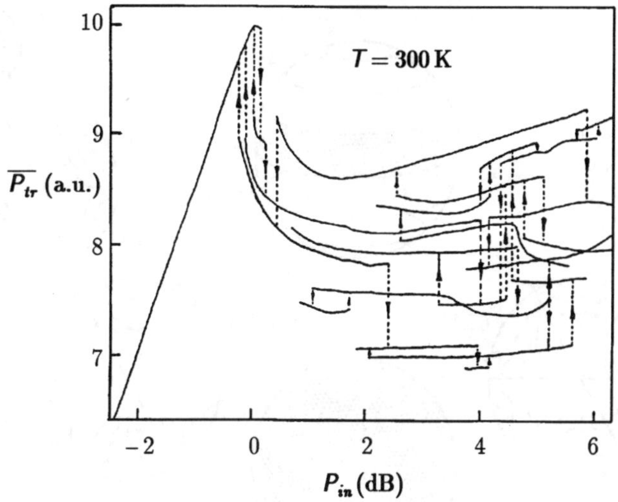

Fig. 2: Transmitted FMR signal with respect to microwave power at room temperature ($H = 995$ Oe; $\omega_p/2\pi = 2.56$ GHz; $\vec{H} \parallel (100)$). The first-order Suhl threshold corresponds to 0 dB.

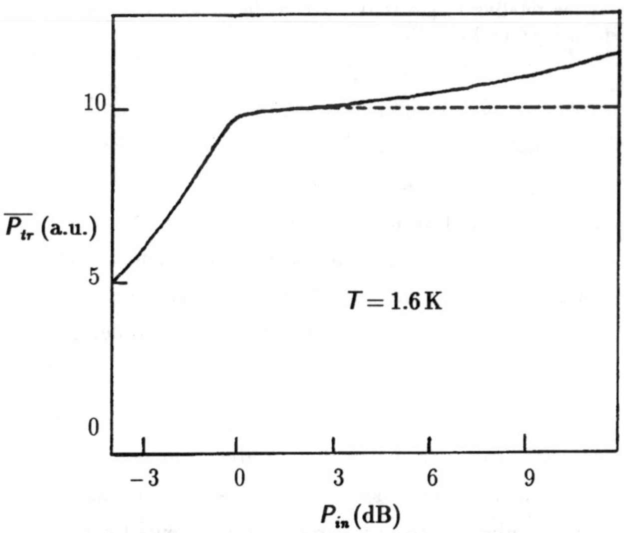

Fig. 3: Transmitted FMR signal at low temperature ($H = 1300$ Oe; $\omega_p/2\pi = 3.01$ GHz; same orientation). Threshold corresponds to 0 dB.

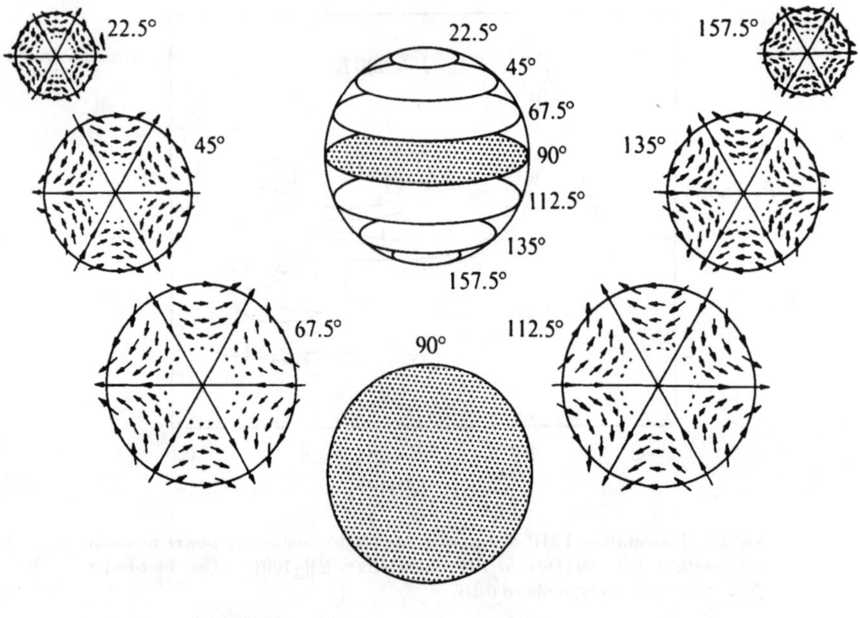

Fig. 4: Magnetostatic (4,3,0) mode in a sphere showing the instantaneous positions of the rotating magnetization vectors in planes transverse to the dc magnetic field (from Ref. 43).

For a YIG sphere of $d = 0.71\,\mathrm{mm}$ diameter, $\omega_0/2\pi = 2.3\,\mathrm{GHz}$, and room temperature the detuning amounts to $1.2\,\mathrm{MHz}$ [46]. Because of its inhomogeneous magnetization (see Fig. 4) the (4, 3, 0) mode cannot directly be excited by the homogeneous microwave field, but as it is nearly degenerate with the uniform mode, it can be excited on resonance by a three-magnon confluence process $(\rho^*_{430,\kappa\kappa'}\, c^*_{430}\, c_\kappa\, c_{\kappa'})$ with two parametrically excited spherical modes. Note that this process can only occur within the coincidence regime of first-order Suhl instability $(\omega_\mathrm{p} = 2\,\omega_\kappa \simeq \omega_{430})$.

For computing the coupling coefficients $\rho_{430,\kappa\kappa'}$ it is necessary to consider boundary conditions for the parametrically excited modes c_κ. The detailed discussion of this problem in terms of spherical modes has been presented in Refs. 17 and 47. Here, for simplicity, we only present some heuristic approach to our explanation by considering a cube of size d instead of a sphere. The corresponding

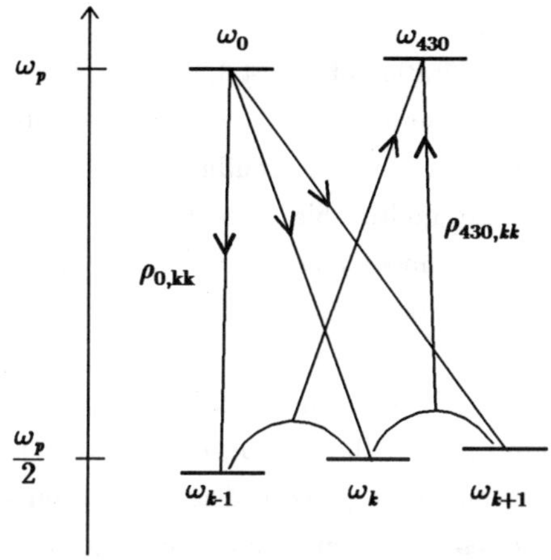

Fig. 5: Indirect excitation of the $(4,3,0)$ mode by standing spin waves which are neighbouring in k-space. The uniform mode (ω_0) is driven by the microwave field (ω_p) and decays into standing spin waves at half the pumping frequency $\omega_p/2$ through three-magnon decay processes $(\rho_{0,\,kk}\, c_0\, c_k^*\, c_k^*)$.

eigenmodes are standing spin waves with wavevector components[4] quantized in multiples of π/d ($k_{eff} = n\cdot\pi/d,\ n \gg 1$), whereas the related longwave magnetostatic mode ($\lambda_{eff} \simeq d$) is characterized by an effective wavevector in the order of only π/d. Conservation of quasi-momentum implies that the coupling coefficients $\rho_{430,kk'}$ remain non-zero only between neighbouring spin waves in k-space, which is schematically indicated in Fig. 5 by using the labels $k-1$, k, and $k+1$. Since in a real experiment not only a single critical spin wave is excited but several spin waves, which are neighbouring in k-space, an indirect excitation of the (4,3,0) mode by resonant confluence processes of parametrically excited spin waves becomes possible. Neglecting four-magnon interactions between the spin waves, the system is described by the following Hamiltonian

$$\mathcal{H} = \sum_{i=0,430,k} \omega_i c_i c_i^* + \sum_k \left(\rho_{0,kk} c_0 c_k^* c_k^* + \text{c.c.}\right) + \sum_{kk'} \left(\rho_{430,\,kk'}\, c_{430}\, c_k^* c_{k'}^* + \text{c.c.}\right)$$
$$+ \left(\gamma h_{eff}^+ c_0^* e^{-i\omega_p t} + \text{c.c.}\right) \quad . \tag{2.2}$$

[4] Accordingly, we will use the label k rather than κ.

There are two experiments which indicate that the multistable behaviour is related to the resonant excitation of the (4,3,0) mode. In the first one, the uniform mode was excited off resonance which implies that the (4,3,0) mode, too, is no longer driven on resonance by the confluence processes. For a detuning of $|\omega_0 - \omega_p| > 2\pi \cdot 5\,\mathrm{MHz}$ the multistable behaviour vanishes (for comparison, the linewidth η_{430} of the (4,3,0) mode amounts to roughly $2\pi \cdot 0.8\,\mathrm{MHz}$). In the second experiment, a large sphere of 1.75 mm in diameter was used, and according to Eq. (2.1) the detuning between ω_0 and ω_{430} increases to roughly $|\omega_0 - \omega_{430}| \simeq 2\pi \cdot 7.6\,\mathrm{MHz}$. Even if the uniform mode is driven on resonance the multistability does not occur in the big sphere since the (4,3,0) mode again is not excited by a resonant confluence process. These two results can be interpreted in the sense that only the resonant excitation of the (4,3,0) mode is responsible for the multistable behaviour.

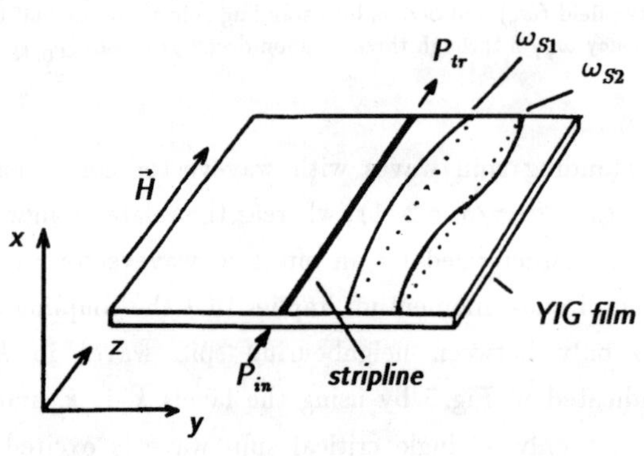

Fig. 6: Experimental set-up with a microstrip line on the top of a YIG film. The lines corresponding to ω_{S1} and ω_{S2} show the variation of the amplitude in z-direction of the two surfaces modes ω_{S1} and ω_{S2} shown in Fig. 9.

2.2. Multistability in YIG Films

Similar effects have been observed in films. The experimental setup applied in these experiments is shown in Fig. 6. A YIG film of $L_y = 3.5\,\mathrm{mm}$ and $L_z = 2.8\,\mathrm{mm}$ lateral size and of $d = 12.8\,\mu\mathrm{m}$ thickness grown on a (111) gadolinium gallium garnet (GGG) substrate is in contact with a 23 μm wide micro-stripline fixed on an aluminum ground plane (which is not shown in Fig. 6). The external magnetic field is applied parallel to the stripline. By means of microwave excitation at 3 GHz with an input power P_{in} magnetostatic waves are launched in the YIG film. This, in turn, results in an increased attenuation of the microstrip signal and in a reduction of the output power P_{tr}. Figure 7 shows the ratio of transmitted power P_{tr}/P_{in} as a function of H for different input powers. Following the theory of Damon and Eshbach [25] for magnetostatic waves in tangentially magnetized films, the resonances in the field range between 220 and 530 Oe correspond to the excitation of surface waves propagating perpendicularly to the stripline with a wavevector component k_y quantized in multiples of π/L_y and a component $k_z = \pi/L_z$ due to pinning at the edges of the film (see Fig. 6). The resonances above 530 Oe correspond to volume waves. As long as the system remains linear, the curves of P_{tr}/P_{in} versus H should not depend on the input power P_{in}, which, in fact, is observed above 500 Oe (see Fig. 7). Below 500 Oe the transmission ratio P_{tr}/P_{in} increases with increasing power. This effect, already observed in earlier studies [48], results from the three-magnon decay of surface waves, the amplitude of which remains exactly constant on variation of P_{in} above threshold due to the decay process. Brillouin light

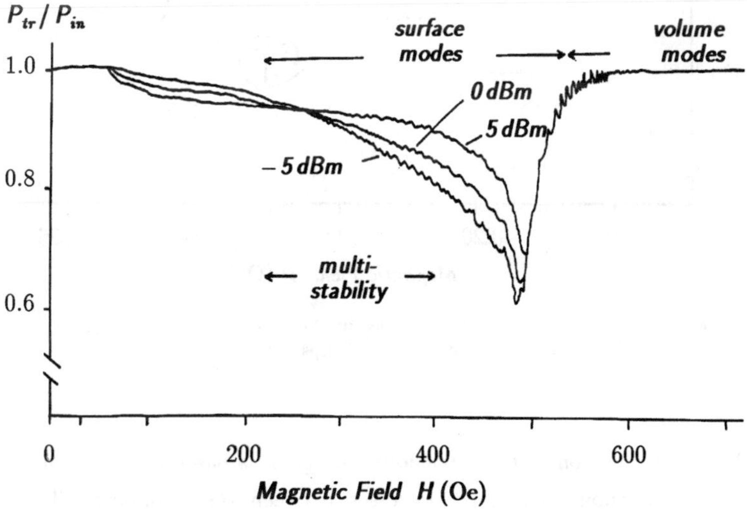

Fig. 7: Transmission ratio P_{tr}/P_{in} as a function of H for input powers of – 5, 0 and 5 dBm. The field regions which correspond to the excitation of surface and volume waves and in which multistable behaviour occurs are indicated.

scattering experiments[48] in the same experimental set-up yield direct evidence for the three-magnon decay of a surface mode into spin waves at $\omega_p/2$. Since the transmission ratio increases with P_{in} the experimental set-up can be used as a signal-to-noise enhancer device. Below 250 Oe the transmission ratio, however, decreases slightly with increasing P_{in}. The interpretation of this opposite behaviour is not yet clear.

In the field range from 220 to 400 Oe where only surface waves are excited, multistable behaviour is observed at some power level between 0 and 5 dBm depending on the field (Fig. 8) The output signal P_{tr} shows abrupt jumps on variation of H, leading to well reproducible coexisting levels of $P_{tr}(H)$. At certain points (A, B) the signal jumps intermittently between two well defined levels, which is probably induced by external noise.

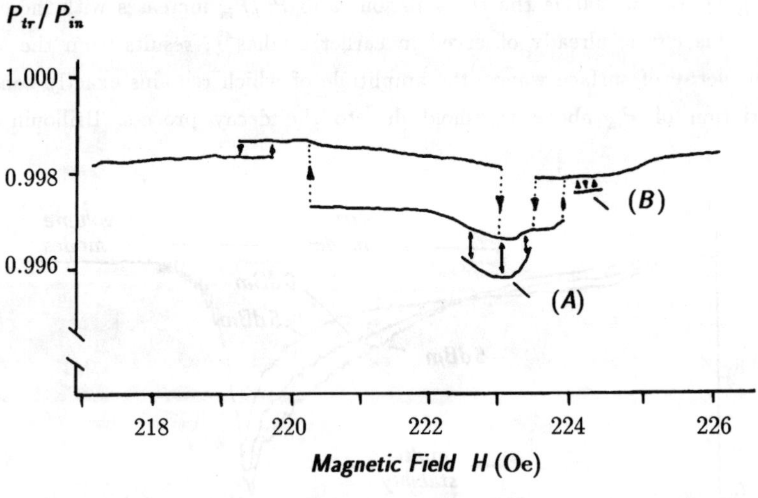

Fig. 8: Multistable behaviour occuring in the transmission ratio P_{tr}/P_{in} versus H. At (A) and (B) intermittent jumps between different levels are observed.

Our interpretation of these phenomena, again, is based on the occurrence of resonant three-magnon confluence processes between the parametrically excited spin waves and special magnetostatic surface waves. Figure 9 shows the dispersion relation of surface modes[30], the wavevectors of which are quantized in integer

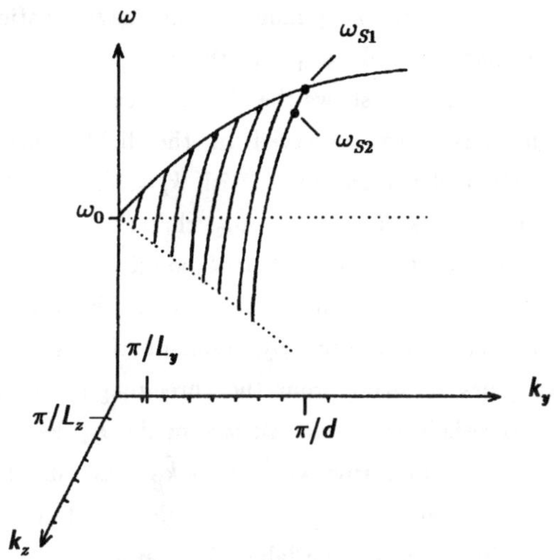

Fig. 9: Dispersion of the surface modes. The wavevector components k_y and k_z are quantized in integer multiples of π/L_y and π/L_z, respectively, due to pinning of the magnetization field at the edges of the film.

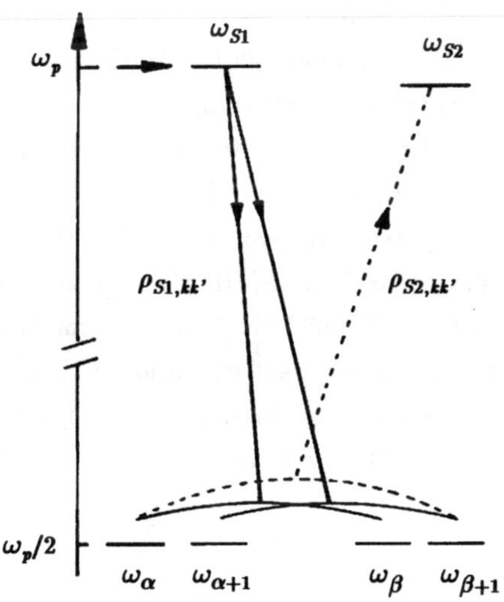

Fig. 10: Indirect excitation of the surface mode ω_{S2} by resonant three-magnon confluence processes of the parametrically excited spin waves.

multiples of π/L_y and π/L_z due to the pinning of the magnetization field at the edges of the film. The surface mode ω_{S1} has the smallest possible wavevector component $k_z = \pi/L_z$ and, therefore, shows the strongest excitation by the stripline (Fig. 6). Therefore, the resonances observed in the field range $220-530$ Oe correspond to the excitation of surface waves with $k_z = \pi/L_z$ and k_y quantized in integer multiples of π/L_y. The surface mode ω_{S2} characterized by a wavevector component $k_z = 2\pi/L_z$, on the other hand, cannot be excited by the stripline for symmetry reasons. However, since the modes ω_{S1} and ω_{S2} have roughly the same frequency, ω_{S2} can indirectly be excited by resonant three-magnon confluence processes of spin-wave pairs resulting from the three-magnon decay of ω_{S1}, as indicated in Fig. 10. At threshold the driven surface mode ω_{S1} decays into a pair of spin waves following the selection rule $\vec{k}_{S1} = \vec{k}_\alpha + \vec{k}_\beta$. As already discussed in section 2.2 we expect the excitation not only of a single pair, but of several neighbouring pairs in k-space (indicated by the labels $\vec{k}_{\alpha+1}$ and $\vec{k}_{\beta+1}$ in Fig. 10). Since the wavevectors of ω_{S1} and ω_{S2} differ by only $\Delta k_z = \pi/L_z$, resonant confluence processes of parametrically excited spin waves ($\vec{k}_{S2} = \vec{k}_\alpha + \vec{k}_{\beta+1}$) become possible and, this way, the indirect resonant excitation of ω_{S2}.

Like in the case of a sphere there are analog experiments supporting the presented idea: (i) Multistability is restricted to a limited range of the magnetic field and vanishes below 220 Oe and above 400 Oe. With increasing magnetic field the 'surface of dispersion' (Fig. 9) is shifted up with respect to the pumping frequency, which means that the detuning between ω_{S1} and ω_{S2} becomes smaller. At 220 Oe, where the pumping frequency is on the top of the surface mode branch (Fig.9) the detuning $|\omega_{S1} - \omega_{S2}|$ tends to zero. At 400 Oe, where the multistability vanishes (Fig. 7), the detuning exceeds $2\pi \cdot 5$ MHz (for comparison, the linewidth of the surface-mode η_{S2} amounts to $2\pi \cdot 0.8$ MHz.). This is consistent with the idea that multistability is related to the resonant excitation of ω_{S2}. (ii) Measurements on a YIG film of $d = 105\,\mu$m thickness have shown that the multistability is completely missing. From the Damon-Eshbach theory [25] it follows that the detuning scales like

$$|\omega_{S1} - \omega_{S2}| \propto d^2 / L_z^2 \quad , \tag{2.3}$$

which means that in this sample the detuning is about 100 times larger than in the previous film with $d = 12.8\,\mu$m. This again supports our idea that multistability is related to the indirect excitation of a second resonant mode ω_{S2} through three-magnon confluence processes of the parametrically excited spin waves. A detailed description of multistability in films will be published elsewhere[19].

2.3. Discussion

As already discussed above three-magnon confluence processes of parametrically excited spin waves are usually out of resonance. In the preceding sections we presented two systems where the special properties of the magnetostatic mode spectrum, however, allows a resonant excitation. In this special case we can no longer eliminate the resonant magnetostatic mode adiabatically from the equations of motion, but have to consider the full set of equations corresponding to the Hamiltonian (2.2):

$$\begin{aligned}
\dot{b}_0^+ &= -\,[\eta_0 + \mathrm{i}(\omega_0 - \omega_p)]\; b_0^+ \;-\; \sum_k \mathrm{i}\,\rho_{0,kk}^*\, b_k\, b_k \;-\; \mathrm{i}\,\gamma\, h_{eff}^+ \\
\dot{b}_{430}^+ &= -\,[\eta_{430} + \mathrm{i}\,(\omega_{430} - \omega_p)]\; b_{430}^+ \;-\; \sum_{kk'} \mathrm{i}\,\rho_{430,kk'}^*\, b_k\, b_{k'} \\
\dot{b}_k &= -\,[\eta_k + \mathrm{i}(\omega_k - \omega_p/2)]\; b_k \;-\; \sum_{k'} 2\mathrm{i}\,(\rho_{0,kk'}\, b_0^+ + \rho_{430,kk'}\, b_{430}^+)\; b_{k'}^* \,.
\end{aligned} \qquad (2.4)$$

Nevertheless it is suggestive to consider an expression corresponding to Eq. (1.40), which would refer to the adiabatic elimination of the (4,3,0) mode, since this simpler approach gives us some idea how resonant confluence processes into a certain magnetostatic mode act back to the spin waves at $\omega_p/2$. If we simply include the effect of the (4,3,0) mode just by introducing the renormalization (1.40) into the equation of motion for the spin waves, as described in section 1.4, we obtain - among other terms - the following renormalizations of dampings and eigenfrequencies:

$$\eta'_k = \eta_k + \beta_{kk'} |\, b_{k'}|^2 \;, \quad \beta_{kk'} = \frac{4\,\eta_{430}\; |\,\rho_{430,\,kk'}|^2}{(\omega_k + \omega_{k'} - \omega_{430})^2 + \eta_{430}^2} \qquad (2.5)$$

$$\omega'_k = \omega_k + \alpha_{kk'} |\, b_{k'}|^2 \;, \quad \alpha_{kk'} = \frac{(\omega_k + \omega_{k'} - \omega_{430})\; |\,\rho_{430,\,kk'}|^2}{(\omega_k + \omega_{k'} - \omega_{430})^2 + \eta_{430}^2} \qquad (2.6)$$

The resonance behaviour of the coefficients $\alpha_{kk'}$ and $\beta_{kk'}$ implies that especially the damping renormalization (2.5) vanishes very soon if the (4,3,0) mode is excited out of resonance by the spin waves ω_k and $\omega_{k'}$.

The following example will show that this kind of renormalization of dampings (2.5) is capable to produce multistable behaviour: We consider a simplified system with the uniform mode driven on resonance and two degenerate standing spin waves $\omega_k = \omega_{k+1} = \omega_p/2$ with equal couplings $\rho_{0,kk} = \rho_{0,\,k+1\;k+1} \equiv \rho$, but slightly different dampings $\eta_{k+1} > \eta_k$. They are supposed to be neighbouring in k-space, so, damping renormalizations can occur. This system is described by the following set of equations:

$$
\begin{aligned}
\dot{b}_0^+ &= -\eta_0\, b_0^+ - \mathrm{i}\,\rho^* \left(b_k\, b_k + b_{k+1}\, b_{k+1}\right) - \mathrm{i}\,\gamma\, h_{eff}^+ \\
\dot{b}_k &= -\left(\eta_k + \beta_{k,k+1}\, |\, b_{k+1}|^2\right)\, b_k - 2\,\mathrm{i}\,\rho\, b_0^+\, b_k^* \\
\dot{b}_{k+1} &= -\left(\eta_{k+1} + \beta_{k,k+1}\, |\, b_k|^2\right)\, b_{k+1} - 2\,\mathrm{i}\,\rho\, b_0^+\, b_{k+1}^* \quad .
\end{aligned}
\tag{2.7}
$$

These equations have been solved numerically, and the solutions for omitting the spin wave b_{k+1} (solid lines) or b_k (dashed lines) are represented in Fig. 11.

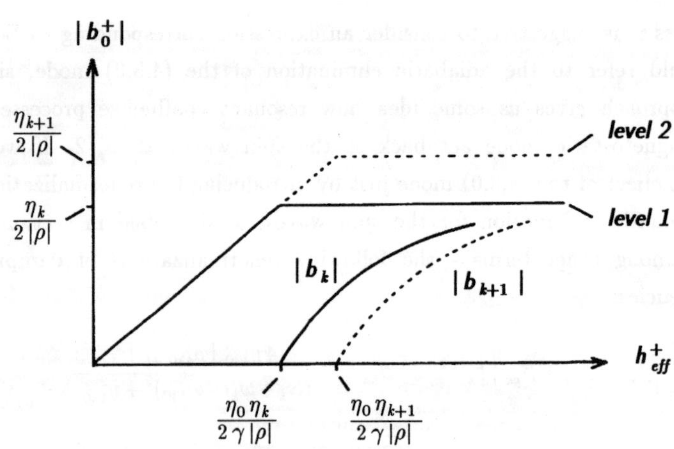

Fig. 11: Simulation of 2-surface mode model with unequal dampings, Eqs. (2.7). Solutions for vanishing amplitude b_k (dashed line) and for vanishing amplitude b_k (solid line).

This way, the well known behaviour is reproduced that the amplitude of the uniform mode becomes exactly constant above the threshold of parametric excitation. Considering the complete system (2.7) for a vanishing renormalization of damping, $\beta_{k,k+1} = 0$, we obtain the solid lines as solution. This means that above threshold only spin wave b_k is excited, and spin wave b_{k+1} with its slightly larger damping will never be excited even for very high pumping powers, since the amplitude of the uniform mode is frozen at the level $\eta_k / 2\,|\rho|$ and cannot reach the critical level $\eta_{k+1} / 2\,|\rho|$ necessary to excite spin wave b_{k+1}. For this case the dashed line represents an unstable solution of the system. If we introduce the damping renormalization, both solutions (solid and dashed lines), become stable depending on the previous state of the system. This can be veryfied in the following way: If the system is on level 2 (dashed line in Fig. 11), the critical amplitude of b_0^+ for the excitation of the mode b_k is now given by $(\eta_k + \beta_{k,k+1} |\, b_{k+1}|^2) / 2\rho$ and therefore higher than $\eta_{k+1} / 2\,|\rho|$. This explains that level 2 now represents a stable solution of the system. The fact that the damping of b_k increases when b_{k+1} is excited and vice versa, leads to the bistable behaviour that either b_k or b_{k+1} is excited above threshold.

If instead of (2.7) we use the correct equations of motions (2.4) including explicitly the (4,3,0) mode we observe qualitatively the same behaviour. The presence of three-magnon confluence processes $c_k + c_{k+1} \rightarrow c_{430}$ leads to the bistability that either c_k or c_{k+1} is excited above threshold, and both solutions are stable. This discussion was specified to experiments on spheres (section 2.2) but can also be transferred to experiments on films (section 2.3) by replacing standing spin waves ω_k and ω_{k+1} by the pairs of modes $(\vec{k}_\alpha, \vec{k}_\beta)$ and $(\vec{k}_{\alpha+1}, \vec{k}_{\beta+1})$ (see Figs. 5 and 10).

Considering the more realistic case of more than two spin waves, each of them coupled by the matrix $\rho_{430,\,kk'}$ to several neighbouring ones in three-dimensional k-space, we obtain many coexisting solutions consisting of different sets of excited modes. Therefore, we believe that the coexisting levels in Fig. 2 and Fig. 8 correspond to different sets of excited modes. The vanishing of these renormalizations when the (4,3,0) mode is no longer driven on resonance supports our idea that resonant confluence processes are responsible for the observed multistability.

Although the excitation of the (4,3,0) mode in spheres or of corresponding modes in films through resonant three-magnon confluence processes leads to a lot of coexisting solutions, i.e. multistable behaviour, the presented equations of motion still cannot completely describe the experimental results for the following reason: If we slowly increase the driving microwave field h^{+}_{eff} in the numerical simulations of Eqs. (2.7) and follow the corresponding solutions for the amplitudes – which is the situation in experiment when slowly increasing the microwave power – the system keeps staying on level 1 (solid line in Fig. 11) but never jumps to level 2 (dashed line). Also in more realistic simulations based on several interacting spin waves we get the same problem that the present solution is maintained and jumps to other coexisting solutions, as observed in experiment, do not occur. Numerical simulations for slowly increasing h^{+}_{eff} show more or less the same behaviour as observed at $T = 1.6\,\mathrm{K}$ (see Fig. 3) without any jumps of b^{+}_{0}. This may be taken as an indication that some additional mechanism responsible for the jumps is missing. The experimental fact that jumps between different levels only occur at room temperature (Fig. 2), but vanish at low temperatures (Fig. 3), suggests that this missing mechanism might be induced by the nonzero temperature of the system.

The role of temperature in spin-wave instabilities so far was theoretically investigated only by L'vov *et al.*[49]. They showed that, in contrast to $T=0$, where only one critical standing spin wave $\vec{k}_{crit}$ is excited, for $T>0$ a whole packet of spin waves is excited with finite widths $(\Delta\omega_{th}, \Delta k_{th})$ in ω and k. This corresponds to the idea that temperature causes a scattering of energy from the critical mode into neighbouring modes in k-space which have a frequency detuning $|\omega_k - \omega_{k,crit}|$ smaller than $\Delta\omega_{th}$. Considering in addition the frequency renormalizations due to neighbouring modes (2.6) one can interpret the occurence of jumps between different solutions as follows: If we consider two neighbouring modes c_k and c_{k+1} with a frequency distance $|\omega_{k+1} - \omega_k|$ larger than $\Delta\omega_{th}$, an increase of the amplitude b_{k+1} results in such a shift of ω'_k that the detuning becomes smaller than $\Delta\omega_{th}$ and the 'thermal scattering of energy' between these modes is switched on (Fig. 12). Estimates for $\Delta\omega_{th}$, for the frequency shift $\Delta\omega'_k = \alpha_{k,k+1}\,|\,b_{k+1}|^2$ and for typical frequency distances between neighbouring modes are given by $\Delta\omega_{th}/2\pi \simeq 4\,\mathrm{kHz}$, $\Delta\omega'_k/2\pi \simeq 300\,\mathrm{kHz}$ and $|\omega_{k+1} - \omega_k|\,/\,2\pi \simeq 0 - 200\,\mathrm{kHz}$ [47]. These values show that the frequency renormalization $\Delta\omega'_k$ is in the order of the frequency detuning between neighbouring modes $|\omega_{k+1} - \omega_k|$ and that thermal

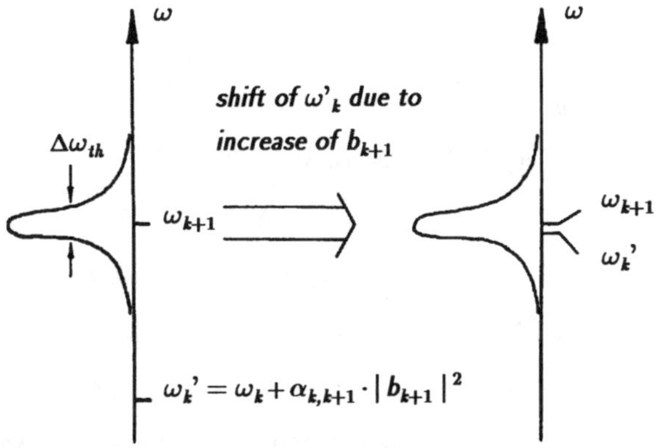

Fig. 12: Switching of thermal couplings between neighbouring modes (b_k, b_{k+1}) due to the renormalization of frequency.

scattering of energy can be switched on rather abruptly because of $\Delta\omega_{th} \ll \Delta\omega'_k$. The abrupt switching of thermal couplings between neighbouring modes might be the reason for the jumps between the different states of the system. A quantitative description of this temperature-induced effect, however, requires a more sophisticated treatment, which we hope to stimulate by these ideas.

Acknowledgements

This work was supported in part as a project of SFB 185 'Nichtlineare Dynamik', by the *Deutsche Forschungsgemeinschaft* and by Grant DMR-8921761 of the US National Science Foundation. One of us (G.W.) was supported on a postdoctoral fellowship from the *Deutsche Forschungsgemeinschaft*. We gratefully acknowledge helpful discussions with Dr. Ming Chen, Dr. P. Kabos, H.-A. Krug von Nidda, Prof. Patton and Dr. F. Rödelsperger.

References

1. H. Suhl, *J. Phys. Chem. Solids* **1** (1957) 209.
2. F. R. Morgenthaler, *J. Appl. Phys.* **31** (1960) 95 S.
3. G. Gibson and C. Jeffries, *Phys. Rev.* **A 29** (1984) 811.
4. F. Waldner, D. R. Barberis, and H. Yamazaki, *Phys. Rev.* **A 31** (1985) 420.
5. H. Yamazaki and M. Warden, *J. Phys. Soc. Jap.* **55** (1986) 4477.
6. F. M. de Aguiar and S. M. Rezende, *Phys. Rev. Lett.* **56** (1986) 1070.
7. P. H. Bryant, C. D. Jeffries, and K. Nakamura, *Phys. Rev.* **A38** (1988) 4223.
8. M. Warden and F. Waldner, *J. Appl. Phys.* **64** (1988) 5386.
9. T. L. Caroll, L. M. Pecora, and F.J. Rachford, *J. Appl. Phys.* **64** (1988) 5396.
10. G. Wiese and H. Benner, *Z. Phys.* **B 79** (1990) 119.
11. G. Wiese, *Z. Phys.* **B 91** (1993) 57.
12. G. Wiese, P. Kabos, and C. E. Patton, *J. Appl. Phys.* **74** (1993) 1218.
13. G. Wiese, *Z. Phys.* **B 82** (1991) 453.
14. W. Jantz and J. Schneider, *Phys. Stat. Sol. (a)* **31** (1975) 595.
15. H.-A. Krug v. Nidda, G. Wiese, and H. Benner, *Z. Phys. B* (in press).
16. G. Wiese, H.-A. Krug von Nidda, and H. Benner, *Europhys. Lett.* **15** (1991) 585.
17. G. Wiese, *J. Appl. Phys.* **73** (1993) 6819.
18. H. Benner, F. Rödelsperger, and G. Wiese, in *Nonlinear Dynamics in Solids*, ed. H. Thomas (Springer, Berlin-Heidelberg, 1992), p. 129.
19. G. Wiese, P. Kabos, and C. E. Patton, to be published in *J. Appl. Phys.*
20. B. A. Kalinikos, N. G. Kovshikov, and N. V. Kozhus', *Sov. Phys. Sol. State* **27** (1985) 1681.
21. R. D. McMichael and P. E. Wigen, *Phys. Rev. Lett.* **64** (1990) 64.
22. F. Holstein and H. Primakov, *Phys. Rev.* **58** (1940) 1048.
23. L. R. Walker, *Phys. Rev.* **105** (1957) 390.
24. P. C. Fletcher and R. O. Bell, *J. Appl. Phys.* **30** (1959) 687.

25. R. W. Damon and J. R. Eshbach, *J. Phys. Chem. Solids* **19** (1961) 308.

26. A.I. Akhiezer, V.G. Bar'yakhtar, and S.V. Peletminskii, *Spin Waves* (North-Holland, Amsterdam 1969).

27. P.H. Bryant, *Phys. Rev. B* **39** (1989) 4363.

28. C. Kittel, *Phys. Rev.* **73** (1948) 155.

29. E. Schlömann and R.I. Joseph, *Appl. Phys.* **32** (1961) 1006.

30. C.E. Patton, *Phys. Reports* **103** (1984) 251.

31. C.E. Patton, *Phys. Stat. Sol. (b)* **92** (1979) 211.

32. Y.H. Liu and C.E. Patton, *J. Appl. Phys.* **53** (1982) 5116.

33. W. Jantz, J. Schneider, and B. Andlauer, *Solid State Commun.* **10** (1972) 937.

34. G. Wiese, P. Kabos, and C.E. Patton, to be published in *J. Appl. Phys.*

35. V.E. Zakharov, V.S. L'vov, and S.S. Starobinets, *Sov. Phys. Usp.* **17** (1975) 896.

36. T.L. Caroll, L.M. Pecora, and F.J. Rachford, *Phys. Rev.* **A40** (1989) 377.

37. X.Y. Zhang and H. Suhl, *Phys. Rev.* **A32**, 2530 (1985)

38. S.M. Rezende, O.F. de Alcantara Bonfirm, and F.M. de Aguiar, *Phys. Rev.* **B33** (1986) 5153.

39. K. Nakamura, S. Ohta, and K. Kawasaki, *J. Phys.* **C15** (1982) L143.

40. H. Suhl and X.Y. Zhang, *Phys. Rev. Lett.* **57** (1986) 1480.

41. X.Y. Zhang and H. Suhl, *Phys. Rev.* **B38** (1988) 4893.

42. S.M. Rezende, F.M. de Aguiar, and A. Azevedo, *J. Appl. Phys.* **67** (1990) 5624.

43. L.R. Walker, *J. Appl. Phys.* **29** (1958) 318.

44. J.E. Mercereau, *J. Appl. Phys.* **30** (1959) 184S.

45. R.L. White, *J. Appl. Phys.* **31** (1960) 86S.

46. P. Fletcher, I.H. Solt, and R. Bell, *Phys. Rev.* **114** (1959) 739.

47. G. Wiese, Ph.D. thesis, TH Darmstadt (1991)

48. G. Srinivasan and C.E. Patton, *J. Appl. Phys.* **61** (1987) 2318.

49. V.S. L'vov and L.A. Prozorova, in *Spin Waves and Magnetic Excitations, Vol. 1*, eds. A.S. Borovik-Romanov and S.K. Sinha (Elsevier, Amsterdam, 1988) p. 233.

PART III

INTERACTION OF MICROWAVE MAGNETIC EXCITATIONS WITH LIGHT

DIFFRACTION OF GUIDED LIGHT BY SPIN WAVES IN FERROMAGNETIC FILMS

A. A. STASHKEVICH*

Centre for Chemical Physics, University of Western Ontario, London, Ontario, Canada N6A 3K7

1. Introduction

Investigations of light scattering by magnetic excitations in ferromagnets: bulk samples and single crystal films, are carried out since nineteen sixties.[1-7] However, only in early eighties it was shown by Fisher[8,9] et al that it is in the case of waveguide propagation of both waves in a ferromagnetic film that a really effective interaction of light with spin waves can occur. Such regime of interaction will be henceforth called the waveguide interaction. In the investigations that followed it has been demonstrated that the waveguide interactions hold promise for both applications with optical information processing devices operating with microwave frequencies as high as 5- 20 GHz[10-16] and as a unique tool for optical probing of linear and nonlinear magnetic excitations in thin ferromagnetic films.[17-19]

It must be noted that there exists a strong analogy between the acoustooptics and magnetooptical interaction of light and spin waves. In both cases light is essentially scattered by a diffraction phase grating, which is induced in the medium by the corresponding wave process and which has the same distribution in space and time as the wave. Hence the waveguide interaction of light and spin waves can be treated as a diffraction process in a periodically inhomogeneous medium. For this reason the terms interaction and diffraction are used as synonyms in this chapter. The rigorous solution of the problem of light propagation in a phase diffraction grating given by Brillouin[20] and Nomoto[21] involves finding the eigenfunctions of the perturbed wave equation. The computational difficulties associated with summing the

*Permanent address: Department of Radio Engineering, Electrotechnical University, St-Petersburg, 197376, Russia.

resulting series in Mathieu functions makes this method unsuitable for many cases of practical interest. There are similar difficulties in other theoretical approaches such as the geometric optics method[22,23] and Rytov's method of smooth perturbations.[24]

The most suitable method for the problem is the coupled wave method, where the solution is constructed from the eigenfunctions of the unperturbed wave equation. The resulting system of differential equations can be solved approximately in a relatively simple way for the range of values of the basic parameters encountered in situations of practical interest. Fourier representation of the electromagnetic field through the angular spectrum of plane waves simplifies application of the analysis of the optical interaction to an entire optical system in which the magnetooptical element functions: the formalism of Fourier optics is widely used in physics and in applied optics.[25-27]

The system of coupled wave equations can be obtained in two ways: through the wave equation[25], or directly through Maxwell's equations as done in Ref.[26] or in Ref.[28] for the case of the diffraction of a plane (in the plane of the film) light wave by a harmonic spin wave. The second method is preferable; as a rigorous approach, it leads to fewer mathematical difficulties than the first one.

Below this method will be extended to the more general situation of diffraction of an optical mode by a spin wave with an arbitrary spectrum. We will also touch upon the subject of a strong magnetooptical interaction accompanied by the creation of several diffraction orders. In all previous publications the formulation of the problem was restricted to the classic Bragg diffraction, with only two diffraction orders involved. At the same time, the aforementioned aspects are typical of the situations often occurring in practice, for example, during the signal analysis in an optical processor or in the course of diffraction by a nonlinear spin wave characterized by the modulation instability.[29,30] The theory constructed was in its form similar to that already developed in acoustooptics.[31,32]

The solution of the problem is constructed according to the perturbation method which was selected on the basis of the following considerations.

First, it allows to achieve a reasonable and flexible compromise between the generality of the formulation of the problem being solved and the complexity of the solution obtained. Thus, in the case of the two-magnon interaction with one spin wave harmonic involved the solution can be extended to regimes of interaction characterized by high diffraction efficiency which requires to retain higher terms in the perturbation series. On the contrary, in the case of a complex spectrum (multifrequency diffraction) higher order terms in the solution can be neglected except the first two ones.

On the other hand, the perturbation method allows the magnetooptical diffraction to be considered as a multiple-scattering process and be graphically illustrated by Feynman-type multiple-transition diagrams like it is done in acoustooptics.

2. Generalized Coupled Wave Equation

The geometry of the waveguide interaction is as follows. We considered a layered structure which is infinite in the x_2x_3 plane. A ferromagnetic film magnetized to saturation occupies the layer $0<x_1<h$. A spin wave with a complex spectrum propagates along x_2. The spin wave is localized in the coordinate x_3 inside the interval $0<x_3<L$. The film acts like a plane dielectric waveguide to optical waves: $n^f>n^s>n^c$, where n^f, n^s, n^c are the indices of refraction of the film, substrate, and covering layer, respectively. A monochromatic electromagnetic field with frequency $\omega^{(0)}_{opt}$ propagating in such a structure can be represented as a superposition of N pairs of optical waveguide modes (radiation modes are not considered):

$$\vec{E} = \sum_{s=0}^{2N-1} C_s \vec{E}^e_s = \sum_{s=0}^{2N-1} C_s \vec{E}_s(x_1)\, exp(-i\beta_s(\vec{e},\vec{R}))\, exp(i\omega^{(0)}_{opt}t), \tag{1a}$$

$$\vec{H} = \sum_{s=0}^{2N-1} C_s \vec{H}^e_s = \sum_{s=0}^{2N-1} C_s \vec{H}_s(x_1)\, exp(-i\beta_s(\vec{e},\vec{R}))\, exp(i\omega^{(0)}_{opt}t), \tag{1b}$$

The functions $\vec{E}^{e}_{s}$ and $\vec{H}^{e}_{s}$ satisfy Maxwell's equations

$$rot\vec{H}^{e}_{s} = \varepsilon^{i}_{1}\varepsilon_{0} \frac{\partial \vec{E}^{e}_{s}}{\partial t}, \quad rot\vec{E}^{e}_{s} = -\mu_{0} \frac{\partial \vec{H}^{e}_{s}}{\partial t}, \tag{2}$$

$$\varepsilon^{1}_{1} = \varepsilon^{s}_{1} = (n^{s})^{2}, \quad \varepsilon^{2}_{1} = \varepsilon^{f}_{1} = (n^{f})^{2}, \quad \varepsilon^{3}_{1} = \varepsilon^{c}_{1} = (n^{c})^{2},$$

where ε^{s}_{1}, ε^{f}_{1}, ε^{c}_{1} are the dielectric constants of the substrate, film, and covering layer, respectively; C_s are numerical coefficients; s is the integer index of the optical mode (odd s corresponds to transverse electric modes (TE modes), even s corresponds to transverse magnetic modes (TM modes)); $\vec{e}$ is a unit vector determining the direction of propagation of the optical mode in the plane of the film; $\vec{R}$ is the position vector in the plane of the film; $(\vec{e},\vec{R})$ is the scalar product of the vectors $\vec{e}$ and $\vec{R}$. The magnitudes of the wavevectors $\vec{\beta}$ in (1) are determined by the dispersion relation, which is obtained with the help of the electrodynamic boundary conditions. The vector functions $\vec{E}_s(x_1)$ and $\vec{H}_s(x_1)$ characterize the electric and magnetic field distribution in the optical mode over a cross section of the film. For TE modes

$$\vec{E}^{TE}_{s} = E^{y}_{s}\vec{e}^{y}, \quad \vec{H}^{TE}_{s} = H^{x}_{s}\vec{e}^{x} + H^{z}_{s}\vec{e}^{z}. \tag{3a}$$

For TM modes

$$\vec{E}^{TM}_{s} = E^{x}_{s}\vec{e}^{x} + E^{z}_{s}\vec{e}^{z}, \quad \vec{H}^{TM}_{s} = H^{y}_{s}\vec{e}^{y}. \tag{3b}$$

Optical waveguide modes propagating along a given direction $\vec{e}$, are orthonormal in the sense of the cross power.[26] It can be shown that optical modes propagating in different directions with wavevectors $\vec{\beta}_s$ and $\vec{\beta}_{s'}$ are orthogonal in the sense of the cross power

$$P_{ss'} = \int_{-\infty}^{\infty} dx_1(([\vec{E}_s,\vec{H}^{*}_{s'}] + [\vec{E}^{*}_{s'},\vec{H}_s]), \vec{e}^{\Delta}) = 0. \tag{4}$$

Here and after $\Delta\beta\vec{e}^{\Delta} = \vec{\beta}_s - \vec{\beta}_{s'}$, $*$ corresponds to a complex conjugate, and $[\vec{a},\vec{b}]$ denotes a vector product of two vectors.

For convenience of notation an auxiliary coordinate system x,y,z

is introduced in (3). It is defined with respect to the optical wave : the z axis is along the direction of propagation of the optical mode, the x axis is perpendicular to the plane of the film (and coincides with the x_1 axis), and the y axis is perpendicular to the projection of the propagation direction of the optical mode in the plane of the film.

We consider now the perturbations of the dielectric permittivity of a ferromagnetic film carrying spin waves. The problem of magnetic permeability modulation at optical frequencies is discussed elsewhere[33-35] and will not be considered here. The magnetization of a spin wave having a complex spectrum $\tilde{m}(\omega)$ and propagating along x_2 has the form

$$\vec{m}(x_1,x_2,t) = s(x_1,K_0)\,\frac{1}{\sqrt{2\pi}}\int_{-\infty}^{\infty}\Big(\tilde{m}^e(K(\omega)-K_0(\omega_0))\,\vec{e}(x_1,K_0(\omega_0)) + \tilde{m}^{e*}(-K(\omega)-K_0(\omega_0))\,\vec{e}(x_1,-K_0(\omega_0))\Big)\exp(i(K(\omega)x_2-\omega t))\,dK, \qquad (5)$$

where the functions $s(x_1,K_0)$ and $\vec{e}(x_1,K_0)$ describe the distribution of the amplitude and polarization of the magnetization with film thickness. The dispersion relation $K = K(\omega)$ is derived through integration of the equation of motion for magnetization (Landau-Lifschitz equation) together with the equations of magnetostatics.[36,37]

The following practical considerations are reflected in the structure of (5). The spin wave signal has a limited frequency band with the spectral components concentrating around the central temporal frequency ω_0 and, correspondingly, spatial frequency K_0. We can assume that the functions $s(x_1,K)$ and $\vec{e}(x_1,K)$ retain their form in the vicinity of $K_0(\omega_0)$. However, the type of polarization does depend on the sign of K_0. Because the function $\vec{m}(x,t)$ is real its spectrum can be represented as $\vec{m}(K) = \tilde{m}^e(K-K_0)\,\vec{e}(x_1,K_0) + \tilde{m}^{e*}(-K-K_0)\,\vec{e}(x_1,-K_0)$, where $\tilde{m}^e(K)$ is the spectrum of the spin wave envelope.

The presence of the spin wave leads to small tensor corrections to the dielectric constant of the film. These corrections have the same space-time dependence as the magnetization of the spin wave

$$\hat{\Delta\varepsilon}(x_1,x_2,t) = s(x_1)\,\frac{1}{\sqrt{2\pi}}\int_{-\infty}^{\infty}\hat{\Delta\varepsilon}'(x_1,K(\omega))\,exp(i(K(\omega)x_2-\omega t))\,dK, \qquad (6)$$

where $\hat{\Delta\varepsilon}'(x_1,K) = \hat{\Delta\varepsilon}(x_1,K_0)\,\tilde{m}^e(K-K_0) + \hat{\Delta\varepsilon}(x_1,-K_0)\,\tilde{m}^{e*}(-K-K_0).$

The structure of the tensor $\hat{\Delta\varepsilon}$ will be revealed below. In the presence of the perturbing influence of the spin wave, Maxwell's equations take the form

$$rot\vec{H} = \varepsilon_1^i\varepsilon_0\,\frac{\partial\vec{E}}{\partial t} + \varepsilon_0\frac{\partial}{\partial t}\,(\hat{\Delta\varepsilon}\vec{E}), \qquad (7a)$$

$$rot\vec{E} = -\mu_0\,\frac{\partial\vec{H}}{\partial t}\,. \qquad (7b)$$

Suppose initially an optical wave with frequency $\omega_{opt}^{(0)}$ is propagating in the half-space $x_3 < 0$ at an angle ϑ_r to the axis x_3. This mode will be called the incident wave and all other optical modes will be called diffracted waves.

The field of the incident wave is characterized by the vector functions $\vec{E}_r(x_1)$, $\vec{H}_r(x_1)$ and the wave vector

$$\vec{\beta}_r = \beta_r\vec{e}^r = \beta_r sin\,\vartheta_r\,\vec{e}^2 + \beta_r cos\,\vartheta_r\,\vec{e}^3.$$

In the case of a continuous spatial frequency spectrum for the spin wave magnetic disturbance the angular spectrum of the diffracted light is also continuous. Therefore, the conventional concept of spectral diffraction orders[38] is meaningless, and the discrete index p in the system of coupled wave equations is replaced by a continuous variable K, which has the sense of a spatial frequency.

Thus, the fields in the waveguide perturbed by a spin wave ($x_3>0$) are sought in the form of superpositions of coupled wave continua

$$\vec{E} = \sum_{s=0}^{2N-1}\frac{1}{\sqrt{2\pi}}\int_{-\infty}^{\infty}C_s(K,x_3)\,\vec{E}_s^e(K)\,dK + \tilde{E}\vec{e}^3, \qquad (8a)$$

$$\vec{H} = \sum_{s=0}^{2N-1} \frac{1}{\sqrt{2\pi}} \int_{-\infty}^{\infty} C_s(K,x_3)\, \vec{H}^e_s(K)\, dK \;+ \tilde{H}\vec{e}^3, \tag{8b}$$

where here, unlike (1), C_s are functions of the coordinate x_3. We have adopted the following notation in (8):

$$E^e_s(K) = E_s(x_1)\, exp(-i(\vec{\beta}_s(K),\vec{R}))\, exp(i\omega(K)t), \tag{9a}$$

$$H^e_s(K) = H_s(x_1)\, exp(-i(\vec{\beta}_s(K),\vec{R}))\, exp(i\omega(K)t), \tag{9b}$$

$$\vec{\beta}_s(K) = (\beta_r sin\vartheta_r + K)\, \vec{e}^2 + \sqrt{(\beta_s(K))^2 - (\beta_r sin\vartheta_r + K)^2}\, \vec{e}^3. \tag{9c}$$

The quantity $\beta_s(K)$ corresponds to the wave number of the optical mode with an index s at the optical frequency $\omega_{opt}(K) = \omega^{(0)}_{opt} + \omega(K)$

$$\beta_s(K) = \beta_s(\omega_{opt} = \omega^{(0)}_{opt}) + \frac{\omega}{V_g}, \tag{9d}$$

$V_g = \dfrac{d\omega}{d\beta_s}$ is the group velocity of the optical mode in the waveguide.

Expressions (8a) and (8b) are double superpositions of the solutions of the unperturbed problem, the functions $\vec{E}^e_s(K)$ and $\vec{H}^e_s(K)$ satisfying Eq.(2). Summation with respect to the index s goes over a finite number $2N$ of optical waveguide eigenmodes. For each optical eigenmode integration goes over a continuum of the plane wave angular spectrum. The angle of propagation of a plane wave component ϑ is set in accordance with the spatial frequency $K' = \beta_s sin\vartheta$ and is correspondingly Doppler shifted by $\omega(K' - \beta_r sin\vartheta_r)$. For convenience of notation we will in fact use spatial frequency $K = K' - \beta_r sin\vartheta_r$.

The relations (8a,b) are constructed in such a way that the two wavevectors $\vec{\beta}_s(K)$ and $\vec{\beta}_{s'}(K)$ (optical modes with different numbers s and s' but corresponding to the same spatial frequency K) differ by a vector parallel to the x_3 axis, i.e.

$$\vec{\beta}_s(K) - \vec{\beta}_{s'}(K) = \Delta\beta\vec{e}^3.$$

This means that the modes with indices (discrete and continuous) s,K and s',K are orthogonal in the sense of (4) and $\vec{e}^\Delta = \vec{e}^3$. Because the orthogonality does not extend to the components of the fields $\vec{E}^e_s(K)$; $\vec{H}^e_s(K)$ parallel to x_3, strictly speaking, additional vector functions $\tilde{\vec{Ee}}^3$, $\tilde{\vec{He}}^3$ must be introduced in (8), by analogy with the procedure of Ref. 39.

The series (8) are constructed from functions which are not eigenfunctions of the differential operator describing light propagation in an optical waveguide perturbed by spin waves. This representation reduces the diffraction problem to a system of first-order integro-differential equations for the amplitudes of the expansions in Eq. (8). Substituting Eq. (8) into the perturbed Maxwell's equations (7) and taking account of the orthogonality relation (4), we finally obtain the system of coupled wave equations, which describes the diffraction of a light wave by a spin wave with an arbitrary spectrum and which will be referred to as the generalized coupled wave equations:

$$\frac{dC_s(K)}{dx_3} = -i \sum_{s'=0}^{2N-1} \int_{-\infty}^{\infty} (W^{s,s'}(K_0)\, \tilde{m}^e(K-K_0-K') + W^{s,s'}(-K_0)\, \tilde{m}^{e*}(-K-K_0-K'))$$

$$C_{s'}(K')\, exp(i\, \Delta\beta^{s,s'}(K,K')\, x_3)\, dK', \qquad (10a)$$

$$W^{s,s'}(\pm K_0) = \omega_{opt}\, \varepsilon_0 \int_0^h s(x_1)\, (\vec{E}^*_s(x_1),\, \hat{\Delta\varepsilon}(\pm K_0)\, \vec{E}_{s'}(x_1))\, dx_1, \qquad (10b)$$

$$\Delta\beta^{s,s'}(K,K') = ((\vec{\beta}_s(K)-\vec{\beta}_{s'}(K')),\, \vec{e}^3). \qquad (10c)$$

The basis for the interaction of light and spin waves is the perturbation of the dielectric constant of the medium due to magnetooptical effects: the linear (in the magnetic field) Faraday effect (magnetic optical activity) and the quadratic Cotton-Mouton effect (magnetic birefringence). Our analysis will apply to single-crystal ferrite-garnet ferromagnetic films (e.g., yttrium iron garnet (YIG) $Y_3Fe_5O_{12}$ or bismuth substituted garnet $(BiLu)_3Fe_5O_{12}$) grown on nonmagnetic optically isotropic dielectric substrates (gadolinium -gallium garnet). This structure is the one most often used in waveguide magnetooptics and in spin-wave technology. For these two garnet crystals, as well as for other ferrite-garnets belonging to the cubic system, the tensors describing magnetooptical effects are relatively simple in form. For the Faraday effect

$$\hat{\Delta\varepsilon}^{F}_{ij} = if\ \delta_{ijk} M_k. \tag{11a}$$

For the Cotton-Mouton effect

$$\hat{\Delta\varepsilon}^{CM}_{ij} = g_{12} M^2 \delta_{ij} + 2g_{44} M_i M_j + (g_{11} - g_{12} - 2g_{44})\, M^2\, \hat{\nu}_{ij},$$

$$M_i = Me^M_i, \quad \hat{\nu} = \begin{vmatrix} (e^M_1)^2 & 0 & 0 \\ 0 & (e^M_2)^2 & 0 \\ 0 & 0 & (e^M_3)^2 \end{vmatrix}. \tag{11b}$$

In Eq. (11a), corresponding to the Faraday effect, δ_{ijk} is the Levi-Cevita unit antisymmetric tensor of the third rank; M_k is the magnetization vector characterizing the total magnetic field in the film; f is a scalar constant. The representation of the tensor $\hat{\Delta\varepsilon}^{CM}_{ij}$ describing the Cotton-Mouton effect in the form (11b) is constructed according to the method suggested by Fedorov for the Christoffel tensor in crystal acoustics.[40] In Eq. (11b) the first two terms characterize the magnetooptical properties of an isotropic ferromagnetic medium, while the third term gives a correction for the anisotropy of the cubic crystal. The first term does not contribute to the magnetooptical birefringence and it involves the unit isotropic tensor of the second rank δ_{ij}. The magnetooptical anisotropy parameter for yttrium-iron garnet is[41]

$$a = \frac{2g_{44}}{g_{11} - g_{12}} \approx 1{,}335.$$

Hence the second term is approximately four times larger than the third, i.e. in the first approximation the dyad $2g_{44} M_i M_j$ determines the magnetooptical birefringence in yttrium iron garnet. The representation of the magnetooptical birefringence in the form of the dyad $2g_{44}M_iM_j$ will be called the dyadic approximation. The indices on g_{11}, g_{12}, g_{44} correspond to the abbreviated six-dimensional form of the representation of three-dimensional tensors of the second rank which are symmetric in two pairs of indices;[40] the tensor ν_{ij} is written in crystallographic coordinates and e_i^M denotes the direction cosines of the total magnetization vector. Therefore the complete correction to the tensor $\hat{\varepsilon}$ from both magnetooptical effects can, in the first approximation, be written in a compact form which is independent of the coordinate system

$$\Delta\hat{\varepsilon} = \Delta\hat{\varepsilon}^F + \Delta\hat{\varepsilon}^{CM} \approx -if\hat{M}^{\times} + g_{12}(M)^2\,\hat{1} + 2g_{44}\vec{M}\cdot\vec{M}. \tag{12}$$

In Eq. (12) $\hat{M}^{\times}$ is an antisymmetric tensor of the second rank dual to the vector $\vec{M}$;[40] $\hat{1}$, $\vec{M}\cdot\vec{M}$ are coordinate-independent forms of the second-rank unit tensor and the outer product of two vectors $\vec{M}$.

In correspondence with the previous equation the bilinear form $A = (\vec{E}^*_{s'}, \Delta\hat{\varepsilon}\,\vec{E}_s)$, determining the strength of interaction in Eq. (10), can be expressed simply in terms of scalar and mixed products of the vectors appearing in this form

$$A = if\,(\vec{E}^*_s, [\vec{E}_{s'}, \vec{M}]) + g_{12}(M)^2(\vec{E}^*_s, \vec{E}_{s'}) + 2g_{44}\,(\vec{E}^*_s, \vec{M})\,(\vec{E}_{s'}, \vec{M}).$$

The forms this expression assumes for specific geometries will be considered below.

3. General Solution

Equation (10) describes the propagation of an optical mode in a dielectric waveguide whose dielectric constant is disturbed by a spin wave of arbitrary form. As a matter of fact, it is a generalization of a conventional system of difference-differential equations of coupled modes in the theory of sinusoidal phase diffraction gratings, derived for the first time in acoustooptics by Raman and Nath[42,43] and then extensively used in optics and optoelectronics.[25-27,38]

We construct the solution of equation (10) with the help of the perturbation method. In practice, due to the Bragg selectivity, an effective magnetooptical interaction usually occurs only between two optical waveguide modes which means that the summation over the index s in Eq. (8) is no longer necessary. The strength of the interaction of the two optical modes s and s' directly depends on the magnitude of the coupling coefficient $W^{s,s'}(\pm K_0)$ whose physical meaning is the variation of the wavenumber of an optical mode due to the presence of the spin wave.

After replacing Eq. (10) with its Laplace transform:

$$(p-i\gamma(k))\ S_s(K,p-i\gamma(K)) - \delta(K) = i[S_{s'}(K,p-i\gamma(K)) \otimes \tilde{u}(K)],$$

we substitute in it the perturbation series

$$S_s(K,p) = \sum_{m=0}^{\infty} S_s^m(K,p).$$

Here $S_s(K,p) = \tilde{L}(C_s(K,x_3))$ is the Laplace transform of $C_s(K,x_3)$ and the second term on the left hand side is due to the boundary condition $C_s(K,0) = \delta(K)$ (for simplicity of notation the amplitude of the incident optical wave is taken equal to unity); the operator symbol $\otimes$ denotes a convolution integral of the type that occurs in Eq. (10) and $\tilde{u}(K) = W^{s,s'}(K_0)\ \tilde{m}^e(K-K_0) + W^{s,s'}(-K_0)\ \tilde{m}^{e*}(-K-K_0)$.

The quantity $\gamma(K) = \beta_r \cos\vartheta_r - \beta_3(K)$ is an important parameter of the interaction which describes the phase mismatch between the spatial (angular) harmonic K and the incident optical wave (zero spatial harmonic $K = 0$).

The small parameter in the expansion for $S_s(K,p)$ is $W^{s,s'}(\pm K_0)/\beta_s$ or, in other words, the ratio of the perturbation of the optical wavenumber due to the presence of the spin wave to its unperturbed value. The recursion relation

$$S'^m(K,p) = \frac{-i}{p - i\gamma(K)}(S'^{m-1}(K,p) \otimes \tilde{u}(K)),$$

where $S'^m(K,p) = S(K,p-i\gamma(K))$, allows us to obtain the expressions for the specific terms of the perturbation series

$$S'^{0}(K,p) = \frac{\delta(K)}{p}, \tag{13a}$$

$$S'^{1}(K,p) = \frac{-i}{p(p - i\gamma(K))}\,\tilde{u}(K), \tag{13b}$$

$$S'^{2}(K,p) = \frac{-1}{p(p - i\gamma(K))}\int_{-\infty}^{\infty} \frac{\tilde{u}(K_1)\,\tilde{u}(K-K_1)}{p - i\gamma(K-K_1)}\,dK_1, \tag{13c}$$

$$S'^{3}(K,p) = \frac{i}{p(p-i\gamma(K))}\int_{-\infty}^{\infty}\int_{-\infty}^{\infty} \frac{\tilde{u}(K_1)\,\tilde{u}(K_2)\tilde{u}(K-K_1-K_2)}{(p-i\gamma(K-K_1))(p-i\gamma(K-K_1-K_2))}\,dK_1 dK_2, \tag{13d}$$

and for the m-th approximation

$$S'^{m}(K,p) = \frac{i^{m}}{p(p-i\gamma(K))}\int_{-\infty}^{\infty}\!\!\cdots\int_{-\infty}^{\infty} \frac{\prod_{i=1}^{m-1}\tilde{u}(K_i)\tilde{u}(K-\sum_{j=1}^{m-1}K_j)}{\prod_{i=1}^{m-1}(p-i\gamma(K-\sum_{j=1}^{i}K_j))}\,dK_1\ldots dK_{m-1}. \tag{13e}$$

The transition from region p to region x_3 by means of the inverse Laplace transform and the residue theorem does not involve any serious difficulties. As a result, for example, the first- and the third-order approximations of the perturbation series are finally obtained in the form

$$C'^{1}(K,x_3) = \frac{\tilde{u}(K)}{\Gamma 1}\,(1 - exp(i\Gamma 1\; x_3)) = -ix_3\,\tilde{u}(K)\;exp(i\Gamma 1\frac{x_3}{2})\;sinc(\Gamma 1\frac{x_3}{2}), \tag{14a}$$

$$C'^{3}(K,x_3) = \int_{-\infty}^{\infty}\int_{-\infty}^{\infty}\left(\frac{1}{\Gamma 1\;\Gamma 2\;\Gamma 3} - \frac{exp(i\Gamma 1x_3)}{\Gamma 1\;\Gamma 12\;\Gamma 13} + \frac{exp(i\Gamma 2x_3)}{\Gamma 2\;\Gamma 12\;\Gamma 23} - \frac{exp(i\Gamma 3x_3)}{\Gamma 3\;\Gamma 13\;\Gamma 23}\right)\tilde{u}(K_1)\tilde{u}(K_2)\tilde{u}(K-K_1-K_2)\;dK_1\;dK_2. \tag{14b}$$

The following notation has been used in Eq. (14)

$$\Gamma 1 = \gamma(K); \qquad \Gamma 2 = \gamma(K-K_1); \qquad \Gamma 3 = \gamma(K-K_1-K_2);$$

$$\Gamma 12 = \Gamma 1-\Gamma 2; \qquad \Gamma 13 = \Gamma 1-\Gamma 3; \qquad \Gamma 23 = \Gamma 2-\Gamma 3;$$

$$C'^{m}(K,x_3) = C^{m}(K,x_3)\ exp(i\gamma(K)x_3)).$$

For brevity of the notation we omitted the subscript "s" designating the number of the optical mode involved in the interaction.

The conditions $K = K_1$; $K_1 = -K_2$ correspond to the appearance in Eq. (13d) of second-order poles, which are not taken into account by Eq. (14b). In the numerical calculations special equations must be constructed for these points or they should be bypassed.

4. Diffraction by One Spin-Wave Harmonic

In the case of a purely sinusoidal spin wave with the wavenumber K_0 the perturbation spectrum is

$$\tilde{m}(K) = \frac{m}{2}\,[(\delta(K - K_0) + \delta(K + K_0)]. \tag{15}$$

The substitution of Eq. (15) into Eq. (13) or Eq. (14) indicates that the solution obtained is meaningful only for a discrete set of values of the argument $K_n = nK_0$ (n = 0, ±1, ±2, ...) which corresponds to conventional diffraction orders in the angular spectrum of the light wave scattered by a periodic diffraction grating. For these K_n the generalized Eq. (10) reduces to a traditional system of difference-differential equations of coupled modes. Consequently, the integrals describing the structure of the diffracted field in Eq. (13) and Eq. (14) reduce to finite sums.

As it was for the first time shown by Berry[22] for the acoustooptical case, the mechanism for the diffraction of light by a sinusoidal phase grating can be interpreted in terms of the Feynman-type multiple scattering diagrams. Then the formation of the diffracted wave is considered as a set of transitions between the allowed levels $K_n = nK_0$. Within the framework of this theoretical approach the summation in the aforementioned sums goes over all the transitions from the zero level (n = 0 or the incident light wave) to a given n-th level. Extensive calculations of such transitions, or the partial waves in terms of physical optics, have been carried out by Martynov and Mirer.[44] The approach was generalized to the case of diffraction by a multifrequency grating, comprised of N spatial harmonics in Refs.[31,32].

A multiple scattering diagram for the case $N = 1$ is given in Fig. 1a. The order m of the perturbation theory which amounts to the multiplicity of the scattering process is plotted along the horizontal axis of the diagram, and the system of levels along the vertical axis corresponds to the space-time harmonics or, which is equivalent, diffraction orders. As it is seen from Fig. 1a, only partial waves of odd multiplicity $2m + 1$ contribute to the first diffraction order (the first allowed level). Being of most practical interest, the first diffraction orders will become a point of particular interest in further discussion. For a regime of weak diffraction it is sufficient to take into account only a single transition $\hat{\hat{01}}$. Because the amplitude of the diffracted light wave in such a case is linear in the amplitude of the perturbation, this regime is called linear. In such mode of diffraction the transfer of energy from the zero order to the first one does not exceed a few percent.

In place of the coefficient W it is expedient to introduce the dimensionless quantity $v = WL$ (L is the interaction length of the magnetooptical interaction), which is called a phase-modulation index. In this sense, the criterion for the linear regime of diffraction can be formulated as

$$v < 1. \tag{16}$$

The linear diffraction/weak interaction approximation holds for a wide class of optical information processing devices such as spectrum analyzers, correlators, and convolvers. Unless the inequality (16) is broken the amplitude in the first diffraction order is described by a simple expression

$$C_1^1 = C_1 = \frac{-i}{2} v \exp\left(-\frac{i\gamma(K_0)L}{2}\right) \operatorname{sinc}\left(\frac{\gamma(Ko)L}{2}\right). \tag{17}$$

Eq. (17) is very convenient for estimating the frequency properties of magnetooptical interactions of optical waveguide modes and spin waves in information processing devices. Moreover, the weak interaction regime is characteristic of most experimental studies carried out so far.

When the phase-modulation index exceeds the level determined by inequality (16) we must move further upward the perturbation series which means that the contributions of the threefold transitions to the amplitude of the first diffraction order can not be any longer neglected. These are the three transitions $\hat{0}\hat{1}\hat{0}\hat{1}$, $\hat{0}\hat{1}\hat{2}\hat{1}$, and $\hat{0}\,\widehat{-1}\,\hat{0}\hat{1}$. Their contributions are described by the following expressions written in region p

$$S'^{3}_{1}(\hat{0}\hat{1}\hat{2}\hat{1}) = i\, v^{+}v^{+}v^{-}\, \frac{1}{(2L)^{3}}\, \frac{1}{p}\, \frac{1}{p-i\gamma(K_0)}\, \frac{1}{p-i\gamma(2K_0)}\, \frac{1}{p-i\gamma(K_0)}, \qquad (18a)$$

$$S'^{3}_{1}(\hat{0}\hat{1}\hat{0}\hat{1}) = i\, v^{+}v^{-}v^{+}\, \frac{1}{(2L)^{3}}\, \frac{1}{p}\, \frac{1}{p-i\gamma(K_0)}\, \frac{1}{p}\, \frac{1}{p-i\gamma(K_0)}, \qquad (18b)$$

$$S'^{3}_{1}(\hat{0}\,\widehat{-1}\,\hat{0}\hat{1}) = i\, v^{-}v^{+}v^{+}\, \frac{1}{(2L)^{3}}\, \frac{1}{p}\, \frac{1}{p-i\gamma(-K_0)}\, \frac{1}{p}\, \frac{1}{p-i\gamma(K_0)}, \qquad (18c)$$

and in region x_3

$$C'^{3}_{1}(\hat{0}\hat{1}\hat{2}\hat{1}) = i\, v^{+}v^{+}v^{-}\, \frac{1}{(2L)^{3}} \left(\frac{-i}{\Gamma_2(\Gamma_1)^2} + \frac{i\, exp(i\Gamma_2 x_3)}{\Gamma_2(\Gamma_{12})^2} - exp(i\Gamma_1 x_3) \left(\frac{x_3}{\Gamma_1\Gamma_{12}} + \frac{i}{(\Gamma_1)^2\Gamma_{12}} + \frac{i}{(\Gamma_{12})^2\Gamma_1} \right) \right), \qquad (19a)$$

$$C'^{3}_{1}(\hat{0}\hat{1}\hat{0}\hat{1}) = i\, v^{+}v^{-}v^{+}\, \frac{1}{(2L)^{3}} \left(-\frac{x_3}{(\Gamma_1)^2}\,(1 + exp(i\Gamma_1 x_3)) + \frac{2i}{(\Gamma_1)^3}\,(1 - exp(i\Gamma_1 x_3)) \right), \qquad (19b)$$

$$C'^{3}_{1}(\hat{0}\hat{-1}\hat{0}\hat{1}) = i\, v^{-}v^{+}v^{+}\,\frac{1}{(2L)^{3}}\left(\frac{-x_{3}}{\Gamma_{1}\Gamma_{-1}} + \frac{i}{(\Gamma_{1})^{2}\Gamma_{-1}} + \frac{i}{(\Gamma_{-1})^{2}\Gamma_{1}} + \frac{i\,exp(i\Gamma_{1}x_{3})}{(\Gamma_{1})^{2}\Gamma_{1-1}} - \frac{i\,exp(i\Gamma_{-1}x_{3})}{(\Gamma_{-1})^{2}\Gamma_{1-1}}\right), \tag{19c}$$

$$\Gamma_{1} = \gamma(K_{0}); \qquad \Gamma_{2} = \gamma(2K_{0}); \qquad \Gamma_{-1} = \gamma(-K_{0}); \tag{19d}$$
$$\Gamma_{12} = \Gamma_{1} - \Gamma_{2}; \qquad \Gamma_{1-1} = \Gamma_{1} - \Gamma_{-1}.$$

The expressions (18) are constructed according to a simple and obvious principle. The sequence of factors in the denominator describes

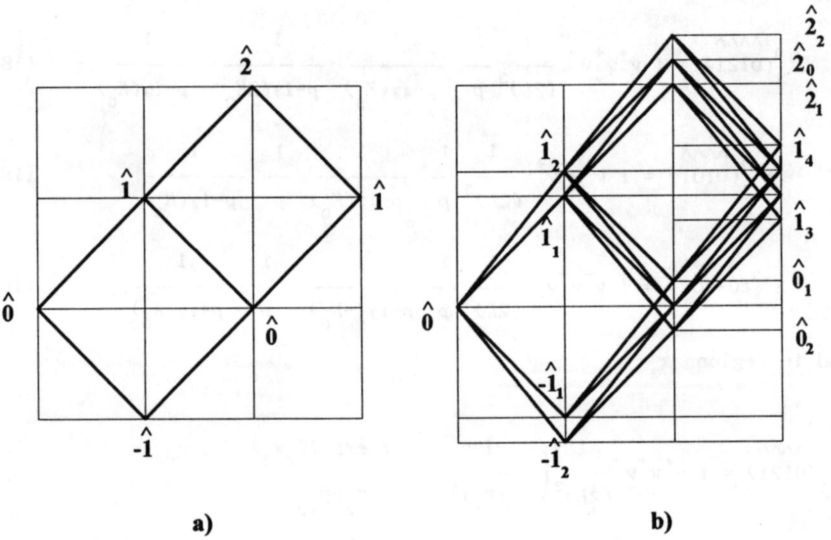

Fig.1. Multiple scattering diagrams for diffraction of light a) by one spin-wave harmonic; b) by two spin-wave harmonics.

the sequence of elementary transitions in a corresponding diagram in Fig. 1a, while the right hand side term of the type $\gamma(nK_0)$ in each factor is equal to the phase mismatch of the optical wave in the nth order (nth allowed level) with respect to the incident light wave (0th level).

Since the amplitudes of partial waves in Eq. (19) are cubic in the phase-modulation index v, and therefore the spin wave harmonic amplitude, the corresponding threefold transitions are referred to as nonlinear. The superscripts "+" and "-" are ascribed to the phase-modulation index v to distinguish between the elementary transitions accompanied by an increase in frequency $v^{+}=v(W(K_0))$ and the elementary transitions accompanied by a decrease in frequency $v^{-}=v(W(-K_0))$ (Stokes and anti-Stokes transitions). It must be noted here that, in contrast with acoustooptics[38], in spin wave magnetooptics $v(W(K_0)) \neq v(W(-K_0))$.

In addition, to make the parallel between the structure of Eq. (19) and that of the system of transitions in Fig. 1a even more graphic the Eq. (19) is rewritten in such a form that the sequence of scalar factors v^{+} and v^{-} repeats the upward and downward "jumps" along a corresponding trajectory in Fig. 1a.

5. Diffraction by Two Spin-Wave Harmonics

In the case of diffraction of a plane wave by two spin-wave harmonics K_{01} and $K_{02}+\Delta K$ the system of transitions gets far more complicated. This is illustrated in Fig. 1b. Along with the fundamental levels of the form $K_n=nK_{01}$, $K_m=mK_{02}$, there are sublevels due to alternate scattering by both harmonics $K_{nm}=nK_{01}+mK_{02}$.

The single linear transitions contribute to both first-order fundamental levels K_{01} and K_{02}. The nonlinear threefold transitions

-transfer energy to fundamental levels through scattering by one of the spin-wave harmonics: the transitions $\hat{0}\hat{1}_1\hat{0}\hat{1}_1$, $\hat{0}\hat{1}_1\hat{2}_1\hat{1}_1$, $\hat{0}\text{-}\hat{1}_1\hat{0}\hat{1}_1$ for the K_{01} level and the transitions $\hat{0}\hat{1}_2\hat{0}\hat{1}_2$, $\hat{0}\hat{1}_2\hat{2}_2\hat{1}_2$, $\hat{0}\text{-}\hat{1}_2\hat{0}\hat{1}_2$ for the level K_{02};

-transfer energy to fundamental levels through alternate scattering by the two harmonics K_{01} and K_{02}, for example, the transition $\hat{0}\hat{1}_2\hat{2}_0\hat{1}_1$ for the K_{01} level and the transition $\hat{0}\text{-}\hat{1}_1\hat{0}\hat{1}_2$ for the K_{02}

level;

-transfer energy to the sublevels $K_{01}-\Delta K$ and $K_{02}+\Delta K$ due to alternate scattering by both harmonics: for example, the transition $\hat{0}\hat{1}_{1}\hat{2}_{1}\hat{1}_{3}$ for the sublevel $K_{01}-\Delta K$ and the transition $\hat{0}\hat{1}_{2}\hat{2}_{2}\hat{1}_{4}$ for the sublevel $K_{02}+\Delta K$.

The aforementioned transitions are described, like in the case of one spin wave harmonic, by one of the expressions (19). The only exception lies in the fact that there can be transitions in which no level is repeated twice, for example $\hat{0}\hat{1}_{2}\hat{2}_{0}\hat{1}_{1}$. This means that all three poles are of the first order and therefore the corresponding contribution is determined by the expression

$$C'^{3}_{1}(\hat{0}\hat{1}_{2}\hat{2}_{0}\hat{1}_{1}) = i\, v^{+}v^{+}v^{-}\,\frac{1}{(2L)^{3}}\left(\frac{1}{\Gamma_{1}\Gamma_{2}\Gamma_{3}} - \frac{exp(i\Gamma_{1}x_{3})}{\Gamma_{1}\Gamma_{12}\Gamma_{13}} + \frac{exp(i\Gamma_{2}x_{3})}{\Gamma_{2}\Gamma_{12}\Gamma_{23}} - \frac{exp(i\Gamma_{3}x_{3})}{\Gamma_{3}\Gamma_{13}\Gamma_{23}}\right), \tag{19e}$$

$$\Gamma_{1} = \gamma(K_{02}), \quad \Gamma_{2} = \gamma(K_{01} + K_{02}), \quad \Gamma_{3} = \gamma(K_{01}),$$

$$\Gamma_{12} = \Gamma_{1} - \Gamma_{2}, \quad \Gamma_{13} = \Gamma_{1} - \Gamma_{3}, \quad \Gamma_{23} = \Gamma_{2} - \Gamma_{3}.$$

With the diffraction of light by N spin wave harmonics the system of levels and transitions in a Feynman diagram gets even more complicated. At the same time, the fundamental relationships characterizing the previous cases are retained and the corresponding partial amplitudes are described by one of the Eq. (19).

Thus, in this section we considered the general structure of expressions, which describe the magnetooptical diffraction of an optical waveguide mode by a spin wave with a complex spectrum as a multiple scattering process. However, in specific situations occurring in practice some of the transitions are "prohibited" due to the Bragg selectivity and the system of transitions is substantially simplified. In the next section we will dwell on another parameter of the interaction: the phase-modulation index v.

6. Phase-Modulation Index for Specific Geometries

The coupling between two optical modes depends on

-geometric factor which includes a) the orientation of the bias constant saturating magnetic field with respect to the plane of the film and the direction of spin wave propagation, i.e. the type of the spin wave propagating in the film and b) mutual orientation of wavevectors of the interacting waves and therefore the shape of the corresponding vector triangle in the Fourier space;

-polarization factor which is determined by the polarization of the incident and diffracted optical waves.

In accordance with the first principle of classification, we will consider the peculiarities of the scattering mechanism for the diffraction of light by the three fundamental types of spin waves: forward volume spin waves in a normally magnetized film and backward volume and surface spin waves in a tangentially magnetized magnetic film, the angles of incidence and diffraction being taken into account. Within each such regime of diffraction we will distinguish between two distinct types of diffraction process: anisotropic and isotropic. In the anisotropic case the incident and the diffracted optical waves are of orthogonal polarization while in the isotropic case, on the contrary, these waves are of identical polarization.

6.1. Diffraction in a normally magnetized film.

The vector of total magnetization $\vec{M}$ consists of a constant component $\vec{M}^0$ and a component variable in time and space due to the magnetic field of the spin wave and represented by the complex amplitude $\vec{m}$. Then

$$a)\ (\vec{m}, \vec{M}^0) = 0, \quad b)\ |\vec{m}| \ll |\vec{M}^0|. \tag{20}$$

In the coordinate system adopted here

$$\vec{M} = \begin{vmatrix} M^0 \\ \pm imb \\ m \end{vmatrix} \tag{21}$$

where $b < 1$ is the ellipticity of the spin wave polarization.

Anisotropic diffraction TE⇒TM. The fields of the incident and the diffracted wave have the form

$$\vec{E}^{TE} = \begin{vmatrix} 0 \\ E^y \cos\vartheta_0 \\ -E^y \sin\vartheta_0 \end{vmatrix}, \quad \vec{E}^{TM} = \begin{vmatrix} E^x \\ E^z \sin\vartheta_1 \\ E^z \cos\vartheta_1 \end{vmatrix}, \tag{22}$$

where ϑ_0 is the angle of incidence of the TE optical mode and ϑ_1 is the angle of diffraction of the TM optical mode. Substitution of Eq. (21) and Eq. (22) into Eq. (10b) for the coefficient W gives the following relation for the phase-modulation index v:

$$v^{\alpha} = \beta L\ (\Delta n_{\alpha}^{yx} I^{yx} - \Delta n_{\alpha}^{yz} I^{yz}),$$

$$\Delta n_{\alpha}^{yx} = \frac{\Delta\varepsilon_{\alpha}^{yx}}{2N^{TE}}, \quad \Delta n_{\alpha}^{yz} = \frac{\Delta\varepsilon_{\alpha}^{yz}}{2N^{TE}},$$

$$\Delta\varepsilon_{\alpha}^{yx} = (\vec{e}^{y}, \hat{\Delta\varepsilon}^{\alpha}\, \vec{e}^{x}), \quad \Delta\varepsilon_{\alpha}^{yz} = (\vec{e}^{y}, \hat{\Delta\varepsilon}^{\alpha}\, \vec{e}^{z}). \tag{23}$$

In Eq. (23) $\beta = 2\pi/\lambda$ is the wavenumber in vacuum; $N^{TE} = \beta^{TE}/\beta$ is the effective index of refraction of the incident wave (in the case considered here this is the TE mode); the binary index α assumes the values "+" for the diffraction process accompanied by an increase in the optical frequency and "-" for those accompanied by a decrease in frequency. The bilinear forms $\Delta\varepsilon_{\alpha}^{yx}$ and $\Delta\varepsilon_{\alpha}^{yz}$ are scalars and give the effective perturbation of $\hat{\varepsilon}$ tensor for this diffraction geometry. Then Δn_{α}^{yx} and Δn_{α}^{yz} can be considered as effective perturbations of the index of refraction.

The overlap integrals I^{yx} and I^{yz} give the degree of correlation of the spatial distribution (along x_1) of the fields of the interacting wave processes. Figure 2 shows (in logarithmic scale) the dependence of these quantities on the film thickness h. The decibel scale $I(\text{dB}) = 20\ lg|I|$ is used on the ordinates of Fig. 2.

According to Kalinikos[37] the distribution of magnetization in a dipole-exchange spin wave can be represented as an infinite superposition of spin-wave resonance (SWR) modes

$$\vec{m}(x_1, K) = \sum_{n=0}^{\infty} m^n \, S^n(x_1) \, \vec{e}^n . \tag{24}$$

Scalar functions $S^n(x_1)$ represent a full set of mutually orthogonal

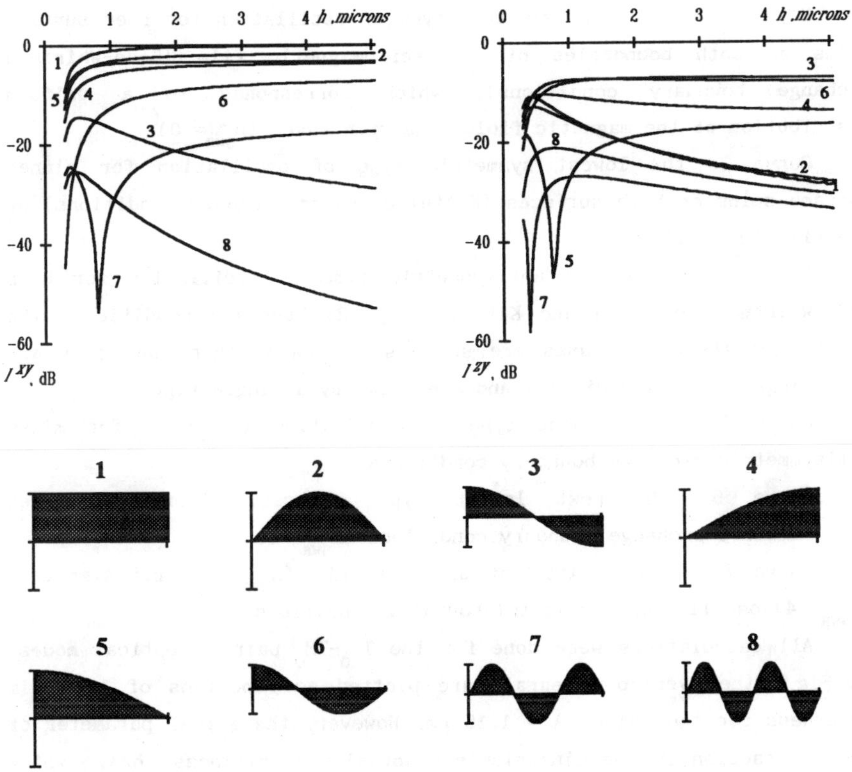

Fig.2. Dependences of the overlap integrals on the film thickness.

sine/cosine functions. Their form depends on the exchange boundary conditions on the two surfaces of the film. Vectors $\vec{e}^n$ characterize the ellipticity of each SWR mode.

Eq.(21) is written in the so called diagonal approximation.[37] It holds for a practically important situation when the spins on both film surfaces are totally unpinned. Then the series reduces to the lowest mode, which amounts to the case of a non-exchange or magnetostatic wave.

The curves in Fig.2 correspond to the following symmetry types of the spin-wave magnetization distribution with film thickness:

Curve 1: the lowest symmetric type of oscillation for free surface spins on both boundaries of the ferromagnetic film (Rado-Whirtman exchange boundary conditions), which corresponds to a uniform distribution of the magnetic field of a spin wave ($n_{SWR} = 0$).

Curve 2: the lowest symmetric type of oscillation for pinned surface spins on both surfaces (Kittel exchange boundary conditions) of the film ($n_{SWR} = 1$).

Curve 3: the lowest antisymmetric type of oscillation for both Rado-Whirtman ($n_{SWR} = 1$) and Kittel ($n_{SWR} = 2$) boundary conditions; the curves for these two cases are so close to each other that they are indistinguishable graphically and are shown by a single figure.

Curve 4,5: the lowest type of oscillation ($n_{SWR} = 1$) for mixed antisymmetric exchange boundary conditions

Curve 6: the next lowest type of oscillation for mixed antisymmetric exchange boundary conditions ($n_{SWR} = 3$).

Curve 7,8: higher type of antisymmetric ($n_{SWR} = 3$) and symmetric ($n_{SWR} = 4$) oscillation for Kittel boundary conditions.

All calculations were done for the TE_0-TM_0 pair of optical modes. In Fig.2 the overlap integrals are plotted as functions of the film thickness for the value $\lambda = 1.15$ μm. However, the actual parameter of the interaction is the dimensionless normalized thickness h/λ, which means that the results can be converted to apply to a different wavelength of light by simply renormalizing the abscissa.

The results shown in Fig.2 essentially reflect the symmetry of the integrand in Eq.(10b). The behavior of curve 1 in Fig.2a is typical in this sense. It corresponds to the interaction of the lowest pair of optical modes TE_0 and TM_0 in the case when the spin-wave magnetization is uniform with film thickness, i.e., only the symmetry of the optical modes affects the quantities I. In the case of I^{yx} both of the distributions $E^x(x_1)$ and $E^y(x_1)$ are "almost" symmetric and the magnitude of the overlap integral I^{yx} is almost equal to unity in the interval of h from 1 to 5 μm. When the thickness h decreases to submicron levels there is a redistribution of light inside the layered structure: the field of the optical waveguide mode is expelled into the substrate to a greater extent, the smaller the thickness h. Hence the power of the light carried within the film decreases and the symmetry of the optical field distribution is broken. As a result, there is observed sharp decrease in I^{yx}. The zero TM mode is cut off at $h \approx$ 0.3 μm and I^{yx} is no longer a physically meaningful quantity.

The dependence $I^{zy}(h)$ (see curve 1 in Fig.2b) is completely different. In this case the optical field $E^y(x_1)$ is nearly symmetric, while the second field $E^z(x_1)$ is nearly asymmetric. Therefore $I^{yz} \ll I^{yx}$. In addition, the longitudinal component E^z depends strongly on thickness h. It increases rapidly with decreasing h and therefore I^{yz} also rapidly increases. For submicron thicknesses redistribution processes of the optical field in the layered structure begin to show up near the cutoff and a break is observed in the function $I^{yz}(h)$. Similar behavior is demonstrated by curve 3 in Fig.2a for I^{yx} in the case of "purely" asymmetric SWR mode. As in the previous situation, the product of two symmetric optical distributions and an asymmetric magnetic one gives an asymmetric integrand.

Curves 4 and 5 in Fig.2a demonstrate an interesting behavior. They are calculated for the case of antisymmetric exchange boundary conditions (the lowest mode). Their behavior at submicron thicknesses significantly depends on whether the magnetic field of the spin wave presses against the substrate or the covering layer. This is explained by the asymmetry of the optical properties of the wave-carrying structure: as the cutoff point is approached the field of the mode presses against the boundary between the film and the substrate and is "pulled back" from the boundary between the film and the covering layer. Similar considerations explain pronounced dips on curves 5 and 7 in

Fig. 2b.

It should be noted in the end that, on the one hand, the values of both integrals fall sharply with the increase of SWR mode number, on the other hand, the introduction of antisymmetric boundary conditions may, in some cases, noticeably increase the value of an overlap integral (e.g., compare curves 6 and 3 in Fig. 2a).

We consider now the structure of the tensor correction $\hat{\Delta\varepsilon}^{\alpha}$ to the dielectric constant due to the perturbing influence of the spin wave. The quantity $\hat{\Delta\varepsilon}^{\alpha}$ in the approximation of Eq. (12), with the use of Eq. (20) and also the fact that the magnitude of the total magnetization M is constant in time and space, is given by

$$\hat{\Delta\varepsilon}^{\alpha} = if\hat{m}^{\times} + 2\,g_{44}(\vec{M}^{0}\cdot\vec{m} + \vec{m}\cdot\vec{M}^{0}), \tag{25}$$

which leads to the following equation for the phase modulation index

$$v^{iso} = icm(f(\cos\vartheta_{0} \pm ib\sin\vartheta_{0}) \pm 2g_{44}M^{0}(b\cos\vartheta_{0} \pm i\sin\vartheta_{0}))\, I^{yx},$$

$$c = \frac{\beta_{0}L}{2N^{TE}} \tag{26}$$

The ellipticity of the polarization of volume spin waves is nearly unity, therefore we can write

$$v^{iso} = icm\, \exp(\pm i\vartheta_{0})(f \pm 2g_{44}M^{0})\, I^{yx}. \tag{27}$$

For yttrium-iron garnet we have, using the data of[41,45] $fM^{0} = 4.0\cdot10^{-4}$, $2g_{44}(M^{0})^{2} = 2.3\cdot10^{-4}$, i.e., the difference between the intensities of diffraction into the positive and negative first orders is about 10 dB. This type of diffraction asymmetry was observed in[10,11] in the course of a waveguide noncollinear interaction ($\vartheta_{0} \approx 0$) and earlier[4] in bulk samples. Since the forward volume spin wave ellipticity is nearly unity, the same degree of asymmetry must occur for the collinear geometry ($\vartheta_{0}=\pi/2$) that is the distinguishing feature of the waveguide interaction of light and spin waves in a thin ferromagnetic film. However, experimentally this regime has not been realized so far mainly because of the difficulties involved in one-directional excitation of a volume wave.

The above results were obtained assuming isotropic magnetooptical effects in ferrite. We now take into account the contribution of the final term in (11b), which is due to the anisotropy of the cubic crystal. The corresponding correction to the tensor $\Delta\hat{\varepsilon}^{\alpha}$ has the form

$$\Delta\hat{\varepsilon}_{\alpha}^{ani} = 2mM^{0}\,(g_{11} - g_{12} - 2g_{44})(\hat{\nu}' \pm ib\hat{\nu}''), \tag{28}$$

$$\hat{\nu}' = \begin{vmatrix} e_1^0 e_1^{t1} & & 0 \\ 0 & e_2^0 e_2^{t1} & 0 \\ 0 & 0 & e_3^0 e_3^{t1} \end{vmatrix}, \quad \hat{\nu}'' = \begin{vmatrix} e_1^0 e_1^{t2} & & 0 \\ 0 & e_2^0 e_2^{t2} & 0 \\ 0 & 0 & e_3^0 e_3^{t2} \end{vmatrix} \tag{29}$$

where e_i^0, e_i^{t1}, e_i^{t1} are the direction cosines in crystallographic coordinates of the direction of the constant magnetization and two orthogonal directions in the plane perpendicular to the constant magnetization. The magnitude of the anisotropic contribution v^{ani} to the total phase-modulation index $v = v^{iso} + v^{ani}$ depends on the crystallographic direction in which the film was grown.

a. Direction [111]. In this case

$$\hat{\nu}' = \begin{vmatrix} \frac{1}{3\sqrt{2}} & 0 & 0 \\ 0 & \frac{1}{3\sqrt{2}} & 0 \\ 0 & 0 & -\frac{2}{3\sqrt{2}} \end{vmatrix}, \quad \hat{\nu}'' = \begin{vmatrix} -\frac{1}{\sqrt{6}} & 0 & 0 \\ 0 & \frac{1}{\sqrt{6}} & 0 \\ 0 & 0 & 0 \end{vmatrix}.$$

The anisotropic correction is given by the expression

$$v^{ani} = icm\,M^{0}(g_{11} - g_{12} - 2g_{44})(\pm\frac{2}{3}bI^{yx} + \frac{\sqrt{2}}{3}I^{yz}(-\sin3\varphi \pm ib\cos3\varphi)). \tag{30}$$

The angle φ is measured from the direction $[1\bar{1}1]$.

b. Direction $[1\bar{1}0]$. In this case

$$\hat{\nu}' = 0, \quad \hat{\nu}'' = \begin{vmatrix} \frac{1}{2} & 0 & 0 \\ 0 & -\frac{1}{2} & 0 \\ 0 & 0 & 0 \end{vmatrix},$$

$$\nu^{ani} = \pm icb\ mM^0(g_{11} - g_{12} - 2g_{44})\cos^2\varphi \quad I^{yx}. \tag{31}$$

The angle φ is measured from the direction [110].

c. Direction [100]. In this case $\hat{\nu}' = \hat{\nu}'' = 0$, and and there is no anisotropic correction.

Anisotropic diffraction TM⇒TE. The structure of equations is similar for the two anisotropic diffraction processes TE⇒TM and TM⇒TE. The difference is that in relations of the type (21)-(30) which determine the strength of the diffraction TM⇒TE, the term due to the Faraday effect appears with the opposite sign from the preceding case TE⇒TM. In addition, one must take into account that $E^{z*} = -E^z$. Otherwise these relations remain valid. Equations (23) and (27) for the phase modulation index, for example, are rewritten in the form

$$\nu = \beta L\ (\Delta n_\alpha^{xy} I^{xy} + \Delta n_\alpha^{zy} I^{zy}), \tag{32}$$

$$\nu^{iso} = icm\ exp(\pm i\vartheta_0)\ (-f \pm 2g_{44}M^0)I^{xy}. \tag{33}$$

In other words, diffraction into the positive first order has different intensities for TE⇒TM and TM⇒TE. Diffraction into the negative first order is of different intensity than diffraction into the positive first order, as before, but the direction of the "strong" diffraction is the opposite of the preceding case.

6.2. Diffraction in a tangentially magnetized film.

We discuss in detail the two typical situations: longitudinal magnetization, when the direction of the constant part of magnetization is along the wavevector K of the spin waves, which corresponds to backward volume spin waves; transverse magnetization, when the direction of the constant magnetic field is perpendicular to K, which corresponds to surface spin waves.

Diffraction by a longitudinal spin wave. In this case the total magnetization vector is written in the coordinate system adopted here as

$$\vec{M} = \begin{vmatrix} \mp ibm \\ M^0 \\ m \end{vmatrix}.$$

The strength of TE⇒TM anisotropic diffraction is described in the isotropic approximation by the expression

$$v^{iso} = icm\cos\vartheta_0\, I^{yx}(f \mp 2g_{44}bM^0) + icmI^{yz}\cdot$$
$$\cdot(2g_{44}M^0\cos(\vartheta_1 + \vartheta_0) \mp fb\cos(\vartheta_1 - \vartheta_0)). \qquad (34)$$

Contribution of the anisotropic term to the index of modulation depends on the film orientation.

Direction [111]: $v^{ani} = 2cmM^0(g_{11} - g_{12} - 2g_{44})\left(\frac{1}{3\sqrt{2}}\cos3\varphi \mp i\frac{b}{3}\right)I^{yx}.$

Direction [1$\bar{1}$0]: $v^{ani} = 2cmM^0(g_{11} - g_{12} - 2g_{44})\left(\mp i\frac{b}{2}\cos^2\varphi\right)I^{yx}.$

Direction [100]: $v^{ani} = 0.$

The expressions for the efficiency of anisotropic diffraction TM⇒TE are the same, except that the sign in front of the constant f and the overlap integral I^{yz} are switched.

Diffraction by a transverse spin wave. The total magnetization vector is given by

$$\vec{M} = \begin{vmatrix} \pm ibm \\ m \\ M^0 \end{vmatrix}.$$

The isotropic part of the phase modulation index is, for anisotropic TE⇒TM diffraction

$$v^{iso} = -icm\sin\vartheta_0\, I^{yx}(-f \pm 2g_{44}bM^0) + icmI^{yz}(\pm fb\cos(\vartheta_1 - \vartheta_0) +$$
$$2g_{44}M^0\cos(\vartheta_1 + \vartheta_0)). \qquad (35)$$

The corresponding expression for anisotropic TM⇒TE diffraction is the same as Eq.(35) except that the sign in front of the constant f and I^{yz} is switched. The anisotropic contribution, as always, depends on the crystallographic orientation of the film.

Direction [111]: $v^{ani} = 2cmM^0(g_{11} - g_{12} - 2g_{44})\ \frac{1}{3\sqrt{2}} cos3\varphi\ I^{yx}.$

Direction [1$\bar{1}$0]: $v^{ani} = 2cmM^0(g_{11} - g_{12} - 2g_{44}) \mp i\frac{b}{4} sin2\varphi\ I^{yx}.$

Direction [100]: $v^{ani} = 0.$

The regimes of isotropic diffraction are not considred here. They are characterized by low efficiency and therefore are of less interest. One can find relevant information on them in.[28]

7. Two-magnon interaction

The geometry of the two-magnon interaction is illustrated in Fig.3a. It is characterized by the following features: The incident

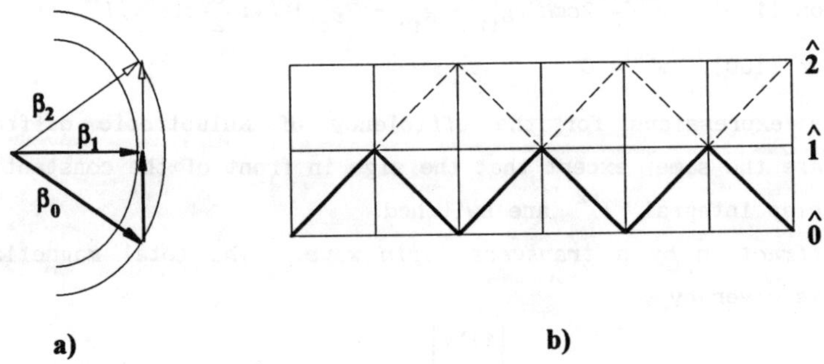

Fig.3. Vector diagram a) and multiple scattering diagram b) for the the two-magnon geometry.

light wave propagates at the double Bragg angle $\vartheta_0 = K/\beta N^{TE}$, while the diffracted wave propagates at zero angle, i.e. perpendicular to the direction of propagation of the spin wave. For such situation to occur, the wavenumber of the incident wave must be larger than that of the diffracted wave, i.e. in yttrium-iron garnet films this is diffraction of the type TE⇒TM, which is not necessarily so in Bi-substituted films.[46,47] The symmetric character of the vector diagram in Fig.3a ensures fulfillment of the Bragg conditions for a double scattering of the type TE⇒TM⇒TE or, in other words, a two-magnon interaction.

The corresponding multiple scattering diagram is given in Fig.3b. The asymmetry of diffraction pattern inherent in the magnetooptical interactions, leads to a considerable difference in the "strength" of the two transitions. While the first diffraction process TE⇒TM (transition $\hat{0}\hat{1}$) is strong, the second one TM⇒TE (transition $\hat{1}\hat{2}$) is necessarily weak. In Fig.3b strong transitions of the type $\hat{0}\hat{1}$ and $\hat{1}\hat{0}$ are given in a solid bold line, weak transitions of the type $\hat{1}\hat{2}$ and $\hat{2}\hat{1}$ in a dashed line.

As it was noted before, the linear diffraction/weak interaction described by single scattering (transitions $\hat{0}\hat{1}$) holds for a wide class of optical information processing devices. It is convenient for estimating the frequency performance of such devices. Since the volume waves exist within a wider band of microwave frequencies, they are of greater interest for applications with information processing systems. On the other hand, the geometry in Fig.3a ensures the fulfillment of the Bragg conditions within a wider range of spin-wave wave numbers, and therefore, wider band of spin-wave frequencies. This is due to the fact that the spin-wave wavevector is tangential to the semicircle, which describes the TM optical modes in Fig.3a.

Fig.4 shows the calculations of the spin-wave frequency band Δf (measured according to the -3dB criterion) of efficient magnetooptical diffraction on the film thickness h. The interaction length (width of the spin-wave beam) L was taken equal to 1 cm and the value of the ferromagnetic resonance frequency f_H was 5 GHz (Fig.4a) and 15 GHz (Fig.4b). Also varied was the value of the parameter $\Delta\beta$ characterizing the birefringence in a pair of optical modes. Curve 1 corresponds to $\Delta\beta$ = 2 cm^{-1}; 2 - 3 cm^{-1}; 3 - 5 cm^{-1}; 4 - 10 cm^{-1}; 5 - 20 cm^{-1}; 6 - 40

cm^{-1}.

It is evident from Fig. 4 that the bandwidth of the magnetooptical interaction is much greater than that of the acoustooptical interaction. This is explained by the fact that the group velocity of spin waves is on average an order of magnitude greater than that of acoustic waves. Therefore analogous relations in wavevector space transform differently in the frequency domain in magnetooptics and acoustooptics. Also there is an optimum (from the point of view of bandwidth) value of the parameter $\Delta\beta$, which is of order $3\ cm^{-1}$. For such values of $\Delta\beta$,

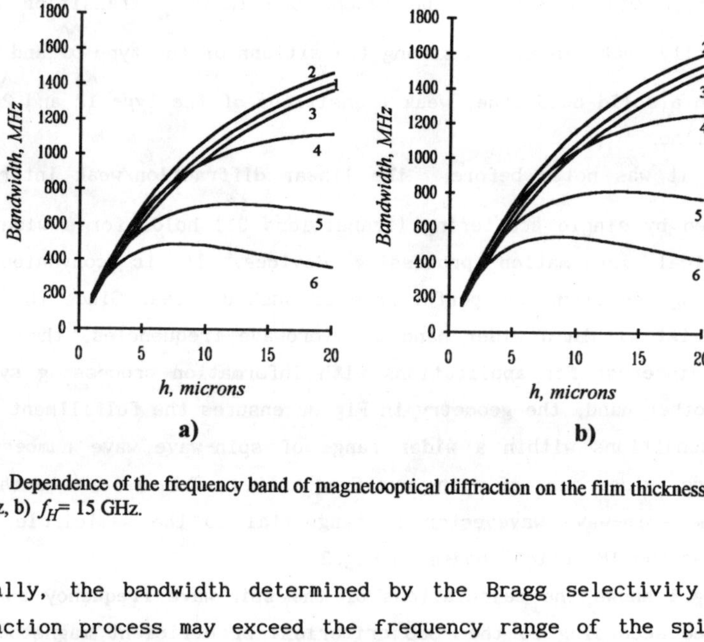

Fig.4. Dependence of the frequency band of magnetooptical diffraction on the film thickness a) f_H = 5 GHz, b) f_H= 15 GHz.

typically, the bandwidth determined by the Bragg selectivity of the diffraction process may exceed the frequency range of the spin waves excited by an antenna of limited width w (here it was taken w = 10 μm). This is taken into account in Fig. 4. We see also that in the large frequency region from 5 GHz to 15 GHz the bandwidth varies by a small amount of less than 10 percent.

To quantify the effects, which accompany the diffraction process when the approximation of weak interaction (16) no longer holds higher terms in the perturbation series should be taken into account or, in terms of the diagrams in Fig. 3b multiple transitions. For a practically important case, when all three interacting optical waves, the zero order (incident wave), the first order, and the second order are in strict synchronism (an isosceles vector triangle in Fig. 3a), the perturbation series (see Eq. (13,14)) for the amplitudes of these orders reduce to

$$E_1 = \frac{v^+}{2}\left(1 - \frac{A}{24} + \frac{(A)^2}{1920} - \frac{(A)^3}{322560} + \frac{(A)^4}{92897280} - \ldots\right), \tag{36a}$$

$$E_2 = -\frac{v^+v^-}{8}\left(1 - \frac{A}{48} + \frac{(A)^2}{5760} - \frac{(A)^3}{1290240} + \frac{(A)^4}{464486400} - \ldots\right), \tag{36b}$$

$$A = v^+v^+ + v^-v^-.$$

It should be noted here that the Eq. (36) can be easily obtained through

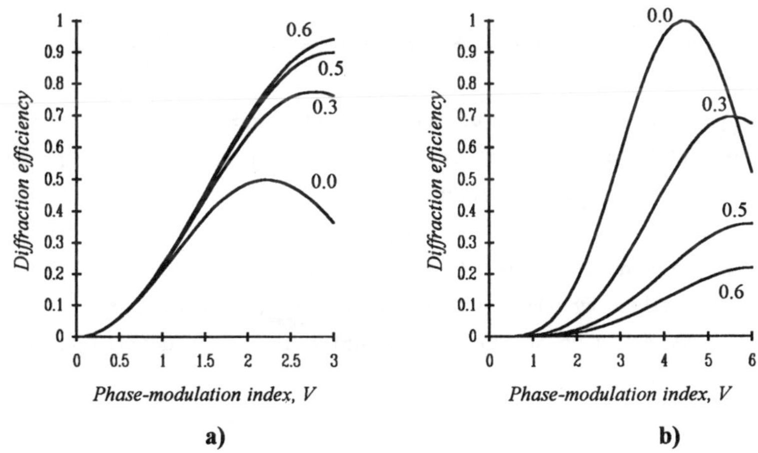

Fig.5. Efficiency of magnetooptical diffraction a) in the first order; b) in the second order.

analysis of transitions in diagrams in Fig. 3b. The calculations of the dependence of the intensities I_1 and I_2 on the phase-modulation index $v = v^+$, which took into account multiple transitions in Fig. 3b up to the tenfold ones, are illustrated in Fig.5. The ratio $R = 2g_{44}/f$ of the magnetooptical constants (Cotton-Mouton and Faraday) is an important parameter here.

For the values of $R \approx 0.5 - 0.6$, typical for YIG crystals, more than 90 percent of light intensity can be concentrated in the first order. By contrast, for $R = 0$, which corresponds to the acoustooptical case,[38] the maximum efficiency of diffraction in the first order can not exceed 50 percent, while a 100 percent efficiency occurs in the second diffraction order. This effect of the so called mid-band degeneracy can severely distort the performance of an optical devices.

Similar analysis of threefold scattering by several spin-wave harmonics makes it possible to estimate the distortions of the angular spectrum of diffracted light with respect to the spectrum of the initial spin wave. For example, in the case of two spin-wave harmonics of equal amplitude we obtain the following expression for the intensity of the diffracted optical wave at the combination frequency ($K_{01}-\Delta K$ or $K_{02}+\Delta K$)

$$I_1^3(K_{01}-\Delta K) = I_{inc}\,(C_1^3(K_{01}-\Delta K))^2 = I_1^3(K_{02}+\Delta K) = I_{inc}\Big(\frac{v^+(v^+v^+ + v^-v^-)}{48}\Big)^2.$$

Here I_{inc} is the intensity of the incident light wave.

The optical component at the combination frequency, being generated in the course of magnetooptical diffraction, is a spurious one; it distorts the spectrum of the initial spin-wave signal. Thus, the upper limit of the dynamic range of a magnetooptical information processing device, described by the maximum value v_{max}, at which the generated spurious components do not exceed the noise level, is determined by

$$\frac{(v^+)^3(1+r)^2}{48} = \sqrt{\frac{I_{noise}}{I_{inc}}} \qquad (37)$$

where I_{noise} is the intensity of the optical noise due to scattering of the incident light wave by static inhomogeneities in the film. In (37) we used the following notation $r = (1-R)/(1+R)$. There is no need in moving further upward along the perturbation series, because already in the course of the threefold scattering unacceptable spurious components are effectively generated.

8. Conclusion

A theory of multiple scattering of light by a non-harmonic spin wave in a ferromagnetic film has been developed. The application of the perturbation method is graphically illustrated with Feynman type multiple scattering diagrams.

Special attention was paid to the analysis of the two-magnon geometry. It was shown that it has some important advantages over its close analog in acoustooptics: the two-phonon geometry. Apart from a superior frequency performance (a wider range of microwave frequencies within which efficient diffraction occurs) it is distinguished by better amplitude chracteristics. Thus, the asymmetry of the magnetooptical diffraction pattern makes it possible to transfer over 90 percent of light intensity into the first diffraction order. This is impossible in principle in acoustoooptical interactions, with the maximum efficiency of the first diffraction order being 50 percent.

Now, with the development of Bi-substituted single crystal films with excellent optical and microwave characteristics,[14,46,47] high levels of diffraction efficiency are achievable. Although the ratio of magnetooptical constants R determining the degree of asymmetry in the diffraction pattern has not been directly measured in Bi-substituted films, there is experimental evidence that it is large enough to influence noticeably the diffraction pattern.[47]

The analysis of threefold scattering of light by two spin-wave harmonics, which generates spurious components in the angular spectrum of the diffracted light, leads to the estimate of the dynamic range of magnetooptical information processing devices. However, the nonlinearity of the spin-wave system proper can put even more serious restrictions on the dynamic characteristics, and therefore further investigation of this problem is needed.

When an optical mode is diffracted by a dipole-exchange spin wave with a complex distribution of magnetization across the film, only the lowest transverse mode contributes significantly to the diffraction process. At the same time, the situation can change for the case of submicron films and for the case of asymmetric boundary conditions. This should be taken into account in the experiment.[48]

References

1. L.K.Anderson, *Appl. Phys. Lett.* **1** (1962) 44.
2. J.F.Dillon and J.S.Remeika, *Appl. Phys. Lett.* **2** (1963) 38.
3. J.H.Collins and D.A.Wilson, *Appl. Phys. Lett.* **12** (1968) 332.
4. H.L.Hu and F.R.Morgenthaller, *Appl. Phys. Lett.* **18** (1971) 307.
5. B.Desormiere and H. Le Gall, *IEEE Trans. Magn.* **MAG-8** (1972) 379.
6. N.N.Kiryukhin, I.N.Kotel'nikov, and F.V.Lisovkiy, *Fiz. Tverd. Tela* **12** (1972) 3686.
7. Yu.F.Gaidai, I.I.Kondilenko, and A.A.Solomko, *Pis'ma Zh. Eksp. Teor. Fiz.* **21** (1975) 578.
8. A.D.Fisher, J.N.Lee, E.S.Gaynor, and A.B.Tveten, *Appl. Phys. Lett.* **41** (1982) 779.
9. A.D.Fisher, *Circuits, Systems, and Signal Processing (USA)* **4** (1985) 265.
10. A.N.Sigaev and A.A.Stashkevich, *Sov. Tech. Phys. Lett.* **14** (1988) 209.
11. D.Young and C.S.Tsai, *Appl. Phys. Lett.* **53** (1988)1696.
12. S.H.Talisa, *IEEE Trans. Magn* **MAG-24** (1988) 2811.
13. C.S.Tsai and D.Young, *Appl. Phys. Lett.* **54** (1989) 196.
14. D.Young and C.S.Tsai, *Appl. Phys. Lett.* **55** (1989) 2242.
15. S.H.Talisa and D.D.Stancil, *IEEE Trans. Magn.* **MAG-25** (1989) 3494.
16. N.Bilaniuk, D.D.Stancil, and S.H.Talisa, *J. Appl. Phys.* **67** (1990) 508.
17. A.A.Stashkevich, B.A.Kalinikos, N.G.Kovshikov et al, *Sov. Tech. Phys. Lett.* **11** (1987) 20.
18. N.Bilaniuk and D.D.Stancil, *J. Appl. Phys.* **67** (1990) 4790.
19. V.V.Matyushev, A.A.Stashkevich, and A.I.Lukyanov, *J. Appl. Phys.* **74** (1993) 3680.
20. L.Brillouin, *La Diffraction de la Lumière par des Ultrasons* (Act.

Sci et Ind. Fasc.,59, Hermann, Paris, 1933).

21 O.Nomoto, *Jap. J. Appl. Phys.* **10** (1971) 611.

22. M.V.Berry, *The Diffraction of Light by Ultrasound* (Academic Press, N.Y.-L., 1966).

23. A.Defebre, *Ultrasonics* **13** (1975) 73.

24. A.M.Martynov, *Radiotekh. Electron.(in Russian)* **18** (1973) 1948.

25. M.Barnoski, *Introduction to Integrated Optics* (Plenum Press, New York, 1974).

26. *Integrated Optics. Topics Appl. Phys., 2nd ed., vol.*7, Ed. T.Tamir (Springer, Berlin, Heidelberg, New-York, 1979).

27. A.Yariv, *Quantum Electronics* (J.Wiley and Sons, New York, London, Sydney, Toronto, 1975).

28. A.A.Stashkevich, *Physics of Magnetic Phenomena, Soviet Physics Journal* **4** (1989) 524.

29. B.A.Kalinikos, N.G.Kovshikov, and A.N.Slavin, *Sov. Phys. JETP,* **67** (1988) 303.

30. A.N.Slavin, B.A.Kalinikos, and Kovshikov, *in "Linear and Nonlinear Spin Waves in Magnetic Films and Superlattices",* ed. M.G.Cottam (World Scientific, Singapore, New Jersey, London, Hong Kong, 1994).

31. A.A.Stashkevich, *Opt. Spectr. (USSR)* **45** (1978) 798.

32. A.A.Stashkevich, *Opt. Spectr. (USSR)* **58** (1985) 793.

33. G.S.Krinchik, *Physics of Magnetic Phenomena (in Russian)* (MGU, Moscow,1985).

34. A.N.Ageev, V.N.Gridnev, O.G.Rutkin, et al., *Fiz. Tverd. Tela (in Russian)* **25** (1983) 478.

35. A.M.Prokhorov, G.A.Smolenskii, and A.N.Ageev, *Sov. Phys. Usp.* **27**, (1984), 339.

36. A.G.Gurevich and G.A.Melkov, *Magnetic Oscillations and Waves (in Russian)* (Nauka, Moscow, 1994).

37. B.A.Kalinikos, *in "Linear and Nonlinear Spin Waves in Magnetic Films and Superlattices",* ed. M.G.Cottam (World Scientific, Singapore, New Jersey, London, Hong Kong, 1994).

38. A.Korpel, *Acousto-Optics* (Marcel Dekker, New York and Basel, 1988).

39. L.A.Vainshtein, *Electromagnetic Waves (in Russian)* (Radio i Sviaz',

Moscow, 1988).

40. F.I.Fedorov, *Theory of Elastic Waves in Chrystals (in Russian)* (Nauka, Moscow, 1965).

41. R.V.Pisarev, I.G.Sinii, N.N.Kolpakova, et al., *Zh. Eksp. Teor. Fiz.* **60** (1971) 2189.

42. C.V.Raman and N.S.S.Nath, Proc. Ind. Acad. Sci. **A3** (1936) 119.

43. C.V.Raman and N.S.S.Nath, Proc. Ind. Acad. Sci. **A3** (1936) 459.

44. A.M.Martynov and I.S.Mirer, *Izv. Vyssh. Uchebn. Zaved., Radiofiz (in Russian)* **18** (1975) 1845.

45. R.V.Pisarev, I.G.Sinii, and G.A.Smolenskii, *Pis'ma Zh. Eksp. Teor. Fiz.* **9** (1969) 112.

46. H.Tamada, M.Kaneko, and T.Okamoto, *J. Appl. Phys.* **64** (1988)

47. V.V.Matyushev, A.A.Stashkevich, and J.M.Desvignes, *J. Appl. Phys.* **69** (1991) 5972.

48. V.V.Matyushev, A.A.Stashkevich, M.P. Kostylev, and J.M.Desvignes, *Technical Digest of the Intermag Conference 1993*, BR-08.

COLLINEAR INTERACTION OF OPTICAL GUIDED MODES WITH MICROWAVE SPIN WAVES IN MAGNETIC FILMS

DANIEL D. STANCIL
Department of Electrical and Computer Engineering, Carnegie Mellon University, Pittsburgh, PA 15213, U.S.A.

NYKOLAI BILANIUK
National Research Council
Ottowa, Canada

1.0 Introduction

1.1 Brief History

Optical waveguides and waveguide devices made with magnetic garnet thin films were first demonstrated by P. K. Tien and coworkers in 1972.[1,2] The low absorption of these materials in the near infrared and the high quality of epitaxial films made them promising components for an emerging integrated optics technology.[3]

At about the same time, Collins and Wilson[4], Hu and Morgenthaler[5], and Desormiere and Le Gall[6] were studying optical scattering from coherent propagating spin waves in bulk single crystals of yttrium iron garnet (YIG). However, these devices were complicated by the fact that the inhomogeneous static magnetic field inside the nonellipsoidal bulk samples made it difficult to control and analyze the spin wave propagation.

As mentioned above, the early seventies also saw the development of high-quality epitaxial thin films of yttrium iron garnet.[7] Using technology developed for bubble memory devices, these films provide low-loss propagation media for magnetostatic spin waves without the problems associated with nonuniform internal magnetic fields. This stimulated considerable interest in magnetostatic wave (MSW) devices.[8-11] This new technology has similarities with both surface acoustic wave devices (SAW), and microwave filters and limiters using single crystal YIG spheres. Potential advantages over SAW include operation at multi-gigahertz frequencies and the flexibility of magnetic tuning. The principal advantage over YIG sphere devices is the economy and convenience of a planar technology.

The maturing of MSW technology and the fact that the epitaxial YIG films possess a higher index of refraction than the commonly-used gadolinium gallium garnet (GGG) substrate made it attractive to take a new look at MSW-optical scattering using dielectric waveguide modes of the thin film rather than gaussian beams in a bulk sample. In the early 1980's, this concept was being pursued by Fisher and coworkers at the Naval Research Laboratories, and by a group led by Tsai at the University of California at Irvine. In 1982 Fisher et al. reported observing the interaction between optical guided modes propagating parallel to magnetostatic surface waves (MSSWs).[12] The following year Tsai[13] reported observations of the transverse interaction with MSSWs, and Fisher et al.

reported observations of the transverse interaction with magnetostatic forward volume waves (MSFVWs).[14] Interest in this new device technology has grown steadily since these reports and about a dozen laboratories have become involved in some aspect of MSW-optical device research.

As with MSW devices, research on MSW-optical interactions has initially been motivated by device analogs using acoustic waves. Advantages over acousto-optical devices include operation with microwave frequencies perhaps as high as 30 GHz[15], electronic tunability via a static magnetic bias field, and greater modulation and beam deflection speeds owing to the higher velocity of magnetostatic waves. Fortunately, many of the difficulties encountered with MSW analogs of SAW devices do not appear with MSW-optical devices; complicated transducer structures are not required, and the presence of dispersion does not appear to be a problem with the devices proposed. Microwave signal processing devices such as filters, correlators, and spectrum analyzers as well as optical components such as switches, modulators, frequency shifters, and tunable filters are possible with this technology.

The demonstration of MSW-optical guided wave interactions is not the only reason for reconsidering Tien's proposal for a magnetic garnet-based integrated optics device technology. As optical communications systems have matured, there has been a growing need for compact optical isolators.[16-19] Magnetic garnet films are the most promising materials for this application. Other integrated optical building blocks that have been demonstrated include channel waveguides,[20] phase shifters,[21] mode converters,[22-24] and lenses.[25] Stimulated emission has been observed in doped films suggesting the possibility of pumped lasers,[26,27] and new photo-induced effects[28,29] may lead to practical ways of making detectors.

1.2 Collinear Interactions

If the optical guided modes and the MSW are traveling at angles close to parallel or antiparallel, the configuration is referred to as the collinear interaction geometry. In this chapter, we will concentrate our attention on the collinear interaction geometry shown in Fig. 1. An epitaxial film of YIG grown on a GGG substrate serves as the optical and MSW waveguide.The presence of the MSW perturbs the optical index of refraction through the Faraday and Cotton-Mouton magneto-optic effects. The MSW thus creates a moving diffraction grating for the optical modes. Efficient diffraction takes place when the TE wavenumber is equal to the sum or difference of the MSW and TM wave numbers, depending on whether the MSW and optical propagation directions are parallel or antiparallel, respectively. This is referred to as the phase-matching condition.

This chapter is intended to provide an overview of collinear MSW-optical interactions and applications. In Section 2, the physical phenomena responsible for magneto-optical interactions are discussed phenomenologically. Section 3 provides a brief review of MSW and optical modes and the pertinent aspects of coupled-mode theory. The effects of microwave and optical losses are discussed in Section 4, and microwave nonlinear phe-

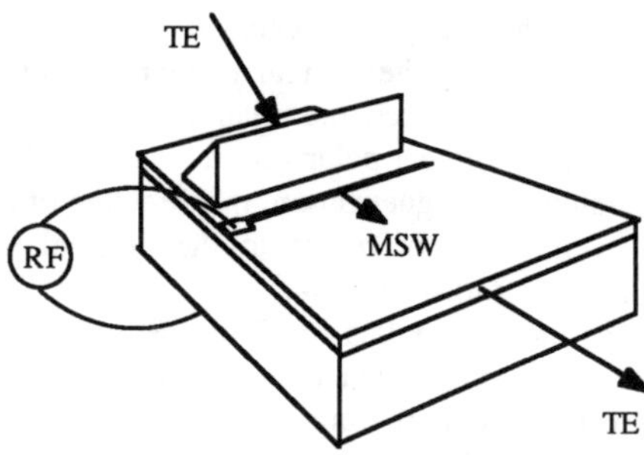

FIGURE 1. Basic collinear configuration for optical-MSW interactions.[30] Coupling to the optical modes can be either through a prism or through the edge of the film. In this figure, the microwave excitation is applied between a microstrip transducer in contact with the film and a ground plane beneath the substrate.

nomena are discussed briefly in Section 5. Finally, the stringent material requirements and possible ways they can be met are described in Section 6.

2.0 Magneto-optic Interactions

2.1 Magneto-optic Coupling Mechanisms

The presence of magnetostatic waves in a magnetic garnet film causes small time-varying perturbations in the direction of the magnetization. These perturbations can affect both the optical permittivity and permeability of the material, in principle. However, from a phenomenological point of view, the polarization **P** and magnetization **M** cannot be uniquely defined in the presence of time-varying fields.[31-33] Consequently, magneto-optical effects can be modeled either with an effective permittivity, an effective permeability, or a combination of both (although modifications of the electromagnetic boundary conditions may be required[33]). The most common approach is to model the effects with an effective permittivity, set the permeability to that of free space, and use the standard electromagnetic boundary conditions. This is equivalent to neglecting the direct interaction of the magnetic field of the wave with the medium. I believe it is likely that careful experiments will eventually identify inadequacies with this approach to describing the MSW-optical interaction. In the absence of such data, we will proceed with the common approach.

The largest perturbations in the permittivity are linear in the magnetization and give rise to the Faraday effect. Perturbations of second order in the magnetization are responsible for the Cotton-Mouton effect. Although the Faraday effect is the stronger of the two, in certain geometries the MSW-optical interaction is almost entirely due to the Cotton-Mouton effect.

In a magnetic garnet, the optical normal modes of propagation parallel to the magnetization are right- and left-circularly polarized waves. These waves travel with different phase velocities and therefore have different effective indices of refraction. This phenomenon is called *magnetic circular birefringence*. A linearly polarized wave incident upon such a medium can be decomposed into the circularly-polarized normal modes. As these modes travel with different phase velocities, the relative phase between them gradually changes. A rotation of the equivalent linear polarization results. If the direction of propagation reverses without reversing the magnetization, the direction of the polarization rotation remains the same (i.e., it does not return to the original incident state). This nonreciprocal effect is known as *Faraday rotation.*

The situation becomes more complicated if the medium also exhibits (nonmagnetic) *linear birefringence*; i.e., if orthogonal linearly polarized modes in the medium travel with different phase velocities apart from the magnetic perturbation. In this case, rotations in linear polarization caused by the magnetic perturbation can be thought of as mode conversion between the unperturbed linearly-polarized modes. To achieve large angles of polarization rotation, significant mode conversion must occur before the modes go out of phase from the linear birefringence.

In optical thin-film waveguides, Faraday rotation manifests itself as mode conversion between TE and TM optical modes. As in the case of a birefringent bulk crystal, the phase velocities of the two unperturbed modes must be very nearly equal for efficient mode conversion to occur. Since optical modes of a thin film exhibit shape birefringence (and possibly growth and stress birefringence), significant mode conversion is ordinarily prevented. This difficulty can be overcome by periodically reversing the magnetization so that the sign of the coupling coefficient is reversed to compensate for the phase difference between the modes. Such a periodic variation in the magnetization is caused by a propagating magnetostatic wave. Mode conversion from Faraday rotation results when there is a component of the time-varying magnetization parallel to the direction of optical mode propagation.

If the magnetization in a garnet is perpendicular to the direction of light propagation, the optical normal modes will be linearly polarized waves either parallel or perpendicular to **M**. Again, these normal modes travel with different phase velocities resulting in *magnetic linear birefringence*. No rotation of the polarization will occur in this case.

In a thin-film waveguide, orienting the magnetization perpendicular to the film plane or in the plane perpendicular to the direction of light propagation will alter the linear birefringence but will not cause mode conversion. However, orienting the magnetization at an oblique angle creates real off-diagonal permittivity tensor elements resulting in reciprocal mode conversion. As before, significant mode conversion can occur only when there is a periodic reversal of the magnetization in step with the phase beats of the TE and TM waves. Again, a propagating magnetostatic wave can provide the required periodic variation in **M**.

2.2 *Form of the Optical Permittivity Tensor with Magnetization*

In the presence of magnetostatic waves, the permittivity will have time-varying components with microwave frequencies. Since these variations are slow compared to optical frequencies, we can write the dielectric constitutive relation in a mixed-domain representation:

$$D_m(\omega_L) = \varepsilon_0 \varepsilon_{mn}(\omega_L, t) E_n(\omega_L), \tag{1}$$

where ω_L is an optical frequency, the time variations described by t are slow compared with the optical period, and summation over repeated indices is assumed. To second order in the magnetization, the relative permittivity tensor can be expressed in the crystal coordinate system as[33]

$$\begin{aligned}\varepsilon_{mn} &= \varepsilon_{mn}^{(0)} + \varepsilon_{mn}^{F} + \varepsilon_{mn}^{CM} \\ &= \varepsilon_{mn}^{(0)} + je_{mnp}f_{pq}M_q + g_{mnpq}M_pM_q\end{aligned} \quad (2)$$

where e_{mnp} is the Levi-Civita pseudotensor, the forms of the magneto-optic tensors f_{pq} and g_{mnpq} are determined by the symmetry of the crystal, and a time dependence of the form $\exp(j\omega t)$ is assumed. The first term represents the nonmagnetic component of the permittivity, the second term gives rise to the Faraday effect, and the last term gives rise to the Cotton-Mouton effect. For the cubic magnetic garnets of class *m3m*,

$$\varepsilon_{mn}^{(0)} = K\delta_{mn}, \quad (3)$$

$$f_{pq} = f\delta_{pq}, \quad (4)$$

and

$$g_{mnpq} = g_{nmpq} = g_{mnqp} = g_{nmqp}, \quad (5)$$

where δ_{pq} is the Kronecker delta function and K is the isotropic relative permittivity. Because of the symmetry of g_{mnpq}, the indices mn may be replaced with a single index r, and the indices pq may be replaced with the single index s. The elements may then be represented in matrix form using the Voigt notation:

$$g_{rs} = g_{mnpq} \quad (6)$$

where the indices r and s take on values from 1 to 6 according to the following table:

TABLE 1. Voigt notation for contraction of tensor indices

r* or *s	***mn* or *pq***
1	11
2	22
3	33
4	23, 32
5	13, 31
6	12, 21

Further, in cubic crystals, there are only three independent components of g_{rs} denoted g_{11}, g_{12}, and g_{44}. Consequently, g_{rs} can be represented in matrix form as[33]

$$\left[g_{rs}\right] = \begin{bmatrix} g_{11} & g_{12} & g_{12} & 0 & 0 & 0 \\ g_{12} & g_{11} & g_{12} & 0 & 0 & 0 \\ g_{12} & g_{12} & g_{11} & 0 & 0 & 0 \\ 0 & 0 & 0 & g_{44} & 0 & 0 \\ 0 & 0 & 0 & 0 & g_{44} & 0 \\ 0 & 0 & 0 & 0 & 0 & g_{44} \end{bmatrix} . \tag{7}$$

The elements of the Cotton-Mouton permittivity tensor can also be expressed in this notation:

$$\left[\varepsilon_{mn}^{CM}\right] = \begin{bmatrix} \varepsilon_{11}^{CM} & \varepsilon_{12}^{CM} & \varepsilon_{13}^{CM} \\ \varepsilon_{21}^{CM} & \varepsilon_{22}^{CM} & \varepsilon_{23}^{CM} \\ \varepsilon_{31}^{CM} & \varepsilon_{32}^{CM} & \varepsilon_{33}^{CM} \end{bmatrix} = \begin{bmatrix} \varepsilon_{1}^{CM} & \varepsilon_{6}^{CM} & \varepsilon_{5}^{CM} \\ \varepsilon_{6}^{CM} & \varepsilon_{2}^{CM} & \varepsilon_{4}^{CM} \\ \varepsilon_{5}^{CM} & \varepsilon_{4}^{CM} & \varepsilon_{3}^{CM} \end{bmatrix} . \tag{8}$$

Combining this notation with Eq. (2) and Eq. (7) enables us to express the permittivity tensor elements in the form

$$\begin{bmatrix} \varepsilon_{1}^{CM} \\ \varepsilon_{2}^{CM} \\ \varepsilon_{3}^{CM} \\ \varepsilon_{4}^{CM} \\ \varepsilon_{5}^{CM} \\ \varepsilon_{6}^{CM} \end{bmatrix} = \begin{bmatrix} g_{11} & g_{12} & g_{12} & 0 & 0 & 0 \\ g_{12} & g_{11} & g_{12} & 0 & 0 & 0 \\ g_{12} & g_{12} & g_{11} & 0 & 0 & 0 \\ 0 & 0 & 0 & g_{44} & 0 & 0 \\ 0 & 0 & 0 & 0 & g_{44} & 0 \\ 0 & 0 & 0 & 0 & 0 & g_{44} \end{bmatrix} \begin{bmatrix} M_1M_1 \\ M_2M_2 \\ M_3M_3 \\ 2M_2M_3 \\ 2M_1M_3 \\ 2M_1M_2 \end{bmatrix} \tag{9}$$

The factors of two in the lower three elements of the column vector containing the magnetization components result from the sum of the symmetric off-diagonal elements M_iM_j and M_jM_i in the implied summation of Eq. (2).

We can also expand the Faraday component of the relative permittivity using the symmetry property given in Eq. (4). The result is

$$\left[\varepsilon_{mn}^{F}\right] = jf \begin{bmatrix} 0 & M_3 & -M_2 \\ -M_3 & 0 & M_1 \\ M_2 & -M_1 & 0 \end{bmatrix} . \tag{10}$$

All of the expressions up to this point have been expressed in the principal coordinate system of the crystal. However, most epitaxial garnet films are grown with the [111] axis normal to the plane, and it is convenient experimentally to choose a coordinate system that uses this direction as one of its axes. Following Prokhorov *et al.*[33], we define the *xyz* system so that $x \parallel [111]$, $y \parallel [11\bar{2}]$, and $z \parallel [\bar{1}10]$ (see Fig. 2). Vectors and tensors may be expressed in the new coordinate system using the transformation matrix S_{mn}:

$$\hat{M}_m = S_{mn} M_n, \tag{11}$$

$$\hat{\varepsilon}_{mn} = S_{mp} \varepsilon_{pq} S^T_{qn}. \tag{12}$$

Here quantities expressed in the *xyz* system are indicated by a hat (^). For the two coordinate systems under consideration, S_{mn} is given by

$$\left[S_{mn}\right] = \begin{bmatrix} \frac{1}{\sqrt{3}} & \frac{1}{\sqrt{3}} & \frac{1}{\sqrt{3}} \\ \frac{1}{\sqrt{6}} & \frac{1}{\sqrt{6}} & -\frac{2}{\sqrt{6}} \\ -\frac{1}{\sqrt{2}} & \frac{1}{\sqrt{2}} & 0 \end{bmatrix}. \tag{13}$$

The form of the antisymmetric tensor is unchanged by such a transformation:

$$\left[\hat{\varepsilon}^F_{mn}\right] = jf \begin{bmatrix} 0 & M_z & -M_y \\ -M_z & 0 & M_x \\ M_y & -M_x & 0 \end{bmatrix}. \tag{14}$$

The Cotton-Mouton tensor transformation has been worked out by Prokhorov, *et al.*[33] The result is

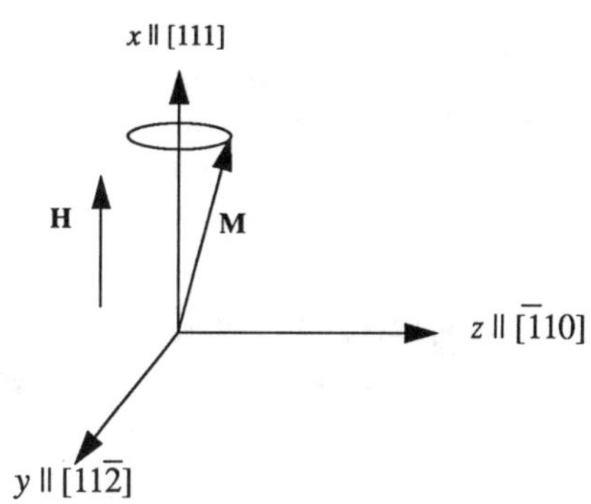

FIGURE 2. Coordinate system for a perpendicularly magnetized garnet film.

$$\begin{bmatrix} \hat{\varepsilon}_1^{CM} \\ \hat{\varepsilon}_2^{CM} \\ \hat{\varepsilon}_3^{CM} \\ \hat{\varepsilon}_4^{CM} \\ \hat{\varepsilon}_5^{CM} \\ \hat{\varepsilon}_6^{CM} \end{bmatrix} = \begin{bmatrix} g_{11}-2\Delta g/3 & g_{12}+\Delta g/3 & g_{12}+\Delta g/3 & 0 & 0 & 0 \\ g_{12}+\Delta g/3 & g_{11}-\Delta g/2 & g_{12}+\Delta g/6 & 0 & 0 & -\Delta g/(3\sqrt{2}) \\ g_{12}+\Delta g/3 & g_{12}+\Delta g/6 & g_{11}-\Delta g/3 & 0 & 0 & \Delta g/(3\sqrt{2}) \\ 0 & 0 & 0 & g_{44}+\Delta g/6 & \Delta g/(3\sqrt{2}) & 0 \\ 0 & 0 & 0 & \Delta g/(3\sqrt{2}) & g_{44}+\Delta g/3 & 0 \\ 0 & -\Delta g/(3\sqrt{2}) & \Delta g/(3\sqrt{2}) & 0 & 0 & g_{44}+\Delta g/3 \end{bmatrix} \begin{bmatrix} M_x M_x \\ M_y M_y \\ M_z M_z \\ 2M_y M_z \\ 2M_x M_z \\ 2M_x M_y \end{bmatrix} \quad (15)$$

where $\Delta g = g_{11} - g_{12} - 2g_{44}$.

In the case of forward volume waves biased along the x direction, the magnetiztion can be approximated by

$$\mathbf{M} = \hat{x}M_S + \hat{y}m_y(t) + \hat{z}m_z(t) \quad (16)$$

where M_S is the static saturation magnetization and m_y and m_z are small-signal microwave perturbations of the direction of $\mathbf{M}$. Under these circumstances, m_y^2, m_z^2, and $m_y m_z$ are of second-order in small quantities and can be neglected. Making these approximations in Eq. (15) yields the Cotton-Mouton permittivity tensor to first-order in the small-signal magnetization:

$$\left[\hat{\varepsilon}_{mn}^{CM}\right] = \begin{bmatrix} (g_{12}+2g_{44}+\Delta g/3)M_S^2 & 2(g_{44}+\Delta g/3)M_S m_y & 2(g_{44}+\Delta g/3)M_S m_z \\ 2(g_{44}+\Delta g/3)M_S m_y & (g_{12}+\Delta g/3)M_S^2 - \sqrt{2}\Delta g M_S m_y/3 & \sqrt{2}\Delta g M_S m_z/3 \\ 2(g_{44}+\Delta g/3)M_S m_z & \sqrt{2}\Delta g M_S m_z/3 & (g_{12}+\Delta g/3)M_S^2 + \sqrt{2}\Delta g M_S m_y/3 \end{bmatrix}. \quad (17)$$

The diagonal element $(g_{12}+\Delta g/3)M_S^2$ is present regardless of the direction of the magnetization and is thus indistinguishable from an isotropic contribution to the nonmagnetic permittivity.[33,34] We will therefore absorb $(g_{12}+\Delta g/3)M_S^2$ into the nonmagnetic permittivity but retain the remaining terms explicitly in the Cotton-Mouton tensor. Combining Eq. (14) and Eq. (17) and allowing for the possibility of growth and/or stress-induced nonmagnetic anisotropy in the film yields the total relative permittivity tensor:

$$\left[\hat{\varepsilon}_{mn}\right] = \begin{bmatrix} K_{\parallel} & 0 & 0 \\ 0 & K_{\perp} & 0 \\ 0 & 0 & K_{\perp} \end{bmatrix} + \begin{bmatrix} 2g_{44}M_S^2 & 0 & 0 \\ 0 & 0 & jfM_S \\ 0 & -jfM_S & 0 \end{bmatrix}$$

$$+ \begin{bmatrix} 0 & jfm_z + 2(g_{44}+\Delta g/3)M_S m_y & -jfm_y + 2(g_{44}+\Delta g/3)M_S m_z \\ -jfm_z + 2(g_{44}+\Delta g/3)M_S m_y & -\sqrt{2}\Delta g M_S m_y/3 & \sqrt{2}\Delta g M_S m_z/3 \\ jfm_y + 2(g_{44}+\Delta g/3)M_S m_z & \sqrt{2}\Delta g M_S m_z/3 & \sqrt{2}\Delta g M_S m_y/3 \end{bmatrix} \quad (18)$$

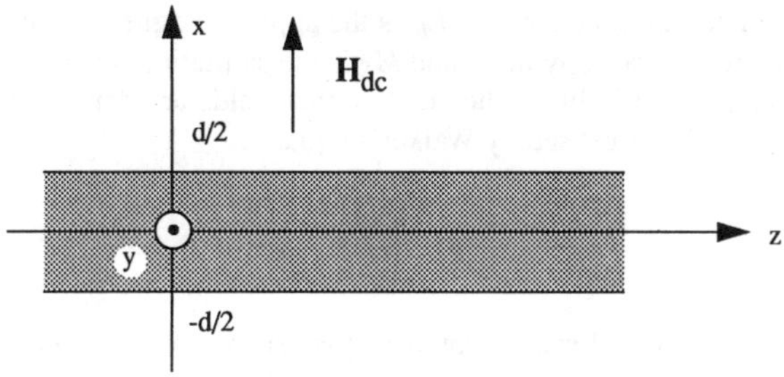

FIGURE 3. Geometry for the analysis of the guided modes of a magnetic thin film.

The first term on the right-hand-side represents the nonmagnetic component of the permittivity, the second term represents the static magneto-optic component, and the last term represents the dynamic magneto-optic effect.

3.0 Basic Optical-Magnetostatic Wave Interactions

3.1 Magnetostatic Waves in a Perpendicularly Magnetized Film

Let us first focus our attention on the propagation of microwave guided modes in the thin-film structure shown in Fig. 3. Assuming the applied magnetic field $\mathbf{H}_{dc}$ is large enough to saturate the film, the small-signal magnetization and magnetic fields in the film are related by:

$$\begin{bmatrix} m_y \\ m_z \end{bmatrix} = \begin{bmatrix} \chi & j\kappa \\ -j\kappa & \chi \end{bmatrix} \begin{bmatrix} h_y \\ h_z \end{bmatrix} \tag{19}$$

where

$$\chi = \frac{\omega_0 \omega_M}{\omega_0^2 - \omega_q^2}, \tag{20}$$

$$\kappa = \frac{\omega_q \omega_M}{\omega_0^2 - \omega_q^2}, \tag{21}$$

$$\omega_0 = -\gamma\mu_0 (H_{dc} + H_k - M_S), \tag{22}$$

$$\omega_M = -\gamma\mu_0 M_S, \tag{23}$$

γ is the gyromagnetic ratio (negative), H_{dc} is the applied magnetic field intensity (see Fig. 3), H_k is the effective anisotropy field, and M_S is the saturation magnetization of the film.

In the magnetostatic limit, the small signal fields are derivable from a magnetic scalar potential ψ which must satisfy Walker's equation

$$\frac{\partial^2\psi}{\partial x^2} + (1+\chi)\left(\frac{\partial^2\psi}{\partial y^2} + \frac{\partial^2\psi}{\partial z^2}\right) = 0. \tag{24}$$

This equation is valid for all space provided χ is set to zero in the nonmagnetic regions. The complete potential is obtained by finding the solutions to Eq. (24) in each region and matching boundary conditions between regions. Once the potential is found, the small signal magnetic field intensity can be obtained from

$$\boldsymbol{h} = -\nabla\psi. \tag{25}$$

Since the electric field is neglected in the magnetostatic approximation, the most appropriate boundary conditions are to require tangential **h** and normal **b** to be continuous at each boundary. It is easily shown that the boundary condition on tangential **h** is equivalent to requiring ψ to be continuous. Likewise, the absence of a susceptibility along the x direction reduces the boundary condition on normal **b** to requiring $d\psi/dx$ to be continuous.

For this geometry, the potential function has the following form:

$$\psi = \phi(x)\, e^{j(\omega_q t - \beta_q z)}, \tag{26}$$

where the function $\phi(x)$ is given explicitly in Reference (35). To normalized the fields to P mW/mm along the y direction we require

$$-\frac{\omega_q \mu_0 \beta_q}{2}\int_{-\infty}^{\infty}[1+\chi(x)]\,\phi^2(x)\,dx = P. \tag{27}$$

The small-signal magnetization components obtained from Eqs. (19) and (25), and (26) are given by

$$m_y = -\kappa\beta_q\psi \tag{28}$$

$$m_z = j\chi\beta_q\psi. \tag{29}$$

For the lowest-order thickness mode, the potential ψ is approximately constant across the film in the limit of small β_q ($\beta_q d$ <<1). In this case, imposing the normalization condition (27) leads to

$$\psi \approx \psi_0 e^{j(\omega_q t - \beta_q z)}, \tag{30}$$

where

$$\psi_0 = \sqrt{\frac{2P}{\omega_q \mu_0}}. \tag{31}$$

Further, in the long wavelength limit the dispersion relation may also be approximated as[35]

$$\beta_q d \approx -\frac{2}{\chi}. \tag{32}$$

Substituting Eqs. (31) and (32) into Eqs. (28) and (29) gives the following simplified expressions for the magnetization components:

$$m_y = m_{0y} e^{j(\omega_q t - \beta_q z)}, \tag{33}$$

$$m_z = m_{0z} e^{j(\omega_q t - \beta_q z)}, \tag{34}$$

where

$$m_{0y} = \frac{2}{d}\left(\frac{\omega_q}{\omega_0}\right)\sqrt{\frac{2P}{\omega_q \mu_0}}, \tag{35}$$

$$m_{0z} = -j\frac{2}{d}\sqrt{\frac{2P}{\omega_q \mu_0}}. \tag{36}$$

The power per unit length P can be related to the current I flowing in a transducer of width w by the approximate expression[35]

$$P = \frac{\omega_q \mu_0}{4[-(1+\chi)]\beta_q d}\left(\cos\left[\frac{\beta_q d}{2}\sqrt{-(1+\chi)}\right]\right)^2 |F|^2 \tag{37}$$

were F is given by

$$F = I e^{-\beta_q s} J_0(\beta_q w/2) \tag{38}$$

and s is the spacing of the transducer above the film.

These expressions are valid for small β_q and the absence of a ground plane. Rigorous calculations of MSW excitation in the presence of a ground plane are given by Weinberg,[36] Adam et al.,[11] and Pu and Tsai.[37]

3.2 Optical Guided Modes of a Thin Film

If the optical index of refraction of the magnetic film is greater than the indices of the surrounding layers, then total internal reflection can occur at the boundaries of the film. In this case, optical guided waves also exist, provided the film is sufficiently thick. In the absence of magneto-optical coupling, such a film supports both TE and TM modes. For TE modes propagating along the z direction, there is a single component of the electric field along y from which the magnetic field components along x and z can be derived. For the lowest-order modes far from cutoff, the dominant component of the magnetic field intensity is h_x. Consequently, the two dominant optical fields for TE modes are

$$E_y = e_y(x)\, e^{j(\omega_{TE}t - \beta_{TE}z)}, \tag{39}$$

$$H_x = h_x(x)\, e^{j(\omega_{TE}t - \beta_{TE}z)}, \tag{40}$$

where the transverse mode function h_x is obtained from e_y using Maxwell's equations:

$$h_x = -\frac{\beta_{TE}}{\omega_{TE}\mu_0} e_y. \tag{41}$$

Note that for the optical wave, we use the free space permeability rather than the susceptibility used for microwave frequencies.

For TM waves propagating along z, there is a single y component of the magnetic field from which the two components of the electric field can be derived. The dominant fields in this case are:

$$H_y = h_y(x)\, e^{j(\omega_{TM}t - \beta_{TM}z)}, \tag{42}$$

$$E_x = e_x(x)\, e^{j(\omega_{TM}t - \beta_{TM}z)}, \tag{43}$$

where

$$e_x = \frac{\beta_{TM}}{\omega_{TM}\varepsilon(x)} h_y, \tag{44}$$

and $\varepsilon(x)$ is the optical permittivity appropriate for each region.

The transverse mode functions e_y and h_y are found by writing appropriate solutions to Maxwell's equations in each region, then requiring the tangential components of **E** and **H** to be continuous at both film surfaces. Explicit expressions for these functions are available in the literature.[35,38] Far from cutoff, however, the mode fields are tightly confined to the film. A large index difference between the film and the surrounding layers further contributes to this confinement. Under this approximation, the fields for the lowest-order TE and TM modes inside the film simplify to

$$e_y^{(0)} = 2\sqrt{\frac{\omega_{TE}\mu_0}{\beta_{TE}d}} \cos\left(\frac{\pi x}{d}\right) \qquad |x| \le \frac{d}{2} \tag{45}$$

$$h_y^{(0)} = 2\sqrt{\frac{\omega_{TM}\varepsilon}{\beta_{TM}d}} \cos\left(\frac{\pi x}{d}\right) \qquad |x| \le \frac{d}{2} \tag{46}$$

where ε is the permittivity of the guiding layer, and the mode amplitudes are normalized to correspond to a power of 1 W/m.

The process of matching the boundary conditions also gives the dispersion relations for the modes. For TE modes, the dispersion relation is obtained from[38]

$$\tan(hd) = \frac{2p}{h[1-(p/h)^2]} \tag{47}$$

where

$$h = \sqrt{\varepsilon(\omega/c)^2 - \beta^2}, \tag{48}$$

$$p = \sqrt{\beta^2 - (\omega/c)^2}, \tag{49}$$

and c is the velocity of light in free space.

Similarly, the dispersion relation for TM modes is given by[38]

$$\tan(hd) = \frac{2\tilde{p}}{h[1-(\tilde{p}/h)^2]}, \tag{50}$$

where

$$\tilde{p} = \frac{\varepsilon}{\varepsilon_c} p, \tag{51}$$

and ε_c is the permittivity in the cladding layers ($|x|>d/2$).

3.3 *Coupled-mode Interactions*

The optical TE and TM modes described in the previous section are uncoupled and orthogonal. However, the presence of a magnetostatic wave introduces off-diagonal components of the permittivity tensor that can cause mode conversion between TE and TM modes. The details of this interaction have been considered theoretically by a number of authors,[12,35,39-48] and are similar to the theories developed for acousto-optics.[49,50] The work of Stashkevich[46] is of particular note because of its completeness. Here we briefly review the key features of the theory using forward volume waves in the collinear interaction geometry as an example.

For definiteness, consider the case of TM→TE conversion. The TE wave only has a y component to the electric field, while the dominant electric field component of the TM wave is along the x direction. The presence of the TM wave could result in the excitation of the TE wave if the permittivity component ε_{yx} is nonzero. In this case, the x component of the electric field from the TM wave would generate a y component of the electric polarization that would, in turn, act as a source for a y component of the electric field and hence excite the TE wave. In a similar way, the presence of a TE wave will excite a TM wave if ε_{xy} is nonzero. From Eq. (18), we see that $\varepsilon_{xy}=(\varepsilon_{yx})^*$ and mode conversion in both directions will occur. The Hermitian form of the permittivity is a general result valid for all lossless media.

If the coupling is weak so that the changes in mode amplitudes occur on length scales much longer than an optical wavelength, then the mode amplitudes are described by the following coupled-mode equations[38]

$$\frac{\partial A}{\partial z} e^{j(\omega_{TE}t - \beta_{TE}z)} = \kappa_{ab} B e^{j(\omega_{TM}t - \beta_{TM}z)} \tag{52}$$

$$\frac{\partial B}{\partial z} e^{j(\omega_{TM}t - \beta_{TM}z)} = \kappa_{ba} A e^{j(\omega_{TE}t - \beta_{TE}z)} \tag{53}$$

where A and B are the normalized amplitudes of the TE and TM modes, respectively, and $\kappa_{ba}=-(\kappa_{ab})^*$ to ensure power conservation. As described above, the coupling coefficient κ_{ab} arises from the tensor element ε_{yx} along with the overlap between the mode functions e_y and e_x:

$$\kappa_{ab} = -j\frac{\varepsilon_0 \omega_{TM}}{4} \int_{-\infty}^{\infty} dx \;\; e_y(x)\, \varepsilon_{yx}(x)\, e_x(x)\,. \tag{54}$$

Alternatively, e_x can be expressed in terms of h_y using Eq. (44):

$$\kappa_{ab} = -j\frac{\varepsilon_0 \beta_{TM}}{4} \int_{-\infty}^{\infty} dx \frac{e_y(x)\, \varepsilon_{yx}(x)\, h_y(x)}{\varepsilon(x)}, \tag{55}$$

where $\varepsilon(x)$ is the diagonal element of the permittivity (assumed isotropic).

If ε_{yx} is not time-varying, then Eqs. (52) and (53) can only be satisfied if $\omega_{TE}=\omega_{TM}$. Further, if ε_{yx} is constant with z, then significant mode conversion can only occur if the phase matching condition $\beta_{TE}=\beta_{TM}$ is satisfied. However, TM and TE modes with the same frequency usually have different wave numbers because the operative boundary conditions for the two modes are different (this is sometimes referred to as *shape anisotropy*). Consequently, significant mode conversion will not occur unless additional steps are taken to cause the two modes to be degenerate.

If ε_{yx} depends on both t and z, then significant mode conversion can occur when both the frequencies and the wave numbers of the TE and TM modes differ. To see this, let us write ε_{yx} explicitly in terms of the magnetization using Eq. (18):

$$\begin{aligned} \varepsilon_{yx}(x,z,t) &= -jfm_z(x,z,t) + 2\,(g_{44}+\Delta g/3)\, M_S m_y(x,z,t) \\ &= -jf\left(\frac{1}{2} m_{0z}(x)\, e^{j(\omega_q t - \beta_q z)} + cc\right) \\ &\quad + 2\,(g_{44}+\Delta g/3)\, M_S \left(\frac{1}{2} m_{0y}(x)\, e^{j(\omega_q t - \beta_q z)} + cc\right) \end{aligned} \tag{56}$$

where cc represents the complex conjugate. The first term on the right-hand-side of Eq. (56) results from the Faraday effect, and the second term results from the Cotton-Mouton effect. Substituting Eqs. (55) and (56) into Eq. (52) allows us to rewrite the first coupled equation (52) as

$$\frac{\partial A}{\partial z} e^{j(\omega_{TE}t - \beta_{TE}z)} = -j\frac{\varepsilon_0 \beta_{TM}}{8} B \int_{-d/2}^{d/2} dx \; \frac{e_y(x) h_y(x)}{\varepsilon(x)}$$

$$\times \{ [-jfm_{0z}(x) + 2(g_{44} + \Delta g/3) M_S m_{0y}(x)] e^{j[(\omega_{TM} + \omega_q)t - (\beta_{TM} + \beta_q)z]}$$

$$+ [-jfm_{0z}{}^*(x) + 2(g_{44} + \Delta g/3) M_S m_{0y}{}^*(x)] e^{j[(\omega_{TM} - \omega_q)t - (\beta_{TM} - \beta_q)z]} \} \quad (57)$$

Note that we have limited the range of integration to $\pm d/2$ since the magnetization vanishes outside the film. Since the two terms on the right-hand-side of Eq. (57) have different frequencies, both cannot be simultaneously synchronous with the TE wave on the left-hand-side. The term that is capable of synchronously driving the TE wave depends on the specific geometry. For example, it is usually the case that $\beta_{TE} > \beta_{TM}$. If the magnetostatic wave is propagating parallel to both optical modes, then the phase matching condition needed for maximum conversion will be satisfied by the first term on the right-hand-side and

$$\omega_{TE} = \omega_{TM} + \omega_q \quad (58)$$

$$\beta_{TE} = \beta_{TM} + \beta_q. \quad (59)$$

Similarly, if the magnetostatic wave is propagating antiparallel to both optical beams, then the phase matching condition is satisfied by the second term yielding

$$\omega_{TE} = \omega_{TM} - \omega_q \quad (60)$$

$$\beta_{TE} = \beta_{TM} - \beta_q. \quad (61)$$

If we assume that only the terms with frequency ω_{TE} are kept, then the coupled-mode equations (52) and (53) can be written in the following form:

$$\frac{\partial A}{\partial z} = \kappa_{ab} B e^{-j\Delta z} \quad (62)$$

$$\frac{\partial B}{\partial z} = -\kappa_{ab}{}^* A e^{j\Delta z} \quad (63)$$

where

$$\Delta = \beta_{TM} - \beta_{TE} \pm \beta_q, \quad (64)$$

$$\kappa_{ab} = \kappa^F + \kappa^{CM}, \quad (65)$$

$$\kappa^F = -\frac{f\beta_{TM}}{8\varepsilon_r} \int_{-d/2}^{d/2} dx \; e_y(x) h_y(x) \begin{bmatrix} m_{0z}(x) \\ m_{0z}{}^*(x) \end{bmatrix}, \quad (66)$$

and

$$\kappa^{CM} = -j\frac{\beta_{TM}(g_{44}+\Delta g/3)M_S}{4\varepsilon_r}\int_{-d/2}^{d/2} dx\ e_y(x)\,h_y(x)\begin{bmatrix} m_{0y}(x) \\ m_{0y}^*(x) \end{bmatrix}. \tag{67}$$

where ε_r is the (diagonal) relative permittivity in the guiding layer. The upper terms in the brackets in Eqs. (66) and (67) go with the upper sign in Eq. (64), and correspond with parallel propagation of the optical and magnetostatic wave modes. In writing Eqs. (66) and (67) we have assumed that the film permittivity is uniform through the thickness.

The coupling coefficients (66) and (67) can be simplified considerably if we assume the magnetization is approximately constant through the film and that the optical modes are tightly confined. Substituting Eqs. (45) and (46) into (66) and (67) and performing the indicated integrations gives

$$\kappa^F = -\frac{\Phi_F}{2}\left(\frac{m_{0z}}{M_S}\right), \tag{68}$$

$$\kappa^{CM} = -j\frac{\Phi_{CM}}{2}\left(\frac{m_{0y}}{M_S}\right), \tag{69}$$

where Φ_F and Φ_{CM} are the Faraday rotation and the Cotton-Mouton rotation in the film, respectively, and are given by[45,51]

$$\Phi_F = \frac{fM_S k_0}{2n}, \tag{70}$$

$$\Phi_{CM} = \left(g_{44}+\frac{\Delta g}{3}\right)\frac{M_S^2 k_0}{n}, \tag{71}$$

where k_0 is the free-space optical wave number, and n is the isotropic index of refraction.

Since $\omega_q \approx \omega_0$ for long wavelength magnetostatic forward volume waves, Eqs. (35) and (36) show that the small-signal magnetization is approximately circularly polarized. It follows that $m_{0y} \approx jm_{0z}$ for propagation along z (parallel propagation) and $m_{0y}^* \approx -jm_{0z}^*$ for propagation along $-z$ (antiparallel propagation). Making this approximation allows the total coupling coefficient to be written

$$\kappa_{ab}^{\pm} \approx j(\Phi_F \mp \Phi_{CM})\frac{\sin\theta}{2} \tag{72}$$

where $\sin\theta = |m_{0z}/M_S|$, θ is the precession angle of the small-signal magnetization, and the upper and lower signs correspond to parallel and antiparallel propagation, respectively. Clearly the coupling strength depends on the relative signs of the Faraday and Cotton-Mouton coefficients as well as the direction of propagation.[5,6,52]

To obtain a simple physical interpretation of (72), let us first consider only coupling via the Faraday effect. For a static magnetization, the amount of Faraday rotation is proportional to the component of **M** along the direction of light propagation. Thus, if **M** is tilted so that it makes an angle θ with respect to the film normal, the magnitude of the coupling constant is $|\kappa_{ab}|=|\Phi_F| \sin\theta$. (Clearly, if **M** is parallel to the direction of light propagation, the mode conversion is just a description of Faraday rotation and $|\kappa_{ab}|=|\Phi_F|$.) The additional factor of 1/2 in Eq. (72) results from the time variations of the magnetization. A similar interpretation can be given to the Cotton-Mouton term in Eq. (72), except mode conversion arises from the component of **M** along y rather than z.

Eq. (72) can be expressed in terms of the power per unit width in the magnetostatic wave using Eq. (36):

$$\kappa_{ab}^{\pm} \approx j\frac{(\Phi_F \mp \Phi_{CM})}{M_S d}\sqrt{\frac{2P}{\omega_q \mu_0}}\,. \tag{73}$$

The complete solutions to Eqs. (62) and (63) are found to be[35]

$$A(z) = e^{j\Delta z/2}\left[A(0)\cos\beta_0 z + \frac{\kappa_{ab}B(0) - j(\Delta/2)A(0)}{\beta_0}\sin\beta_0 z\right], \tag{74}$$

$$B(z) = e^{-j\Delta z/2}\left[B(0)\cos\beta_0 z + \frac{\kappa_{ba}A(0) + j(\Delta/2)B(0)}{\beta_0}\sin\beta_0 z\right], \tag{75}$$

where $A(0)$ and $B(0)$ are the mode amplitudes at z=0, and

$$\beta_0 = \sqrt{|\kappa|^2 + (\Delta/2)^2}. \tag{76}$$

As an example, suppose the incident optical mode is TM so that $A(0)=0$ and $B(0)=B_0$. The normalized power in the converted mode is given by

$$\left|\frac{A(z)}{B_0}\right|^2 = \frac{|\kappa|^2}{|\kappa|^2 + (\Delta/2)^2}\sin^2\left(z\sqrt{|\kappa|^2 + (\Delta/2)^2}\right). \tag{77}$$

The optical power is periodically transferred between the TE and TM modes, resulting in the repeating peaks and zeros of Eq. (77). It is clear from (77) that total mode conversion cannot occur unless the modes are phased matched, i.e., unless Δ=0. If the modes are phased matched, then the optical power will be totally tranferred from mode B to mode A in a length

$$L_c = \frac{\pi}{2|\kappa|}. \tag{78}$$

3.4 Tunable Microwave Filter

When the collinear interaction is observed at low power levels, the diffraction takes place over a relatively narrow microwave bandwidth. From Eq. (77), the bandwidth between first nulls is determined approximately by the condition $\Delta=4\pi/L$, if L is the interaction distance and $\Delta \gg \kappa$ near the zero (valid for weak interactions). If $\beta_q d \ll 1$, the frequency of the forward volume waves is $\omega_q \approx \omega_0$, and the bandwidth is approximately[53,54]

$$\Delta f = v_g \frac{\Delta}{2\pi} \approx \frac{\pi f_M d}{L}, \tag{79}$$

where $f_M=|\gamma|\mu_0 M_S/(2\pi)$, v_g is the group velocity, and we have used the approximation $v_g \approx \pi f_M d/2$. For YIG, $f_M \approx 5$ GHz and Δf is typically 5-10 MHz. Thus the bandwidth is constant with frequency and the frequency can be tuned continuously by varying the applied bias field. This phenomenon can be used to realize a type of tunable microwave filter.[55,56] If the diffracted beam is sensed with diode detectors, the output will be the modulation envelope of the microwave signal rather than the microwave signal itself. Such a device could be called a demodulating filter. On the other hand, optical heterodyne detection of the diffracted beam would permit complete recovery of the microwave signal. Since the recovered microwave signal would be proportional to the amplitudes of the optical beams used in the up and down conversion steps, Talisa, et al.,[55] have argued that such a filter could, in principle, realize zero insertion loss.

A filter bank can be made by placing several of these collinear MSW-optical filter elements in a linearly-graded bias field and connecting the inputs in parallel.[55] Such a structure has the potential for a higher physical channel density (chan/cm^3) than either coupled-sphere resonators or an equivalent MSW multichannel filter without the optical interactions.

3.5 Optical Frequency Shifter

As discussed in connection with Eqs. (58) and (60), the frequency of the diffracted optical beam is shifted by the frequency of the microwave signal. By simultaneously varying both the bias field and the microwave frequency so that the MSW wavelength remains constant, the interaction can be maintained as the frequency is tuned continuously over many GHz in frequency. The frequency shift is single sideband and no deflection of the output beam occurs as the frequency is varied. The multigigahertz frequency range and the absence of beam deflection are advantages over acousto-optical techniques, in which the absence of beam deflection can be achieved only over limited frequency ranges.[57,58] The optical frequency shift resulting from the collinear MSFVW-optical interaction has been directly measured over the microwave frequency range of 2-12 GHz by Bilaniuk, and coworkers.[59] A frequency shifter such as this may find applications in coherent optical communications systems,[60] optical heterodyne spectroscopy,[61] and the optical generation of microwave signals. As an example of the latter application, control signals for the elements of phased array antennas could be generated by nonlinear mixing between a frequency-shifted optical signal and an optical reference signal. These optical signals could be conveyed to the array elements using fibers. Use of an MSW-optical frequency shifter would permit continuous amplitude and phase control of the array element excitations.

3.6 Other Device Possibilities

A number of additional applications of the collinear interaction have been proposed by Fisher.[41] As examples, we will briefly mention two of these: an optical modulator and a tunable optical filter.

Since the diffracted optical beam amplitude is proportional to the amplitude of the MSW, the collinear interaction can be used as an optical modulator. The switching speed is faster than an analogous acousto-optical device owing to the higher MSW propagation velocity. However, it is unlikely that modulation rates competitive with electro-optical techniques can be achieved with MSWs.

The interaction can also be used as a tunable optical filter owing to the dependence of $\beta_{TE}-\beta_{TM}$ on optical wavelength. If the wavelength of the light is varied and the MSW wavelength held constant, it is clear that the interaction would take place only when $\beta_{TE}-\beta_{TM}$ satisfies the phase-matching condition $\Delta=0$. Changing the MSW wavelength would cause the interaction to take place at a different optical wavelength. Tuning can be accomplished by varying either the bias field or the microwave frequency.

4.0 Effects of Microwave and Optical Attenuation[62,63]

The presentation in the previous section was based on the assumption of lossless media. If both microwave and optical losses are included, the coupled-mode equations take the form

$$\frac{\partial A}{\partial z} = -\alpha_o A + \kappa_0 e^{-\alpha_M z} B e^{-j\Delta z} \tag{80}$$

$$\frac{\partial B}{\partial z} = -\alpha_o B - \kappa_0{}^* e^{-\alpha_M z} A e^{j\Delta z} \tag{81}$$

where the optical and magnetostatic mode amplitudes are assumed to decay according to the factors $\exp(-\alpha_o z)$ and $\exp(-\alpha_M z)$, respectively, and κ_0 is the maximum coupling coefficient at the point of magnetostatic wave excitation. Under conditions of perfect phase matching $\Delta=0$ these equations can be solved exactly. If κ_0 is assumed to be real, the solutions take a particularly convenient form. Taking κ_0 to be real causes no loss of generality since this just establishes the phase reference for the microwave signal. Making these assumptions, the solutions for $A(0)=0$ and $B(0)=B_0$ are:

$$A(z) = B_0 e^{-\alpha_o z} \sin\left[\frac{\kappa_0}{\alpha_M}(1-e^{-\alpha_M z})\right] \tag{82}$$

$$B(z) = B_0 e^{-\alpha_o z} \cos\left[\frac{\kappa_0}{\alpha_M}(1-e^{-\alpha_M z})\right]. \tag{83}$$

The effect of magnetic damping is to cause the MSW to exponentially decay in amplitude with propagation. This results in a continuous weakening of the coupling coefficient as the waves move away from the point of MSW excitation. Eventually, the coupling becomes sufficiently weak that no further mode conversion takes place. If the loss is large enough, this condition may be reached before total mode conversion can occur. Examination of these equations shows that for total conversion to occur, the arguments of the sinusoids must exceed $\pi/2$. Thus we have the following condition for total mode conversion:

$$\frac{\kappa_0}{\alpha_M} \geq \frac{\pi}{2}. \tag{84}$$

When optical attenuation is taken into account, it is clear that even when total mode conversion takes place, the intensity of the output beam will be less than that of the input beam. In such a case, it may be more desirable to choose the length to maximize the intensity of the converted beam rather than to insist on total mode conversion.[64] Thus, in the presence of both optical and microwave magnetic losses, the optimum conversion length L is obtained by finding the maximum of Eq. (82) with respect to z. This leads to the following transcendental equation for the optimum length:

$$\tan\left[\frac{\kappa_0}{\alpha_M}\left(1-e^{-\alpha_M L}\right)\right] = \frac{\kappa_0}{\alpha_o} e^{-\alpha_M L} . \tag{85}$$

Another effect of microwave loss is the broadening of the interaction bandwidth. Referring to Eq. (79), the bandwidth depends inversely on length if the MSW amplitude is constant along the interaction path. In the presence of microwave attenuation, the decay of the MSW mode amplitude will result in a lower total conversion than would otherwise be expected. This can be thought of as resulting from an effective interaction length[54] that is shorter than the actual interaction length. It follows that the bandwidth will be boadened as well.

5.0 High Power Effects

As indicated by Eqs. (72) and (73), efficient optical mode conversion requires large cone angles and consequently high microwave power. Under these circumstances, we must closely examine the linearization approximations that lead to the small-signal susceptibility appearing in Eq. (19). To identify the first nonlinear effects and how they effect the MSW-optical interaction, let us return to the Landau-Lifshitz equation of motion:

$$\frac{\partial \mathbf{M}}{\partial t} = \gamma\mu_0 \mathbf{M}\times\mathbf{H} . \tag{86}$$

Assuming the static bias field and the equilibrium magnetization are along the x direction, it is convenient to follow the usual procedure of separating the fields into static and time-varying components:

$$\mathbf{M} = \mathbf{M}_0 + \mathbf{m}(t) , \tag{87}$$

$$\mathbf{H} = \mathbf{H}_0 + \mathbf{h}(t) . \tag{88}$$

On the right-hand-side of these equations we have used upper-case symbols for static components and lower-case symbols for dynamic components. Substituting these components into the Landau-Lifshitz equation (86) leads to

$$\frac{\partial \mathbf{m}}{\partial t} = \gamma\mu_0 \left[\mathbf{M}_0\times\mathbf{H}_0 + \mathbf{M}_0\times\mathbf{h} - \mathbf{H}_0\times\mathbf{m} + \mathbf{m}\times\mathbf{h}\right] . \tag{89}$$

Assuming the static components are parallel and point along the equilibrium direction of the magnetization, the first term on the right-hand-side vanishes. In the usual linearization procedure, both **m** and **h** are taken to be small in magnitude so that their product can be neglected to first-order in small quantities. After neglecting the last term in Eq. (89), the magnitude of the static magnetization $\mathbf{M}_0$ is approximated by the saturation magnetization M_s since the precession cone angle is taken to be small. The linearized susceptibility in Eq. (19) then results.

We are now interested in the case where **m** is circularly polarized (or nearly so), but not necessarily small. Since the magnitude of **M** is conserved, the component M_0 along the equilibrium direction will be reduced to $M_s\cos\theta$. Further, if **m** is not small, then the magnitude of the last term may no longer be negligible. To examine the effects of this term, let us consider the frequency components that may be present in Eq. (89). For definiteness, let us assume that the magnetic system is being driven by a sinusoidal field **h** with frequency ω. We expect the dominant response to be a precession of **m** at the frequency ω, although nonlinearities may generate additional frequency components. Our primary interest is the effect of a large precession amplitude on a propagating wave at frequency ω rather than the additional frequency components pumped by nonlinearities. Consequently, we wish to keep only the terms in Eq. (89) that contribute components at frequency ω. The first two non-vanishing terms on the right-hand-side of Eq. (89) are linear in the time-varying fields **m** and **h** and are therefore of frequency ω. However, the last term involves products of the time-varying fields and must be considered more carefully. For definiteness, let us take the equilibrium direction to lie along x and separate Eq. (89) into components parallel and transverse to x:

$$\frac{\partial \mathbf{m}_t}{\partial t} = \gamma\mu_0\,[\mathbf{M}_0 \times \mathbf{h}_t - \mathbf{H}_0 \times \mathbf{m}_t + \mathbf{m}_x \times \mathbf{h}_t + \mathbf{m}_t \times \mathbf{h}_x] \tag{90}$$

$$\frac{\partial \mathbf{m}_x}{\partial t} = \gamma\mu_0\,[\mathbf{m}_t \times \mathbf{h}_t]\,. \tag{91}$$

If the fields are approximately circularly polarized, it follows that $|\mathbf{m}_x| \ll |\mathbf{m}_t|$. Consequently we neglect the third term on the right-hand-side of Eq. (90). Eq. (91) could be used to obtain an approximation to $\mathbf{m}_x$, if desired. However, since it is a product of two fields with frequency ω, $\mathbf{m}_x$ will be at the frequency 2ω and is therefore not of primary interest to us. Similarly, the last term in Eq. (90) will have frequency 2ω and will therefore not contribute to the wave fields with frequency ω. We conclude, then, that the term $\mathbf{m}\times\mathbf{h}$ can again be neglected–but for a different reason: the magnitude may not be negligible, but it makes a negligible contribution to the field components with frequency ω. Dropping the last two terms in Eq. (90) leads to the equation of motion in the limit of large precession angle but nearly-circular polarization:

$$\frac{\partial \mathbf{m}_t}{\partial t} = \gamma\mu_0\,[\mathbf{M}_0 \times \mathbf{h}_t - \mathbf{H}_0 \times \mathbf{m}_t] \tag{92}$$

with

$$M_0 = M_S\cos\theta. \tag{93}$$

The equation of motion (92) can be rewritten in the same form as Eq. (19) but with a non-linear susceptibility obtained by substituting M_0 for M_S in the elements of the susceptibility tensor according to Eqs. (22) and (23). It should be pointed out, however, that simply substituting M_0 for M_S in Eq. (17) will *not* give the required corrections to the Cotton-Mouton tensor at large cone angles. Instead, Eq. (15) must be used.

5.1 MSW Wavenumber Shift at Large Cone Angles

From the discussion of the previous section, we conclude that when the fields are approximately circularly polarized, the effects of a large cone angle can be obtained by reducing the saturation magnetization by $\cos\theta$ in the equations describing the behavior of magnetostatic waves. Of particular interest to the MSW-optical interaction is any shift in MSW wavenumber that would destroy phase matching between the MSW and optical modes. For small shifts we can write

$$\Delta\beta_q = \frac{\partial\beta_q}{\partial M}\Delta M, \tag{94}$$

where

$$\Delta M = M_S(\cos\theta - 1). \tag{95}$$

To obtain the derivative in Eq. (94), we can approximate the dispersion relation in the long-wavelength limit as

$$\beta_q = (\omega_q - \omega_0)\frac{\partial\beta_q}{\partial\omega_q} \tag{96}$$

where

$$\frac{\partial\beta_q}{\partial\omega_q} = \frac{4}{\omega_M d} \tag{97}$$

and ω_0 is given by Eq. (22). Substituting Eqs. (22) and (97) into the dispersion relation (96) and simplifying gives

$$\frac{\partial\beta_q}{\partial M} = \frac{4}{M_S d}\frac{(\omega_{dc} - \omega_q)}{\omega_M} \tag{98}$$

where $\omega_{dc}=-\gamma\mu_0 H_{dc}$. Combining Eqs. (94) and (95), and (98) leads to the following expression for the shift in MSW wavenumber:

$$\Delta\beta_q = \frac{4}{d}\frac{(\omega_{dc} - \omega_q)}{\omega_M}(\cos\theta - 1). \tag{99}$$

If this shift is large compared with the interaction bandwidth, then the phase-matching condition will be spoiled as the cone angle is increased resulting in diminished mode conversion. As shown in Fig. 4, the phase-matching condition can be restored at large cone angles (high power) by shifting the frequency by the amount

$$\Delta\omega_q = -\left.\frac{\partial\omega_q}{\partial\beta_q}\right|_\theta \Delta\beta_q, \tag{100}$$

where

$$\left.\frac{\partial\omega_q}{\partial\beta_q}\right|_\theta = \frac{\omega_M d}{4}\cos\theta. \tag{101}$$

Using Eqs. (99) and (101), the frequency shift (100) becomes

$$\Delta\omega_q = (\omega_{dc} - \omega_q)\cos\theta\,(1 - \cos\theta). \tag{102}$$

For small θ, the frequency shift can be approximated by

$$\Delta\omega_q \approx (\omega_{dc} - \omega_q)\frac{\theta^2}{2}. \tag{103}$$

Using Eq. (36), the precession angle in this limit is

$$\theta \approx \frac{2}{M_S d}\sqrt{\frac{2P}{\omega_q \mu_0}}. \tag{104}$$

The frequency shift can finally be written:

$$\Delta\omega_q \approx \frac{(\omega_{dc} - \omega_q)}{\omega_q}\frac{4P}{\mu_0 M_S^2 d^2}. \tag{105}$$

If the magnetostatic waves were lossless so that the amplitude was uniform along the interaction distance, then the interaction peak would simply be shifted in frequency by the amount given by Eq. (105). However, in the presence of loss, the amplitude will decay exponentially away from the point of excitation. Consequently the wave number shift will vary with position; it will have its largest value at the excitation antenna, and will decay asymptotically to zero away from the antenna. In this situation the interaction peak would not simply be shifted, but would be smeared out over a frequency range determined by the maximum precession angle beneath the antenna. This effect may explain the broadened bandwidth at high power observed by several researchers.[65,66]

5.2 Thermal Effects

The dissipation of microwave power within the film will cause temperature increases. Although the detailed temperature distribution will depend on lateral thermal diffusion, we expect the rise in temperature to be highest where the magnetostatic wave

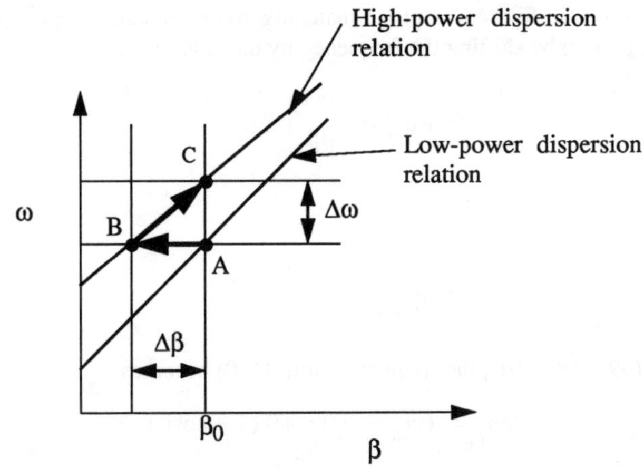

FIGURE 4. Changes in the MSW dispersion relation at high power. Point A represents the phase-matching point at low power. When the power is increased, the operating point shifts to B and phase-matching is destroyed. Phase-matching can be restored at high power by increasing the frequency thereby moving the operating point to C.

amplitude is greatest. Since a rise in temperature reduces the saturation magnetization, effects qualitatively similar to an increase in precession angle are expected. A change in temperature would also modify the optical index of refraction. This could cause distortion of the optical wavefront and spreading of the beam. The introduction of additional optical wave vector components in this way could also broaden the interaction peak and affect the spatial distribution of the output optical beams. Further work is needed to fully understand these effects.

6.0 Garnet Heterostructures

6.1 Basic Principles

All of the experimental investigations of MSW-optical guided-wave interactions to date have involved a single film of YIG. Significant improvements in diffraction efficiency can be obtained by using Bi-substitued films[45,67] and by using nonuniform bias fields to focus the MSWs.[64] However, use of a single layer places rather stringent material requirements upon the film since it must simultaneously exhibit low optical absorption, low ferromagnetic resonance line width, and high Faraday rotation. In addition, microwave and optical modes place conflicting restrictions on the thickness of the guiding layer. To achieve optical single mode operation, the film should be on the order of 2 microns or less in thickness. However, spin wave resonance absorption lines prevent low-loss magnetostatic wave propagation in films thinner than about 5 microns.[68] The microstrip radiation impedance is also more favorable for thicker films permitting more efficient coupling to magnetostatic waves. In response to this difficulty, several authors have suggested that the desired qualities can be more easily realized in a structure consisting of two or more layers.[48,56,69,70] In such a case, alternating layers could be optimized for magnetic and optical properties. In addition to allowing separate minimizations of optical and magnetic losses, independent control of the optical and magnetostatic wave dispersion relations would be possible by varying the thicknesses of the layers. In a recent paper,[30] a design procedure was presented that includes MSW losses in an approximate way. In the present discussion, we will briefly review the concepts underlying enhanced interactions in heterostructures and the design of these structures. The improvements to the coupled-mode analysis described in Section 4 will be used to estimate the optimum device length and the resulting diffraction efficiency in the presence of both optical and microwave losses.

Since the desired properties are difficult to achieve in a single layer, the first obvious advantage of the heterostructure is that the demands on any given layer are relaxed. For example, the microwave layer can be YIG which has low microwave losses but modest Faraday rotation, and the optical layer can be (Bi,Lu)IG which has high Faraday rotation but (typically) higher microwave losses than pure YIG. (As discussed previously, large Faraday rotation is required for a strong MSW-optical interaction.) Both layers can be epitaxially grown on GGG substrates.[45] An example of the proposed geometry is shown in Fig. 5. In this figure, layer 1 is the optical layer and layer 3 is the microwave layer. All other layers are nonmagnetic. We will restrict our consideration to the case of optical interactions with magnetostatic forward volume waves.

To ensure that the microwave propagation is dominated by the microwave layer, the magnetic properties of the layers are chosen so that a frequency range exists where magnetostatic waves propagate in layer 3 only (Fig. 6). when the operating point is chosen to lie in this range, the microwave energy is concentrated in layer 3 where the losses are small.

The decrease in microwave coupling between the layers resulting from their separation can be recovered by a proper choice of the magnetic parameters of the two layers as follows. There is a resonance in the susceptibility at the bottom of the forward volume wave band. As shown in Fig. 6, the resonance in the optical layer susceptibility can be approached from below in such a way as to enhance the response of the magnetization to the fringing fields from layer 3. The decrease in the magnetic response in the optical layer owing to the separation can therefore be recovered.

The optical mode analysis is simplified by the presence of the spacing layer (layer 2). When this layer is greater than about 0.5 μm, the optical mode fields do not extend into the microwave layer. As a result, the optical attenuation of the microwave layer is not important, and the optical modes are determined only by the three-layer structure consisting of the cover, layer 1, and layer 2. In contrast, because of the much longer MSW wave length, the presence of layer 2 results in only a modest decrease in microwave coupling. Although we want most of the microwave energy to be concentrated in layer 3, some coupling between the layers is required for MSW-induced optical mode conversion to occur in layer 1.

Let us now discuss these ideas in a more quantitative manner. If both the optical and MSW modes are propagating along the z direction, the optical mode coupling constant in layer 1 is given approximately by (cf. Eq. (72))

$$|\kappa| \approx \left|\frac{(\Phi_F \mp \Phi_{CM})}{2}\right|\left|\frac{m_{1z}}{M_{S1}}\right| \tag{106}$$

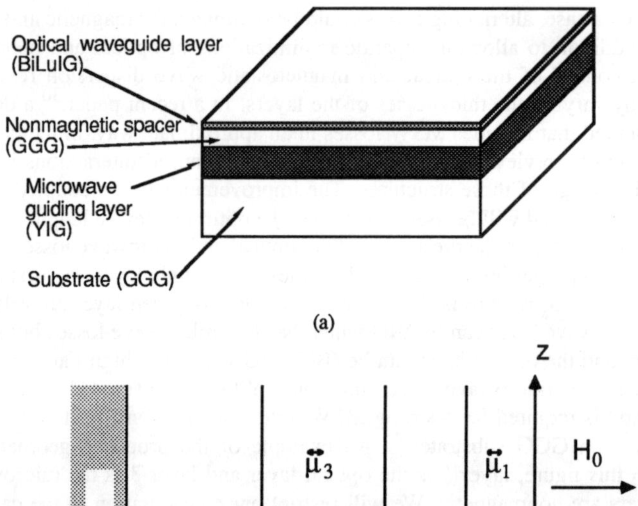

FIGURE 5. Five-layer dielectric structure for optical and magnetostatic mode analysis. (a) Geometry of the epitaxial layers. (b) Coordinate system used in the analysis.

where the subscript "1" refers to layer 1. For this geometry, the small-signal magnetization in both layers is related to the small-signal magnetic field intensity by (cf. Eq. (19)):

$$m_{iz} = \chi_i h_{iz} \qquad i = 1, 3 \tag{107}$$

where the susceptibility in the ith layer is given by Eq. (20) evaluated with the appropriate material parameters for the layer under consideration. Eq. (20) is strictly valid only in the lossless case, but is a reasonable approximation if the resonance in the denominator is not approached too closely. Using Eq. (107) and the fact that h_{iz} decays exponentially away from layer 3 leads us to the approximate expression

$$\frac{m_{1z}}{m_{3z}} \approx \frac{\chi_1}{\chi_3} e^{-\beta_q d_{sep}}, \tag{108}$$

where the effective separation distance is approximated by

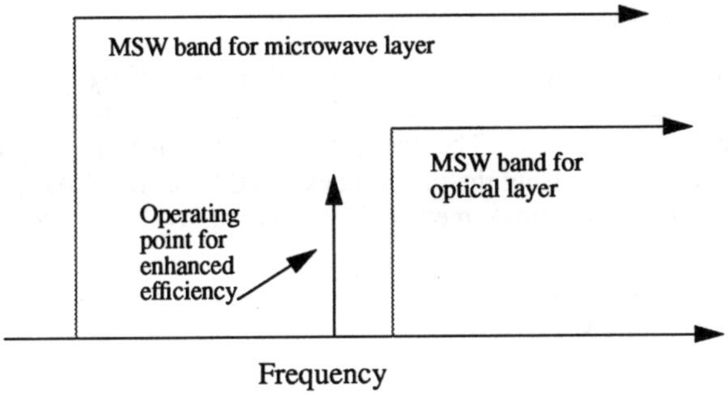

FIGURE 6. The relationship between the operating point and the MSW passbands in the two magnetic layers.[48,56]

$$d_{sep} \approx \frac{d_1}{2} + d_2 + \frac{d_3}{2}. \tag{109}$$

For near-circular precession, we can write

$$\left|\frac{m_{1z}}{m_{3z}}\right| \approx \frac{|\mathbf{m}_1|}{|\mathbf{m}_3|} \equiv \eta. \tag{110}$$

To recover the separation loss, we choose the magnetic parameters so that the ratio of the susceptibilites in Eq. (108) compensates for the separation factor $\exp(-kd_{sep})$:

$$\eta \geq 1. \tag{111}$$

Alternatively, one can impose a minimum absolute value on m_{1z} necessary to achieve a desired coupling strength as given by Eq. (106). This is clearly the stronger of the two conditions.

The ratio of the magnetizations η cannot be arbitrarily large, however. If the operating point is too close to the susceptibility resonance in layer 1, then Eq. (20) becomes invalid, and layer 1 begins to dominate the loss characteristics. We must therefore establish a criterion for ensuring that the energy is concentrated in layer 3. This will effectively limit the allowed value of η. We will return to this topic in the next section.

6.2 Microwave Loss in a Heterostructure

As mentioned previously, the MSW amplitude decays as exp($-\alpha_M z$) in the presence of magnetic damping. For a heterostructure, the MSW attenuation constant α_M depends on the properties of each layer. The attenuation constant for the magnetostatic mode can be rigorously calculated using the method described by Buris.[71] As an alternative, we present the following approximate method for estimating α_M based upon the lossless dispersion relation.

The attenuation constant for the magnetostatic mode can be obtained from the mode relaxation time T using the expression[72]

$$\alpha_M = \frac{1}{T}\frac{\partial k}{\partial \omega_0}. \tag{112}$$

The derivative may be estimated numerically with any computer program capable of calculating the dispersion relation. The propagation loss is then given by

$$P_L = -8.686\alpha_M \qquad (dB/m). \tag{113}$$

We now turn our attention to obtaining an approximation for the heterostructure relaxation time T. The following derivation uses the approximations that the small-signal magnetization is circularly polarized and approximately constant within each magnetic film. As previously remarked, these approximations are valid in the long-wavelength regime most often employed with MSW-optical interactions. Our approach will be to obtain the inverse relaxation time from the ratio of the time-averaged power dissipated per unit volume to the time-averaged energy density.

We begin by calculating the time-averaged power dissipated per unit volume P_d by magnetic loss mechanisms. From the complex Poynting theorem,

$$P_d = \frac{1}{2}\mathrm{Re}\,\{j\omega\mathbf{h}^* \bullet \mathbf{b}\}\ . \tag{114}$$

Using the constitutive relations $b_i = \mu_0(h_i + m_i)$ and $h_i = \chi^{-1}{}_{ij}m_j$, the dissipated power density can be written

$$P_d = \frac{\omega\mu_0}{2}Re\,\{-jm^*{}_k\chi^{-1}_{kl}m_l\}\,, \tag{115}$$

where the inverse magnetic susceptibility obtained from the Gilbert form of the Landau-Lifshitz equation in the present coordinate system is[35]

$$[\chi^{-1}_{kl}] = \frac{1}{\omega_0}\begin{bmatrix}\omega_0 + j\omega\alpha & -j\omega \\ j\omega & \omega_0 + j\omega\alpha\end{bmatrix}. \tag{116}$$

Substituting the inverse susceptibility (116) into the dissipated power density (115) and simplifying gives

$$P_d = \frac{\mu_0\omega}{2\omega_M} Re\{(-j\omega_0 + \omega\alpha)(|m_y|^2 + |m_z|^2) + \omega(m_y m_z^* - m_y^* m_z)\}. \tag{117}$$

Noting that the last term in parentheses is imaginary, the real part of the total expression can be explicitly taken. The dissipated power density then reduces to

$$P_d = \frac{\mu_0}{2}\frac{\omega}{\omega_M}\frac{|\mathbf{m}|^2}{T_0} \quad (W/m^3) \tag{118}$$

where

$$|\mathbf{m}|^2 = |m_y|^2 + |m_z|^2, \tag{119}$$

and the uniform precession relaxation time T_O is related to the Gilbert damping parameter α by[72]

$$\frac{1}{T_0} = \omega\alpha. \tag{120}$$

Next, we need to compute the energy density stored in the mode fields. Although the stored energy extends to regions away from the film because of magnetic fringing fields, the total time-averaged mode energy can be more conveniently obtained from the *pseudo-kinetic energy density* [35,73] which is nonzero only within the magnetic layers:

$$W = -j\frac{\mu_0}{4}\frac{\omega}{\omega_M}(\mathbf{m} \times \mathbf{m}^*) \bullet \hat{\mathbf{x}}. \tag{121}$$

Assuming the precession is in the resonant sense and is approximately circular, $m_Z = -jm_Y$, and W becomes

$$W = \frac{\mu_0}{4}\frac{\omega}{\omega_M}|\mathbf{m}|^2. \tag{122}$$

The mode relaxation time is given by

$$\frac{1}{T} = \frac{1}{2}\frac{\int_{-\infty}^{\infty} P_d dx}{\int_{-\infty}^{\infty} W dx}. \tag{123}$$

This expression can be evaluated using the dissipated power density (118) and the stored energy density (122). If the small-signal magnetization is taken to be uniform through the thickness of each layer, and the ratio of the magnitudes of the small-signal magnetization is denoted

$$\frac{|\mathbf{m}_1|}{|\mathbf{m}_3|} \equiv \eta, \tag{124}$$

the inverse relaxation time (123) can be written

$$\frac{1}{T} = \frac{\left[\dfrac{d_3}{\omega_{M3}T_3} + \dfrac{\eta^2 d_1}{\omega_{M1}T_1}\right]}{\left[\dfrac{d_3}{\omega_{M3}} + \dfrac{\eta^2 d_1}{\omega_{M1}}\right]}, \tag{125}$$

where T_1, T_3 are the uniform precession relaxation times in layers 1 and 3, respectively.

Recall from the previous section that the objective is to localize most of the mode energy to layer 3 where the losses are small. If the properties of layer 3 are to dominate, then Eq. (125) indicates that we must require

$$\frac{d_3}{d_1}\frac{\omega_{M1}}{\omega_{M3}}\frac{T_1}{T_3}\frac{1}{\eta^2} \gg 1. \tag{126}$$

Typically, the saturation magnetizations of the two layers will be comparable. If we further take η=1, then the ratio of the film thicknesses must be chosen to overcome the ratio of the relaxation times. Under these conditions the energy will be concentrated in layer 3 as desired.

The propagation loss P calculated with the approximate relaxation time method is compared with a rigorous calculation in Fig. 7.[30] The approximate solution (solid circles) agrees well with the rigorous calculation (curve). The curve was obtained using the method reported by Buris.[71] The sharp dip in the curve marks the resonance in the susceptibility of layer 1. Above this frequency magnetostatic waves can propagate in both film layers. To ensure that the energy is concentrated in layer 3, operation at a frequency below this dip is required.

6.3 Design Example

As an example, consider two layers with equal saturation magnetizations and magnetic losses characterized by ferromagnetic resonance (FMR) line widths (FWHM) ΔH_1=2 Oe, and ΔH_3=0.5 Oe. The FMR line width is related to the uniform precession relaxation time by[72]

$$\frac{1}{T_0} = -\frac{\gamma\mu_0\Delta H}{2}. \tag{127}$$

The design can be divided into three steps as follows.

1. Choose film thicknesses to offset the difference in FMR line width. To make sure we do not get too close to the susceptibility resonance in layer 1, we make the conservative choice η=1. For example, with d_1=1.9 μm and d_3=50, Eq. (126) reduces to

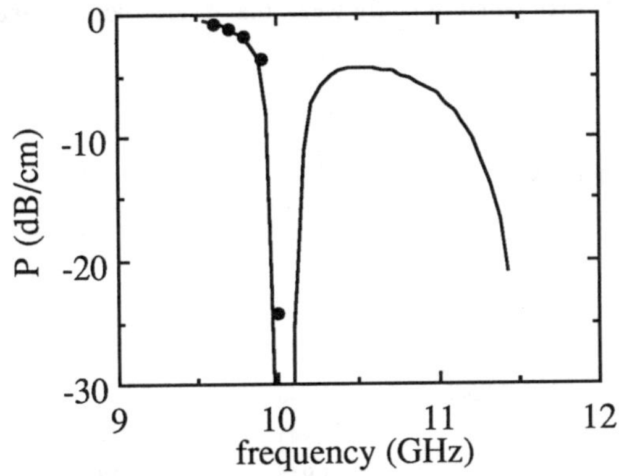

FIGURE 7. Propagation loss for the lowest-order heterostructure mode. The solid curve was calculated using the method of Buris.[71] The solid circles were calculated using the approximate relaxation time method from Eqs. (108)- (110), (112), (113), and (125). The parameters for the heterostructure are d_1=1.8 µm, d_2=5 µm, d_3= 50 µm, d_4=254 µm, H_k=0 in all layers, $-\gamma/(2\pi)$ = 28 GHz/T in all layers, M_1= 127.3 kA/m ($4\pi M_1$= 1600 G), M_3= 141.6 kA/m ($4\pi M_3$ = 1780 G), ΔH_1 (FWHM)= 398 A/m (5 Oe) at 9 GHz, ΔH_3 (FWHM)= 39.8 A/m (0.5 Oe) at 9 GHz, and H_{dc}= 412.8 kA/m (H_{dc} = 5186.8 Oe).

$$\frac{d_3}{d_1}\frac{\Delta H_3}{\Delta H_1} = 6.6 \gg 1. \tag{128}$$

2. Once the film thicknesses are chosen, optical and MSW dispersion relations can be calculated. Using these calculations, we must choose the magnetic anisotropy of the optical layer so as to achieve the assumed η=1 at the frequency and field needed for the interaction.

3. Finally, we must determine the microwave power necessary to achieve the desired coupling constant κ_0.

The results of an example calculation[30] are given in Tables 2 and 3. For these calculations, the coupling coefficient was obtained using the expressions given in Reference (48) (rather than the approximate expression given in Eq. (106)). If the MSW power/width is 1 W/m, then a coupling constant of $|\kappa_0|$=141 m^{-1} results. Neglecting the attenuation of the magnetostatic waves, this would result in a 100% conversion length of about 1.1 cm. When both optical and microwave magnetic losses are included, the optimum conversion length increases to 1.21 cm with a conversion efficiency of R = 71.8%.

It should be noted that the power level assumed is higher than the threshold for nonlinear effects,[74,75] although the cone angle is still relatively small owing to the large film thickness. The consequences of nonlinear effects for a given device would need to be carefully considered. Spurious frequencies resulting from the nonlinearities could adversely affect devices such as spectrum analyzers and filters. However, in many cases, these spurious components may be sufficiently weak that device performance will not be significantly degraded.

TABLE 2. INPUT PARAMETERS

Layer:	0	1	2	3	4
d (μm)	-	1.9	5	50	254
n_x	1	2.2502	1.945	2.212	1.945
n_y	1	2.25	1.945	2.212	1.945
n_z	1	2.25	1.945	2.212	1.945
Φ_F (°/cm)	-	-2300	-	210	-
$g_{44}M_S^2$ (SI)	-	-0.0004	-	-0.000114	-
$\Delta g M_S^2$ (SI)	-	0.0002	-	5.73e-005	-
$\gamma/(2\pi)$ (MHz/G)	-	2.8	-	2.8	-
$4\pi M_S$ (G)	-	1780	-	1780	-
H_k (Oe)	-	425	-	65	-
ΔH (FWHM) (Oe)	-	2	-	0.5	-

optical wavelength = 1.32 (μm)
optical atten. const. (α_o)= 0.13 (1/cm)
optical mode number = 0
microwave frequency = 10 (GHz)
MSW mode number = 0
MSW power = 1000 (mW/mm)

7.0 Summary and Conclusions

Optical-magnetostatic wave devices are promising for a number of microwave and optical signal processing devices such as optical frequency shifters and microwave spectrum analyzers. However, it is difficult to achieve all of the desired optical and microwave properties in a single layer film. A possible way to overcome this difficulty is the use of multiple layer heterostructures. The principles of operation, design procedure, and an example calculation have been presented. The calculations show that power levels of the order of 1 W will be required to get conversion lengths of about 1 cm. At these power levels, nonlinear effects may have a significant influence on device behavior.

TABLE 3. OUTPUT PARAMETERS

β_{TE} =	10613700.2 (1/m)
β_{TM} =	10602123.2 (1/m)
β_q =	11576.9 (1/m)
H_{dc} =	5085.0 (Oe)

TOTAL COLLINEAR COUPLING:
$|\kappa_0|$ = 140.9 (1/m)

Lossless 100% Conversion length:
L = 1.11 (cm)

Optimum Collinear Conversion length with loss:
L = 1.21 (cm)

Normalized output beam intensity:
R = 0.718

Max. Precession angle: θ = 3.4 deg.

ACKNOWLEDGMENTS

This work was supported in part by the National Science Foundation under Grants ECS-8902692 and ECS-9206817 and by the Naval Research Laboratory. Helpful discussions with Professor Andrei Slavin and Professor Yuri Fetisov on nonlinear effects are gratefully acknowledged.

References

1. P. K. Tien, R. J. Martin, S. L. Blank, S. H. Wemple, and L. J. Varnerin, *Appl. Phys. Lett.*, **21** (1972) 207.

2. P. K. Tien, R. J. Martin, R. Wolfe, R. C. Le Craw, and S. L. Blank, *Appl. Phys. Lett..*, **21** (1972) 394.

3. P. K. Tien, *Rev. Mod. Phys.*, **49** (1977) 361.

4. J. H. Collin and D. A. Wilson, *Appl. Phys. Lett.*, **12** (1968) 331.

5. H. L. Hu and F. R. Morgenthaler, *Appl. Phys. Lett.*, **18** (1971) 307.

6. B. Desormiere and H. Le Gall, *IEEE Trans. Magn.*, **MAG-8** (1972) 379.

7. H. L. Glass and M. T. Elliott, *J. Crystal Growth*, **34** (1976) 285.

8. J. D. Adam and J. H. Collins, *Proc. IEEE*, **64** (1976) 794.

9. J.-P. Castera, *J. Appl. Phys.*, **55** (1984) 2506.

10. W. S. Ishak, *Proc. IEEE*, **76** (1988) 171.

11. J. D. Adam, M. R. Daniel, P. R. Emtage, and S. H. Talisa, "Magnetostatic Waves," in *Thin Films for Electronic Devices*, H. Francombe Maurice, Ed. (Vol. 15, Physics of Thin films Series) Academic Press (1991).

12. A. D. Fisher, J. N. Lee, E. S. Gaynor, and A. B. Tveten, *Appl. Phys. Lett.*, **41** (1982) 779.

13. C. S. Tsai, "Hybrid Integrated Optic Modules for Real-Time Signal Processing," Proceedings of NASA Optical Information Processing Conference II, NASA Conference Publication No. 2302, Aug. 30,31, 1983, NASA Langley Research Center, Virginia, pp. 149.

14. A. D. Fisher, E. S. Gaynor, and John N. Lee, Proc. 1983 IEEE Ultrasonics Symp., vol. 1 (1983) 226.

15. J. M. Owens, Y. J. Ataiiyan, G. Avitabile, and R. L. Carter, *Proc. IEEE Ultrasonics Symp.*, (1986) 183.

16. R. Wolfe, J. F. Dillon, Jr., R. A. Lieberman, and V. J. Fratello, *Appl. Phys. Lett.*, **57** (1990) 960.

17. H. Dammann, E. Pross, G. Rabe, and W. Tolksdorf, *Appl. Phys. Lett.*, **56** (1990) 1302.

18. N. Koshizuka and T. Okuda, *IEEE Translation J. Magnetics in Japan*, **TJMJ-1** (1985) 1044.

19. H. Dötsch, P. Hertel, B. Lührmann, S. Sure, and H. P. Winkler, *Ferrites: Proc. Sixth Int. Conf. on Ferrites (ICF6)*, Tokyo and Kyoto, Japan (1992) 1507.

20. Y. Okamura, M. Ishida, and S. Yamamoto, *Appl. Optics*, **23** (1984) 124.

21. Y. Okamura, H. Inuzuka, T. Kikuchi, and S. Yamamoto, *IEEE J. Lightwave Tech.*, **LT-4** (1986) 711.

22. M. Torfeh, L. Courtois, L. Smoczynski, H. Le Gall, and J. M. Desvignes, *Physica* **89B** (1977) 255.

23. G. Hepner, J. P. Castera, and B. Desormiere, *Physica* **89B** (1977) 264.

24. Y. Okamura, M. Chatani, T. Kikuchi, S. Yamamoto, *Optical and Quantum Electronics,* **17** (1985) 427.

25. T. Q. Vu, C. S. Tsai, D. Young, and C. L. Wang, *Appl. Phys. Lett.*, **55** (1989) 2271.

26. L. F. Johnson, J. P. Remeika, and J. F. Dillon, Jr., *Phys. Lett.*, **21** (1966) 37.

27. L. F. Johnson, J. F. Dillon, Jr., and J. P. Remeika, *J. Appl. Phys.*, **40** (1969) 1499.

28. N. G. Nakhodkin and V. V. Voznyuk, *Fizika Taverdogo Tela.* **31** (1989) 114 (Translation in *Sov. Phys.-Sol. State*, **31** (1989) 610).

29. V. G. Veselago, R. A. Doroshenko, R. Z. Khalilov, and V. A. Timofeeva, *Pis'ma v. Zhurnal Tekhnicheskoi Fizika*, **16** (1990) 34 (Translation in *Sov. Tech. Phys. Lett.*, **16** (1990) 257).

30. D. D. Stancil, N. Bilaniuk, and N. E. Buris, *Proc. Sixth Int. Conf. Ferrites (ICF6)*, Tokyo and Kyoto (1992) 1265.

31. L. D. Landau and E. M. Lifshitz, *Electrodynamics of Continuous Media* (Pergamon Press Inc., New York, 1960) pp. 251-253.

32. P. S. Pershan, *J. Appl. Phys.* **38** (1967) 1482.

33. A. M. Prokhorov, G. A. Smolenskii, and A. N. Ageev, *Usp. Fiz. Nauk* **143** (1984) 33 (Translation in *Sov. Phys. Usp.* **27** (1984) 339).

34. M. Torfeh and H. Le Gall, *Phys. Stat. Sol. (a)* **63** (1981) 247.

35. D. D. Stancil, *Theory of Magnetostatic Waves* (Springer-Verlag, New York, 1993) Chapters 4, 5, 7.

36. I. J. Weinberg, *IEEE Trans. Magn.* **MAG-18** (1982) 1607.

37. Y. Pu and C. S. Tsai, *Int. J. High Speed Electron.* **2** (1991) 185.

38. A. Yariv, *IEEE J. Quantum Electron.*, **QE-9** (1973) 919.

39. Y. Miyazaki and R. Bhandari, Proc. of 1984 IEEE Ultrasonics Symposium, (1984) 171.

40. R. Bhandari and Y. Miyaziki, *Trans. IECE of Japan*, **E 67** (1984) 502.

41. A. D. Fisher, *Circuits, Systems and Signal Process.*, **4** (1985) 265.

42. Yu. V. Gulyayev, I. A. Ignat'yev, V. G. Plekhanov, and A. F. Popkov, *Radiotekhnika i Elektronika*, No. 8, (1985) 1522. (Translation in *Sov. J. Commun. Technol. Electron.* **30** (1985) 44.

43. C. S. Tsai, D. Young, W. Chen, L. Adkins, C. C. Lee and H. Glass, *Appl. Phys. Lett.*, **47** (1985) 651.

44. He Huahui, Lee Yibing, and Zhou Shichang, *IEEE Trans. Magn.* **MAG-23** (1987) 3500.

45. H. Tamada, M. Kaneko, and T. Okamoto, *J. Appl. Phys.*, **64** (1988) 554.

46. A. A. Stashkevich, *Izvestiya Vysshikh Uchebnykh Zavednii, Fizika*, No. 4, (1989) 5 (Translation in *Sov. Phys. J.*, **32** (1989) 241).

47. C. S. Tsai and D. Young, *IEEE Trans. Microwave Theory Tech.* **38** (1990) 560.

48. D. D. Stancil, *IEEE J. Quant. Elect.*, **27** (1991) 61.

49. A. Korpel, *Acousto-Optics* (Marcel Dekker, Inc., New York, 1988).

50. A. Yariv and P. Yeh, *Optical Waves in Crystals* (John Wiley & Sons, New York, 1984) Chap. 9.

51. B. Neite and H. Dotsch, *J. Appl. Phys.*, **62** (1987) 648.

52. D. Young and C. S. Tsai, *Appl. Phys. Lett.*, **53** (1988) 1696.

53. A. A. Stashkevich, B. A. Kalinikos, N. G. Kovshikov, O. G. Rutkin, A. N. Sigaev, and A. N. Ageev, *Pis'ma Zh. Tekh. Fiz.* **13** (1987) 49 (Translation in *Sov. Tech. Phys. Lett.* **13** (1987) 20).

54. N. Bilaniuk and D. D. Stancil, SPIE Vol. 1177 Integrated Optics and Optoelectronics (1989) 365.

55. S. H. Talisa, J. D. Adam, and M. R. Daniel, *Microwave J.* **33** No. 11 (1990) 105.

56. S. H. Talisa and D. D. Stancil, *IEEE Trans. Magn.* **25** (1989) 3494.

57. C. S. Tsai and Z. Y. Cheng, *Appl. Phys. Lett.* **54** (1989) 1616.

58. J. Neev and F. V. Kowalski, IEEE J. Quant. Elect. QE-26 (1990) 1682.

59. N. Bilaniuk, D. D. Stancil, and S. H. Talisa, *J. Appl. Phys.* **67** (1990) 508.

60. W. A. Stallard, A. R. Beaumont, and R. C. Booth, *IEEE J. Lightwave Tech.* **LT-4** (1986) 852.

61. J. A. R. Griffith, Phil. Trans. Royal Soc. London **A 307** (1982) 563.

62. N. Bilaniuk, *A Study of the Collinear Interaction between Magnetostatic Waves and Optical guided Modes in Garnet Thin Films*, Ph.D. Thesis, Carnegie Mellon University (1992).

63. N. Bilaniuk and D. D. Stancil, *IEEE J. Quant. Elect.*, submitted for publication.

64. A. A. Klimov, V. L. Preobrazhenskii, and Yu. K. Fetisov, *Pis'ma Zh. Tekh. Fiz.* **16** (1990) 18 (Translation in *Sov. Tech. Phys. Lett.* **16** (1990) 649).

65. O. G. Rutkin, N. G. Kovshikov, A. A. Stashkevich, B. A. Kalinikos, A. N. Ageev, A. S. Trifonov, V. B. Kravchenko, and L. M. Filimonova, *Pis'ma Zh. Tekh, Fiz.* **11** (1985) 933 (Translation in *Sov. Tech. Phys. Lett.* **11** (1985) 386).

66. S. H. Talisa, *IEEE Trans. Magn.* **24** (1988) 2811.

67. D. Young and C. S. Tsai, *Appl. Phys. Lett.* **55** (1989) 2242.

68. J. D. Adam, T. W. O'Keeffe, and R. W. Patterson, *J. Appl. Phys.,* **50** (1979) 2446.

69. K. Hasegawa and Y. Miyazaki, Proc. 11th Symposium on Ultrasonic Electronics, Tokyo 1990; *Jap. J. Appl. Phys.* **30** Supplement 30-1 (1991) 188.

70. K. Hasegawa and Y. Miyazaki, Proc. 12th Symposium on Ultrasonic Electronics, Tokyo 1991; *Jap. J. Appl. Phys.* **31** Supplement 31-1 (1992) 230.

71. N. E. Buris, *J. Appl. Phys.,* **64** (1988) 4726.

72. D. D. Stancil, *J. Appl. Phys.,* **59** (1986) 218.

73. D. A. Fishman and F. R. Morgenthaler, *J. Appl. Phys.,***54** (1983) 3387.

74. B. A. Kalinikos, N. G. Kovshikov, and A. N. Slavin, *Piz'ma Zh. Tekh. Fiz.*, **10** (1984) 936 (Trans. in *Sov. Tech. Phys. Lett.*, **10** (1984) 392).

75. P. E. Zil'berman, S. A. Nikitov, and A. G. Temiryazev, Piz'ma Zh. Eksp. Teor. Fiz. **42** (1985) 92 (Translation in *JETP Lett.*, **42** (1985) 110).

BRILLOUIN LIGHT SCATTERING FROM DIPOLE-EXCHANGE MICROWAVE SPIN WAVES IN MAGNETIC FILMS

M.G. COTTAM[*], **I.V. ROJDESTVENSKI**[*], and **A.N. SLAVIN**[†]

[*] *Physics Department, University of Western Ontario, London, Ontario N6A 3K7, Canada*

[†] *Physics Department, Oakland University, Rochester, Michigan 48309-4401, USA*

1. Introduction

The first observation of spin waves by light scattering was made in 1966, when Fleury et al[1] used the Raman scattering technique to study the excitation of spin waves and spin-wave pairs in the antiferromagnet FeF_2. Extensive studies of Raman light scattering in many other antiferromagnets followed shortly afterwards. However, it was several years later before the innovation of a multipass Fabry-Perot spectrometer made possible the observation of ferromagnetic and ferrimagnetic spin waves by the technique of Brillouin light scattering (BLS). The first BLS measurements of spin waves[2,3] were reported in 1973 for the ferrimagnet yttrium iron garnet (YIG) and in 1974 for the ferromagnet $CrBr_3$.

More recently there have been BLS studies of spin waves in opaque ferromagnets, and on thin films and multilayers, with a view to studying spin waves at surfaces. The first experiments[4,5] were performed for EuO and for Fe and Ni, using BLS at a single surface of a thick sample. They showed scattering from localized surface spin waves, as well as a modification of the bulk spin-wave properties. Subsequent BLS studies were reported for ferromagnetic thin films[6], double-layer systems[7], and superlattices[8]. BLS spectroscopy has now developed into a convenient and sensitive technique for studying the spin-wave spectra and the magneto-optical properties of films and multilayers.

A general description of the progress (up to about 1986) in light scattering from spin waves can be found in the book by Cottam and Lockwood[9]. Some review accounts, dealing mainly with BLS measurements in magnetic films and multilayers, have been given (for example) by Grünberg[10,11] and Dutcher[12].

From the kinematics of BLS it is easily deduced[9] that the spin waves participating in the light scattering have wave vectors $\mathbf{q}$ close to the Brillouin zone center. Typically $|\mathbf{q}|$ is of order 10^6 cm^{-1}, although the value depends on the laser wavelength, the scattering geometry, and the refractive index of the magnetic medium. Usually this correspond to the region of *dipole-exchange spin waves*, in which the long-range magnetic dipole-dipole interactions and the short-range exchange interactions both influence the spin-wave properties. Approximate descriptions of BLS in ferromagnets can be obtained in some cases by using magnetostatic theory; for example, the early experiments[4,5] on BLS from EuO, Fe, and Ni showed strong scattering from a surface spin wave that approximated to the Damon-Eshbach surface magnetostatic mode[13]. However, light scattering experiments performed in different magnetic materials in thin-film and multilayer geometries have shown conclusively[10,11] that a full theory of BLS from magnetic excitations should take into account both the dipole-dipole and exchange interactions in the material, as well as the anisotropies at the surfaces and interfaces of the sample.

The earlier theories of BLS from dipole-exchange spin waves (SW) in ferromagnetic films[14,15] give, in general, a good overall picture of the process and are in fair agreement with the experimental data. These light-scattering theories use essentially the same description of the dipole-exchange SW spectrum of the film as developed originally by De Wames and Wolfram[16] (see also Refs. 17-19). In this approach the initial system (consisting of a linearized Landau-Lifshitz equation for the magnetization, Maxwell's equations in the magnetostatic limit, and electrodynamic and exchange boundary conditions) is reduced to a bicubic equation for the values of SW wave number. This equation can then be solved numerically to obtain the frequencies of the SW modes. Consequently the theory of BLS from dipole-exchange SW modes involves lengthy numerical calculations, whether one uses a treatment based on Green functions[14] or on the application of statistical mechanics to the SW normal modes of the film[15].

In this chapter we elaborate a theory of light scattering from magnetic excitations in ferromagnetic films which takes into account the relevant interactions and, moreover, yields explicit *analytical* expressions for the light-scattering cross sections from surface and bulk SW modes[20-23].

In order to achieve this we use the tensorial Green-function method introduced by Vendik and Chartorizhskii[24] and later further developed by Kalinikos and Slavin[25,26] (see also Refs. 27,28) for the dipole-exchange SW spectrum. The advantage of this method lies in the fact that it yields a simple approximate analytical solution for the SW frequencies in films, while still giving sufficient accuracy for practical applications. The method also allows one to obtain explicit analytical expressions for the SW linear-response functions which can then be employed to obtain analytically the BLS cross section. We illustrate the advantages of our approach by examining in detail the particular case of a perpendicularly magnetized ferromagnetic film with antisymmetric exchange boundary conditions. However, we also for completeness include the generalizations to other magnetization directions and to other boundary conditions. Also we develop BLS results appropriate to double-layer magnetic systems.

2. Fundamentals of light scattering from spin waves

In this section we provide an account of the kinematics of BLS from the surface of a magnetic film. In particular, we obtain an expression for the differential scattering cross section for a general geometry and for general orientation of the static magnetization in the film, and we discuss the form of the magneto-optical coupling. Then we consider in more detail the results for the case of a 180° backscattering geometry and a perpendicularly magnetized film.

2.1. The differential cross section for BLS

We assume a general geometry (see Fig. 1) of backscattering from the upper surface of a ferromagnetic film of thickness L deposited on a nonmagnetic substrate. The axes X, Y, and Z are related to the film surface as shown, and do not necessarily coincide with the direction of static magnetization. The incident (I) and scattered (S) light beams are at angles θ_I and θ_S, respectively, to the Z axis and are in different vertical planes if angle $\phi \neq 0$. The special cases of a 180° backscattering geometry ($\theta_I = -\theta_S$, $\phi = 0$) and a 90° scattering geometry ($\theta_I + \theta_S = 90°$, $\phi = 0$) are typical of many surface BLS experiments.

From the translational symmetry in the XY plane, it follows that the excitation wave vector **q** parallel to the surface has components given by

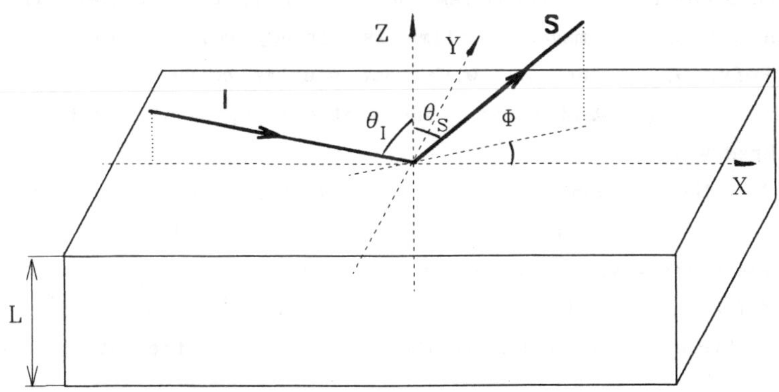

FIG. 1. The ferromagnetic film of thickness L, showing the coordinate axes and the directions of the incident (I) and scattered (S) light beams for a general BLS backscattering geometry.

$$q_X = (\omega_I/c)\sin\theta_I - (\omega_S/c)\sin\theta_S\cos\phi \ , \quad q_Y = -(\omega_S/c)\sin\theta_S\sin\phi \tag{1}$$

where ω_I and $\omega_S = (\omega_I - \omega)$ are the frequencies of the incident and scattered light, c is the velocity of light in vacuum, and the medium above the film ($Z > 0$) is assumed to have refractive index equal to 1. The cases of Stokes and anti-Stokes scattering correspond to $\omega > 0$ and $\omega < 0$ respectively. Denoting the wave vector of the light *inside* the film (with complex dielectric function ε for an optically absorptive medium) by $\mathbf{k}$, we have

$$k_I^Z = (\omega_I/c)[\varepsilon - \sin^2\theta_I]^{1/2} \ , \quad k_S^Z = (\omega_S/c)[\varepsilon - \sin^2\theta_S]^{1/2} \tag{2}$$

as the complex components perpendicular to the surface. From the property of translational symmetry parallel to the surface, the in-plane components of the incident and scattered light wave vector inside the film are real and correspond to $k_I^X = (\omega_I/c)\sin\theta_I$, $k_I^Y = 0$, $k_S^X = (\omega_S/c)\sin\theta_S$, and $k_S^Y = (\omega_S/c)\sin\theta_S\sin\phi$.

The Green-function formalism for calculating the differential cross section for BLS in magnetic films is already well established (see, e.g., Refs. 9, 14, and 29) with most studies being for the case of tangentially-magnetized films. The scattering process may be divided into three stages:

Stage 1. The incident light with wave vector k_I and frequency ω_I is partially transmitted from the vacuum region ($Z > 0$) to the magnetic film, where its wave vector becomes $\mathbf{k}_I$. The electric-field component μ ($\mu = X,Y,Z$) in the film is given simply by $f^\mu E_I^\mu$, where E_I^μ is the electric field in the vacuum and the the Fresnel coefficients f^μ are

$$f^X = 2k_I^Z/(k_I^Z + \varepsilon k_I^Z) \ , \quad f^Y = 2k_I^Z/(k_I^Z + k_I^Z) \ , \quad f_I^Z = f^X k_I^Z/k_I^Z \ . \tag{3}$$

Stage 2. The incident light in the film interacts with a spin wave with wave vector $\mathbf{Q} = (\mathbf{q},\kappa_n)$ and frequency ω. The components of $\mathbf{q}$ are real and given by Eq. (1), but the κ_n may be complex in general. Either κ_n is real for bulk modes, taking a series of discrete values as discussed later, or κ_n is complex (usually pure imaginary) for surface modes. The interaction produces a polarization in the film, and we denote its Fourier components by $\mathbf{P}_{int}\exp[i(\mathbf{K}\cdot\mathbf{r}-\omega_S t)$ where $\omega_S = \omega_I - \omega$ and $\mathbf{K} = \mathbf{k}_I - \mathbf{q}$.

Stage 3. The induced polarization in the film produces a scattered beam with wave vector k_S and frequency ω_S in the vacuum region (Z > 0). The solution for the corresponding electric field gives

$$E_S^{\mu} = (\omega_S/\varepsilon_o c)\,\exp(ik_S.\mathbf{r}) \sum_{\mathbf{q},\nu} \psi(B + \kappa_n) g^{\mu\nu} P^{\nu}_{int} \quad . \tag{4}$$

Here $B = (k_I^z + k_S^z)$ is a complex quantity depending on the optical wave vectors in the film, and we define

$$\psi(x) \quad = \quad [1 - \exp(ixL)]/x \quad . \tag{5}$$

The matrix $\mathbf{g}$ describes the transmission of the scattered beam through the surface at Z = 0:

$$\mathbf{g} = \mathbf{R} \begin{pmatrix} \dfrac{-(c/\omega_S)k_S^Z k_S^Z}{\varepsilon k_S^Z + k_S^Z} & 0 & \dfrac{-(c/\omega_S)k_S^Z k_S^X}{\varepsilon k_S^Z + k_S^Z} \\ 0 & \dfrac{\omega_S/c}{k_S^Z + k_S^Z} & 0 \\ \dfrac{(c/\omega_S)k_S^Z k_S^X}{\varepsilon k_S^Z + k_S^Z} & 0 & \dfrac{(c/\omega_S)(k_S^Z)^2}{\varepsilon k_S^Z + k_S^Z} \end{pmatrix} \mathbf{R}^{-1} \tag{6}$$

where

$$\mathbf{R} = \begin{pmatrix} \cos\phi & -\sin\phi & 0 \\ \sin\phi & \cos\phi & 0 \\ 0 & 0 & 1 \end{pmatrix} \tag{7}$$

Using the above results for the scattered electric field, it is now straightforward to deduce the BLS cross section, which is proportional to $|\mathbf{E}_S|^2$. The general expression for the differential cross section for scattering into an elementary solid angle $d\Omega$ with frequency between ω_S

and $\omega_S + d\omega_S$ has the form[9]

$$\frac{d^2h}{d\Omega d\omega_S} = [n(\omega) + 1]\ \mathrm{Im}\left[\sum_{n,n'} \psi(B+\kappa_n)^* \psi(B+\kappa_{n'}) \times \sum_{\nu,\nu',\delta,\delta'} F(\nu,\delta,\nu',\delta')\ \langle\langle (\chi^{\delta\nu})^* ; \chi^{\delta'\nu'} \rangle\rangle \right] \tag{8}$$

where the n and n′ refer to summations over the discrete SW modes of the film, $n(\omega)$ is the Bose-Einstein thermal factor, and the indices ν, δ, ν', δ' refer to Cartesian components. The function F depends on the magneto-optical coupling, the optical transmission coefficients at the surface, and the scattering geometry. In terms of the quantities defined above, it is given by

$$F(\nu,\delta,\nu',\delta') = -\frac{\bar{A}\omega_I\omega_S^3\cos^2\theta_S}{2\pi^2c^4} \sum_{\mu,\mu'} e_S e_S (f^\nu)^* f^{\nu'} e_I^\nu e_I^{\nu'} (g^{\mu\delta})^* g^{\mu'\delta'} \tag{9}$$

where $\mathbf{e}_I$ is the unit polarization of the incident beam, and we assume that the polarization $\mathbf{e}_S$ is selected in the scattered beam. Also $\bar{A}$ is the area of the film surface through which the scattered beam emerges. Finally the term in Eq. (8) of the form $\langle\langle\chi^*;\chi\rangle\rangle$ is a Green function, where χ is a component of the magnetization-dependent susceptibility tensor, which corresponds to the polarization $\mathbf{P}_{int}$.

The susceptibility χ can be expanded[9] in terms of components of the total magnetization **M** in the film. If we suppose that the film has its static magnetization M_o along the z axis of a xyz coordinate system, the light scattering from single spin waves can occur in two ways: (i) *linear magneto-optical coupling*, involving terms in χ proportional to M^x and M^y, and (ii) *quadratic magneto-optical coupling*, involving terms $M^xM^z + M^zM^x$ and $M^yM^z + M^zM^y$. Usually the quadratic parts can be simplified by replacing terms like M^xM^z by M^xM_o (ignoring quantum fluctuations in the z component), and M^x and M^y can then be replaced by the fluctuating magnetizations m^x and m^y, respectively. For a ferromagnetic material with cubic symmetry, the appropriate form of χ for substitution into Eq. (8) is

$$\chi = \begin{pmatrix} 0 & 0 & 2M_oGm^x - Km^y \\ 0 & 0 & 2M_oGm^y + Km^x \\ 2M_oGm^x + Km^y & 2M_oGm^y - Km^x & 0 \end{pmatrix} \qquad (10)$$

where K and G are the coefficients of linear and quadratic magneto-optical coupling, respectively. They are proportional to the Faraday rotation and the magnetic linear birefringence, respectively, in the ferromagnet. We note from Eq. (10) that χ is antisymmetric in the case of linear magneto-optic coupling only (G = 0) and symmetric in the case of quadratic coupling only (K = 0); this property can be used in BLS experiments to distiguish between the two contributions[9]. From Eqs. (8) and (10) it follows that the light-scattering cross section is expressible in terms of the four basic Green functions $\langle\langle (m^x)^*; m^x \rangle\rangle$, $\langle\langle (m^x)^*; m^y \rangle\rangle$, $\langle\langle (m^y)^*; m^x \rangle\rangle$, and $\langle\langle (m^y)^*; m^y \rangle\rangle$. Their evaluation in terms of the dipole-exchange SW theory will be given later in this chapter.

It should be emphasized that Eq. (3)-(9) apply provided the magnetic film is sufficiently thick and/or opaque that light scattering processes involving multiple reflections at the lower film surface (at Z = -L) can be neglected. The generalized formalism to include such effects is known[29], and we shall later make use of this.

2.2. 180° Backscattering from a Perpendicularly Magnetized Film

We now consider the specific form of the expression for $d^2h/d\Omega d\omega_S$ in the case of 180° backscattering ($\theta_I = -\theta_S \equiv \theta$, $\phi = 0$) with crossed polarizations ($\mathbf{e}_I \cdot \mathbf{e}_S = 0$). This would be typical of many BLS experiments. Also it is assumed that the film is magnetized perpendicular to its surfaces, and so the XYZ and xyz axes coincide.

Consider first the case where the incident polarization is in the vertical plane, so that $\mathbf{e}_I = (\cos\theta, 0, \sin\theta)$, and therefore the scattered polarization is $\mathbf{e}_S = (0, 1, 0)$. The differential cross section is found from Eqs. (8)-(10) to be

$$\frac{d^2h}{d\Omega d\omega_S} = - \frac{\bar{A}\omega_I\omega_S^3\cos^2\theta\sin^2\theta}{2\pi^2c^4} [n(\omega) + 1] \; |f^z g^{yy}|^2 \; \mathrm{Im}\Big[\sum_{n,n'} \psi(B + \kappa_n)^*$$

$$\times \; \psi(B + \kappa_{n'}) \; \{ \; |K|^2 \; \langle\langle (m^x)^*; \; m^x \rangle\rangle \; + \; 2M_o G^* K \; \langle\langle (m^y)^*; \; m^x \rangle\rangle$$

$$+ \; 2M_o G \; K^* \langle\langle (m^x)^*; \; m^y \rangle\rangle \; + \; 4M_o^2 |G|^2 \; \langle\langle (m^y)^*; \; m^y \rangle\rangle \; \} \Big] \quad . \tag{11}$$

On the other hand, if the polarizations are $\mathbf{e}_I = (0, 1, 0)$ and $\mathbf{e}_S = (\cos\theta, 0, \sin\theta)$, the required expression is

$$\frac{d^2h}{d\Omega d\omega_S} = - \frac{\bar{A}\omega_I\omega_S^3\cos^2\theta}{2\pi^2c^4} [n(\omega) + 1] \; |f^y|^2 \; [\cos^2\theta |g^{xz}|^2 + \sin^2\theta |g^{zz}|^2$$

$$+ \sin\theta\cos\theta\{(g^{xz})^* g^{zz} + (g^{zz})^* g^{xz}\}] \; \mathrm{Im}\Big[\sum_{n,n'} \psi(B + \kappa_n)^* \; \psi(B + \kappa_{n'})$$

$$\times \{ \; |K|^2 \; \langle\langle (m^x)^*; \; m^x \rangle\rangle \; - \; 2M_o G^* K \; \langle\langle (m^y)^*; \; m^x \rangle\rangle \; - \; 2M_o G \; K^* \langle\langle (m^x)^*; \; m^y \rangle\rangle$$

$$+ \; 4M_o^2 |G|^2 \; \langle\langle (m^y)^*; \; m^y \rangle\rangle \; \} \Big] . \tag{12}$$

3. Non-hybridized spin waves and green functions

To obtain the equations for the dipole-exchange SW frequencies and the Green functions, we consider a ferromagnetic film of finite thickness L magnetized to saturation along the z-axis by an external bias field of magnitude H_e. Specifically we take the case of the static magnetization (denoted by M_o) perpendicular to the film surfaces. Hence the xyz and XYZ axes coincide, and the film surfaces correspond to the planes z = 0 and z = -L (see Fig. 1).

Following Refs. 25 and 26, we represent the variable magnetization $\mathbf{m}(z,\mathbf{r},t)$ of the film in the form

$$\mathbf{m}(z,\mathbf{r},t) \; = \; M_o \sum_n \mathbf{m}_{nq} \; \Phi_n(z) \; \exp[i(\omega t - \mathbf{q}\cdot\mathbf{r})] \tag{13}$$

where $\mathbf{q}$ and $\mathbf{r}$ are respectively the two-dimensional SW wave vector and position vector in the film plane, and $\Phi_n(z)$ (with n denoting a mode index) are the SW modes defined in Ref. 25. The variable driving field

$\mathbf{h}(z,\mathbf{r},t)$ can be written in a similar form to Eq. (13). The SW modes form a complete orthonormal set of functions that satisfy a differential operator of the second order (an operator of the inhomogeneous exchange interaction) and the exchange boundary conditions[25]. In the present case of perpendicular magnetization the boundary condition at $z = 0$ has the simple form

$$\frac{\partial \mathbf{m}^x}{\partial z} + d_1 \mathbf{m}^x = 0 , \qquad \frac{\partial \mathbf{m}^y}{\partial z} + d_2 \mathbf{m}^y = 0 \tag{14}$$

where d_1 is a pinning parameter (related to the surface anisotropy constant, as explained below), and there is an analogous boundary condition at $z = -L$ involving another pinning parameter d_2. The SW modes are the modes of spin wave resonance (SWR), and they correspond to

$$\Phi_n(z) = A_n \{\cos[\kappa_n(z+L)] + (d_1/\kappa_n)\sin[\kappa_n(z+L)]\} . \tag{15}$$

Here κ_n is a transverse SW wave number which takes discrete values determined by the condition[25]

$$[(\kappa_n)^2 - d_1 d_2] \tan(\kappa_n L) = \kappa_n (d_1 + d_2) . \tag{16}$$

Negative and positive pinning parameters correspond to "easy-plane" and "easy-axis" types of surface anisotropy, respectively.

The SW mode frequencies and the corresponding magnetic Green functions (linear response functions) can readily be obtained using the formalism of Ref. 25 and the expansion in Eq. (13) in the case of a perpendicularly-magnetized film. The results are found to be different for the non-degenerate (see below) and degenerate (see Sec. 5) cases.

In the case of non-degenerate SW modes (corresponding to $\omega_{nq} \neq \omega_{mq}$ for $n \neq m$) the SW frequencies are given approximately by

$$\omega_{nq} = [\Omega_{nq} (\Omega_{nq} + \omega_M P_{nn})]^{1/2} \tag{17}$$

where we have defined

$$\Omega_{nq} = \omega_H + \alpha\omega_M(q^2 + \kappa_n^2) . \tag{18}$$

Here $\omega_H = \gamma(H_e - 4\pi M_0)$, $\omega_M = \gamma(4\pi M_0)$, γ is the gyromagnetic ratio, α is an exchange constant, and $P_{nn}(qL)$ is the diagonal matrix element of the dipole-dipole interaction[25]. We note that our exchange parameter α and pinning parameter d are related to the exchange term A and surface anisotropy constant K_s, which are sometimes used by other authors, by $\alpha = A/(2\pi M_0^2)$ and $d = K_s/A$. As q varies from zero to infinity P_{nn} satisfies $0 \le P_{nn} \le 1$. The spin-wave Green functions (response functions) $\hat{C}_{nn}$ in the wave-vector representation are defined by

$$\begin{bmatrix} m^x_{nq} \\ m^y_{nq} \end{bmatrix} = \hat{C}_{nn} \begin{bmatrix} h^x_{nq} \\ h^y_{nq} \end{bmatrix} . \qquad (19)$$

We find[25,30] that these take the simple form $\hat{C}_{nn} = \hat{Y}_{nn}\, \omega_M/(\omega^2 - \omega^2_{nq})$, where the elements of the 2 x 2 matrix $\hat{Y}_{nn}$ are $Y^{xx}_{nn} = \Omega_{nq} + \omega_M P_{nn}$, $Y^{yy}_{nn} = \Omega_{nq}$, and $Y^{xy}_{nn} = -Y^{yx}_{nn} = i\omega$.

We next examine the above results in the particular case of antisymmetric exchange boundary conditions when the pinning parameters have equal magnitudes , but opposite signs, at the film surfaces[30]:

$$d_1 = - d_2 \equiv d , \qquad (d > 0) . \qquad (20)$$

It is then clear from Eq. (16) that the SW spectrum of the film consists of an infinite number of bulk SW dispersion branches ($\kappa_n = n\pi/L$ with n = 1,2,..) and one surface SW branch ($\kappa_n = id$). We note that, in this case of perpendicular magnetization, the surface branch for $q \to 0$ lies *below* the spectrum of bulk SW modes. For general pinning, the solutions for κ_n can be obtained numerically from Eq. (16).

4. Brillouin light scattering from non-hybridized spin waves

The response functions $C^{\alpha\beta}_{nn}$ defined in Eq. (19) can be identified with the magnetic Green functions $\langle\langle (m^\alpha)^*; m^\beta \rangle\rangle$ for mode n which enter into the BLS formalism (α,β = x,y). When the imaginary parts of these Green functions are taken in the expressions for the differential cross section $d^2h/d\Omega d\omega_S$ [e.g., see Eqs. (11) and (12)], the predicted BLS spectrum consists of a series of δ-function peaks at the discrete SW frequencies found in Sec. 3. The integrated intensity $(dh/d\Omega)_n$ for mode n is proportional to the strength of the corresponding δ-function and behaves for Stokes scattering at $\omega_S = \omega_I - \omega_{nq}$ as

$$\left(\frac{dh}{d\Omega}\right)_n \propto \frac{|\psi(B+\kappa_n)|^2}{\omega_{nq}^2}\left[K(\Omega_{nq}+\omega_M P_{nn})^{1/2} + 2M_0 G(\Omega_{nq})^{1/2}\right]^2 \tag{21}$$

where ω_n and Ω_{nq} are defined in Eqs. (17) and (18), and we have approximated the Bose-Einstein factor by assuming $\hbar\omega_{nq} \ll k_B T$ where T is the temperature, since this inequality usually applies in BLS experiments.

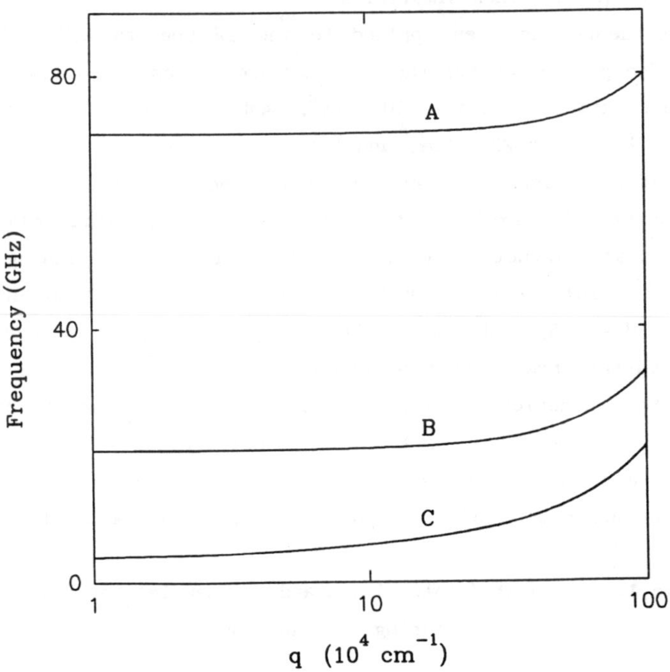

FIG. 2. The SW frequencies plotted against the in-plane wave vector q for a Fe film in the nondegenerate case; see text for the parameters. The n = 1 and n = 2 bulk modes (labeled B and A, respectively) and the surface mode (labeled C) are shown.

We have taken polarizations $\mathbf{e}_I = (\cos\theta, 0, \sin\theta)$ and $\mathbf{e}_S = (0, 1, 0)$, the overall proportionality factor in Eq. (21) is a function of the angle θ, the in-plane wave vector q, and the optical parameters, as can be deduced from Eq. (11). However, this factor is independent of κ_n (equal to id for the surface SW mode and $n\pi/L$ for a bulk SW mode). The corresponding result in the case of anti-Stokes scattering is obtained by the replacement $K \longrightarrow -K$. Hence we predict a Stokes/anti-Stokes asymmetry when both K and G contribute to the light scattering; this is analogous to the effect observed[9,31] in bulk BLS measurements on YIG. In many materials the dominant magneto-optical coupling corresponds to the K term[9], and Eq. (21) then simplifies.

The theory has been applied to the SW spectra and BLS from Fe films. The parameters for the Fe films are chosen as follows[20,29]: the exchange constant $\alpha = 1.15 \times 10^{-13}$ cm^2, $4\pi M_o = 21$ kG, the external bias magnetic field $H_e = 22.5$ kOe, and the complex refractive index $\eta = 2.72 + 2.86i$ at a wavelength $\lambda = 488$ nm for the incident light.

A numerical example of the predicted SW dispersion relations for perpendicularly-magnetized Fe films with antisymmetric pinning is given in Fig. 2. This corresponds to film thickness L = 20 nm and pinning parameter $d = 2.5 \times 10^5$ cm^{-1}. There is no appreciable mode degeneracy and the surface branch lies below the first two (n = 1 and n = 2) bulk modes. Some numerical results are shown in Table 1 for the integrated BLS intensity $I = dh/d\Omega$ for the four lowest-frequency SW modes of a Fe film (surface mode S and n = 1, 2, 3 bulk modes) in this same case. We have taken angle $\theta = 45^0$, implying $q = 1.82 \times 10^5$ cm^{-1}. These results

Table 1. Integrated intensities and frequencies for Brillouin scattering from an iron film

SW Mode	S	1	2	3
I/I_1	14.6	1	0.09	0.03
$\omega/2\pi$ (GHz)	7.20	21.5	71.2	154

have been obtained by employing Eq. (21) for the case of linear magneto-optical coupling only ($G = 0$), and the intensities are expressed relative to I_1 of the $n = 1$ bulk mode. The surface mode scattering is predicted to dominate in this example, and by analogy with experimental results in the parallel-magnetization geometry[11] this would be expected for such a film thickness.

For Fe films, in the case of different thickness or with different pinning parameters, we find that hybridization effects can become important. This is discussed in later sections.

The BLS results described so far are for cases where the light scattering processes do not involve *multiple* reflections of the incident or scattered light at the film surfaces. They are appropriate whenever the penetration depth of the light is less than the film thickness or if the (nonmagnetic) substrate has the same complex refractive index as the film itself. The formalism to include the multiple reflections in a very thin magnetic film has been given by Cottam[29] and applied to transverse-magnetized films in a magnetostatic approximation. Now we summarize the method applied to perpendicular-magnetized films in our dipole-exchange theory and, in particular, we generalize Eq. (8) to include these effects, taking only the case of linear magneto-optical coupling ($G = 0$).

If the light incident from vacuum (in the region where $z > 0$) has electric field components $E_I^{\nu}\exp(-ik_I^z z)$, then the incident electric field components in the film are[29]

$$E_I^{\nu}\left[f^{\nu}(k_I^z)\exp(-ik_I^z z) + F^{\nu}(k_I^z)\exp(ik_I^z z)\right] \tag{22}$$

where, for convenience, we omit the common factor $\exp[i(\omega t - \mathbf{q}.\mathbf{r})]$. The coefficients f and F are, for example, given by

$$f^z(k_I^z) = -F^z(-k_I^z) = 2\mathit{k}_I^z(\varepsilon K_I^z + \varepsilon'\mathit{k}_I^z)/[(k_I^z + \varepsilon\mathit{k}_I^z)(\varepsilon K_I^z + \varepsilon'\mathit{k}_I^z) - (k_I^z - \varepsilon\mathit{k}_I^z)(\varepsilon K_I^z - \varepsilon'\mathit{k}_I^z)\exp(2ik_I^z L)] \quad . \tag{23}$$

Here we follow the notation of Sec. 2 that k_I^z and k_I^z refer to wavevector components in the vacuum and film respectively, and we denote the corresponding optical wave vector in the substrate by K_I^z. Also ε and ε' denote the complex dielectric functions of the film and substrate respectively. We note that, when $\mathrm{Im}(k_I^z L) \gg 1$, $f^z(k_I^z)$ becomes equal to the Fresnel coefficient in Eq. (3) and $F^z(k_I^z)$ vanishes.

For the choice of electric-field polarizations $\mathbf{e}_I = (\cos\theta, 0, \sin\theta)$ and $\mathbf{e}_S = (0, 1, 0)$, the required contribution to the electric field of the scattered light in the vacuum region that arises from magneto-optical coupling in the film with the $\exp(-ik_I^z z)$ term of Eq. (22) is[29]

$$E_S^{(-)} = -(\mathit{k}_S^z)^2 P_{int} \exp(ik_S^z) \Big[g_{yy}(k_S^z)\, \psi(k_I^z + k_S^z + \kappa_n) + g^{yy}(-k_S^z)\, \psi(k_I^z - k_S^z + \kappa_n) \Big] . \tag{24}$$

Here P_{int} is the polarization produced at the scattered light frequency due the coupling between the incident light and a SW with wave-vector component κ_n. The function ψ is defined in Eq. (5) and

$$g^{yy}(k_S^z) = (\omega_S/c)(K_S^z + k_S^z)/[(\mathit{k}_S^z + k_S^z)(K_S^z + k_S^z) - (\mathit{k}_S^z + k_S^z)(K_S^z + k_S^z)\exp(2ik_S^z L)] . \tag{25}$$

The contribution $E_S^{(+)}$, which corresponds to the scattering produced from the coupling with the $\exp(ik_I^z z)$ term in Eq. (22), is given by a similar expression with k_I^z replaced by $-k_I^z$ in Eq. (24).

Finally, following the approach of Ref. 26 adapted to the present geometry, we conclude that Eq. (8) for the differential BLS cross section is generalized (in the case of linear magneto-optical coupling) by making the following replacement of factors:

$$|f^z g_{yy}|^2\, [\psi(B + \kappa_n)]^*\, \psi(B + \kappa_n') \;\rightarrow \sum_{\sigma_I, \sigma_I', \sigma_S, \sigma_S'} \sigma_I \sigma_I' \,[f^z(\sigma_I k_I^z)]^* \times f^z(\sigma_I' k_I^z)\, [g_{yy}(\sigma_S k_S^z)]^*\, g^{yy}(\sigma_S' k_S^z)\, [\psi(\sigma_I k_I^z + \sigma_S k_S^z + \kappa_n)]^* \times \psi(\sigma_I' k_I^z + \sigma_S' k_S^z + \kappa_n) \tag{26}$$

where the σ indices are each summed over the values +1 and -1.

We have made further BLS calculations for EuO films, for which the exchange effects are weaker than for Fe. Also the optical penetration is larger (about 150 nm) in EuO, and so the multiple-reflection effects (including a dependence on the optical constants of the *substrate* as well as those of the magnetic film itself) become more important. For EuO films we take[11,20] $\alpha = 4.6 \times 10^{-15}$ cm^2, $4\pi M_o = 23.8$ kG, $H_e = 25.3$ kOe, and $\eta = 1.6 + 0.25i$ at $\lambda = 514.5$ nm. However, it turns out that SW hybridization effects are significant for these parameter values in EuO, and so we discuss the BLS results later in Sec. 6.

5. Hybridized spin waves and green functions

In the previous section we pointed out certain cases where the hybridization of the SW modes should be taken into account. It is very difficult to treat analytically the hybridization of more than two modes at once. However, in most cases we encounter only the situation where the modes are degenerate in pairs. The occurrence of three or more dispersion branches of the spectrum almost crossing is very rare.

In the case of degeneracy of a pair of SW modes with numbers n and m ($\omega_{nq} = \omega_{mq}$) the hybridized SW dispersion branches ω_{nm} are found, following Ref. 25, to satisfy

$$2\,\omega_{nm}^2 = (\omega_{nq}^2 + \omega_{mq}^2) \pm \left[(\omega_{nq}^2 - \omega_{mq}^2)^2 + 4\,\Omega_{nq}\,\Omega_{mq}\,\omega_M^2\,P_{nm}^2\right]^{1/2} \qquad (27)$$

where P_{nm} is a non-diagonal matrix element of the dipole-dipole interaction[25]. We note that Eq. (27) is a secular equation which removes the degeneracy of "diagonal" dipole-exchange modes ω_{nq} and ω_{mq} by taking into account the *non-diagonal* dipole-dipole hybridization of these modes in the first order of standard perturbation theory. The small parameter of this perturbation theory is proportional to the ratio of the *non-diagonal* matrix element P_{nm} of the dipole-dipole interaction to the difference between the eigenfrequencies of the unperturbed modes, as discussed in detail in Ref. 25.

The spin-wave Green functions in the degenerate case are now expressible in terms of a 4 x 4 matrix, due to "mixing" between the modes. They can be derived following earlier work[25,26], except that it is more convenient here to use a representation in terms of the quantities $a_{nq} \equiv m^{+}_{nq} = m^{x}_{nq} + im^{y}_{nq}$ and $a^{*}_{nq} \equiv m^{-}_{nq} = m^{x}_{nq} - im^{y}_{nq}$ instead of m^{x}_{nq} and m^{y}_{nq}; these are the circular polarizations of the spin-wave mode with the label n and in-plane wave vector q. The corresponding result to Eq. (19) in the degenerate case is

$$\left(\hat{D}_{nm} + (\omega/\omega_M)\hat{L}_{nm} \right) \mathbf{a}_{nmq} = \mathbf{h}_{nmq} \tag{28}$$

where $\hat{L} = \mathrm{diag}(-1,1,-1,1)$ is a diagonal matrix and

$$\mathbf{a}_{nmq} = \begin{bmatrix} a_{nq} \\ a^{*}_{n,-q} \\ a_{m,q} \\ a^{*}_{m,-q} \end{bmatrix}, \qquad \mathbf{h}_{nmq} = \begin{bmatrix} h^{+}_{nq} \\ h^{-}_{n,-q} \\ h^{+}_{mq} \\ h^{-}_{m,-q} \end{bmatrix}. \tag{29}$$

Here $h^{\pm}_{nq}$ are the polarizations of the external magnetic field (along the film normal) in the Fourier representation, and the 4x4 matrix $\hat{D}_{nm}$ (which contains the dipole-exchange and pinning effects) is given by

$$\hat{D}_{nm} = \frac{1}{2} \begin{bmatrix} 2N_n + P_{nn} & P_{nn} & P_{nm} & P_{nm} \\ P_{nn} & 2N_n + P_{nn} & P_{nm} & P_{nm} \\ P_{mn} & P_{mn} & 2N_m + P_{mm} & P_{mm} \\ P_{mn} & P_{mn} & P_{mm} & 2N_m + P_{mm} \end{bmatrix} \tag{30}$$

where $N_n = \Omega_{nq}/\omega_M$ and expressions for the dipole-dipole interaction matrix elements P_{nm} (which depend on the pinning conditions) are written explicitly in Ref. 25. From Eq. (28), the tensorial Green function for the linear-response relationship $\mathbf{a}_{nmq} = \hat{G}_{nm}\,\mathbf{h}_{nmq}$ can be written as

$$\hat{G}_{nm} = \left(\hat{D}_{nm} + (\omega/\omega_M)\hat{L}_{nm} \right)^{-1}. \tag{31}$$

The hybridization (or mixing) of the SW dispersion branches is caused by the long-range dipolar interaction; it can be ignored only if the branches are well separated. To find the representation in which the matrix $\hat{G}_{nm}$ becomes diagonal we need to perform a general Bogolyubov transformation which takes into account the interaction between the two modes[32]:

$$a_{nq} = U_{n1,q}\alpha_{1q} + U_{n2,q}\alpha_{2q} + V^*_{n1,-q}\alpha^*_{1,-q} + V^*_{n2,-q}\alpha^*_{2,-q}$$

$$a^*_{nq} = U^*_{n1,q}\alpha^*_{1q} + U^*_{n2,q}\alpha^*_{2q} + V_{n1,-q}\alpha_{1,-q} + V_{n2,-q}\alpha_{2,-q} \tag{32}$$

and there are similar expressions for a_{mq} and a^*_{mq}. Here α_{gq} and α^*_{gq} (with g = 1,2) are the new canonical variables for the nm and mn hybridized modes (labeled arbitrarily above as 1 and 2). They are chosen for their Green functions to have the form:

$$<<\alpha_{gq};\alpha^*_{g'q}>> = \delta_{gg'}/[\omega - \omega_g(q)] , \tag{33}$$

$$<<\alpha^*_{gq};\alpha_{g'q}>> = -\delta_{gg'}/[\omega + \omega_g(q)], \tag{34}$$

with $\omega_g(q)$ being the hybridized frequencies as given by Eq. (27), or equivalently obtained from the roots of $\det[\hat{\mathbf{D}}_{nm} + (\omega/\omega_M)\hat{\mathbf{L}}_{nm}] = 0$. Note that, in the present geometry of a perpendicularly magnetized film, we have $\omega_g(q) = \omega_g(-q)$.

To calculate the light-scattering (BLS) spectra we need (see Sec. 2) the Green functions $« m^\mu_{nq};\ m^\nu_{mq} »$ with $\mu,\nu = x,y$. We can rewrite them in the terms of a and a^*; for example

$$«(m^x_{nq})^*;\ m^x_{mq}» = \frac{1}{4}\left[G^{++}_{nm} + G^{+-}_{nm} + G^{-+}_{nm} + G^{--}_{nm}\right] \tag{35}$$

where $G^{++}_{nm} = « a^*_{nq};\ a^*_{m,-q}»$, $G^{+-}_{nm} = « a^*_{nq};\ a_{mq}»$, $G^{-+}_{nm} = « a_{n,-q};\ a^*_{m,-q}»$, and $G^{--}_{nm} = « a_{n,-q};\ a_{mq}»$.

Assuming a transformation as in Eq. (32), the coefficients U_{nm} and V_{nm} can be found from the requirement that the Hamiltonian has a diagonal form in α and α^*. This corresponds to the Green functions having the form of Eqs. (33) and (34). Then, using Eqs. (32)-(34) and the definitions following Eq. (35), we can write

$$G^{++}_{nm} = \frac{V_{n1,-q}U^*_{m1,-q}}{\omega - \omega_1(-q)} + \frac{V_{n2,-q}U^*_{m2,-q}}{\omega - \omega_2(-q)} - \frac{U^*_{n1,q}V_{m1,q}}{\omega + \omega_1(q)} - \frac{U^*_{n2,q}V_{m2,q}}{\omega + \omega_2(q)}$$

$$G^{--}_{nm} = \frac{U_{n1,-q}V^*_{m1,-q}}{\omega - \omega_1(-q)} + \frac{U_{n2,-q}V^*_{m2,-q}}{\omega - \omega_2(-q)} - \frac{V^*_{n1,q}U_{m1,q}}{\omega + \omega_1(q)} - \frac{V^*_{n2,q}U_{m2,q}}{\omega + \omega_2(q)}$$

$$G^{+-}_{nm} = \frac{V_{n1,-q}V^*_{m1,-q}}{\omega - \omega_1(-q)} + \frac{V_{n2,-q}V^*_{m2,-q}}{\omega - \omega_2(-q)} - \frac{U^*_{n1,q}U_{m1,q}}{\omega + \omega_1(q)} - \frac{U^*_{n2,q}U_{m2,q}}{\omega + \omega_2(q)}$$

$$G^{-+}_{nm} = \frac{U_{n1,-q}U^*_{m1,-q}}{\omega - \omega_1(-q)} + \frac{U_{n2,-q}U^*_{m2,-q}}{\omega - \omega_2(-q)} - \frac{V^*_{n1,q}V_{m1,q}}{\omega + \omega_1(q)} - \frac{V^*_{n2,q}V_{m2,q}}{\omega + \omega_2(q)} \quad . \qquad (36)$$

On the other hand, the above Green functions are also given by the elements of the matrix reciprocal to $[\hat{D}_{nm} + (\omega/\omega_M)\hat{L}_{nm}]$. Using Cramer's rule for the elements of the reciprocal matrix, we obtain:

$$G^{-+}_{nn} = \frac{\Delta_{11}(q,\omega)}{\Delta(q,\omega)} \; , \qquad G^{-+}_{mm} = \frac{\Delta_{33}(q,\omega)}{\Delta(q,\omega)} \qquad (37)$$

where $\Delta(q,\omega) = \det[\hat{D}_{nm} + (\omega/\omega_M)\hat{L}_{nm}]$ and $\Delta_{11}(q,\omega)$ and $\Delta_{33}(q,\omega)$ are the minors of the (1,1) and (3,3) elements, respectively. Next, taking into account that

$$\Delta(q,\omega) = \big(\omega - \omega_1(q)\big)\big(\omega - \omega_2(q)\big)\big(\omega + \omega_1(-q)\big)\big(\omega + \omega_2(-q)\big) \qquad (38)$$

and making the expressions in Eqs. (36) and (37) equal for $\omega = \pm\omega_g(\pm q)$ for $g = 1,2$, we obtain the following expressions for the U and V coefficients:

$$U_{n1,q} = \left[\frac{\Delta_{11}(\omega_1^+)}{(\omega_1^+ - \omega_2^+)(\omega_1^+ + \omega_1^-)(\omega_1^+ + \omega_2^-)}\right]^{1/2} e^{i\psi_1(q)}$$

$$U_{n2,q} = \left[\frac{\Delta_{11}(\omega_2^+)}{(\omega_2^+ - \omega_1^+)(\omega_2^+ + \omega_2^-)(\omega_2^+ + \omega_1^-)}\right]^{1/2} e^{i\psi_2(q)}$$

$$V_{n1,q} = \left[\frac{\Delta_{11}(-\omega_1^+)}{(\omega_2^+ - \omega_1^+)\cdot(\omega_1^+ + \omega_2^-)(\omega_1^+ + \omega_1^-)} \right]^{1/2} e^{i\psi_1(q)}$$

$$V_{n2,q} = \left[\frac{\Delta_{11}(-\omega_2^+)}{(\omega_1^+ - \omega_2^+)(\omega_1^- + \omega_2^+)(\omega_2^+ + \omega_2^-)} \right]^{1/2} e^{i\psi_2(q)}$$

$$U_{m1,q} = \left[\frac{\Delta_{33}(\omega_1^+)}{(\omega_1^+ - \omega_2^+)(\omega_1^+ + \omega_1^-)(\omega_1^+ + \omega_2^-)} \right]^{1/2} e^{i\psi_1(q)}$$

$$U_{m2,q} = \left[\frac{\Delta_{33}(\omega_2^+)}{(\omega_2^+ - \omega_1^+)(\omega_2^+ + \omega_2^-)(\omega_2^+ + \omega_1^-)} \right]^{1/2} e^{i\psi_2(q)}$$

$$V_{m1,q} = \left[\frac{\Delta_{33}(-\omega_1^+)}{(\omega_2^+ - \omega_1^+)(\omega_1^+ + \omega_2^-)(\omega_1^+ + \omega_1^-)} \right]^{1/2} e^{i\psi_1(q)}$$

$$V_{m2,q} = \left[\frac{\Delta_{33}(-\omega_2^+)}{(\omega_1^+ - \omega_2^+)(\omega_1^- + \omega_2^+)(\omega_2^+ + \omega_2^-)} \right]^{1/2} e^{i\psi_2(q)} \quad . \tag{39}$$

The phases $\psi_1(q)$ and $\psi_2(q)$ cannot be obtained using this approach. To determine them we can use the condition for the Bogolyubov transformation to be canonical, i.e., to keep unchanged the form of Hamilton's equations of motion:

$$\sum_j \left[U_{nj}U_{mj}^* - V_{nj}V_{mj}^* \right] = \delta_{nm} \ , \quad \sum_j \left[U_{nj}V_{mj} - V_{nj}U_{mj} \right] = 0 \quad . \tag{40}$$

The analytical expressions for the phases $\psi_1(q)$ and $\psi_2(q)$ are rather cumbersome. However, when examining the conditions in Eq. (40) numerically, we found out that one can set the phases equal to either 0 or π, as appropriate, without any appreciable change of accuracy.

The expressions derived in Eq. (39) differ slightly from those in Ref. 32. This difference is due to the different choice of the Green functions in Eq. (37). However, our choice seems more appropriate, because it leads to expressions that are more symmetric with respect to the mode indices.

The Green functions in Eq. (35) can eventually be written, using the above Bogolyubov transformation and the solutions for the U and V coefficients, as

$$\langle\langle (m^x_{nq})^*;\ m^x_{mq} \rangle\rangle = \frac{1}{2} \sum_{g=1,2} \left[\frac{A_{gnm}(q)}{\omega - \omega_g(q)} - \frac{\left(A_{gnm}(q)\right)^*}{\omega + \omega_g(q)} \right] \tag{41}$$

where

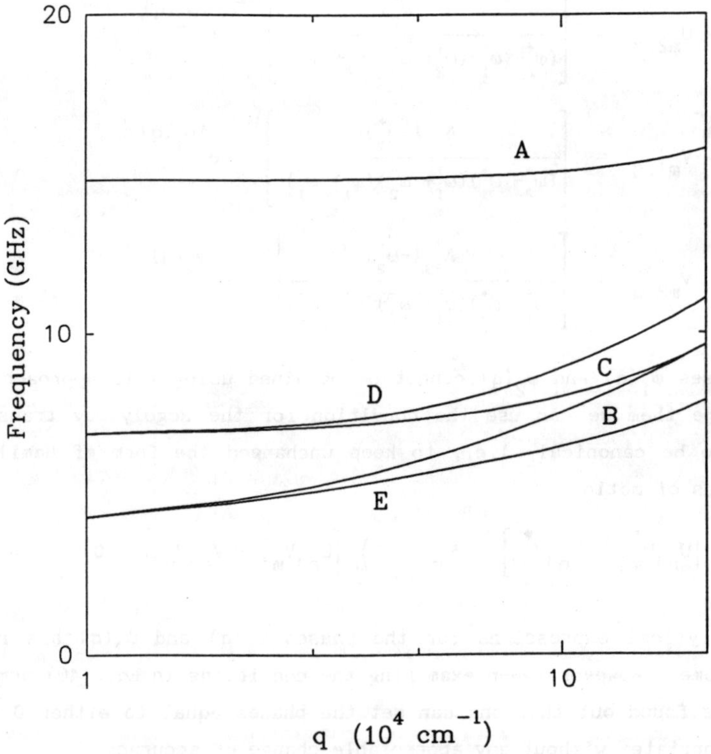

FIG. 3. The SW dispersion relations for a Fe film; see text for the parameters. The notations are: A - the n = 2 bulk mode; D, E - the hybridized n = 1 bulk and surface modes; B, C, - the non-hybridized n = 1 bulk and surface modes.

$$A_{gnm}(q) = \frac{S_{nm}(q,\omega)\ S^*_{mn}(q,\omega)}{2\omega_g(q)\ [\omega_g^2(q) - \omega_{3-g}^2(q)]}, \qquad (g = 1,2)\ , \tag{42}$$

$$S_{nm}(q,\omega) = \left(\Delta_{11}(q,\omega)\right)^{1/2} + \left(\Delta_{11}(q,-\omega)\right)^{1/2} \tag{43}$$

where $\Delta_{11}(q,\omega)$ has been defined earlier.

It is worthwhile mentioning that all the above-written expressions reduce to those for the non-degenerate case for those values of wave vector q that are far from the region of repulsion of the SW branches. This is due to the fact that, far from the region of repulsion, the small parameter of the perturbation theory, which is proportional to $P_{nm}/(\omega_n^2 - \omega_m^2)$, approaches zero.

6. Brillouin light scattering from hybridized spin waves

We now assume, as in Sec. 5, that for any given in-plane wave number **q** we have only two SW modes (labeled n and m) that are appreciably hybridized. For simplicity, we restrict attention to the usual case of linear magneto-optical coupling, so that only the Green function $\langle\langle (m^x)^*; m^x \rangle\rangle$ is required [see Eqs. (11) and (12)]. Using results from Sec. 5, we eventually find that the contribution to $d^2h/d\Omega d\omega_S$ from the hybridized modes is proportional to

$$\sum_{g=1,2} \frac{1}{\omega_g(q)} \Big[|\psi(B + \kappa_n)|^2 A_{gnn}(q) + |\psi(B + \kappa_m)|^2 A_{gmm}(q) + 2\ \mathrm{Re}[\psi^*(B + \kappa_n)\psi(B + \kappa_m)\ A_{gnm}(q)] \Big] \Big[\delta(\omega - \omega_g) + \delta(\omega + \omega_g) \Big] \cdot \tag{44}$$

The integrated intensities for BLS are given by strengths of the δ-functions in Eq. (44) multiplied by the appropriate Fresnel coefficients and magneto-optical constants.

In Fig. 3, which corresponds to L = 50 nm and d = 2.5×10^5 cm^{-1} for a Fe film, we have an example of a degeneracy near q $\sim 1.1 \times 10^5$ cm^{-1} between the surface and n = 1 bulk modes, and the resulting hybridized modes (full lines) are calculated using Eq. (27). In Fig. 4 we display the dependences of the SW eigenfrequencies on the pinning parameter d, taking fixed in-plane wave vector $q = 1.82 \times 10^5\ cm^{-1}$. The surface-mode frequency decreases with increasing d, going to zero at a characteristic value d_o, which lies far from the hybridization region of the dispersion curves and can be approximately deduced from the non-degenerate dispersion relationship $\omega_S = \omega_H + \alpha\omega_M(k^2 - d_o^2) = 0$; it is of interest for the light scattering behavior. The softening of this mode (labeled C in the non-hybridized case and E in the hydridized case) occurs because the pinning at one of the surfaces behaves like an effective anisotropy field that opposes the applied magnetic field.

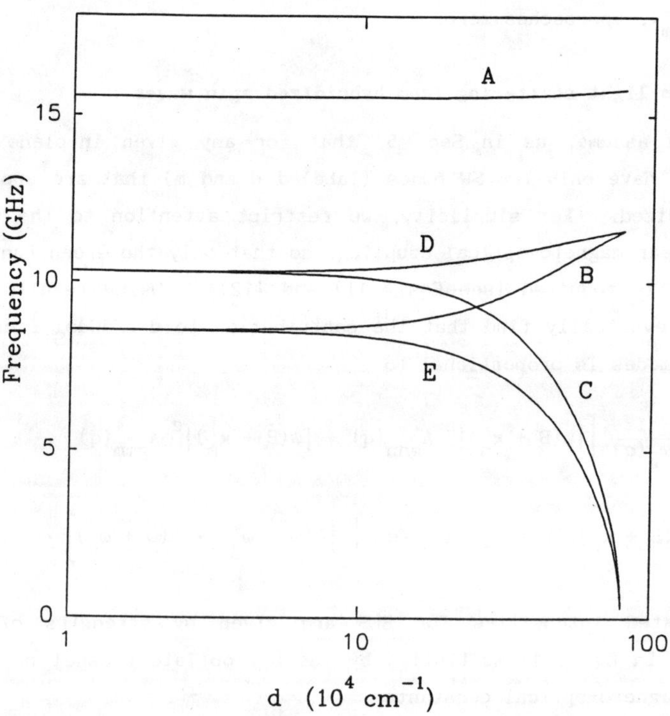

FIG. 4. The SW eigenfrequencies versus pinning d for a Fe film; see text for the parameters. The mode hybridization occurs around $d \cong 2 \times 10^5\ cm^{-1}$. The notations are the same as for Fig. 3.

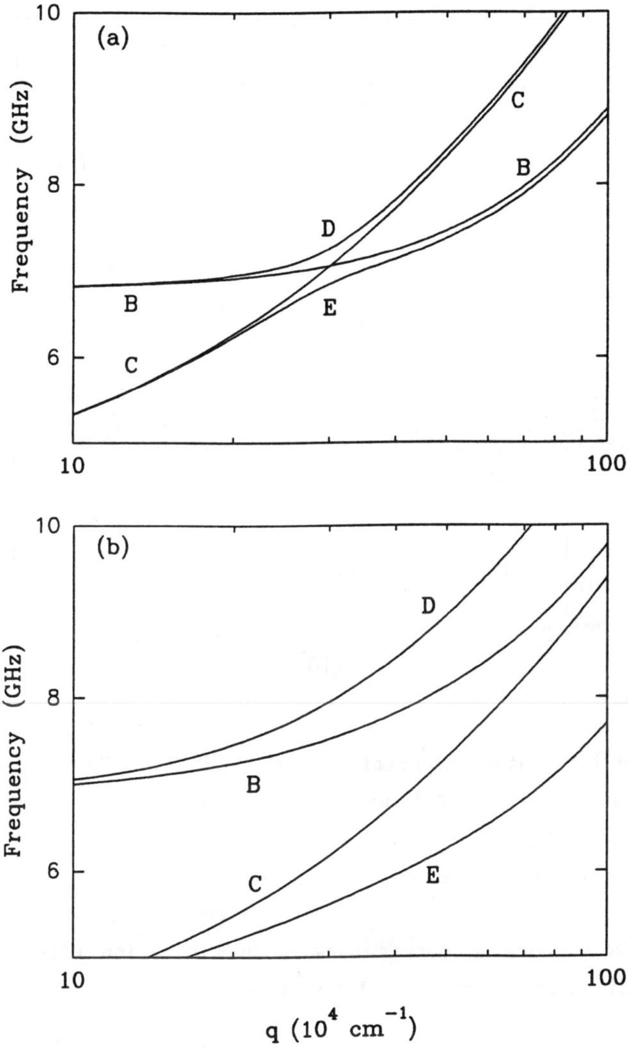

FIG. 5. The SW dispersion relations for a EuO film for two different values of the pinning parameter: (a) d = 2.5×10^5 cm^{-1}, (b) d = 1.4×10^5 cm^{-1}; see text for the other parameters. The notations are: D, E - the hybridized n = 1 bulk and surface modes; B, C, - the non-hybridized n = 1 bulk and surface modes.

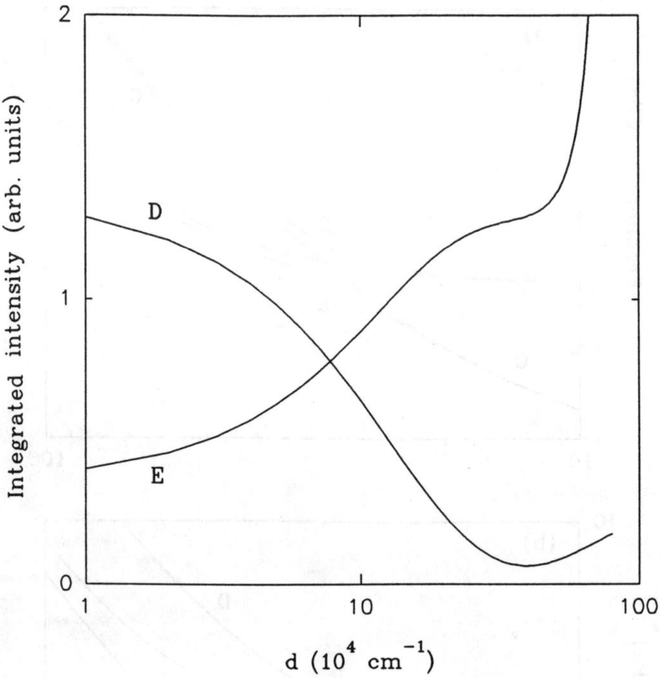

FIG. 6. The integrated intensities of the BLS for a Fe film with back-scattering angle equal to 45° versus pinning parameter d, using the same parameters as in Fig. 3. The curves D and E correspond to the hybridized n=1 bulk and surface modes.

Also some SW dispersion curves showing hybridization effects for EuO films of thickness 10 nm are shown in Fig. 5, taking two different values of d.

For other values of the parameters, the hybridization effects may strongly influence the BLS intensities. A plot of the integrated intensities of the scattered light from the hybridized n = 1 bulk and surface modes versus d is given in Fig. 6 for the same Fe film as in Fig. 3 with angle $\theta = 45^0$. The specific features in the behavior of both curves around $d \cong 2 \times 10^5$ cm^{-1} become clear when account is taken that this is the region of the hybridization of the dispersion relationships. The increase of the integrated BLS intensity from the mode labeled D corresponds to d approaching the value d_o (implying that the unhybridized surface SW frequency becomes very small). Because the optical penetration depth of Fe is relatively small (about 15 nm), we can neglect the multiple reflection of the scattered light from the lower surface of the film for the given film thicknesses.

We have made further BLS calculations for EuO films, for which multiple-reflection effects and a dependence on the (nonmagnetic) substrate material become important, as discussed at the end of Sec. 4. Using the parameter values quoted earlier, we show some numerical calculations in Fig. 7. In the main figure the BLS integrated intensity is plotted against film thickness L for two different choices of substrate material, namely silicon (with complex refractive index 4.3 + 0.1i) and sapphire (1.77 + 0.01i). We have used the BLS formalism described in Sec. 4, and it should be noted that the surface SW mode and the n = 1 bulk SW mode are strongly hybridized for part of the range of L being considered (see the inset to Fig. 7). For larger values of L, the surface mode becomes hybridized, in turn, with the bulk SW modes of larger n.

7. Modification of pinning conditions

The case of antisymmetric pinning is of physical interest, because there are effective ways to control the magnitude and sign of the pinning at each film surface[30]. Nevertheless, experimentally it is hard to achieve *purely* antisymmetric pinning, and so we discuss the situation when the pinning conditions are not exactly antisymmetric.

The general equation for the SW perpendicular wavenumber κ_n in the case of pinning d_1 and d_2 at the two surfaces is given in Eq. (16). In the purely antisymmetric case, the solutions are just the bulk waves with $\kappa_n = n\pi/L$ and the surface wave with $\kappa_n = \pm id$, as discussed earlier. If κ^* denotes any one of these solutions , then for the case of not purely antisymmetric pinning we may define $\kappa_n = \kappa^*(1 + \delta)$, with $d_1 \equiv d$ and $d_2 \equiv -d(1 + \varepsilon)$, assuming $|\varepsilon| \ll 1$ and $|\delta| \ll 1$. We then find, by

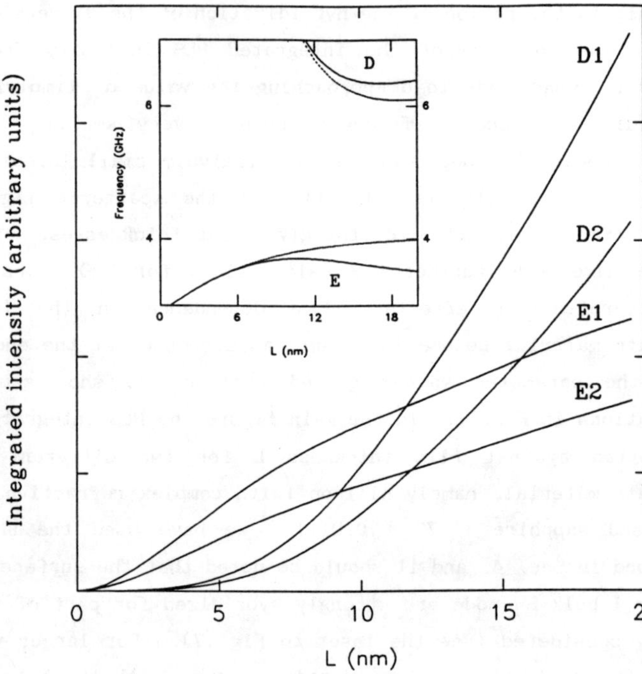

FIG. 7. The integrated BLS intensities and SW frequencies (see inset) for a EuO film for the hybridized surface and n = 1 bulk SW modes. The BLS results are shown as a function of film thickness L for two choices of substrate corresponding to silicon (curves D1 and E1 corresponding to SW modes D and E, respectively) and to sapphire (curves D2 and E2). See the text for parameter values.

linearizing Eq. (16), that the approximate solutions correspond to (in terms of δ):

$$\delta \simeq \begin{cases} \varepsilon[1 - \exp(-2dL)]^{-1}, & \text{for a surface wave} \\ -\varepsilon(d/L)[(n\pi/L)^2 + d^2]^{-1}, & \text{for bulk wave } n\,. \end{cases} \tag{45}$$

For any value of ε in the linear regime, we see that *real* solutions for δ occur in both above cases. Hence the wavenumber κ_n corresponding to the surface wave remains imaginary, and the bulk wave wavenumber remains real.

A numerical analysis shows that, within the given range of the parameters, the ratio δ/ε for the bulk wave is rather small. This leads us to conclude that the bulk wave is relatively stable with respect to deviations from the antisymmetric case. For a surface wave, we conclude from Eq. (45) that δ/ε becomes larger as the pinning parameter d becomes smaller. However, the dispersion relationships in the regime of small d are rather flat, and so even a significant shift in κ_n does not change the frequency appreciably (e.g., the change is only a few percent or

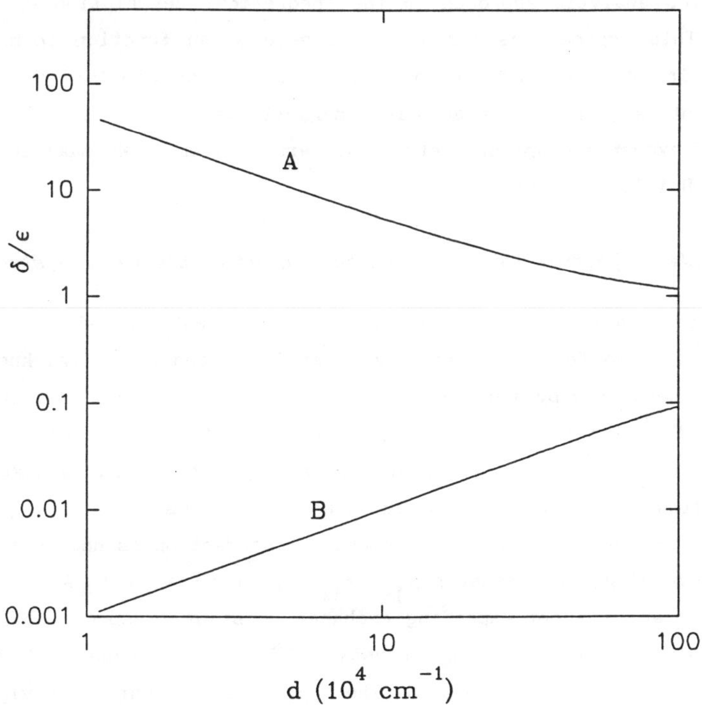

FIG. 8. The shift in the perpendicular SW wave vector (expressed in terms of the dimensionless parameter δ/ε for a fixed ε) as a function of d for the n = 1 bulk mode (curve A) and the surface mode (curve B). We have taken ε = 0.3 and L = 10 nm.

less). Some numerical results for the dependence of δ/ε on d are shown in Fig. 8, taking as an example $\varepsilon = 0.3$.

8. Magnetic double-layer structures

The problem of SW excitations in double-layer systems, i.e., layered structures consisting of two ferromagnetic films separated by a thin nonmagnetic layer, has been addressed by several authors[33-36] and, although theoretical curves obtained adequately describe the existing experimental SW data, the theory involves lengthy numerical calculations of a rather cumbersome system of initial equations. Here we develop a more simple, analytic approach to the theoretical description of this system. This approach is based on the same Green function formalism discussed in the previous sections and accounts for dipole-dipole and exchange interactions within each ferromagnetic layer as well as dipole-dipole and exchange couplings between different layers. We make applications to Brillouin light scattering.

8.1. Spin Wave Spectrum for the Exchange-Coupled Double-Layer System

Let us consider a perpendicularly magnetized layered structure consisting of two ferromagnetic films (labeled 1 and 2) of thicknesses L_1 and L_2, separated by a nonmagnetic layer of thickness b. The pinning constants $d_1^{s,i}$ and $d_2^{s,i}$ characterize pinning on the outer surface (s) and interface (i), respectively, of each film. The z axis is taken to be perpendicular to the planar surfaces and interfaces.

We assume that the interlayer exchange interaction is characterized by a phenomenological parameter $\beta_j = A_{12}/A_j$. Here j and A_j are the film number and the corresponding exchange constant, respectively, and A_{12} is the interlayer exchange constant[33,35]. The parameter β is a function of the spacer thickness b. The typical experimental behavior of β as a function of b is discussed elsewhere[34,36]. It can be positive, negative or oscillatory, being generally of the order of magnitude 10^6 cm^{-1}. In our theory we use β as an empirical parameter already known either from microscopic calculations or from experiment.

The procedure of obtaining the SW spectrum for the multilayered system has been developed[37,38] for the case of purely dipolar coupling between the layers. By analogy to Eq. (13), it is based on the expansion of the variable part of magnetization $\mathbf{m}_j(\mathbf{r},t)$ and the dipolar field $\mathbf{h}^{dip}(\mathbf{r},t)$ of the film j as Fourier series of the following kind:

$$\mathbf{m}_j(\mathbf{r},t) = M_o \sum_{\mathbf{k}} \sum_{n} \mathbf{m}_{\mathbf{k}jn}\, \phi_{jn}(z) \exp[-i(\mathbf{k}\cdot\rho-\omega t)] \tag{46}$$

$$\mathbf{h}_j^{dip}(\mathbf{r},t) = \sum_{\mathbf{k}} \sum_{n} \mathbf{h}_{\mathbf{k}jn}^{dip}\, \phi_{jn}(z) \exp[-i(\mathbf{k}\cdot\rho-\omega t)] \tag{47}$$

where the functions $\phi_{jn}(z)$ for the case of two layers should be taken in the form:

$$\begin{aligned}\phi_{1n}(z) &= A_{1n}\cos\left[\kappa_{1n}\left(z+\frac{b}{2}+L_1\right)\right] + B_{1n}\sin\left[\kappa_{1n}\left(z+\frac{b}{2}+L_1\right)\right]\\ \phi_{2n}(z) &= A_{2n}\cos\left[\kappa_{2n}\left(z-\frac{b}{2}-L_2\right)\right] + B_{2n}\sin\left[\kappa_{2n}\left(z-\frac{b}{2}-L_2\right)\right]\end{aligned} \tag{48}$$

where κ_{jn} are the transverse discrete wave numbers. To incorporate the effects of the interlayer exchange coupling into the theory[37], we require that the eigenfunctions (48) satisfy the linearized Hoffmann boundary conditions[34]:

$$\begin{aligned}&-\frac{\partial\phi_{1n}}{\partial z} + d_1^s\phi_{1n}\Big|_{z=-\left(L_1+\frac{b}{2}\right)} = 0\\ &\frac{\partial\phi_{2n}}{\partial z} + d_2^s\phi_{2n}\Big|_{z=\left(L_2+\frac{b}{2}\right)} = 0\\ &\frac{\partial\phi_{1n}}{\partial z} + \left(d_1^i+\beta_1\right)\phi_{1n}\Big|_{z=-\frac{b}{2}} - \frac{M_{o1}}{M_{o2}}\beta_1\phi_{2n}\Big|_{z=\frac{b}{2}} = 0\\ &-\frac{\partial\phi_{2n}}{\partial z} + \left(d_2^i+\beta_2\right)\phi_{2n}\Big|_{z=\frac{b}{2}} - \frac{M_{o2}}{M_{o1}}\beta_2\phi_{1n}\Big|_{z=-\frac{b}{2}} = 0\end{aligned} \tag{49}$$

where M_{oj} is the static magnetization in the film j. In particular, in the case of a symmetric structure consisting of two identically magnetized layers (i.e., $L_1 = L_2 \equiv L$, $d_1^{i,s} = d_2^{i,s} \equiv d^{i,s}$, $\beta_1 = \beta_2 \equiv \beta$, $M_{o1} = M_{o2} \equiv M_o$), Eqs. (49) lead to

$$\left(\kappa^2 - d^s d^i\right)\tan(\kappa L) = \kappa\left(d^s + d^i\right) \tag{50}$$

for the *symmetric* modes, i.e., the modes with the variable magnetization in the both layers distributed similarly, and

$$\left(\kappa^2 - d^s(d^i+2\beta)\right)\tan(\kappa L) = \kappa\left(d^s + (d^i+2\beta)\right) \tag{51}$$

for the *antisymmetric* modes.

Equation (50) is formally the same as Eq. (16) for a single (uncoupled) layer of thickness L. Equation (51) has a similar form but with a modification of the interface pinning due to the interlayer exchange β, i.e., $d^i \longrightarrow d^i + 2\beta$. Thus we conclude that the interlayer coupling affects only antisymmetric modes. This conclusion is in agreement with the results of Ref. 33. In particular, in the case of unpinned surface spins ($d^i = d^s = 0$) we recover from Eq. (51) all the results for transverse wave numbers of antisymmetric modes obtained in Ref. 33. It also follows from earlier results[25,39] that when $d^i = d^s = 0$ for $\beta > 0$ we get only real solutions for κ (bulk modes). On the other hand, for $\beta < 0$, along with the infinite number of bulk solutions, we obtain one (lowest) surface solution with imaginary κ. By making the replacement $d^i \longrightarrow d^i + 2\beta$, we can apply all the conclusions of the previous sections to describe quantitatively the SW dispersion in exchange coupled magnetic double layers.

Using Eqs. (50) and (51) to calculate the transverse SW wave numbers κ_n in a double layer with exchange coupling, and the formalism[37] developed for dipolar coupling of magnetic layers we can derive an approximate dispersion relation for SW propagating in a magnetic double layer consisting of two identical magnetic films coupled by both dipole-dipole and exchange interactions. In the particular case of the coupling of two identical modes (with label n) in both films we get the following dispersion relation for a double layer:

$$(\omega_n^{\pm})^2 = \Omega_n\left(\Omega_n + \omega_M P_{nn}\right) \pm \left(\omega_M \Omega_n\ P_{nn}^{12}\right) \tag{52}$$

where the "+" sign corresponds to antisymmetric modes, the "-" sign corresponds to symmetric modes, $\Omega_n = \Omega(\kappa_n)$ is the eigenfrequency of the SW mode n of a single film in the purely exchange limit defined by Eq. (18). We do not reproduce here the lengthy expressions for the diagonal $P_{nn} = P_{nn}^{11} = P_{nn}^{22}$ and off-diagonal P_{nn}^{12} dipole matrix elements in coupled magnetic films, as the procedure of their analytical evaluation is well developed[25,37]. We note that, as the expressions for P_{nn} and P_{nn}^{12} depend on the values of transverse wave numbers κ_n (and therefore on the value of the exchange coupling coefficient β), so does the splitting of the spectral branches. This leads us to the conclusion that the dipole-dipole and exchange splitting of the dispersion branches should not be considered separately. Both types of splitting influence the spectrum and, although interlayer exchange mainly affects the frequencies of antisymmetric SW modes near k = 0 [see Eq. (51)], it will also significantly change the dipole-dipole branch repulsion in the short-wavelength ($qL \gg 1$) region probed by BLS.

8.2. Brillouin Light Scattering

The Green function formalism for Brillouin light scattering from multilayered magnetic structures is well established (see Ref. 9 and references therein) and can be considered as a direct generalization of the results obtained in Secs. 4 and 5 for thin films. Since the appropriate Green functions have been calculated from the formalism described earlier, it is straightforward in principle to determine the light scattering intensities associated with the various SW modes. In the general case it is necessary to take account of the internal multiple reflections of the incident and scattered light at the surfaces and interfaces of the sample. In the present section we avoid this difficulty by restricting attention to the case when the optical penetration depth of the light in the sample is smaller than L (e.g., as would be typically in the case of Fe films), thereby making the multiple reflection effects negligible.

8.3. Discussion of Numerical Results for Surface Waves

We are interested here in surface SW modes of a double layer that correspond to the solutions of Eqs. (50)-(52) for imaginary κ_n. The numerical analysis shows, that, depending on the values of d^i, d^s, and β, there exist two, one, or zero such solutions. When both of the effective pinning parameters d^i, d^s are negative and sufficiently large, there are two imaginary solutions for κ_n, while for non-negative values of one of these parameters a single solution exists only if the second parameter is negative. The same kind of analysis can be applied to the dependence of the κ_n values on the exchange coupling coefficient

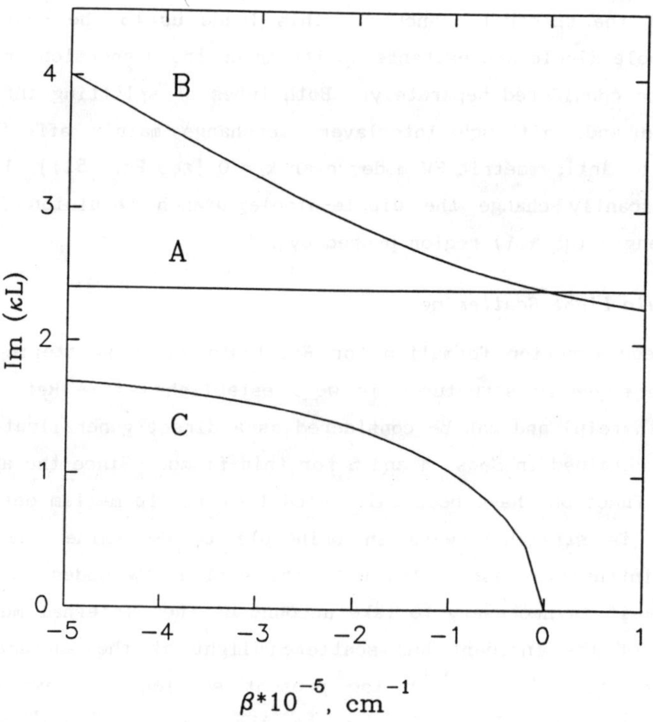

Fig. 9. The dependences of imaginary perpendicular SW wavenumbers on the interlayer exchange coefficient β for the Fe/spacer/Fe structure (see text for the parameters and details).

β. As the symmetric modes are not affected by the interlayer exchange, the κ_n values for the symmetric modes remain the same for all values of β. Contrary to that, for the antisymmetric modes the effective interface pinning changes with β and so do the κ_n values.

We have calculated the dependence of the imaginary transverse wave numbers (corresponding to surface SW modes) on the interlayer exchange parameter β for the system Fe/spacer/Fe. We used the following set of parameters: the exchange parameter α, the field H_e and the static magnetization $4\pi M_0$ are taken as in Sec. 4, the film thickness L = 200 Å, and the pinning parameters are $d^s = d^i = d = K_s/A = -1 \times 10^6$ cm^{-1}. The values of β were taken to vary from -5×10^5 to 1×10^5 cm^{-1}. The results are presented in Fig. 9. For the given set of parameters there

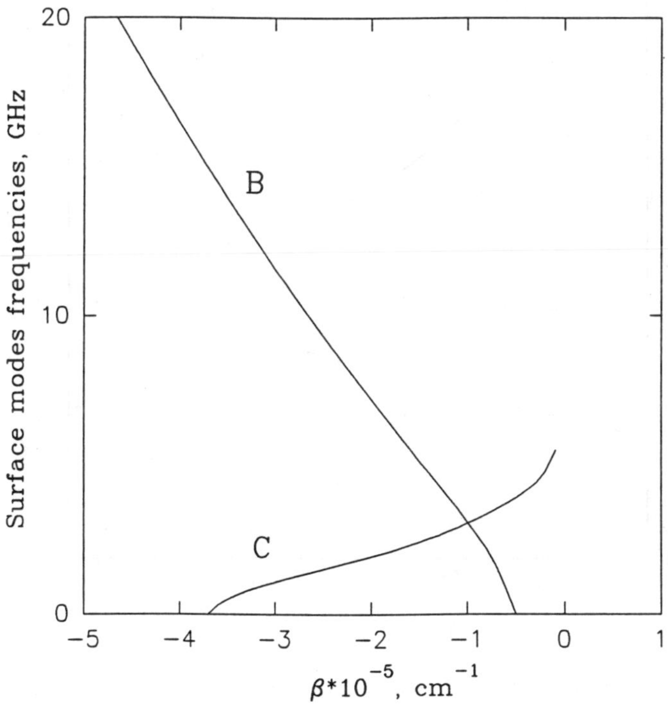

Fig. 10. The dependences of surface SW frequencies on the interlayer exchange coefficient β for the Fe/spacer/Fe structure (see text for the parameters and details).

exist one surface branch for the symmetric mode (horizontal line **A**) and two surface branches for the antisymmetric mode (curves **B** and **C**). Curve **C** ends at the point $\beta = 0$, where the corresponding branch of the SW spectrum meets the bulk continuum.

The existence of the imaginary solutions for κ_n, however, does not guarantee the existence of the corresponding branches in the SW spectrum. This becomes clear from the analysis of the dispersion equation (52). The dipole matrix elements P_{nn}^{ik} are always positive. However, the sign of Ω_n depends on the values of $|q|$ and κ_n, and sometimes becomes negative. In this case if P_{nn}^{ik} are sufficiently large, the

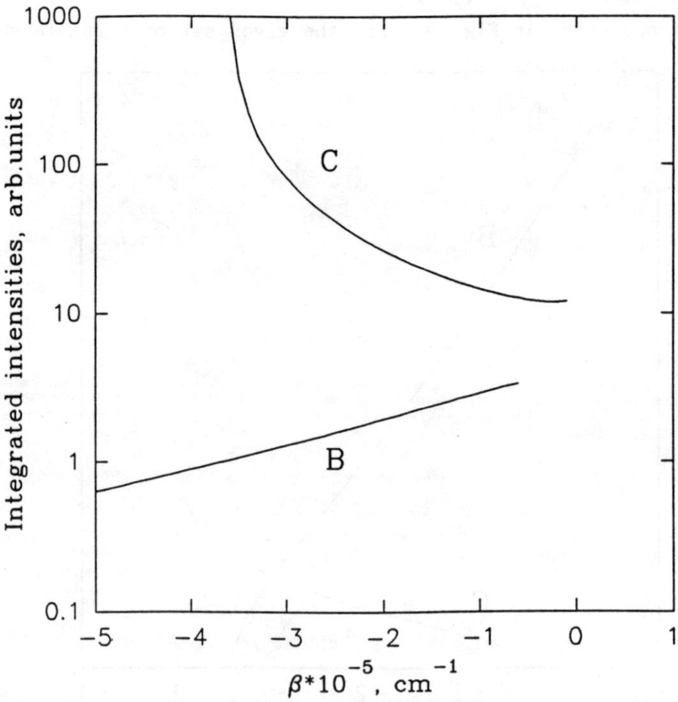

Fig. 11. The dependences of the integrated intensities of Brillouin light scattering from the surface SW on the interlayer exchange coefficient β for the Fe/spacer/Fe structure (see text for the parameters and details).

right hand side of Eq. (52) becomes negative.

The above conclusions are confirmed by the results of our calculations of the SW spectrum and the integrated intensities of the light scattering from the system Fe/spacer/Fe for the given values of the parameters. We used the same 180^0 backscattering geometry and the parameters of the experiment as previously (see Sec. 4), taking the scattering angle $\theta = 45^o$. This provides for the in-plane wavenumber of the SW to be $|q| \cong 1.8 \times 10^5$ cm^{-1}. In Fig. 10. we display the dependences of the surface modes frequencies on β. It can be seen that, although the imaginary κ for symmetric mode exists, there is no corresponding branch in the SW spectrum. As well, the branch **C** in Fig. 10, corresponding to the curve **C** in Fig. 9, appears only when the β values are in the interval from about -3.6×10^5 to 0 cm^{-1}. In the same way, the branch **B** ceases to exist when $\beta \geq -0.5 \times 10^5$cm^{-1} while the corresponding κ dependence (curve **B** in Fig. 9) still shows the existence of the solution. Basically the same characteristic features can be seen in the Fig. 11, where the integrated intensities of the Brillouin light scattering are depicted. The labeling of the branches is the same as for the previous two plots.

9. Conclusions

We have demonstrated in this chapter a simple analytic approach to the calculation of cross sections and intensities for Brillouin light scattering from dipole-exchange spin waves in magnetic thin films. We concentrated on the case of perpendicularly magnetized films with surface anisotropy (pinning) effects included, and extensions were also made to magnetic double-layer structures. An advantage of this approach over previous methods, which were largely numerical, lies in the possibility to carry out the calculations analytically, e.g., in estimating analytically pinning effects or the relative contributions to the spectrum of BLS of such effects as dipole-dipole and exchange interactions (both within a magnetic layer and between magnetic layers in the case of double-layer structures).

The tensorial Green-function formalism for the spin-wave spectrum in magnetic films, which forms the basis of our approach, has recently been generalized for the case of anisotropic magnetic films magnetized in an arbitrary direction to the film normal[40]. Thus, using these results[40], it is straightforward to extend the above BLS formalism to the cases of anisotropic films and different static magnetization directions.

Presently efforts are in progress to extend this formalism also to the cases of magnetic superlattices and non-symmetric multilayered structures where all relevant interactions, including interlayer dipole-dipole and exchange interactions as well as crystallographic and surface anisotropies, are taken into account.

Acknowledgements

This work was supported in part by the grant from NSERC of Canada, by the grant CC-3401 from the Research Corporation, USA and by the grant from the Oakland University Foundation, USA.

REFERENCES

1. P.A. Fleury, S.P.S. Porto, L.E. Cheesman, and H.J. Guggenheim, *Phys. Rev. Lett.* **17** (1966) 84.
2. J.R. Sandercock and W. Wettling, *Solid State Comm.* **13** (1973) 1729.
3. J.R. Sandercock, *Solid State Comm.* **15** (1974) 1715.
4. P. Grünberg and F. Metawe, *Phys. Rev. Lett.* **39** (1977) 1561.
5. J.R. Sandercock and W. Wettling, *IEEE Trans. Magnet.* **14** (1978) 442.
6. M. Grimsditch, A.P. Malozemoff, and A. Brunsch, *Phys. Rev. Lett.* **43** (1979) 711.
7. P. Grünberg, M.G. Cottam, W. Vach, C.M. Mayr, and R.E. Camley, *J. Appl. Phys.* **53** (1982) 2078 .
8. M. Grimsditch, M.R. Khan, A. Kueny, and I.K. Schuller, *Phys. Rev. Lett.* **51** (1983) 498.
9. M.G. Cottam and D.J. Lockwood, *Light Scattering in Magnetic Solids* (Wiley, New York, 1986).
10. P. Grünberg, *Progress in Surf. Science*, **18** (1985) 1.

11. P. Grünberg, in *Light Scattering in Solids V*, eds. M. Cardona and G. Güntherodt (Springer-Verlag, Heidelberg, 1989), p. 303.

12. J.R. Dutcher, in *Linear and Nonlinear Spin Waves in Magnetic Films and Superlattices*, ed. M.G. Cottam (World Scientific, Singapore, 1994), p. 287.

13. R.W. Damon and J. Eshbach, *J. Phys. Chem. Solids* **19** (1961) 308.

14. R.E. Camley, T.S. Rahman, and D.L. Mills, *Phys. Rev. B* **23** (1981) 1226.

15. J.F. Cochran and J.R. Dutcher, *J. Magn. Magn. Mat.* **73** (1988) 299.

16. R.E. De Wames and T. Wolfram, *J. Appl. Phys.* **41** (1970) 987.

17. R.L. Stamps and B. Hillebrands, *Phys. Rev. B* **43** (1991) 3532.

18. V.V. Gann, *Sov. Phys. - Solid State* **8** (1967) 2537 [*Fiz. Tverd. Tela* **9** (1966) 3167].

19. B.N. Filippov, *Phys. Met. Metall.* **32** (1971) 911 [*Fiz. Met. Metalloved* **29** (1970) 1131].

20. M.G. Cottam and A.N. Slavin, *IEEE Trans. Magnetics* **27**, 5489 (1991).

21. I.V. Rojdestvensky, M.G. Cottam, and A.N. Slavin, *J. Appl. Phys.* **73** (1993) 7001.

22. I.V. Rojdestvensky, M.G. Cottam, and A.N. Slavin, *Phys. Rev. B* (1993) in press.

23. A.N. Slavin, I.V. Rojdestvensky, and M.G. Cottam, *J. Appl. Phys.* (to be published).

24. O.G. Vendik and D.N. Chartorizhskii, *Sov. Phys. - Solid State* **12** (1970) 2537 [*Fiz. Tverd. Tela* **12** (1970) 1538].

25. B.A. Kalinikos and A.N. Slavin, *J. Phys. C* **19** (1986) 7013.

26. B.A. Kalinikos and A.N. Slavin, *Acta Phys. Pol.* **A75**, (1989) 541.

27. M.G. Cottam and A.N. Slavin, in *Linear and Nonlinear Spin Waves in Magnetic Films and Superlattices*, ed. M.G. Cottam (World Scientific, Singapore, 1994), p. 1.

28. B.A. Kalinikos, in *Linear and Nonlinear Spin Waves in Magnetic Films and Superlattices*, ed. M.G. Cottam (World Scientific, Singapore, 1994), p. 89.

29. M.G. Cottam, *J. Phys. C* **16** (1983) 1573.

30. N.M. Salanskii and M.S. Erukhimov, *Physical Properties and Applic-*

ation of Magnetic Films (Nauka, Novosibirsk, 1975), p. 38 [in Russian].

31. W. Wettling, M.G. Cottam, and J.R. Sandercock, *J.Phys. C* **8** (1975) 211.
32. V.N. Krivoruchko and D.A. Yablonskii, *Phys. Stat. Sol. (b)* **103** (1981) K41.
33. K. Vayhinger and H. Kronmüller, *J. Mag. Magn. Mat.* **72** (1988) 307.
34. M. Vohl, J. Barnas, and P. Grünberg, *Phys. Rev. B* **39** (1989) 12003.
35. B. Hillebrands, *Phys. Rev. B* **41** (1990) 530.
36. B. Heinrich, Z. Celinski, J.F. Cochran, A.S. Arrott, K. Myrtle, and S.T. Purcell, *Phys Rev. B* **47** (1993) 5077.
37. B.A. Kalinikos and P.A. Kolodin, *Izv. Vuzov-Radiofizika* **32** (1989) 1290 (1989).
38. K. Vayhinger and H. Kronmüller, *J. Mag. Magn. Mat.* **62** (1986) 159.
39. R.C. Moul and M.G. Cottam, *J. Phys. C* **16** (1983) 1307.
40. B.A. Kalinikos, M.P. Kostylev, N.V. Kozhus, and A.N. Slavin, *J. Phys. Cond. Mat.* **1** (1990) 9861.

PHOTOTHERMAL CHARACTERIZATION OF MICROWAVE MAGNETIC EXCITATIONS

O. von Geisau and J. Pelzl

AG Festkörperspektroskopie, Ruhr-Universität Bochum

D-44780 Bochum, F.R.G

1. Introduction

Collective magnetic excitations in ferro- and ferrimagnetic solids with wave numbers below 10^8 m^{-1} can be detected by resonance absorption of microwaves at frequencies in the GHz range (cf. Ch. 1). The most frequently used technique – referred to as conventional ferromagnetic resonance (FMR) here – is based on the detuning of an impedance matched microwave cavity during the resonance absorption. The signal which is detected as a change of the reflected microwave power[1] represents an integral information on the resonance absorption all over the sample. In the last two decades combined optical-microwave techniques have been developed offering the possibility for a locally resolved detection of thermally or coherently excited magnons[2–4] (cf. Ch. 12). Additional alternative FMR detection schemes, photothermal FMR measurement techniques based on the generation and the detection of thermal waves, have been developed in our laboratory. Their common characteristic is the ability of depth discrimination, whereas the lateral resolution available depends on the particular method.

Photothermally based FMR detection techniques are commonly performed in the frequency domain and the magnetic resonance is either modulated by thermal waves excited with a light beam focused onto the sample or heat waves released during the resonance absorption are detected by a contactless method. The amplitude of the photothermal signal is related to the amplitude of the thermal wave and thus characterizes the absorption process and the efficiency of the non-radiative decay channels following the absorption of microwaves. The time delay of the thermal wave with respect to the excitation, the phase angle of the photothermal signal, is a direct measure for the location of the absorbing centres inside the material. Photothermal

FMR detection besides lateral resolution provides depth dependent information also in non-transparent samples. Therefore, the photothermally based methods can be used to investigate the spatial distribution of the resonance absorption in linear and non-linear collective magnetic excitations. The sensitivity to thermal effects and the imaging ability in up to three dimensions distinguish the photothermal FMR techniques from the above mentioned FMR detection methods.

In this report, we first briefly review the basic concepts of the photothermal methods applied to FMR. The experimental arrangements and conditions of the different photothermal detection schemes are summarized and compared with the conventional recording (Sect. 2). Sects. 3 – 6 deal with the theory and the experiment of some recent applications of photothermal FMR techniques devoted to the exploration of the spatial variation of magnetostatic modes in yttrium iron garnet slabs, films and spheres. Results obtained from other materials such as soft ferrites and magnetic tapes are briefly reviewed in Sect. 7. For further reading and more details on the latter topic we would like to refer to preceding reports.[5–7]

2. Experimental Aspects

2.1. Photothermal Methods and Detection Schemes for FMR

At present, three photothermal microwave techniques have been developed: the **p**hoto**a**coustic detection (PA-FMR), the **p**hotothermal laser beam **d**eflection technique (PD-FMR) and the **p**hotothermally **m**odulated detection (PM-FMR).

The common feature of these detection schemes is the generation of oscillatory thermal waves $\vartheta(\mathbf{r},t)$ in the specimen. A time dependent heating can be either achieved by modulating the microwave energy or by locally heating the sample with an intensity modulated laser beam at constant microwave power. When applying a modulated heating process with a modulation frequency ω_{mod} a temperature field consisting of an *ac* part $\vartheta(\mathbf{r},t)$ — oscillatory in space and time — and a *dc* temperature rise is produced in the sample. The spatial component of $\vartheta(\mathbf{r},t)$ is governed by the heat diffusion equation[8]:

$$\alpha\Delta\vartheta(\mathbf{r}) - i\omega_{mod}\vartheta(\mathbf{r}) = Q(\mathbf{r}) \quad . \tag{1}$$

The thermal diffusivity $\alpha = \kappa/\rho c$ is a function of the heat conductivity κ, the mass density ρ and the specific heat capacity c. Q represents the heat source. The solution of the diffusion equation (1) is a strongly damped wave. In the one-dimensional case

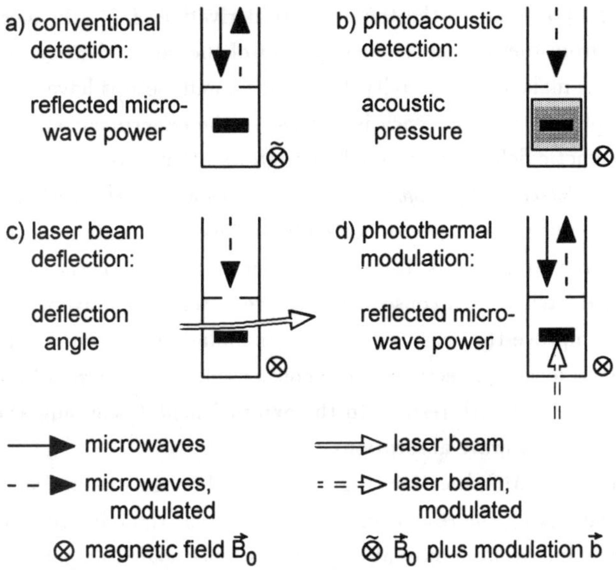

Figure 1: Schematic representation of conventional and photothermal detection methods in microwave resonance absorption and their respective measured quantities.

this thermal wave is given by[8]:

$$\vartheta(x > 0, t) = \vartheta_0 \, e^{-x/\mu} \, e^{i(\omega_{mod} t - x/\mu)} \,. \tag{2}$$

The attenuation length (thermal diffusion length) $\mu = \sqrt{2\alpha/\omega_{mod}}$ decreases with increasing modulation frequency ω_{mod}. The wavenumber of the thermal wave $k = 1/\mu$ and its phase velocity $v_p = \omega_{mod}/k = \sqrt{2\alpha\omega_{mod}}$. Thus, thermal waves are characterized by a strong dispersion of the amplitude and of the phase velocity.

With regard to the thermal waves the photothermal detection schemes can be grouped into two categories. When using the PM-FMR a local periodical heating locally induces small changes in magnetic sample parameters. With a modified conventional setup these are synchronously detected as a part of the conventional diode signal. In the case of PA- and PD-FMR the absorbed incident microwave power – after relaxation to the lattice – produces thermal waves which are detected in an external medium. The local resolution is determined by the particular properties of the measurement technique and the experimental conditions.

The signals of the conventional, the PA- and the PD-FMR are proportional to the imaginary part $\tilde{\chi}''$ of the hf susceptibility tensor

$$\tilde{\chi}(\omega) = \tilde{\chi}'(\omega) - i\tilde{\chi}''(\omega). \tag{3}$$

The PM-FMR signal is proportional to the temperature derivative of $\tilde{\chi}''$. In most cases, in order to benefit from the resonance enhancement of the power density, the mentioned FMR techniques make use of a cavity, but non-resonant setups have been used, too. A microwave of constant frequency is applied to the experiment and by changing the external magnetic field $\tilde{\chi}''$ is scanned through its resonance.

In the *conventional detection technique* (Fig. 1a) the resonance absorption is monitored by observing the microwave power reflected from the cavity containing the sample. At resonance microwave power is dissipated in the sample and reduces the quality factor of the loaded cavity. As a consequence the reflected microwave power is increased. In order to improve the sensitivity a field modulation technique with phase sensitive narrow band lock-in detection is commonly used. The integral of the resulting signal trace $\partial\chi''/\partial H(H)$ with respect to the external field H contains the averaged information on the microwave absorption.

The *photoacoustic (PA) FMR detection* (Fig. 1b) is based on the dissipation of microwave energy into heat during the resonance process. In a gas-tight PA cell the modulation of the incoming microwave power at audio frequencies leads to a periodic heat flow from the sample into the gas. The resulting variation in the gas pressure is detected by a commercial microphone.[9] Besides the possibility of a narrow band detection with a lock-in amplifier the modulation of the signal brings about the most significant advantage inherent to photothermal methods which is the depth sensitivity. When the modulation frequency is varied, heat generated at different depths below the surface can be differentiated because of the changing thermal diffusion length.[5]

The *laser beam deflection (PD-FMR)* or *mirage effect* (Fig. 1c) provides a thermal wave detection which – in addition to the depth resolution – offers a local discrimination, too. The PD technique probes the deflection of a laser beam in the optical density gradient caused by the heat wave in the air adjacent to the sample or in the sample itself.[10] The thermal waves are generated in the same way as discussed in the last paragraph. By scanning the probe beam across the sample a lateral image of the microwave absorption density can be obtained as only those parts of the sample lying below the displaceable laser beam contribute to the PD-signal. The lateral resolution is defined by the diameter of the laser beam and the thermal diffusion lengths of the sample and the surrounding medium. Simultaneously, the depth dependence can be measured by changing the modulation frequency. The laser beam deflection does not require an air tight sample cell but the cavity/waveguide has to provide optical access for the probe beam.

The *photothermally modulated (PM) FMR* (Fig. 1d) is a technique which combines the ability for spatial resolution inherent to the thermal waves and the high sensitivity of the electronic detection circuitry used in the conventional experiments.[11,6] An intensity modulated laser or light beam is focused onto the sample and generates a thermal wave inside the specimen. As a consequence of the local temperature change the resonance absorption is modified giving rise to a reflected microwave power oscillating coherently with the modulation cycle. The detected signal originates from a region inside the sample centred at the laser spot having an extension defined by the size of the laser focus and the thermal diffusion length of the sample. For an opaque sample the modulation frequency controls the depth of the thermal wave penetration into the solid. Therefore, the recording as a function of modulation frequency and laser beam position delivers a real three-dimensional tomographic image of the microwave absorptivity.[12] The main difference as compared to the other photothermal techniques mentioned before arises from the fact that the PM signal is proportional to the temperature derivative of the imaginary part of the hf susceptibility. For that reason the PM signal is governed by the temperature dependent magnetic parameters such as saturation magnetization and anisotropy and magneto-elastic constants.

2.2. Experimental Arrangements and Spatial Resolution

The *conventional FMR measurements* are performed using a VARIAN E-line EPR spectrometer working at X-band frequencies from 8.8 – 9.6 GHz (see Fig. 2). The photothermal experiments are conducted with the conventional spectrometer which is modified in different ways for each technique. The intensity modulation of the incident microwave power for the *PA-FMR* and the *PD-FMR* detection is effected by a PIN-diode modulator. High excitation field strengths are available using an optional travelling wave tube (TWT) amplifier.

For the *PA-FMR detection* (Fig. 3) the sample either placed in a PA waveguide cell[5] or in a quartz tube.[13] For both setups the microphone is placed outside the hf magnetic field and is connected to the sample cell via an air-tight duct.

The *PD-FMR technique* (Fig. 3) measures the deflection of a 1mW-HeNe laser beam passing parallel to the sample surface with a position sensitive quadrant diode. The PD measurement can be effected using a light-access cavity (e. g. a TE_{011} cylindrical cavity[14,11]) or a non-resonant waveguide arrangement with small slits in the short sides of the rectangular waveguide.[15–17,5] The lateral resolution in this experiment is given by the diameter of the laser beam (130 μm) plus two times the thermal

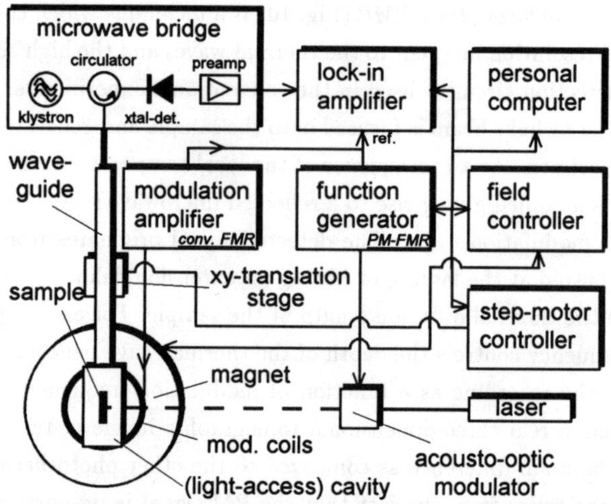

Figure 2: Block diagram of the experimental arrangement for the conventional and the PM-FMR detection.

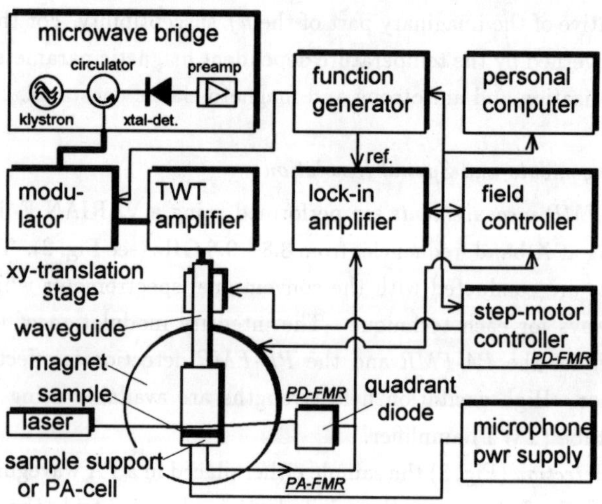

Figure 3: Block diagram of the experimental arrangement for the PA-FMR and the PD-FMR detection.

diffusion length of air (65 μm at 1.5 kHz). Nevertheless, scanning the sample surface with smaller steps of typically 20 μm makes sense as the signal is a superposition of all signal contributions in the area mentioned above.

The *PM-FMR detection* (Fig. 2) works with a HeNe laser beam (10 - 20 mW) which is modulated with an acousto-optic modulator and then focused onto the sample in a light-access cavity. The microwave diode signal is measured synchronously at the frequency of the light modulation by a lock-in amplifier. In the experiments described in this article two types of microwave cavities were used: a commercial optical cavity manufactured by BRUKER (ER 4104 OR) and the home-made wall-less cylindrical TE_{011} cavity.[14,11] The lateral resolution for this case is determined by the laser focus (20 μm) and the thermal diffusion length in YIG (110 μm at 1.5 kHz).

The *depth resolution* of all three photothermal FMR methods is controlled by the thermal diffusion length μ from Eq. 2 which depends on the modulation frequency and on the thermal diffusivity of the material. For a particular sample the depth resolution power is therefore determined by the upper limit of the modulation frequency accessible in the photothermal measurement. The dynamic range of a PA experiment, due to the microphone response, is generally restricted to the audio-frequency range with an upper limit of 20 kHz. For YIG (thermal diffusivity $\alpha = 5.6 \cdot 10^{-5}$ m^{2}s^{-1}) the maximum depth resolution $\mu_{min,PA} = 30$ μm. Modulation frequencies up to 70 kHz could be reached in the PD-FMR experiments on YIG corresponding to $\mu_{min,PD} = 16$ μm. The largest dynamic range and the highest depth resolution is provided by the PM-FMR technique. Up to 2 MHz the PM-FMR signal from YIG is still detectable corresponding to a depth resolution of $\mu_{min,PM} = 3$ μm.

2.3. Sensitivity

The measured quantity in conventional field modulation FMR is the microwave power P which is reflected at an impedance matched cavity. P is connected with the dissipative imaginary part χ'' of the high-frequency susceptibility χ[18]:

$$\sqrt{P} = \frac{1}{2} Q_{u,0} \frac{\int_s \chi''_{xx}(\mathbf{r})\, h_{c,1}^2 \, dV}{\int_c h_{c,1}^2 \, dV} \sqrt{P_0} \tag{4}$$

Here $Q_{u,0}$ denotes the unloaded Q-factor of the cavity and P_0 is the incident microwave power. The subscripts "s" and "c" refer to the sample and the cavity, respectively. If χ''_{xx} is constant within the sample Eq. (4) can be written in the simpler form

$$\sqrt{P} = \frac{1}{2} Q_{u,0} \eta \chi''_{xx} \sqrt{P_0}. \tag{5}$$

For a square law detector the signal to noise ratio of the conventional technique SNR_{conv} being defined as the ratio of the diode voltage ΔU and its equivalent noise voltage ΔU_{noise} (Nyquist noise) takes on the following form:

$$SNR_{conv} = \frac{\Delta U}{\Delta U_{noise}} = \frac{1}{2A_m} Q_{u,0} \eta \chi'' \sqrt{\frac{P_0}{4 k_B T \Delta \nu}} \tag{6}$$

where the factor A_m takes into account the fact that the first Fourier amplitude is measured in the experiment (see e. g.[18]).

The expression for the PM-FMR essentially can be derived from the same formula. Eq. (4) only has to be modified with respect to the signal generation process based on the thermal modulation of χ''_{xx}:

$$SNR_{PM} = \frac{1}{2} Q_{u,0} \cdot \eta \cdot \sqrt{\frac{P_0}{4k_B T \Delta\nu}} \cdot \frac{|\int_s \vartheta_s(\mathbf{r})\, dV|}{V_s} \cdot \frac{\partial \chi''_{xx}}{\partial T} \quad . \tag{7}$$

For a one-dimensional heat diffusion problem[19] the integral in Eq. (7) – the volume averaged temperature $\langle\theta_s\rangle_s$ due to the heating with an optical power P_L – in the limit of low and high modulation frequencies can be approximated by[19]

$$\mu_s \gg s: \quad |\langle\theta_s\rangle_s| \simeq \frac{P_L}{2\sqrt{\omega_{mod}} \cdot \sqrt{\lambda_b \rho_b c_b}} \quad \text{and} \tag{8}$$

$$\mu_s \ll s: \quad |\langle\theta_s\rangle_s| \simeq \frac{P_L}{2\omega_{mod} \rho_s c_s}. \tag{9}$$

The subscripts "s" and "b" refer to the sample and the backing material (substrate) and s is the sample thickness. In the case of a Lorentzian resonance line shape

$$\chi''(H) = \frac{\chi''_{max}}{1 + \left(\frac{H-H_0}{\frac{1}{2}\Delta H_0}\right)^2} \tag{10}$$

the temperature derivative in Eq. (7) phenomenologically can be decomposed into three contributions resulting from the temperature dependences of the susceptibility itself, of the line position H_0 and of the linewidth ΔH[20]:

$$\frac{\partial \chi''}{\partial T} = \frac{\frac{\partial \chi''_{max}}{\partial T}}{1 + (\frac{H-H_0}{\frac{1}{2}\Delta H_0})^2} + \frac{2\chi''_{max} \cdot \frac{\partial H_0}{\partial T} \cdot (H - H_0)}{(\frac{1}{2}\Delta H_0)^2 (1 + (\frac{H-H_0}{\frac{1}{2}\Delta H_0})^2)^2} + \frac{\chi''_{max} \cdot \frac{\partial(\Delta H)}{\partial T} \cdot (H - H_0)^2}{(\frac{1}{2}\Delta H_0)^3 (1 + (\frac{H-H_0}{\frac{1}{2}\Delta H_0})^2)^2}. \tag{11}$$

For YIG at ambient temperatures the second term dominates the PM signal.

Comparison of Conventional FMR and PM-FMR. The sensitivities of the conventional and of the PM-FMR detection schemes are compared for the case of a $590 \times 675 \times 14\ \mu m^3$ YIG film. For this specimen the MSM linewidth ΔH_{pp} is 200 μT leading to a $\chi''_{max} = M_0/\Delta H_{pp}$ of 874. M_0 is the saturation magnetization. The factor $A_m = 40$ for a field modulation of 5 μT. The measurements have been performed in a TE_{011} cylindrical light access cavity with a volume of $6.1 \cdot 10^{-5}$ m^3 and a quality factor of 1100. The filling factor $\eta = 2.1 \cdot 10^{-7}$. The experiments were conducted at 300 K with 10 μW microwave power and a detection bandwidth of 1 Hz.

Insertion of the numbers into Eq. (6) yields a signal to noise ratio SNR_{conv} of the conventional measurement of

$$SNR_{conv} \simeq 6.3 \cdot 10^4. \tag{12}$$

For the comparison of the conventional and the PM-FMR techniques we define the relative sensitivity $G_{PM} = SNR_{PM}/SNR_{conv}$. The combination of (6) and (7) yields

$$G_{PM} = \frac{A_m \langle \theta_s \rangle_s}{V_s} \cdot \frac{\left| \frac{\partial \chi''}{\partial T} \right|}{\chi''_{max}} \tag{13}$$

where, depending on the modulation frequency, $\langle \theta_s \rangle_s$ can be replaced by (8) or (9).

G_{PM} is evaluated at 300 K for modulation frequencies ν_{mod} ranging from 10 Hz to 10 MHz. The temperature derivative of χ''_{max} in YIG is determined by the temperature dependence of the resonance line position. A typical value of $(\partial H_0/\partial T)_{max}$ is 160 μT/K, leading to a maximal value of 700 for the second term in Eq. (11) which dominates $\partial \chi''/\partial T$. For comparison, the temperature dependence of the first term is $\partial \chi''_{max}/\partial T = -3.7$ K^{-1}. The thermal parameters for YIG and the substrate GGG intervening into the calculations according to Eqs. (8) and (9) are: thermal diffusivity $\alpha = \lambda/\rho c = 56 \cdot 10^{-6}$ m^{2}s^{-1} ($72 \cdot 10^{-6}$ m^{2}s^{-1} for GGG) and effusivity $e = \sqrt{\lambda \rho c} = 990$ Ws$^{1/2}$K^{-1}m^{-2} (1060 Ws$^{1/2}$K^{-1}m^{-2}). The laser power P_L was chosen as 10 mW which is a typical value used in the experiments. In Table 1 the relative sensitivity G_{PM} has been listed for different frequencies. Evaluating these numbers it has been taken into account that below 10 kHz the $1/f$-noise determines the noise amplitude of the measurement. The upper frequency limit of PM signals detected up to date is 2 MHz.

Table 1 clearly demonstrates that the PM-FMR possesses the highest sensitivity among the photothermal methods. Depending on the experimental conditions one even obtains sensitivities superior to that of the conventional technique. Experimental results come very close to the values obtained from these theoretical considerations.

Table 1: Relative sensitivity $G_{PM} = SNR_{PM}/SNR_{conv}$ of the PM-FMR technique compared to the conventional field modulation technique

ν_{mod} (Hz)	10^1	10^2	10^3	10^4	10^5	10^6	10^7
G_{PM}	$1.2 \cdot 10^{-1}$	$3.9 \cdot 10^{-1}$	1.2	3.9	$3.6 \cdot 10^{-1}$	$3.6 \cdot 10^{-2}$	$3.6 \cdot 10^{-3}$

Similar considerations show that G_{PM} for YIG can lie in the range of 1 even for temperatures approaching 0 K and the Curie temperature T_C, respectively. A thorough discussion of this point is omitted here because of the temperature dependence of most of the quantities entering the expression for G_{PM} (χ''_{max}, H_0, ΔH, c_s and c_b) the calculation will be rather lengthy. At low temperatures the main contributions to the PM-signal arise from the variation in linewidth (e. g. due to sample impurities) and the decrease in the specific heat of the sample c_s. Near T_C the large temperature coefficient of χ''_{max} strongly enhances the PM-signal.

The minimum temperature oscillation $\vartheta_{PM,min}$ at ambient, averaged over the sample volume can be obtained from the condition $SNR_{PM} = 1$. With the above values and from Eqs. (7–9) we get

$$\vartheta_{PM,min} = \frac{2}{\frac{\partial \chi''}{\partial T} Q_{u,0}\eta} \sqrt{\frac{4k_B T \Delta\nu}{P_0}} \simeq 4 \cdot 10^{-7}\,\mathrm{K}. \tag{14}$$

Comparison of PD-FMR and PM-FMR. The expression for the sensitivity of the laser beam deflection technique is determined by several quantities which have been discussed by NETZELMANN.[16] Above a modulation frequency of 1 kHz the signal to noise ratio is governed by the diode noise and the amplifier noise. The detection limit of the position sensitive detector used in our experiments is $\phi_{n,min} = 10^{-10}$ rad/$\sqrt{\mathrm{Hz}}$.[16] For PD-FMR experiments with YIG bulk samples this limit is normally reached at a modulation frequency of 70 kHz. From an expression for the normal deflection of the laser beam (perpendicular to the sample surface) one yields the smallest temperature oscillation $\vartheta_{PD,min}$ that can be detected

$$\vartheta_{PD,min} = \frac{\mu}{\ell} \frac{n}{\sqrt{2}} \frac{1}{|\frac{\partial n}{\partial T}|} \phi_{n,min}\, e^{-x/\mu} \simeq 6 \cdot 10^{-4}\,\mathrm{K}. \tag{15}$$

Here, $n = 1.0003$, $|\frac{\partial n}{\partial T}| = 10^{-6}\ \mathrm{K}^{-1}$, $\mu = 9.5\,\mu$m, $x = 85\,\mu$m,[7] $\ell = 8.4$ mm and the detection bandwidth is 1 Hz.

3. Theory of Magnetostatic Modes in Slabs and Spheres

The theoretical description of dipolar coupled spin wave modes, so-called magnetostatic modes (MSM), will be treated only briefly as several authors in this book have dealt with this subject. Here, emphasis will be put on the spatial variation of the absorbed microwave power in rectangular specimens which can be detected by the photothermal methods and which allows to unambiguously identify discrete MSM.

3.1. MSM in Slabs

In 1961 DAMON AND ESHBACH (DE)[21] formulated the dispersion relation for the propagation of long wavelength spin wave modes in a tangentially magnetized gyrotropic slab ($\mathbf{H}_0||\mathbf{z}$) which extends infinitely in y and z-direction but has a finite thickness s in x-direction. The physical problem consists of simultaneously solving the equation of motion for the magnetization $\mathbf{M}$ (without damping) and Maxwell's equations $\nabla \cdot \mathbf{B} = 0$ and $\nabla \times \mathbf{H} = 0$ with their respective boundary conditions. In the magnetostatic approximation used here the hf magnetic field $\mathbf{h}$ is expressed by the gradient of the magnetic potential

$$\Psi = A\exp(\pm ik_x x)\exp(\pm ik_y y)\exp(\pm ik_z z). \tag{16}$$

DE obtained a system of three equations in five variables:

$$(1+\kappa)({k_x^i}^2 + k_y^2) + k_z^2 = 0 \tag{17}$$

$$k_x^{e\,2} - k_y^2 - k_z^2 = 0 \tag{18}$$

$$k_x^{e\,2} + 2k_x^e k_x^i (1+\kappa)\cot(k_x^i s) - {k_x^i}^2(1+\kappa)^2 - (\nu k_y)^2 = 0 \tag{19}$$

where k_x^i and k_x^e denote wave vectors inside and outside the sample and

$$\kappa = \frac{\Omega_H}{\Omega_H^2 - \Omega^2}, \quad \nu = \frac{\Omega}{\Omega_H^2 - \Omega^2}, \quad \Omega_H = \frac{H_i}{M_0} \quad \text{and} \quad \Omega = \frac{\omega}{\gamma M_0}. \tag{20}$$

For comparison with experimental results it is convenient to write the dispersion relation in a ω,k_y,k_z-representation:

$$(1+\eta^2) + 2\left|(1+\eta^2)^{1/2}\right|\left(-\frac{1+\eta^2+\kappa}{1+\kappa}\right)^{1/2}(1+\kappa)\cot\left[|k_y|s\left(-\frac{1+\eta^2+\kappa}{1+\kappa}\right)^{1/2}\right]$$
$$+(1+\kappa)^2\left(\frac{1+\eta^2+\kappa}{1+\kappa}\right) - \nu^2 = 0 \tag{21}$$

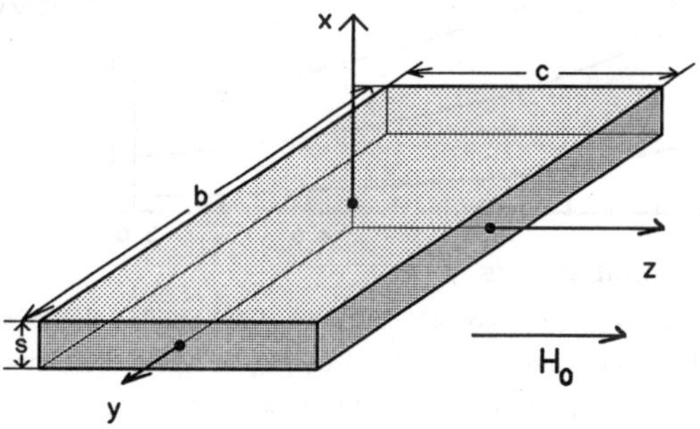

Figure 4: Sample geometry and coordinate system for MSM propagation. In the DE theory b and c tend to infinity.

with $\eta = k_z/k_y$. The solutions of this equation are commonly called "sheets", hyper surfaces in the three dimensional parameter space.

Due to the cotangent term the dispersion relation (21) has two different kinds of solutions depending on the sign of $(1+\kappa)$. For $(1+\kappa) < 0$ there is an infinite number of sheets in the frequency range $\Omega_H < \Omega < (\Omega_H^2 + \Omega_H)^{1/2}$ with purely real k_x^i, i. e. an oscillatory behaviour of the potential Ψ (magnetostatic backward volume waves, MSBVW). For the other case k_x^i is purely imaginary resulting in magnetostatic surface waves (MSSW) with only one sheet between $\Omega = (\Omega_H^2 + \Omega_H)^{1/2}$ and $\Omega = \Omega_H + 1/2$. The special cases of pure mode propagation (see Fig. 5) where either k_y or k_z vanish lead to the following equations

$$\text{MSBVW} \quad (k_y = 0): \qquad 2\cot[|k_z|s\alpha] = \alpha - \frac{1}{\alpha} \quad \text{with} \quad \alpha = \left(\frac{-1}{1+\kappa}\right)^{1/2} \tag{22}$$

$$\text{MSSW} \quad (k_z = 0): \qquad \Omega^2 = \Omega_H^2 + \Omega_H + \frac{1}{2 + 2\coth(|k_y|s)} \tag{23}$$

The introduction of finite sample dimensions can be described by the geometrical quantization of the wave vectors k_y and k_z,

$$k_y = \frac{n_y \pi}{b} \quad \text{and} \quad k_z = \frac{n_z \pi}{c}, \tag{24}$$

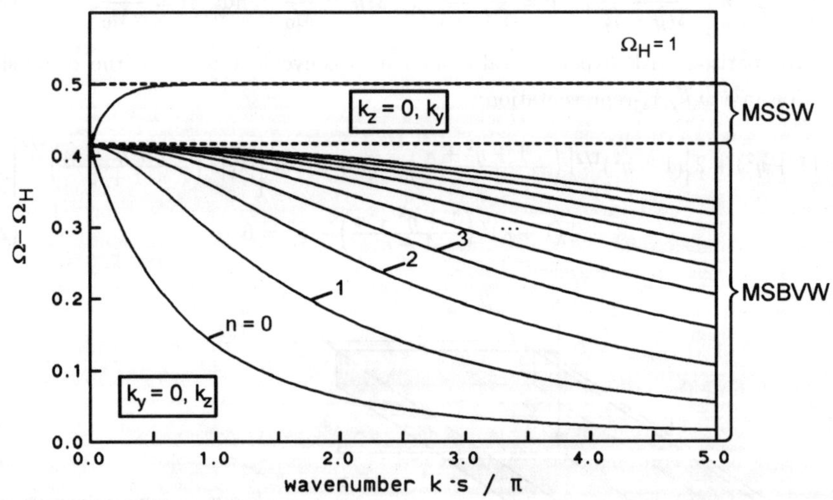

Figure 5: Dispersion of magnetostatic modes in a transversely magnetized unbounded slab of thickness s for pure mode propagation. The volume modes (MSBVW) propagate along the z-direction ($k_y = 0$) and the surface mode (MSSW) propagates along the y-direction ($k_z = 0$).

where b and c are the dimension of the sample according to Fig. 4. For k_x^i a similar simple discretization is not possible as can be seen from Eq. (17). Possible modes in finite samples will be denoted by (n_y, n_z).

This simple condition (24) is only valid if the internal field inside the sample is homogeneous. For thin films this assumption is applicable except for small regions near the sample margins. For bulk samples which will be treated later in this chapter the inhomogeneous demagnetizing field as calculated by JOSEPH AND SCHLÖMANN[22] causes strong variations of the internal magnetic field even in areas far away from the edges. As the potential equation to be solved resembles a Schrödinger equation a WKB-like approximation – with quantization conditions $\int k_i(i)\,di = n_i\pi$, $i = y, z$ as proposed by SCHLÖMANN AND JOSEPH[23] — can be used to correctly take into account this effect. The consideration of demagnetizing effects markedly changes the dispersion relation, especially in thick samples.

Additional influences on the dispersion relation are anisotropy effects which lead to a change in the frequency boundaries for MSSW and MSBVW modes. In addition to these modes anisotropy induced volume modes can be observed between MSSW and MSBVW.[24,25]

In the case of bulk samples the wavelengths of low order modes do no more justify the magnetostatic approximation and propagation effects have to be included into the dispersion relation.[26]

For the comparison of spatially resolved MSM-spectra and theoretical predictions we have performed numerical calculations to simulate such spectra.[19] Here, we would like to discuss the influence of the demagnetizing field on the propagation of MSM in a bulk sample. As the experiments have been performed at a fixed microwave frequency the following spectra have been calculated as a function of the external magnetic field.

For this purpose the dispersion relation (21) has been modified in the following manner: The inhomogeneous demagnetizing field inside a slab has been calculated in a first order approximation. Introducing its diagonal component $N_{zz}M$ in Eq. (21) one can determine the area of the sample in which propagation of MSM is possible, i. e. where the internal field does not exceed the field boundaries for MSM propagation. This area has a complicated curved shape and is approximated by a rectangle. The $y(z)$-dimension of this rectangle is given by the $y(z)$-dimension of the curved surface at $z(y) = 0$. The neglected areas produce only low intensity contributions to the MSM spectrum. The area of propagation as a function of the external magnetic field

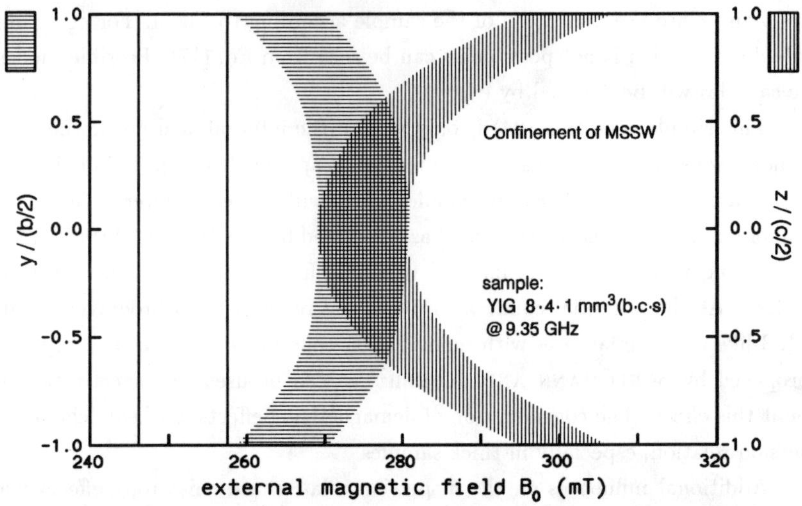

Figure 6: Area of propagation of MSSW in a 8 × 4 × 1 mm^3 YIG-slab at 9.35 GHz. Inside the dashed areas MSM can propagate. The vertical lines are the field limits for MSSW propagation in an infinite sample.

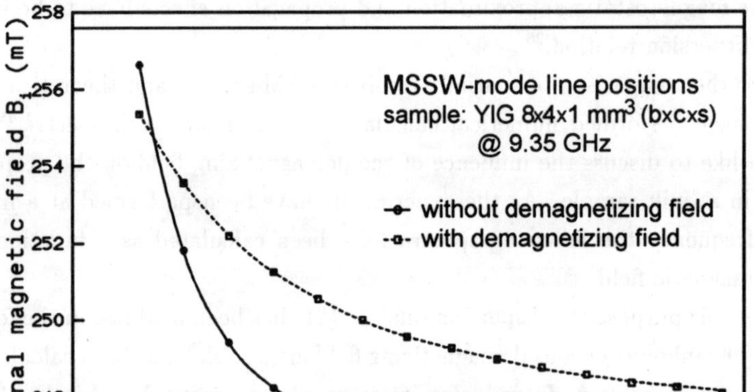

Figure 7: Resonance line positions of MSSW in a 8×4×1 mm^3 YIG sample at 9.35 GHz as a function of the mode index n_y with and without taking into account the influence of the inhomogeneous demagnetizing field.

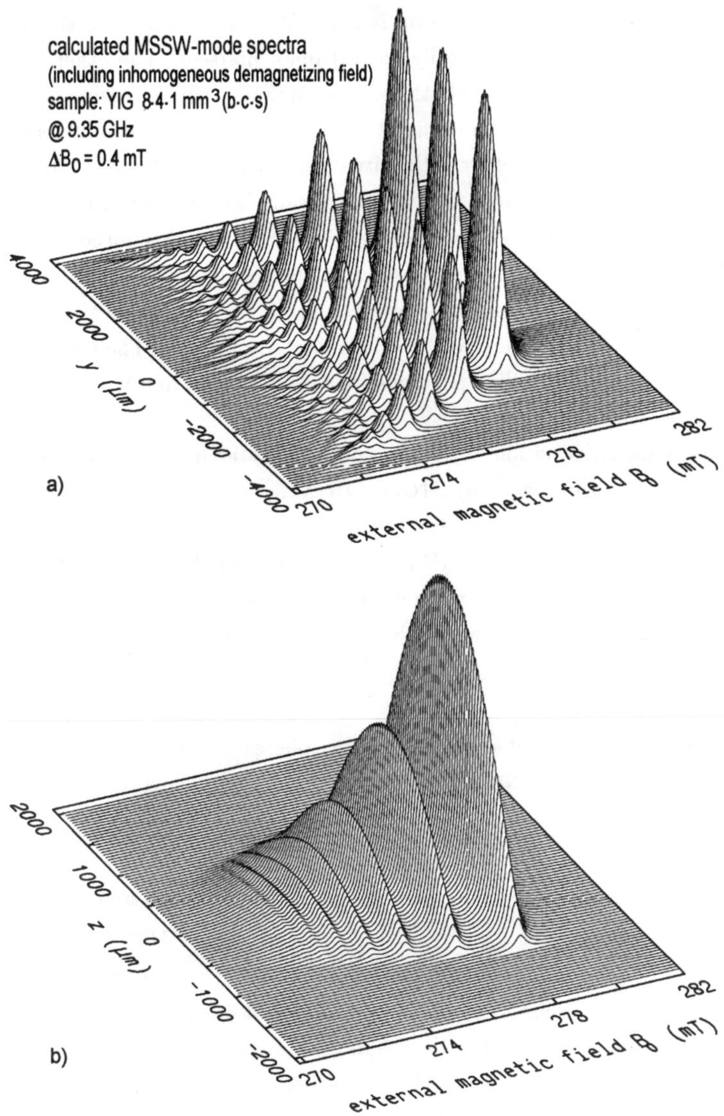

Figure 8: Calculated spatial dependent power absorption of MSSW in a YIG sample ($8 \times 4 \times 1$ mm^3, ΔB_0= 0.4 mT) at 9.35 GHz as a function of the external magnetic field B_0. The influence of the inhomogeneous demagnetizing field has been taken into account.

a) Variation of the absorption perpendicular to the external field B_0

b) Variation of the absorption parallel to the external field B_0

is shown in Fig. 6. MSM propagation is only taken into account in the field range where the horizontally and vertically dashed lines intersect. The other regions also only lead to minor signal contributions. For MSSW with increasing wave vectors (which means decreasing resonance fields) one can state a growing confinement of the propagation in y-direction whereas the situation is vice versa in the z-direction.

With this information on the "effective" sample dimensions the resonance line positions B_0 can be obtained by an iterative procedure. For this purpose B_0 has to be calculated in such a way that after subtracting the demagnetizing field the dispersion relation for the mode in question is solved. During this procedure the wave vectors have to be corrected according to the values plotted in Fig. 6. The result is given in Fig. 7 where the resonance line positions for a $(n_y,1)$ mode in a $8 \times 4 \times 1$ mm^3 YIG sample are shown.

Given the information on the line positions, starting from the expression for the mode intensities suggested by STOREY ET AL.,[27] the spatial dependence of MSM spectra can be calculated assuming a Lorentzian line shape (Fig. 8).

Taking into account a WKB-like approximation the spectra are changed in such a way that the spatial distance between the absorption maxima of MSM at fixed frequencies start to increase towards the sample boundaries.

3.2. MSM in spheres

The calculation of Walker modes[28] in spheres is analogous to that in slabs. According to FLETCHER AND BELL[29] the dispersion relation reads

$$n + 1 + \xi_0 \frac{P_n^{m\,\prime}(\xi_0)}{P_n^m(\xi_0)} \pm m\nu = 0, \tag{25}$$

where P_n^m are associated Legendre polynomials of the first kind and

$$\xi_0^2 = 1 + \frac{1}{\kappa}, \quad H_i = H_0 - \frac{1}{3}M_0 \quad \text{and} \quad P_n^{m\,\prime} = \frac{\mathrm{d}P_n^m(\xi_0)}{\mathrm{d}\xi_0}. \tag{26}$$

A thorough discussion of the spatial distribution of the high-frequency magnetization inside the sphere has been published by RÖSCHMANN AND DÖTSCH.[30] Details on this topic will be discussed when presenting experimental evidence for the identification of Walker modes by PM-FMR in Sect. 6.

4. Magnetostatic Modes in Yttrium Iron Garnet Slabs

The photothermal laser beam deflection (PD) FMR investigations of slabs on which we report in this section have been carried out on three YIG bulk samples with different crystalline orientations. The approximate size of all samples was 8 × 4 × 1 mm^3 ($b \times c \times s$, see Fig. 4). The PD experiments have been performed in a non-resonant slitted waveguide setup with excitation field strengths ranging from 1.3 μT (corresponding to 30 mW of microwave power) to 23 μT. In conventional FMR experiments the lower power limit is approximately three orders of magnitude smaller than for the PD technique.

4.1. MSM spectra recorded by PD-FMR

The MSM spectra shown here in Figs. 9 and 10 have been recorded with 600 mW of microwave power at the position of the sample. The modulation frequency of the incident microwave radiation was chosen as 1.5 kHz corresponding to a thermal diffusion length of 65 μm in air and 110 μm in YIG. The laser beam being aligned parallel to the y-direction was displaced in the z-direction.

The spectra in Fig. 9 originate from a 4.4 mm wide bulk sample. As expected from section 3. (Fig. 8) the MSSW – covering a range of 11 mT above 264 mT – do not propagate over the whole width of the sample but are confined to a region around the centre because of the inhomogeneous demagnetizing field. With the PD-FMR waveguide setup standing MSSW cannot be imaged in the z-direction. This has technical reasons: The laser beam can be displaced along the z-direction only as slits in the long sides of the waveguide would cause severe disturbances in the electromagnetic field inside the waveguide. Experiments with an open cylindrical cavity, of course, allow to obtain the desired mode index n_y if the problem of impedance matching is mastered. The phase information obtained from a photothermal experiment provides some information about the region in the sample where the detected thermal wave has been generated. The experimental results compare well with the calculated ones[19] but the superposition of several modes (including higher order ones) obscure the contribution of a single mode.

The mode at about 274 mT is an anisotropy induced magnetostatic forward volume wave (MSFVW). Its modulation frequency dependence as compared to that of the surface modes (Fig. 11) clearly identifies this mode as a volume mode: For high modulation frequencies only parts of the sample close to the surface contribute to the PD signal because nearly all energy deposited in a surface mode is localized in an area covered by the thermal diffusion length μ. With growing μ (i. e. lower modulation frequency) the signal due to a surface excitation remains about constant whereas the energy of a volume excitation deposited in the bulk of the sample can now contribute

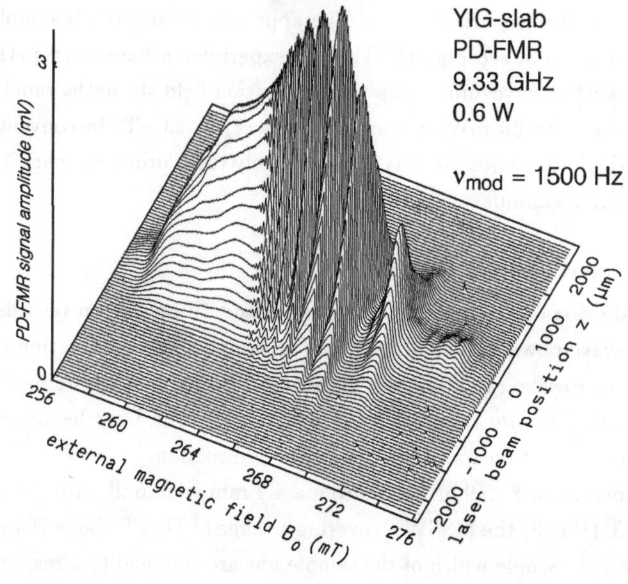

Figure 9: PD-FMR spectra of MSSW in a 8.4×4.4×0.98 mm³ YIG slab.

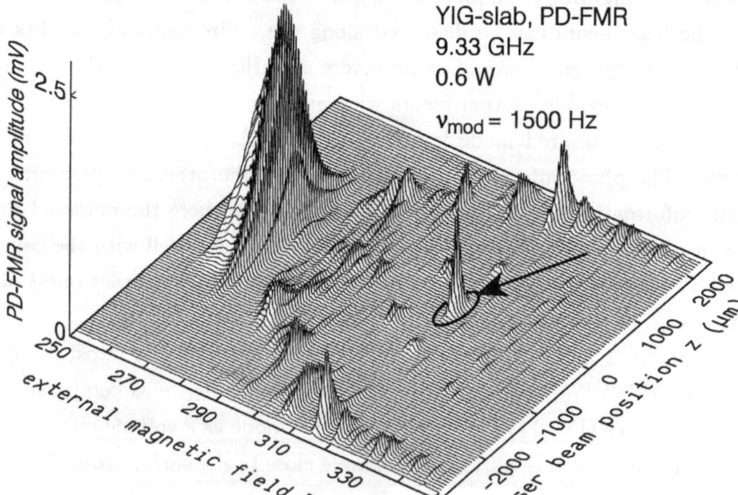

Figure 10: PD-FMR spectra of MSSW and MSBVW in a 8.4×4.4×0.98 mm³ YIG slab. The mode marked by the arrow is the (1,5)-mode.

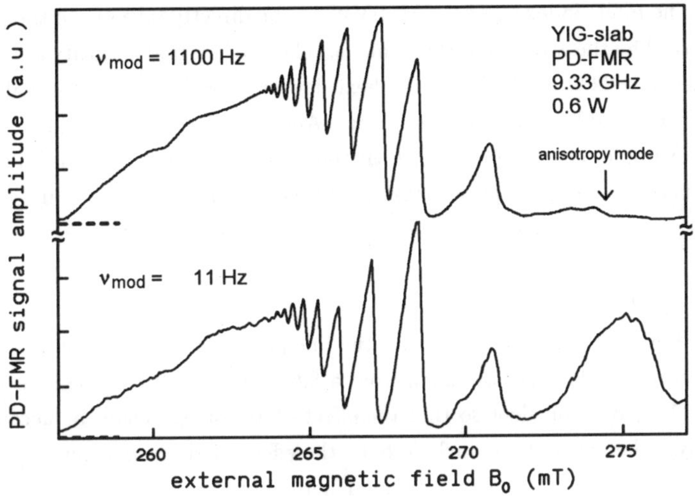

Figure 11: Modulation frequency dependence of PD-FMR spectra of MSSW in a 8.4×4.4×0.98 mm^3 YIG sample microwave power.

to the signal. For decreasing modulation frequencies experiments show the expected rise of the ratio of the PD signals of the anisotropy mode and the first MSSW located at 271 mT.[19]

The unstructured feature below 264 mT is attributed to long-range elastic excitations.

Due to the high microwave power used in this particular experiment the MSSW modes exhibit a sharp decrease in their trailing edges. This effect is linked with the temperature rise in the sample during the magnetic field scan. In the course of the scan the saturation magnetization is lowered leading to a shift of the resonance lines to higher fields. When the field has exceeded the maximum of the resonance lines the temperature drops and with the shift of the resonance lines to lower field the PD-signal decreases instantly. This foldover-like effect is influenced by the scan time and even at higher power can be avoided by either scanning more rapidly or when allowing for deviations of the order of the linewidth by scanning in the reverse direction.

In Fig. 10 the whole spectrum of magnetic excitations as recorded with the PD technique is shown. Above the MSSW modes and the MSFVW volume mode at about 280 mT we find a great number of anisotropy induced MSBVW. At the low field side of the MSBVW mode spectrum the expected confinement of the propagation can be seen. The mode index n_z of the MSBVW can be directly taken from the plot. The mode marked by the arrow is the (1,5)-mode. The rise in signal intensity at the sample boundaries at ± 2200 μm is due to the heat flow leaving the sample through the side faces verified by additional PD experiments. Another effect due to the demagnetizing field, for example, can be observed in the behaviour of the (1,5)-mode: The spatial distance between the absorption peaks grows when moving from $z = 0$ towards the edges of the sample as the wavenumber k_z varies along the z-direction.

4.2. Investigation of Resonance Line Positions

In Fig. 12 the results of a fit for the line positions of MSSW modes is shown. The experimental data are obtained from a $8.6 \times 4.6 \times 1.0$ mm^3 sample where all faces are (100) or equivalent so that the effect of anisotropy could be accounted for very easily. The calculations include both the effect of the demagnetizing field and the retardation effect. The fit presented here shows a considerable improvement concerning the agreement between theoretical and experimental results. The fit for MSBVW is similarly good.[19]

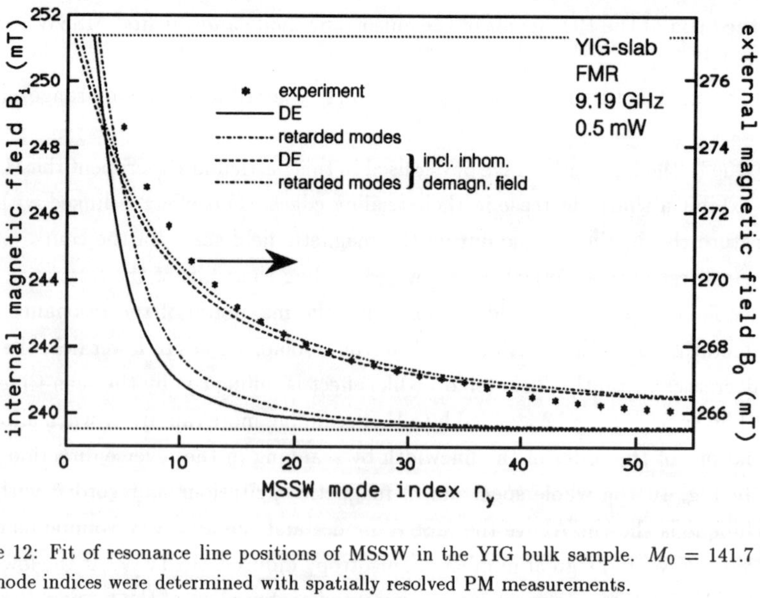

Figure 12: Fit of resonance line positions of MSSW in the YIG bulk sample. $M_0 = 141.7$ kA/m. The mode indices were determined with spatially resolved PM measurements.

4.3. Nonlinear Effects in MSM Spectra

The anomalously high amplitude of the absorption peak of the (1,5)-mode at $z = 0$ visible in Fig. 10 is due to a nonlinear effect. A comparison of the middle peak of the (1,5)-mode and the (1,7)-mode, e. g., for different microwave powers beyond a certain threshold power P_{thr} reveals a stronger growth of the peak of the (1,5)-mode than that of the (1,7)-mode. This is due to a four-magnon effect. The analysis of the matrix element of this process according to NIKITOV[31] shows that the scattering of a pair of different magnons into two magnons of the same mode – this one lying in the middle of the MSBVW mode band – has the highest probability. Therefore the process

$$\mathbf{k}_1 + \mathbf{k}_2 = 2\mathbf{k}_3, \qquad \omega_{\mathbf{k}_1} + \omega_{\mathbf{k}_2} = 2\omega_{\mathbf{k}_3} \tag{27}$$

is observed where the indices "3" refer to the (1,5)-mode and the indices "1" and "2" describe the two interacting mode. The power threshold for this process is[31]

$$P_{thr} \simeq \mu_0 \frac{ca^2(\gamma\mu_0\Delta H)^2 M_0^2}{2(\omega_H^2 + \omega_H\omega_M)^{1/2}}. \tag{28}$$

For a linewidth of 0.2 mT and a saturation magnetization of 176 mT for an internal field of 293 mT for the (1,5)-mode $P_{thr} \simeq 1$ W. The experiments reveal a threshold power somewhere between 300 and 600 mW which is an excellent agreement.

Another nonlinear effect, the subsidiary absorption a three-magnon process, can be observed at lower field values. The upper field limit for this effect is given by

$$\omega_H < \frac{\omega}{2} + \omega_M N_z. \tag{29}$$

For the bulk sample under investigation N_z in the centre of the sample is 0.118 which together with an anisotropy contribution of 1.9 mT results in $H_{max} = 189.8$ mT. With growing microwave power the lower limit could be observed to move to lower fields as expected. The spatially resolved PD-FMR spectra of the subsidiary resonance are shown in Fig. 13. The large signal at the edges is again due to the heat flow through the side faces. The PD-signal in the middle of the sample is markedly higher than towards the boundaries. This effect, in other samples, was even stronger than shown here.

Auto-oscillations in the subsidiary resonance of about 200 kHz could be observed with the conventional technique. Due to the bad signal to noise ratio of 3 – 5 % already using a bandwidth of 0.4 Hz for the modulated measurements it was not possible to detect auto-oscillations as a function of the z-coordinate. However, in the coincidence region such a measurement might be possible.

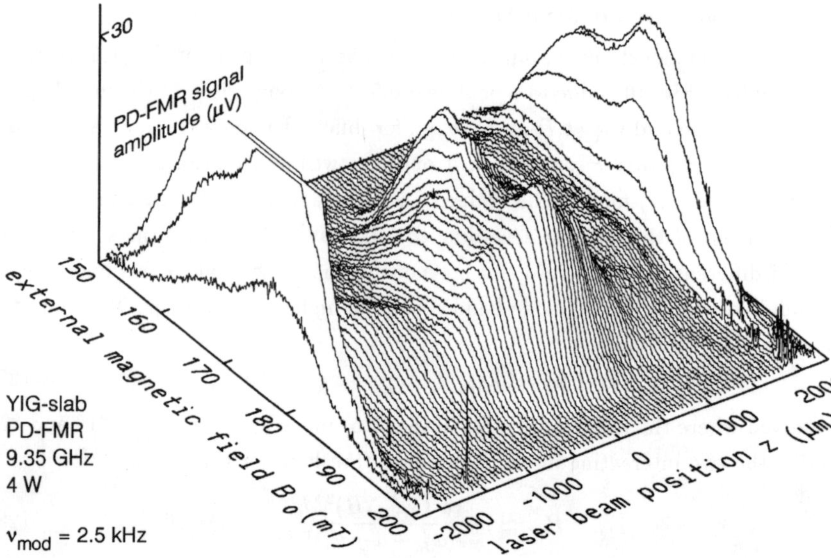

Figure 13: PD-FMR measurement of the subsidiary resonance in a 8.4×4.4×0.98 mm^3 YIG sample at 9.35 GHz and 4 W of microwave power.

5. Magnetostatic Modes in Yttrium Iron Garnet Films

In contrast to the bulk samples of which the experimental results have been presented in Sect. 4. inhomogeneous demagnetization fields in thin films are negligible and only influence the propagation of MSM in a region nearby the edges of the sample (see e. g.[32]). In addition the impedance matching can be improved for a cavity containing only a small amount of sample. For these reasons, film samples are ideally suited for comparative studies of the PM-FMR and the conventional detection technique. Such investigations in particular can serve to elucidate details of the generation process of the PM-FMR signal. The scope of this section is to report on the investigation of anisotropy effects by locally resolved PM-FMR measurements and to discuss conventional and PM-FMR experiments devoted to the temperature dependence of the magnetostatic mode spectra.

Fig. 14 shows the magnetostatic surface wave spectra obtained by PM-FMR from an epitaxial (LPE) 14 μm thick YIG film. The external magnetic field is aligned parallel to the z-direction which was a $[11\overline{2}]$ crystallographic axis. To obtain a two-dimensionally resolved spectrum the laser was scanned parallel (z-direction, $y =$

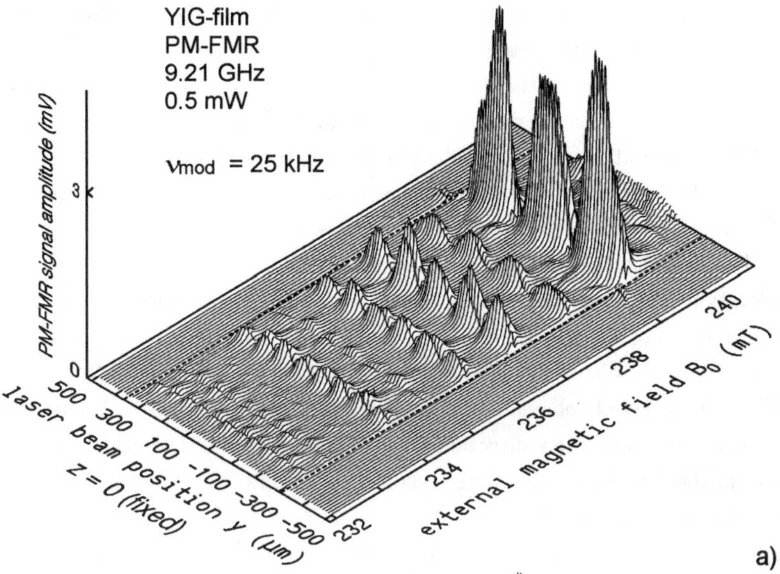

a)

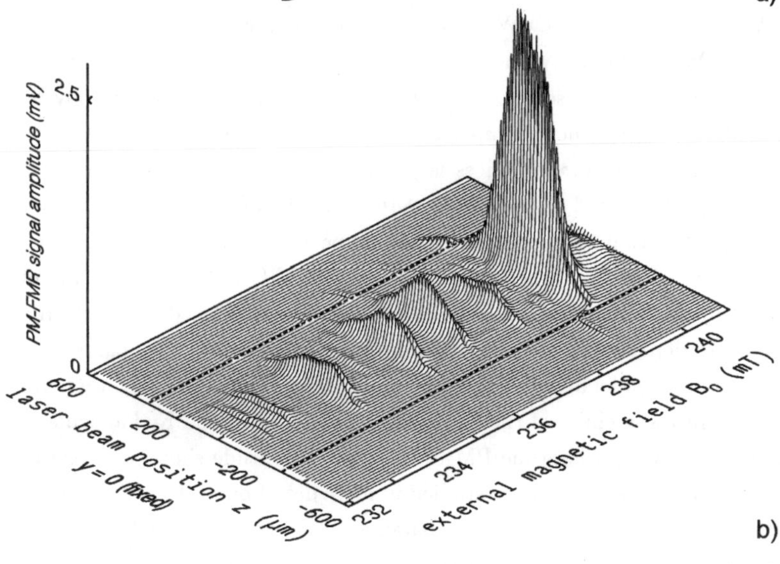

b)

Figure 14: PM-FMR amplitude spectra of MSSW in a 675×590×14 μm^3 YIG film at 9.21 GHz and a microwave power of 0.5 mW. The modulation frequency is 25 kHz. The laser focus was scanned perpendicular (a) and parallel (b) to the external magnetic field. The dashed lines indicate the sample edges.

0, Fig. 14a) and perpendicular to the external magnetic field (y-direction, $z = 0$, Fig. 14b). For the coordinate system see Fig. 4. The double peak structure in the amplitude recording is due to the derivative structure of the photothermal signal. The phase difference between the peak at lower and higher fields of one mode is 180°. The PM-FMR images allow a definite identification of the index numbers of all magnetostatic modes whereas the analysis of conventionally recorded spectra often suffers from an ambiguity in the mode classification (see e. g.[27]).

In the spectra of the MSSW besides the odd modes expected from intensity considerations the even modes are excited, too. This points towards non negligible inhomogeneities of the film sample. Above the mode number $n_y = 14$ the line shape becomes distinctly distorted and simultaneously the line positions deviate from the theoretically expected behaviour. In the Figs. 15 and 16 the measured resonance line positions in the z- and the y-direction are compared with the theoretical values calculated from the DE theory including crystalline and uniaxial anisotropy contributions K_1 and K_U, respectively.[25,33]

The values of the anisotropy constants K_1 and K_U have been deduced from the MSSW, MSBVW and MSFVW results. In order to describe the measured behaviour of the MSSW satisfactorily a much larger magnitude of K_U is required than for the bulk excitations. This discrepancy shows that the treatment of anisotropy has to be refined. At the same time the importance of locally resolved measurements providing some information on mode indices is pointed out, because ambiguities concerning mode labelling occurred when the fits were based on conventional data only.

As a function of temperature the line positions, line shapes and the line intensities of the PM-FMR spectra are changed. In YIG the experimentally observed temperature variations from 4 K to 600 K could be explained satisfactorily by the temperature dependence of the saturation magnetization M_0.[34,12] Larger deviations from the simple model based on the temperature variation of M_0 only are found below 150 K and in the vicinity of the Curie temperature T_C of 560 K. Particularly in the range just below T_C where the PM-FMR signal amplitude shows a strong enhancement these anomalies have been studied in more detail by both the conventional and the PM-FMR.[20] Already below T_C a paramagnetic line is observed superimposed by the MSM spectrum. This line increases in intensity as T_C is approached in order to become again weaker beyond the magnetic transition. Due to this temperature behaviour the paramagnetic excitation is accounted for by long-range fluctuations.[20]

In addition, several degrees below T_C the MSM propagation collapses, i. e.

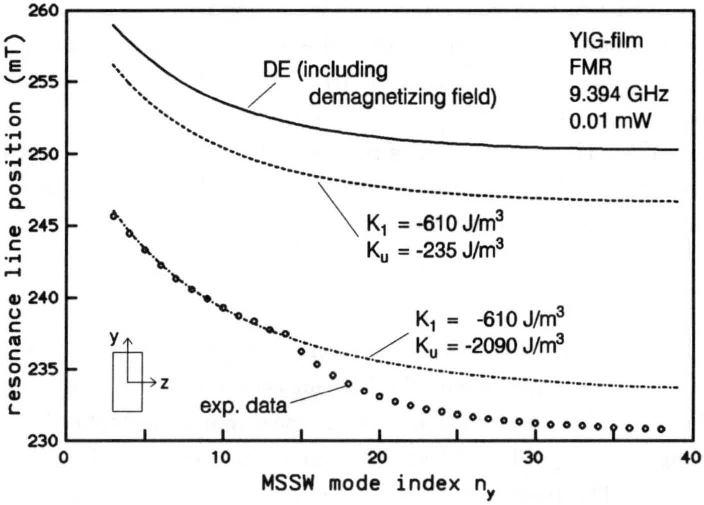

Figure 15: Line positions of MSSW in a 675×590×14 μm^3 YIG film in comparison with theoretical results (see text).

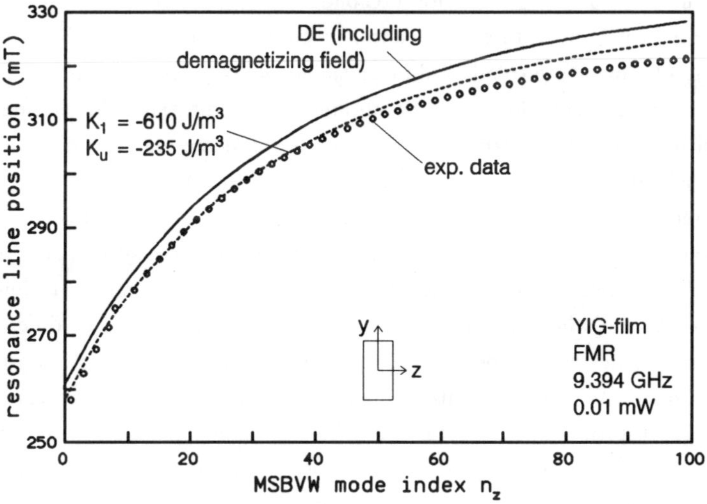

Figure 16: Line positions of MSBVW in a 675×590×14 μm^3 YIG film in comparison with theoretical results (see text).

the number of excited modes decreases with increasing temperature. This has been observed by conventional and PM-FMR[20] and has been explained by the presence of a damping mechanism not included in the DE theory. In order to qualitatively take into account an attenuation effect the authors introduced a criterion for the existence of standing MSM. This criterion states that the attenuation time constant must be larger than the time the wave needs to travel from one boundary to the other and back in order to form a standing wave:

$$t_d > \frac{2 \times \text{lateral sample dimension}}{|v_{gr}|} \tag{30}$$

In the scope of this model the onset for a collapse of the MSM is governed by two temperature dependent parameters: the damping constant and the group velocity v_{gr} of the MSM. From the line width an estimate of the damping constant can be deduced which is roughly of the order 10^{-7} at room temperature an one order of magnitude smaller near T_C. The group velocity has been approximated by the long wavelength limit which for the MSSW becomes:

$$v_{gr}|_{k_{in-plane}\to 0} = \frac{\omega_M^2}{4\omega} \cdot s \tag{31}$$

Inserting the corresponding values for YIG one finds that the group velocity decreases by a factor 9 if the temperature is raised from room temperature to 530 K. Altogether, the expected temperature trends and magnitudes support this simple idea. A further confirmation of the criterion could be obtained experimentally: Measuring a film sample with about 3 times larger dimensions the MSM are found to disappear at a lower temperature as expected from relations (30) and (31).

A methodological aspect has been also approached in the context of the PM-FMR studies on YIG films.[20] The photothermal response of the system only provides the desired information if it is linear in the magnetic and thermal excitations. The linearity of the periodical heating has been proved by experiments at different modulation frequencies with a focused (150 μm diameter) HeNe laser at varying laser powers up to 20 mW. To investigate the influence of the local *dc* temperature change on the magnetostatic modes the HeNe laser beam was focused on the middle of the sample and the conventional signal amplitude was recorded as a function of the heating time. The effect of heating was a shift of the resonance position which after a time delay of about 170 s reached a saturation value of about 0.77 mT. Taking only the temperature variation of the saturation magnetization into account this field value corresponds to a temperature rise of about 6 K in the sample. After switching off the laser the signal recovered with the same time constant. The local heating has two effects on the MSM under investigation: the local effective magnetic field becomes inhomogeneous and the average of the demagnetizing field is decreased. As in the conventional FMR the heat dissipation by spatially confined MSM excitations also leads to localized heating effects, the PM-FMR could offer a tool to quantitatively study the influence of this localized heat source on the MSM spectrum.

6. Magnetostatic Modes in Yttrium Iron Garnet Spheres

In the preceding sections the capabilities of photothermal techniques have been applied to plane geometries. However, PM-FMR can even be used for spatially resolved FMR investigations of samples with complicated surfaces. In this section we will discuss the results of PM-FMR imaging of Walker modes in spherical YIG specimens.[19]

The experiments to be described here have been performed with a broadband as well as with a narrowband setup. For the former a microwave synthesizer together with micro coils for excitation and detection has been used.[35] The narrowband setup consists of the spectrometer and the light-access cylindrical TE_{011} cavity both mentioned in Sect. 2. The optical arrangement, in addition, comprises a mirror which can be rotated around two axes with a high angular resolution. The deflected laser beam is focused onto the sample with a 50 mm lens. For sample diameters not larger than 2 mm there exists a linear relationship between the change in angle and the beam displacement. Towards the poles of the sphere there is a loss of spatial resolution as the size of the focal spot of the laser beam on the sample increases (from 20 μm to 30 μm) and is elliptically distorted both because of the curved sample surface. However, for low order modes such as the (3,1,0)-mode presented later on this effect is negligible.

Here we will restrict the discussion to the experiments performed with the resonant spectrometer. When the YIG sphere (2 mm diameter) is placed on the axis of the cylindrical cavity at 3/8 of its height 28 modes were excited and several of the most intense ones were investigated with the PM technique. For this purpose PM spectra were recorded for laser beam positions on a quadratic grid in y- and z-direction. For the coordinate system see Fig. 17. The image presented below (Fig. 18), e. g., consists of a mesh of 45 by 45 data points which represent the extremum of the PM-signal on the low field side of the resonance lines. Note that the signal may be either positive or negative.

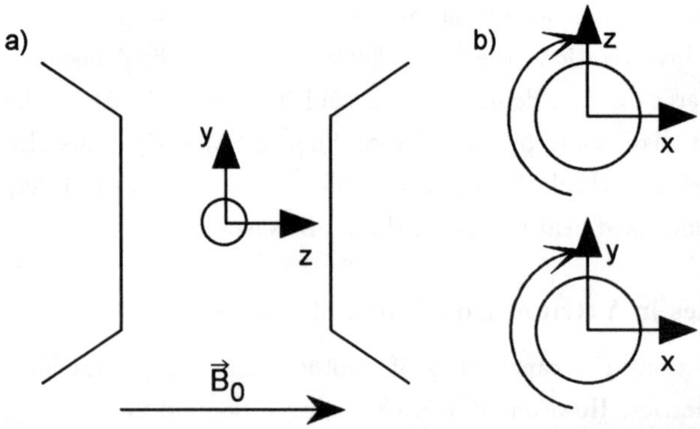

Figure 17:
a) Coordinate system for the spatially resolved recording of Walker modes in spheres. The angle of incidence of the laser beam is perpendicular to the plane of the paper.
b) Coordinate system for the calculated distributions of the high frequency magnetization. The arrows indicate the area scanned with the laser beam.

The interpretation of the measured PM data will be restricted to the behaviour of the (3,1,0)-mode. The following discussion requires some presumptions: (i) the PM signal is measured in the high-frequency side of the resonance line (i. e. the low field side), (ii) a N-type microwave diode is used and (iii) the temperature coefficient $\partial\nu/\partial T$ of the resonance frequency is positive. The following discussion[19] is based on the interpretation of MSM as a demagnetizing effect which results in a deviation of the resonance frequency of inhomogeneous modes from the frequency of the uniform mode. In this picture the resonance frequency of Walker modes can be described to consist of a static and a dynamic demagnetizing contribution, $H_{dem,stat}$ and $H_{dem,dyn}$, respectively. The first one, $-M_0/3$, is constant all over the sphere and can be imagined to result from fictive magnetic charges around the pole caps of the sphere. The dynamic demagnetization field is due to the deviation of the magnetization vectors out of their equilibrium position. Because of the temperature dependence of the saturation magnetization $\partial\nu/\partial T$ related to $H_{dem,stat}$ is positive whereas – for the (3,1,0)-mode – $\partial\nu/\partial T$ related to $H_{dem,dyn}$ is negative.[29] When locally illuminating the spherical sample the change in $H_{dem,stat}$ is constant all over the sphere independent of the laser beam position. In contrast to this $H_{dem,dyn}$ is connected with the high-frequency magnetization $\mathbf{m}(\mathbf{r})$ which is spatially variable. If one uses a focused laser beam and heats a point on the sphere where $\mathbf{m}(\mathbf{r})$ vanishes there is only an increase in $H_{dem,stat}$. Because of this the resonance line is shifted to lower magnetic fields. Scanning the resonance line from low to high magnetic fields one records a derivative curve with negative values below the resonance field. This is due to the modulated heating together with the detection with a N-type diode and is exactly the result of our measurements as can be seen from the comparison of the experimental data in

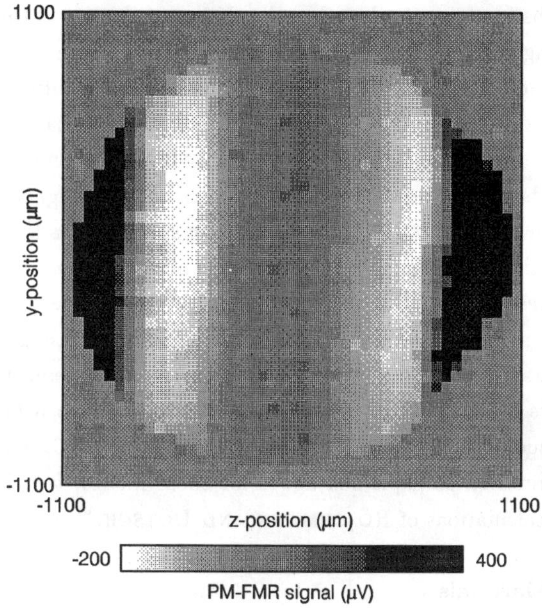

Figure 18: (310)-mode in a YIG-sphere of 2 mm diameter.

frequency ν_0:	9.15 GHz
magnetic field B_0:	363.05 mT
microwave power P_0:	0.5 mW
temperature T:	$\simeq$ 23 °C
side of resonance line:	left
temp. coeff. $\frac{\partial \nu_0}{\partial T}$:	2.05 MHz/K
temp. rise in sample:	1.4 K
mod. frequency ν_{mod}:	221 Hz
LIA time constant τ:	100 ms
laser power $P_{0,L}$:	12 mW
PM signal:	-202.8 – >1000μV
no. of points (z×y):	45 × 45

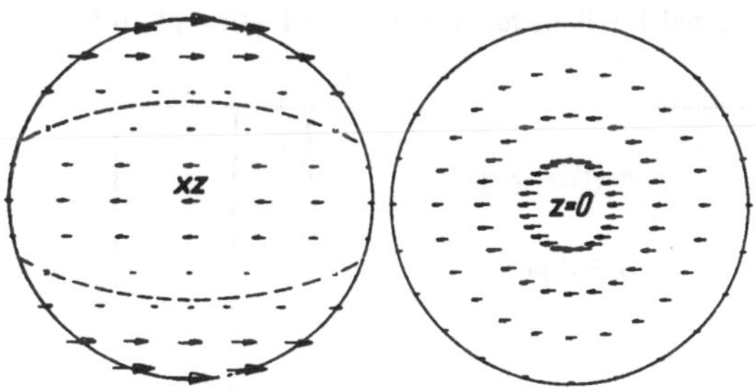

Figure 19: Spatial distribution of the transverse high frequency magnetization (after[30]).
The magnetization vectors precess around the z-axis with a fixed phase relation. On the dashed lines the amplitude of the precession vanishes.

Fig. 20 at $z \approx \pm 600$ μm and the theoretical calculations in Fig. 19. When the laser is moved in z-direction at $y = 0$ starting from the positions of minimum signal the lock-in signal increases and finally becomes positive. From Fig. 19 it can be seen that the amplitude of $\mathbf{m}$ grows towards the poles and towards the equator. This results in a shift of the resonance line due to the contribution of $H_{dem,dyn}$ which counteracts the one induced because of $H_{dem,stat}$. The dynamic contribution can overcompensate the effect of the static part. from Fig. 19 the signal is expected to be constant when the laser beam is moved along the y-axis at fixed z. This is almost the case in the experiment. Slight deviations observed are attributed to the change of the size of the laser focus and the thus resulting change in the temperature oscillations.

Experiments with a non-resonant broad-band setup yield the same results as the ones performed with the resonant setup and the interpretation of the data also essentially is the same. Although the sensitivity of the latter device is lower than that of the cavity-technique it is the better choice as the distortion of the electromagnetic field inside the cavity due to the presence of a large ferrimagnetic specimen leads to problems in the impedance matching, for example. As a consequence of this one has to measure some ten MHz off resonance. With the broad-band setup the line shift due to the *dc*-temperature rise in the sample has been corrected on-line by using a regulating circuit which considerably reduces the time required for the measurement.

In addition to the results presented here, several other modes have been investigated with the PM-technique. Among these are the (2,1,0), (2,2,0), (3,3,0) and the (4,3,0) modes. The interpretation as explained above for these modes, too, is consistent with the theoretical calculations of RÖSCHMANN AND DÖTSCH.[30]

7. Applications to Other Materials

The ability of spatially resolved FMR measurements predestine photothermal

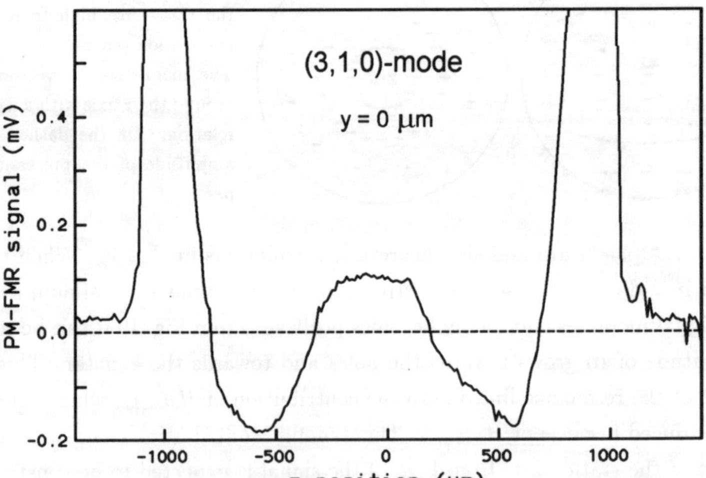

Figure 20: PM-FMR signal of the (3,1,0)-mode in the YIG sphere. The laser beam has been moved along the z-direction at $y = 0$. For the measurement parameters see Fig. 18.

techniques to be utilized for the characterization of materials with lateral and depth dependent magnetic or microwave properties. In this section we briefly discuss a few promising applications to materials other than YIG. For more details on these results we refer the reader to the particular publications or to preceding reviews.[5–7]

Spatial inhomogeneities of microwave magnetic properties have two principal origins: a locally varying magnetization vector in the magnetized material or a spatially dependent microwave power density distribution of a homogeneously magnetized specimen. The latter effect is inherent to microwave materials such as soft ferrites used as "optically" active components in waveguides. The first effect occurs in all types of magnetically ordered materials. Magnetic inhomogeneities can arise from a manifold of sources such as a spatial variation of the chemical composition or of the crystal structure or from inhomogeneous magnetizing and demagnetizing fields acting on the sample.

Using ferromagnetic resonance detected by the photoacoustic effect (PA-FMR) and by laser beam deflection (PD-FMR), for the first time, the spatial distribution of the microwave power density in a Ni-ferrite slab placed directly in front of a short in X-band waveguide has been measured.[17,36] The changes of the FMR traces as a function of the modulation frequency reveal the depth dependence of the FMR absorption. Lateral resolution has been achieved with the laser beam deflection technique. A specially designed waveguide allowed for measurements in the thermal transmission and thermal reflection mode.[36,7] The theoretical model used to interpret the depth dependent data was based on one-dimensional propagation of electromagnetic waves and on one-dimensional heat conduction. A Green's function technique has been applied to take into account the depth dependent heat sources. The main microwave-magnetic features of the model are the "body resonances"[1] which could also be observed by the conventional FMR. Although the principal features of the theoretical model are reproduced by the measurement, the quantitative analysis of the photothermal measurements reveals considerable deviations from the predicted behaviour. This disagreement was attributed to a vast inhomogeneous demagnetizing field in the sample and to the simplicity of the used one-dimensional electrical and thermal models. As to the influence of the demagnetizing field on the resonance line positions, PD-FMR spectra obtained at different sample positions could be explained by the slab geometry.[36,7]

In thin films local detection of the uniform mode of the FMR can be used to explore demagnetizing field effects in the film plane. This has been demonstrated for the first time by PD-FMR studies of an amorphous FeNiB metglass ribbon.[37] The FMR traces of the iron resonance taken at different positions of the ribbon exhibit strong perturbations particularly towards the edges of the ribbons. Shape anisotropy and internal strains are assumed to be the main causes of the observed large lateral variation of the local magnetic field.

Recently, locally resolved FMR studies have been extended to thin metallic films and layers. The photothermally modulated FMR (PM-FMR) has been used as principal method because of its smaller heat impact, its higher sensitivity and its lateral resolution being higher as compared to PD-FMR. In polycrystalline FeNi multilayers and alloys of a few hundred nm thickness thermally modulated resonance spectra from standing spin waves could be recorded as a function of the position of the laser spot on the film surface. The obtained two-dimensional lateral images of the magnetization distribution in alloys reveal weak oscillations of the magnetization along one direction in the plane which have been attributed to growth induced stresses.[38,5] The same technique applied to epitaxial films of iron allowed a mapping of the anisotropy fields providing a new means for a quality control of single crystalline magnetic layers.[39]

Compositional based variations of the magnetization in magnetic layered tapes have been investigated by PA-FMR and PD-FMR. The first photothermal depth profiling measurements have been performed on a commercial ferrochrome tape.[40,5] This tape which is designed for audio recording consists of two layers approx. 3 μm thick containing CrO_2 and Fe_2O_3 particles supported by a 8 μm thick polyester foil. The FMR detected photoacoustically at modulation frequencies of the incoming microwave power between 35 Hz and 23 kHz clearly disclose the layered structure of the magnetization. At high modulation frequencies the amplitude traces of the PA-FMR essentially reproduce the FMR of the upper CrO_2 layer whereas simultaneously the phase traces reflect the FMR of the lower Fe_2O_3 layer. The experimental data have been satisfactorily analysed in the framework of a one-dimensional heat diffusion model with two adjacent layers acting as heat sources. Preceding PD-FMR studies of the same tapes did not reveal any noticeable lateral variation of the compositional magnetic structure.[37]

By the photothermally modulated FMR technique, for the first time, lateral and depth dependent magnetizations profiles recorded with a tape recorder on single layer CrO_2 tape could be deconvolved.[11,5,6] The local thermal modulation was achieved with a HeNe laser with the beam focused down to spot of about 170 μm diameter.

Due to the good signal to noise ratio of the PM-FMR signal from this tape a wide dynamic range up to 100 kHz was available for the depth dependent studies leading to a depth resolution in the micron and submicron range. The PM-FMR images of the recorded magnetic signal tracks simultaneously provide two important pieces of information: the amplitude of the PM-FMR signal is a measure of the magnitude and direction of the local magnetization whereas the phase traces essentially reflect the thermal properties at the same location. In this way, it has been possible – based on the data-set of one measurement – to correlate an observed perturbation of the magnetic track with a delamination at the interface to the substrate.[5] The depth profiling ability, which is the major potentiality of the PM-FMR technique, in this field, has not yet been fully exploited. For samples other than CrO_2 tapes the major drawback of the PM-FMR imaging application is due to the fact that in most cases there is no zero field absorption. FMR generally needs an external magnetic field to drive the spin system into resonance thus probably destroying the information stored on the tape. To overcome this problem we have developed alternative magnetic detection schemes based on the local photothermal modulation of the magnetization. Promising results are already obtained by the photothermally modulated playback process where the depth profiles of sinusoidal signals have been determined for different sorts of tapes.[41,42]

8. Summary

Photothermal methods in microwave resonance offer a tool for laterally and depth resolved ferromagnetic resonance (FMR) investigations. The experimental devices developed and used in photothermal research have already reached a technical standard high enough to approach nowadays' problems in advanced fundamental and applied research.

In this article emphasis is put on the application of photothermal microwave techniques to the study of the spatial behaviour of magnetostatic modes (MSM) in yttrium iron garnet (YIG). Although MSM in this ferrimagnetic compound have been thoroughly studied during the last three decades by ferrimagnetic resonance, only just recently it has become feasible by means of the photothermal microwave resonance techniques to include thick YIG slabs into a quantitative investigation. Using laser beam deflection (PD-FMR) to detect the FMR from a rectangular slab the influence of the inhomogeneous demagnetizing field and retardation effects on the magnetostatic mode pattern could be revealed. Owing to the depth sensitivity of the photothermal detection techniques it was possible to separate experimentally surface and volume modes. In the course of this research work, for the first time, a spatially confined non-linear process has been detected which could be identified and consistently interpreted as a four-magnon process. Applying the photothermally modulated (PM) FMR to thin YIG films two-dimensional laterally resolved MSM spectra were obtained permitting an unambiguous identification of magnetostatic modes. The analysis of the experimental findings within the framework of the Damon-Eshbach theory yields detailed and quantitative information on the anisotropy fields and their influence on MSM. PM-FMR measurements on thin YIG films have been extended to lower and higher temperatures ranging from 4 K up to 600 K which comprises the magnetic ordering temperature.

Among the photothermal techniques in microwave resonance which are able to provide a lateral resolved information the PM-FMR proved to be the most promising method. For YIG it combines a three-dimensional resolution with a sensitivity comparable to that of the conventional FMR. The unique potential of this method is demonstrated by the discussed visualization of Walker modes from a YIG sphere, a result which till now was only accessible to the PM-FMR experiment. Other very recent applications of the photothermally modulated microwave resonance such as the imaging of magnetic anisotropy fields in epitaxial metallic films demonstrate the high potential for the application of this method in the field of magnetism.

Acknowledgements

The authors are indebted to R. Meckenstock for critically reading the manuscript. Partial financial support by the Deutsche Forschungsgemeinschaft (Kennwort: Photothermik) is gratefully acknowledged. Parts of the work have been performed within the framework of the Sonderforschungsbereich 166, project G3.

References

1. B. Lax and K. J. Button, *Microwave Ferrites and Ferrimagnetics* (McGraw-Hill, New York, 1962).
2. C. E. Patton, *Physics Reports* **103** (1984) 251.
3. W. Wettling and W. Jantz, *J. Magn. Magn. Mat.* **45** (1984) 364.
4. S. M. Rezende and A. Azevedo, *Phys. Rev. B* **44** (1991) 7062.
5. J. Pelzl and U. Netzelmann, in *Photoacoustic, Photothermal and Photochemical*

Processes at Surfaces and in Thin Films, ed. P. Hess (Springer, Berlin, 1989).

6. J. Pelzl, U. Netzelmann, T. Orth, and R. Kordecki, in *Photoacoustic and Photothermal Phenomena II*, eds. J. C. Murphy et al. (Springer, Berlin, 1990).
7. J. Pelzl and O. von Geisau, in *Non-Destructive Evaluation*, ed. A. Mandelis, Vol. 2 of *Progress in Photothermal and Photoacoustic Science and Technology* (Prentice Hall, Englewood Cliffs, 1994).
8. H. S. Carslaw and J. C. Jaeger, *Conduction of Heat in Solids* (Clarendon Press, Oxford, 1990).
9. A. Rosencwaig and A. Gersho, *J. Appl. Phys.* **47** (1976) 64.
10. A. C. Boccara, D. Fournier, and J. Badoz, *Appl. Phys. Lett.* **36** (1980) 130.
11. T. Orth, U. Netzelmann, and J. Pelzl, *Appl. Phys. Lett.* **36** (1988) 1979.
12. F. Schreiber, *Untersuchungen zur Signalentstehung und Ortsauflösung der photothermisch modulierten ferromagnetischen Resonanz in YIG.* Master's thesis, Ruhr–Universität, Bochum, 1992.
13. U. Netzelmann, J. Pelzl, H. Vargas, C. Cesar, and L. Miranda, *IEEE Trans. Magn.* **MAG-20** (1984) 1252.
14. M. Chamel, R. Chicault, and Y. M. d'Aubigne, *J. Phys. E* **9** (1976) 87.
15. U. Netzelmann, U. Krebs, and J. Pelzl, *Appl. Phys. Lett.* **44** (1984) 1161.
16. U. Netzelmann, *Ortsaufgelöste Messungen der ferromagnetischen Resonanz mit Hilfe photothermischer Nachweismethoden.* PhD thesis, Ruhr–Universität, Bochum, 1986.
17. O. v. Geisau, U. Netzelmann, and J. Pelzl, *IEEE Trans. Mag.* **24** (1988) 1954.
18. C. P. Poole, *Electron Spin Resonance* (Interscience, New York, 1967).
19. O. von Geisau, *Photothermische Untersuchungen der FMR kollektiver magnetischer Anregungen in YIG.* PhD thesis, Ruhr–Universität, Bochum, 1993.
20. F. Schreiber, M. Hoffmann, O. von Geisau, and J. Pelzl, *Appl. Phys. A* **57** (1993) 545
21. R. W. Damon and J. R. Eshbach, *J. Phys. Chem. Solids* **19** (1961) 308.
22. R. I. Joseph and E. Schlömann, *J. Appl. Phys.* **36** (1965) 1579.
23. E. Schlömann and R. I. Joseph, *J. Appl. Phys.* **35** (1964) 167.
24. R. A. Lemons and B. A. Auld, *J. Appl. Phys.* **52** (1981) 7360.
25. R. Gieniusz and I. Smoczynski, *J. Magn. Magn. Mat.* **66** (1987) 366.

26. R. Ruppin, *J. Appl. Phys.* **62** (1987) 11.
27. B. Storey, A. Tooke, A. Cracknell, and J. Przystawa, *J. Phys. C* **10** (1977) 875.
28. L. R. Walker, *Phys. Rev.* **105** (1957) 390.
29. P. C. Fletcher and R. O. Bell, *J. Appl. Phys* **30** (1959) 687.
30. P. Röschmann and H. Dötsch, *Phys. Stat. Sol. B* **82** (1977) 11.
31. S. A. Nikitov, in *Proceedings of the Congress Ampere* (1991); and priv. comm.
32. J. Barak, R. Ruppin, and J. T. Suss, *J. Appl. Phys.* **63** (1988) 2372.
33. R. Gieniusz, *J. Magn. Magn. Mat.* **119** (1993) 187.
34. M. Hoffmann, T. Orth, O. von Geisau, and J. Pelzl, in *Photoacoustic and Photothermal Phenomena III*, ed. D. Bićanić (Springer, Berlin, 1992).
35. F. Rödelsperger, O. von Geisau, H. Benner, and J. Pelzl, in *Proceedings of the Congress Ampere* (Athens, 1992).
36. O. v. Geisau, U. Netzelmann, and J. Pelzl, *J. Appl. Phys.* **63** (1988) 3347.
37. U. Netzelmann, J. Pelzl, D. Fournier, and A. C. Boccara, *Can. J. Phys.* **64** (1986) 1307.
38. R. Kordecki, *Untersuchungen von Spinwellen in Legierungs- und Stapelschichten aus Eisen und Nickel mit ferromagnetischer Resonanz.* PhD thesis, Ruhr–Universität, Bochum, 1990.
39. R. Meckenstock, F. Schreiber, O. von Geisau, and J. Pelzl, *38th MMM* (Minneapolis, 1993) Poster EQ-08.
40. U. Netzelmann and J. Pelzl, *Appl. Phys. Lett.* **44** (1984) 854.
41. M. Kaack, *Tiefenaufgelöste Messung der Magnetisierung in Audio-Bändern.* Master's thesis, Ruhr–Universität, Bochum, 1993.
42. J. Pelzl, M. Kaack, K. Friedrich, and T. Orth, in *J. de Phys. - Colloque 8th International Topical Meeting of Photoacoustic and Photothermal Phenomena 1994* (in press)